A Textbook of
Engineering Graphics

A Textbook of
Engineering Graphics

Dr. D.A. Hindoliya

B.E. (Mech. Engg.), M.Tech.(IIT Delhi), Ph.D. (IIT Delhi)
Professor in Mechanical Engg.,
Govt. Engineering College Ujjain, (M.P.), India
Formerly - Deputy Registrar, Rajeev Gandhi Technical University, M.P., Bhopal.

BSP BS Publications
A unit of **BSP Books Pvt. Ltd.**

4-4-309/316, Giriraj Lane, Sultan Bazar,
Hyderabad - 500 095
Phone : 040 - 23445605, 23445688

A Textbook of Engineering Graphics/D.A. Hindoliya

Published by :

BSP BS Publications
A unit of **BSP Books Pvt. Ltd.**

4-4-309/316, Giriraj Lane, Sultan Bazar,
Hyderabad - 500 095
Phone : 040 - 23445605, 23445688
e-mail : info@bspbooks.net
www.bspbooks.net

Printed at

Aditya Offset Process (I) Pvt. Ltd.
Hyderabad.

ISBN: 978-93-83635-21-4

Price: Rs. 625.00

Preface

Engineering drawing or graphics is always referred as an international language of engineers. The knowledge of this language is essential for engineers to communicate technical ideas pertaining to design of machine parts, buildings, roads, dams, electrical and electronic devices etc.

The present book is intended for use as a textbook in engineering drawing or graphics course for undergraduate engineering and diploma students. It covers the syllabi prescribed by various technical universities, engineering and polytechnic colleges in the country. An attempt has been made to prepare the entire text in an easy to understand language which will enable the readers to understand various concepts easily. Drawings and illustrations presented in all chapters have been prepared according to latest ISO and BIS standards. A large number of problems asked in the examinations of various universities have been solved and presented. Step-by-step procedure for solving problems will help the students to grasp the concepts in a much faster way.

This book is divided into thirteen chapters. Chapter 1 deals with the introduction of subject. It provides a brief discussion about the essential drawing instruments, lines, lettering, dimensioning and basic geometrical constructions. Chapter 2 describes use and construction of various types of engineering scales. Methods of construction of conic sections such as ellipse, parabola and hyperbola are discussed in Chapter 3. Engineering curves are described in Chapter 4. Chapter 5 deals with introduction of orthographic projection. Projections of points, straight lines, planes and solids are described in Chapters 6 through 9. Chapter 10 discusses sections of solids and intersection of cylinders. Development of surfaces of prism, pyramid, cone, cylinder and sphere are presented in Chapter 11. Chapter 12 deals with isometric projections. An introduction to computer aided drafting (CAD) is included in Chapter 13. Some basic commands of AutoCAD software are discussed. For further and advanced study the readers are advised to refer AutoCAD user manual.

I hope this book will fulfil the need of engineering students very well. Any suggestions and/or criticism from the readers for improving the subject matter of this book will be highly appreciated. The author may be contacted via email:dev_hindoliya@rediffmail.com.

Dr. D. A. Hindoliya

Acknowledgement

I would like to extend my sincere thanks to the Department of Technical Education and Skill Development, Govt. of Madhya Pradesh and Directorate of Technical Education Madhya Pradesh for granting necessary permission to publish this book.

I am indebted to Prof. S.C. Mullick and Prof. T.C. Kandpal, Professors Indian Institute of Technology, Delhi for providing constant encouragement and motivation for writing this book. I express my special thanks to my colleagues of Mechanical Engineering Department, Govt. Engineering College, Ujjain (M.P.) for providing useful suggestions. I am grateful to editorial and production team of M/s BS Publications, Hyderabad for cooperation and support for bringing out this book in time.

This work would have not completed without the blessings of my parents. So, I express my deep sense of gratitude to them. I express my special thanks to my wife Smt. Hemlata Hindoliya, son Shashank and daughter Disha for rendering their full support and cooperation to bring out this book in the present form.

I would like to extend my appreciation to Er. Shashank Hindoliya for providing his assistance in preparation of figures and organizing illustrations and text matter in various chapters of this book.

Dr. D. A. Hindoliya

Contents

CHAPTER 3

Conic Sections

CHAPTER 4

Engineering Curves

CHAPTER 5

Orthographic Projections

CHAPTER 6

Projections of Points

CHAPTER 7

Projections of Straight Lines

CHAPTER 8

Projections of Planes

CHAPTER 9

Projections of Solids

CHAPTER 10

Sections of Solids and Intersection of Cylinders

CHAPTER 11

Development of Surfaces

CHAPTER 12

Isometric Projections

CHAPTER 13

Introduction to Computer Aided Drafting

CHAPTER 1

Introduction

1.1 Graphics: A Tool to Communicate Ideas

Engineering graphics or drawing is the universal language of engineers. An engineer communicate his idea to others with the help of this language. Before actual construction of buildings, bridges, machines components etc., it is required to design them in order to meet technical requirements such as strength, safety etc. After designing, the next step is to prepare detailed drawing which is required during manufacturing or actual construction. A good drawing communicate the message very fast among the people who use it. For example, workers or artisans in the factory, perform manufacturing operations only after understanding the drawing of particular object. This method of communicating ideas is not new. The ancient man also used pictures and symbols to communicate among each other. The engineering drawing helps to provide clear idea about size, shape, internal details of any object.

Internal and complicated details of an object or machine parts can be understood easily. Looking to its high importance, the engineering graphics/drawing is taught to students of almost all branches of engineering. Let us consider some examples where its uses are important. The drawing of a machine component is important for mechanical and electrical engineers. Drawings of maps of buildings, bridges, roads, dams etc., are useful for civil engineers. Drawings of electronic devices are of great importance for electronic engineers.

The "drawing" or "graphics" can be expressed as art of representation of an object by systematic lines on a drawing paper or sheet. It can be broadly classified as artistic drawing and engineering drawing. An artistic drawing deals with the representation of painting, advertisement, pictures etc., whereas engineering drawing deals with the representation of engineering objects such as machine components, dams, roads, buildings, electronic components, computers, motors, generators etc.

An ideal and good drawing should have neat and clean, accurate and beautiful presentation. For such drawing, it is important to use good quality of instruments, pencils, sheets by the skilled draft man.

1.2 Drawing Instruments and Accessories

Knowledge of drawing instruments and other accessories is important for engineers and students. Proper technique for using them is an another aspect to be considered for production of a good drawing. Following is the list of common instruments and accessories/items required by an engineering student.

(i) Drawing board
(ii) T-square
(iii) Set square
(iv) Mini drafter or drafting machine
(v) Instrument box
(vi) Drawing pencil
(vii) Drawing clips/pins
(viii) Sand paper block
(ix) Eraser/Rubber
(x) Scale (Ruler)
(xi) Engineers scales
(xii) French curves
(xiii) Protractor
(xiv) Drawing sheets
(xv) Pencil sharpner
(xvi) Handkerchief or towel cloth

1.2.1 Drawing Board

It is considered as one of the essential instruments. Its top surface should be flat and smooth and edges should be at right angle to each other. Drawing board is usually made of well seasoned soft wood. Five to six narrow strips of soft wood are joined together with the help of suitable glue. To hold and fix the strips firmly, two wooden battens are fixed at the bottom side using screws. One of sides of drawing board is fitted with ebony edge which guides the T-square while sliding. As per the Bureau of Indian Standards (BIS) drawing boards of following (Table 1.1) sizes have been recommended for different use.

Table 1.1

Size in mm (Length × Width)	Designation
1500 × 100	B_0
1000 × 700	B_1
700 × 500	B_2
500 × 350	B_3

Depending upon the size of drawing paper/sheet to be hold the suitable size of drawing board may be selected. Following precautions should be taken in handling of drawing board.

(a) Top flat surface should not be spoiled.

(b) Always use drawing sheet.

(c) Ebony edge (fitted to side edge) should not be spoiled.

A typical design of commonly used drawing board is shown in Fig. 1.1.

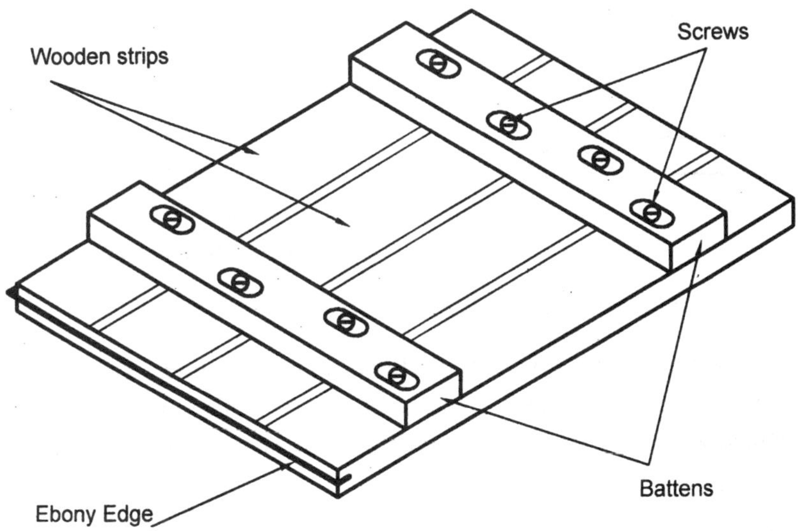

Fig. 1.1 Drawing board.

1.2.2 T-Square

It is used to draw horizontal lines on drawing sheet. It is also used to guide the set square for drawing vertical or inclined lines at 30°, 45° or 60°. It is usually made from hard wood. Two wooden strips e.g., stock and blades (As shown in Fig. 1.2) are joined together at right angles with the help of screw and pins. Stock slides over the working edge of drawing board and blade moves up and down horizontally.

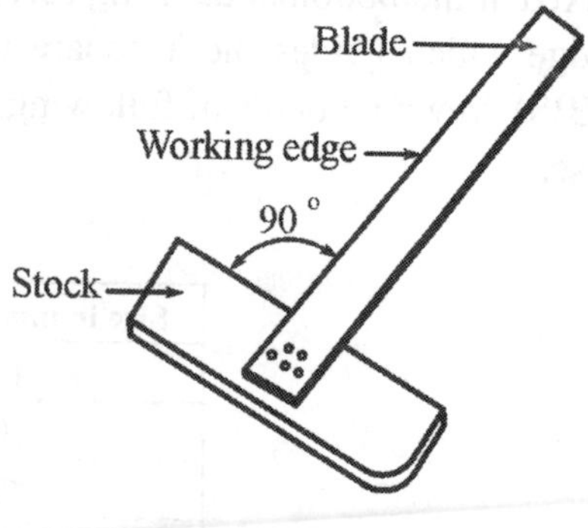

Fig. 1.2 T-Square.

1.2.3 Set Square

Set squares are used to draw vertical and inclined lines at 30°, 45°, 60° etc. Set squares are used with T-square. Set squares are usually made of plastic, wood, etc. A 45° set square and a 30°-60° set square are shown in Figs. 1.3(a) and (b).

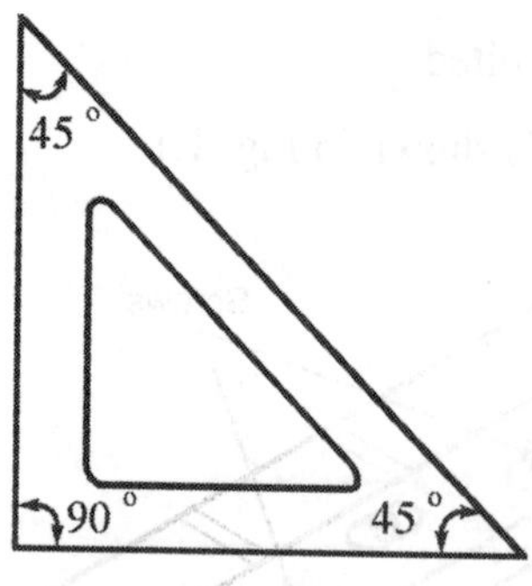

Fig. 1.3(a) 45° set square.

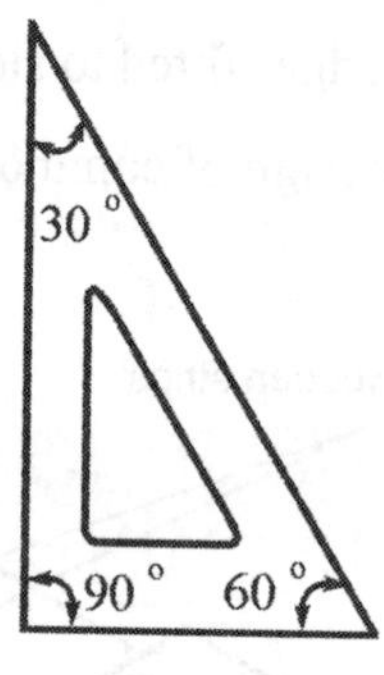

Fig. 1.3(b) 30°-60° set square.

1.2.4 Mini Drafter or Drafting Machine

Mini drafter is used to draw horizontal, inclined or vertical lines on drawing paper or sheet. It has eliminated the use of T-square, set square, scales and protractors. Now-a-days, it has gained wide popularity and is being used by majority of engineering students.

Its one end is provided with clamping screw and clamp. This end can be fixed on one of the edges/corners of drawing board as shown in Fig. 1.4. Other end is fitted with L-type scale and a protractor which can be set at any desired angle.

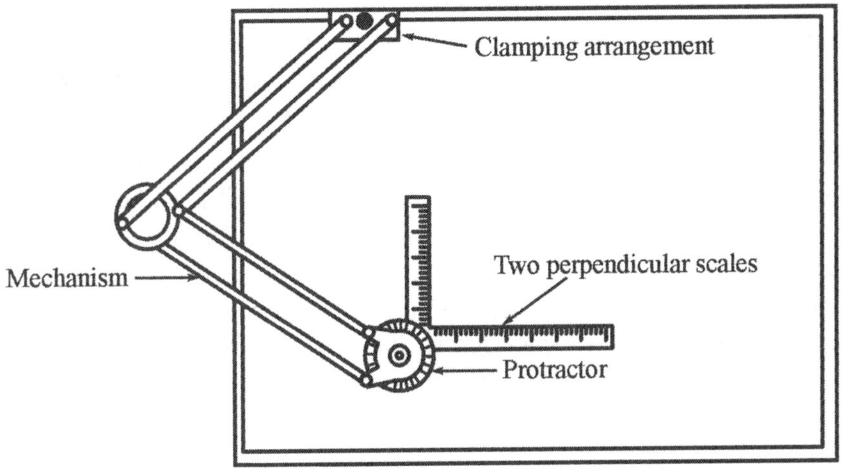

Fig. 1.4 Mini drafter.

1.2.5 Instrument Box

Instrument box contains various instruments for specific work. Usually following instruments are found in a typical instrument box.

(a) Large size compass (b) Large size divider

(c) Small size bow compass (d) Small size bow divider

(e) Inking pen (f) Lengthening bar

(a) ***Large Size Compass:*** The large size compass (as shown in Fig. 1.5) is used to draw circles, semi circles or arcs of required radius. It consists of two metal legs hinged together at upper end. One leg is fitted with a pointed needle whereas other end has a provision/clamp to hold pencil lead. This pencil leg is detachable and can be interchanged with linking pen.

(b) ***Large Size Divider:*** The large size divider is presented in Fig. 1.6. It consists of two legs hinged at the upper end. Both the ends of legs are fitted with steel pins. It is used to divide straight lines into required number of equal parts. It can also be used to transfer dimensions from one part to other part of the drawing.

(c) ***Small Size Bow Compass:*** Small bow compass (Fig. 1.7) is used for drawing circles, semi circles, arcs etc., of small radius (less than 25 mm radius). It is also convenient to use when large number of circles with the same radius are required to be drawn. It has one leg with a steel pin while another leg with a clamp for inserting pencil lead. Small bow compass may be of central adjustment type and side adjustment type. A knurled nut is used to adjust the distance between two legs.

(d) ***Small Size Bow Divider:*** Small bow divider is shown in Fig. 1.8. It is used to divide a line and transfer the distance from one part of the drawing to another part. The distance between legs can be adjusted with the help of knurled nut. This is very convenient for working over small distances.

(e) ***Inking Pen:*** This is used to draw lines in ink. It consists of two steel nibs fitted together in a holder. The distance between two steel nibs can be adjusted with the help of a knurled nut. A typical inking pen is shown in Fig. 1.9.

(f) ***Lengthening Bar:*** Lengthening bar is used as an extension bar with large size compass for making circles or arcs of large radius (more than 75 mm) as shown in Fig. 1.10. For using this, pencil leg is first detached from large compass and then lengthening bar is attached. The pencil leg is now attached to lengthening bar.

Fig. 1.5 Large compass .

Fig. 1.6 Large divider.

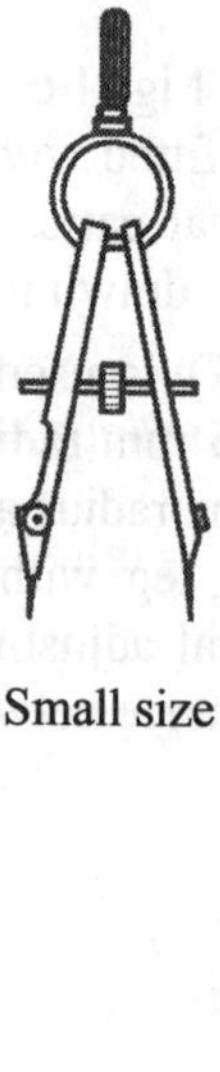

Fig. 1.7 Small size bow compass.

Fig. 1.8 Small size bow divider.

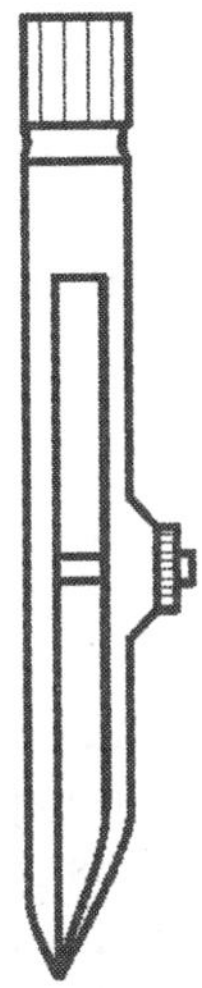

Fig. 1.9 Inking pen.

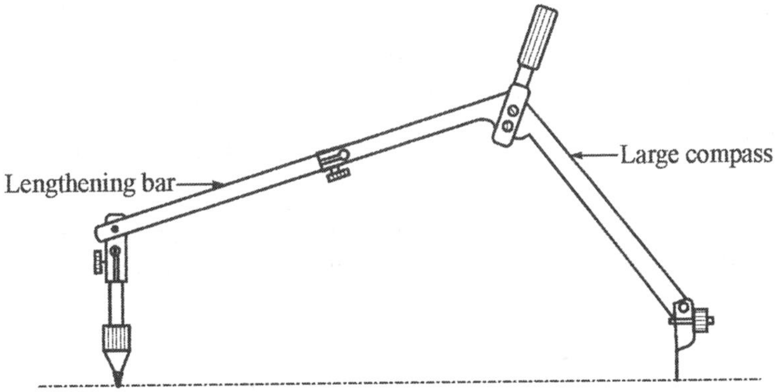

Fig. 1.10 Lengthening bar.

1.2.6 Drawing Pencils

Any kind of drawing work is initiated with drawing pencil. Selection of proper grade of pencil for various components of a drawing is very important in order to achieve accurate and good appearance. To identify pencil, grade is marked on it. Standard way of designating grade is to use some numeral followed by some letter or some letter only e.g., 2H, 3H, B etc. Depending upon hardness of graphite lead, pencils of different grades have been recommended by BIS. Such grades (from hard to soft) are 9H, 8H, 7H, 6H, 5H, 4H, 3H, 2H, H, F, HB, B, 2B, 3B, 4B, 5B, 6B and 7B. Pencils with 9H and B grades are considered as hardest and softest respectively. Hard pencil produces light and thin line whereas soft pencil produces dark and thick line. Pencils with H and 2H grades are most commonly used for drawing various objects. Soft pencils such as H or HB may be used for darkening of outlines, title, lettering etc.

1.2.7 Drawing Clips/Pins

Drawing sheet is placed on the drawing and is fixed using drawing clips or pins. Drawing clips are available in stainless steel or plastic material. Typical drawing clip is shown in Fig. 1.11(a) and (b).

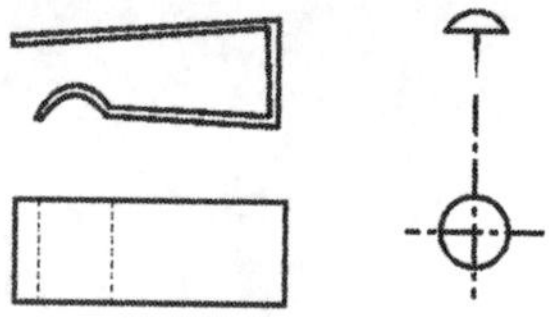

(a) (b)

Fig. 1.11 Drawing clip and pin.

1.2.8 Sand Paper Block

Sand paper block (Fig. 1.12) is used to sharpen the pencil lead to conical point. A piece of sand paper is pasted on a wooden block. The pencil lead is rub on the surface of sand paper to get the required level of sharpness. To draw smooth lines of uniform thickness, it is important to maintain the sharpness of pencil during drawing work because dull pencil always produces fuzzy and unsmooth lines. Pencil can be sharpened to following two types (Fig. 1.13):

(a) Conical point and

(b) Chisel edge

Fig. 1.12 Sand paper block.

Conical point is suitable for drawing lines, dimensioning and general purpose work whereas chisel edge is used to draw straight line of uniform thickness.

(a) (b)

Fig. 1.13(a) Conical point and **(b)** Chisel edge.

1.2.9 Eraser

It is usually made up of rubber and is used to remove pencil lines or marks. It should be made from good quality soft rubber to prevent residue after erasing a drawing.

1.2.10 Scale (Ruler)

A scale is used to measure the distances and to draw straight lines. Scales can be made from wood, plastic or metal. One long edge is marked with divisions of centimetres and millimetres and the other long edge may be marked with division of inches and its sub units.

1.2.11 Engineer's Scale

Drawing of bigger objects can not be prepared with true dimensions. It is always required to reduce the actual dimensions to fit the normal paper size. However, full scale drawing can also be prepared for smaller objects such as watch components, small electronic circuits etc. Engineer's scale can be made from wood, celluloid, card board or metal. On both sides divisions are marked with the reduced or enlarged lengths in a particular ratio. For example, full size scale (1:1) on one side and half size scale (2:1) on other side. As per BIS standard full size scale is designated as 1:1. Reducing scales may be with 1:2, 1:5, 1:10, 1:20, 1:25, 1:50, 1:100 and enlarging scale may be with 10:1, 5:1 and 2:1. The ratio 1:10 means that 1 cm on drawing will represent 10 cm actual distance. Engineer's scales are used to prepare drawings with either enlarging or reducing size.

1.2.12 French Curves

French curves are shown in Fig. 1.14. They are made of transparent plastic or celluloid and are used to draw curved lines or arcs of any irregular shape or of desired curvature.

Fig. 1.14 French curves.

1.2.13 Protractor

Protractor is also made of transparent celluloid material. It is used to measure angles or to construct angles of any required degree. A protractor is shown in Fig. 1.15.

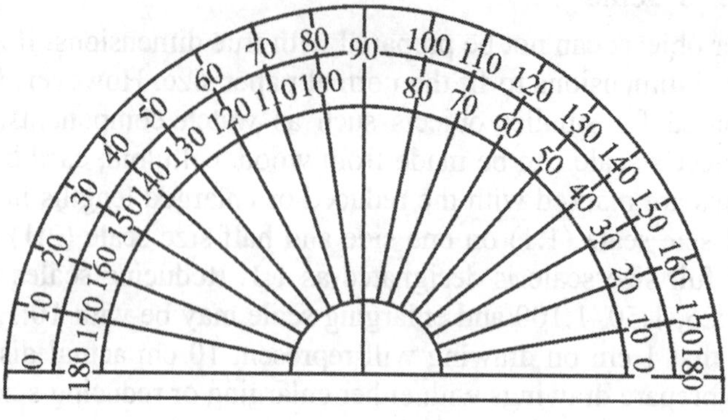

Fig. 1.15 Protractor

1.2.14 Drawing Sheet

Now-a-days, drawing sheet/paper is available in different quality and sizes. The quality of drawing work and its appearance depends on the quality of drawing sheet. A good quality sheet will not leave black marks while erasing. An ideal drawing sheet should meet the following requirements:

(a) It should be tough and strong.

(b) It should have good erasing quality so that, if eraser is used its fibres should not come out.

(c) It should be super white in colour.

As per recommendation of Bureau of Indian standards (BIS), standard size of drawing sheets can be designated as shown in Table 1.2.

Table 1.2

S. No.	Designation	Size of Drawing Sheet Length (mm) × Width (mm)
1	A_0	1189 × 841
2	A_1	841 × 594
3	A_2	594 × 420
4	A_3	420 × 297
5	A_4	297 × 210
6	A_5	210 × 148
7	A_6	148 × 105
8	A_7	105 × 74
9	A_8	74 × 52

1.2.15 Pencil Sharpener

Pencil sharpener is used to sharpen the pencil. It consists of a cutting blade, fitted inside its body. It removes the wood around the lead and produces conical end leaving the lead exposed. The lead can be prepared as conical point or chisel edge.

1.2.16 Handkerchief or Towel Cloth

Handkerchief or towel cloth of suitable size is used as a duster. Before starting the drawing work, all the instruments and accessories should be cleaned thoroughly using duster or towel cloth. This can also be used for sweeping away the crumbs which are formed due to erasing with rubber or eraser.

1.3 Lines

Engineering drawing of any object is systematic combination of lines of different types. Different types of lines are considered as alphabet of this language. As per Indian standards (SP 46-1988) types of lines and their applications are presented in Table 1.3. Figure 1.16 illustrates use of various lines.

Table 1.3

Line	Description	Applications
A	Continuous thick	- Visible outlines - Visible edges
B	Continuous thin	- Imaginary lines of intersection - Dimension lines - Projection lines - Leader lines - Hatching - Outlines of revolved section in place - Short centre lines
C	Continuous thin free hand	- Limits of partial or interrupted views and section, if the limit is not a chain thin
D	Continuous thin (straight with zigzags)	- Long break line
E	Dashed thick	- Hidden outlines - Hidden edges
F	Dashed thin	- Hidden outlines - Hidden edges
G	Chain thin	- Centre line - Lines of symmetry - Trajectories

Table 1.3 *Contd...*

Line	Description	Applications
H	Chain, thick at ends and change of direction	- Cutting planes
J	Chain thick	- Indication of lines or surfaces to which a special requirement applies
K	Chain thin Double Dashed	- outlines of adjacent part - Alternative and extreme positions of moving parts - Centroidal lines - Initial outlines prior to forming - Parts situated in front of the cutting plane.

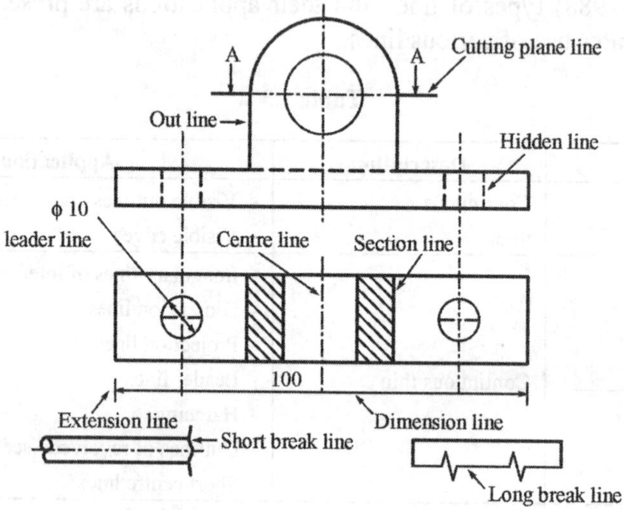

Fig. 1.16 Use of various lines.

1.4 Lettering

Lettering is used to write titles, dimensions and other necessary information on a drawing. All letters should have uniform height and width and should be legible, simple and easy to write. Lettering should be done with dark pencil such as 2H or HB. The lettering should be done in such a manner that it should take minimum time without the use of drawing instruments. Letters used for engineering drawing may broadly be classified into two types i.e., single stroke letters and gothic letters.

1.4.1 Single Stroke Letters

Single stroke letters are simplest form of letters and are used to write titles, dimensions, notes and other information on a drawing. Lettering using single stroke letters has been recommended by Bureau of Indian Standards (BIS) and described in IS:9609-1990. The word "single stroke" indicates that the thickness of lines used in letters should be such as obtained in single stroke of pencil. It does not mean that letter written in single stroke without lifting the pencil. The single stroke letters may be of two types e.g., vertical and inclined letters.

The vertical letters are used for general purpose lettering in engineering drawing. Height to width ratio of such letters varies. But for all practical purposes the letters with height to width ratio may be taken as 7:5 or 6:5 for all capital letters except I, J, L, M and W. For M and W the suitable ratio is 10:8 and for I, this ratio is taken as 10:2. For numerals such ratio may be taken as 7:4 or 6:3. Single stroke vertical upper and lower case letters and numerals are shown in Fig. 1.17. The inclined/italics letters are written

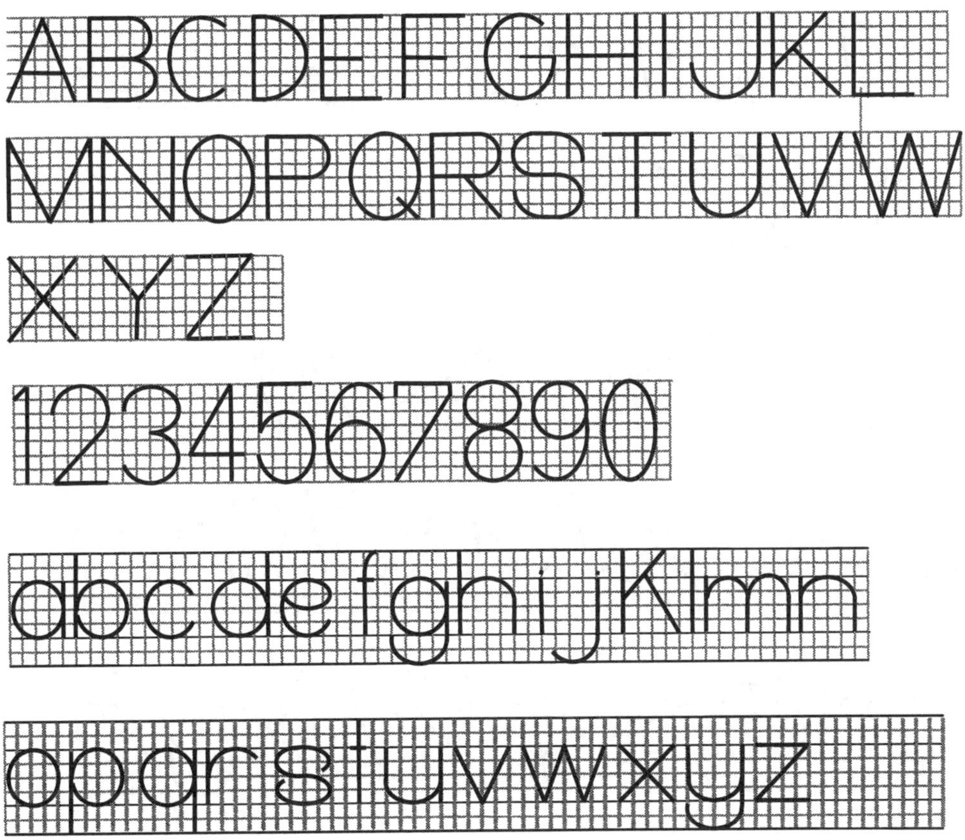

Fig. 1.17 Single stroke vertical letters and numerals.

at an angle of 75° from horizontal towards right. Fig. 1.18 illustrates typical inclined letters and numerals. The size of letters and numerals is specified by their heights 'h'. The standard heights of letters and numerals for practical use are 1.8, 2.5, 3.5, 5.7, 10, 14 and 20 mm as recommended by BIS. The letters size 5 mm-10 mm may be used for main title of drawing. For sub-titles, letter size of 3.5 mm or 5 mm may be used conveniently. The letters and numerals having size 2.5, 3.5 or 5 mm are suitable for writing notes, dimensions etc. The single stroke letters may be of A type or B type. In A type letter, height of letter is divided into fourteen parts whereas in B type letter, the height is divided into 10 parts. The recommended ratio, spacing between letters and thickness of lines (stems) of letters are summarized in Table 1.4 and are explained in Fig.1.19.

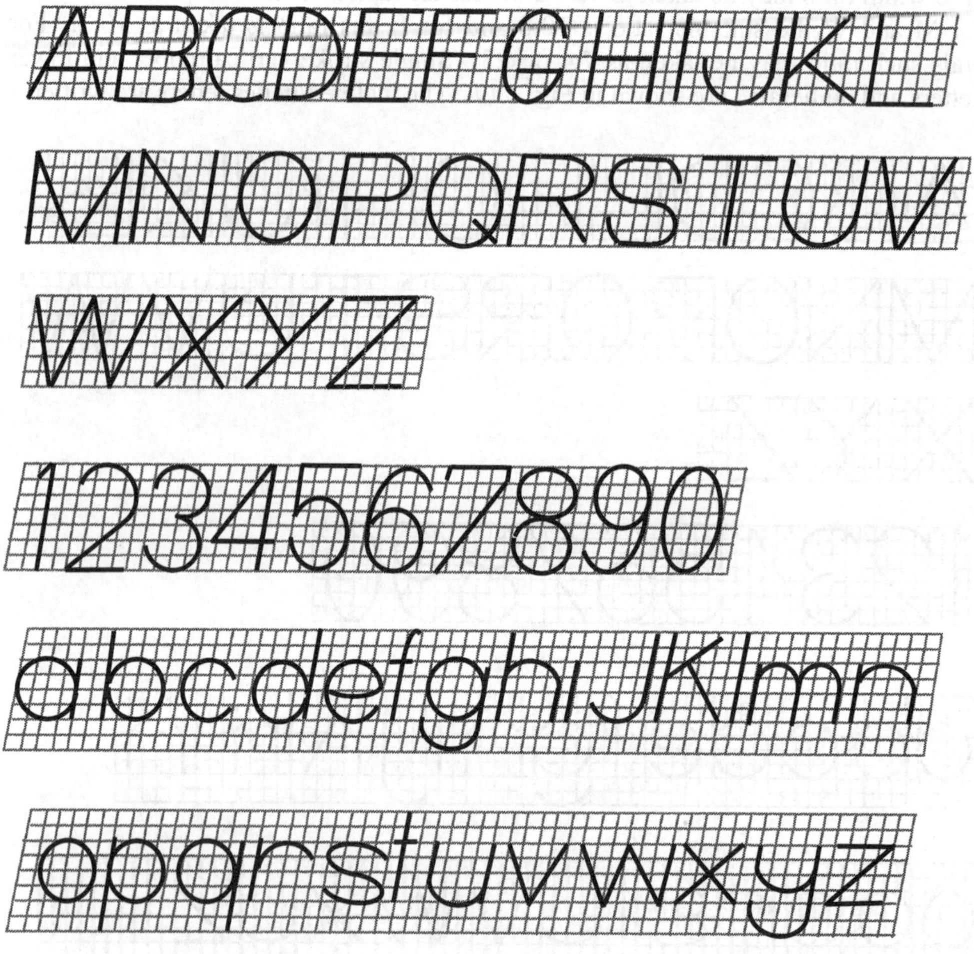

Fig. 1.18 Single stroke inclined letters and numerals.

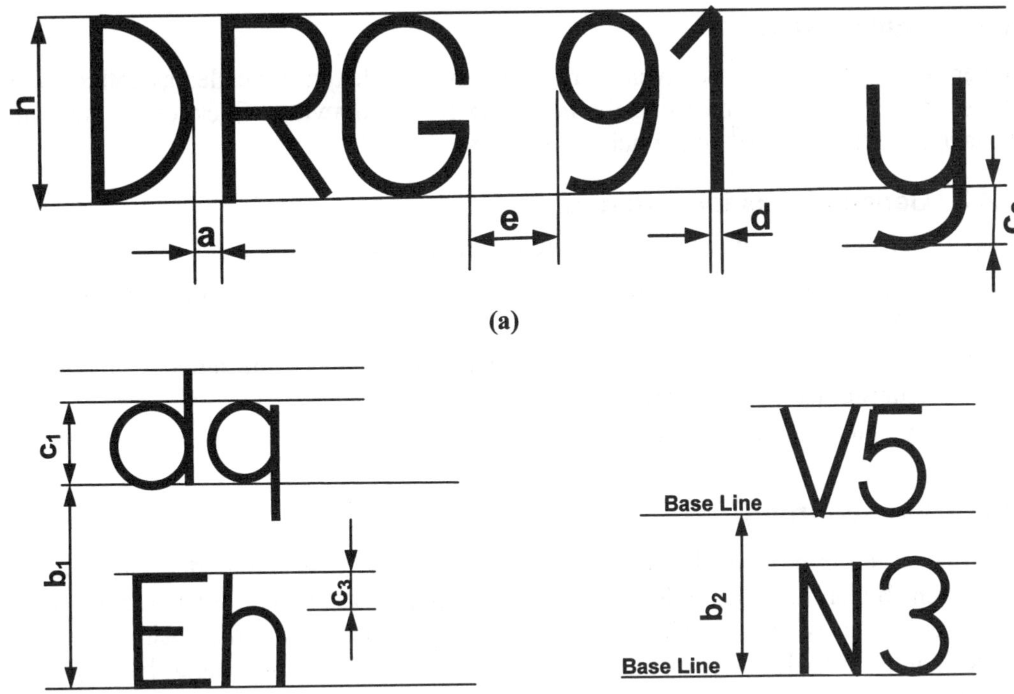

(b) (c)

Fig. 1.19

Table 1.4 Standard dimensions for A type and B type letters

Types of Letters	Height of Upper Case Letters (h)	Height of Lower Case letters (c_1)	Tails of Lower Case Letters (c_2)	Stem of Lower Case Letters (C_3)	Spacing Between Characters (a)	Minimum Spacing Between Base Lines for Upper Case and Lower Case Letters (b_1)	Minimum Spacing Between Base Lines for Upper Case Letters (b_2)	Minimum Spacing Between Words (e)	Line Width (d)
Type A	$\left(\frac{14}{14}h\right)$	$\left(\frac{10}{14}h\right)$	$\left(\frac{4}{14}h\right)$	$\left(\frac{4}{14}h\right)$	$\left(\frac{2}{14}h\right)$	$\left(\frac{21}{14}h\right)$	$\left(\frac{17}{14}h\right)$	$\left(\frac{6}{14}h\right)$	$\left(\frac{1}{14}h\right)$
Type B	$\left(\frac{10}{10}h\right)$	$\left(\frac{7}{10}h\right)$	$\left(\frac{3}{10}h\right)$	$\left(\frac{3}{10}h\right)$	$\left(\frac{2}{10}h\right)$	$\left(\frac{15}{10}h\right)$	$\left(\frac{13}{10}h\right)$	$\left(\frac{6}{10}h\right)$	$\left(\frac{1}{10}h\right)$

1.4.2 Gothic Letters

When the stem of letters are given more thickness, such letters are called as gothic letters. These letters are used for writing the main title for ink drawing. Thickness of stem may be taken as $1/5^{th}$ to $1/10^{th}$ of the height of letters.

1.4.3 General Rules for Lettering

(a) Select the size of letters suitable to given drawing.

(b) Draw horizontal guide lines keeping the distance between them equal to the height of letters.

(c) Width of letters may be taken equal to the height of the letter. Complete the lettering with the standard dimensions.

(d) Do not erase the guide lines.

1.5 Dimensioning

In order to describe any object or drawing technically, it is required to show all the measurements on it. Placing details such as length, breadth, height, thickness of objects, diameter/radius of holes, slots, grooves etc., is known as dimensioning. The dimensioning should be very clear, legible and easy to understand by engineers, technicians etc.

1.5.1 Dimensioning Terminology

Various terms used for dimensioning (Refer Fig. 1.20) are explained as below:

(a) Dimension Line: Dimension line is a thin continuous line used to indicate measurement. The numerical value of a measurement is placed near the middle of this line.

(b) Extension Line: Extended line beyond the outline of any object is known as extension line. The extension line is drawn in such a way that dimension line remains perpendicular to it.

(c) Arrowhead: Arrowheads are placed at the extreme ends of a dimension line in a direction opposite to each other. These arrowheads are used to terminate dimension line. Different types of arrowheads have been recommended by BIS (as shown in Fig. 1.21) such as open 90°, open 30°, closed blank, closed filled etc. The length of an arrowhead is kept 3 times its breadth. The length of an arrowhead is kept about 3 mm for most of the drawings.

(d) Leaders or Pointer Line: Leader lines are thin continuous line which is used to indicate dimensions, notes etc., outside the drawing. Leaders are drawn at any desired angle such as 30°, 45° or 60°. Leaders should not be drawn as vertical, horizontal or curved lines.

(e) **Dimension Figure:** It is a numerical figure which indicates size of the given drawing.

(f) **Symbol:** A symbol is a mark which is used to represent any object, machine component or machining process in order to save time and avoid confusion due to complex detail.

(g) **Notes:** A note on a drawing gives required information regarding specific operation or features.

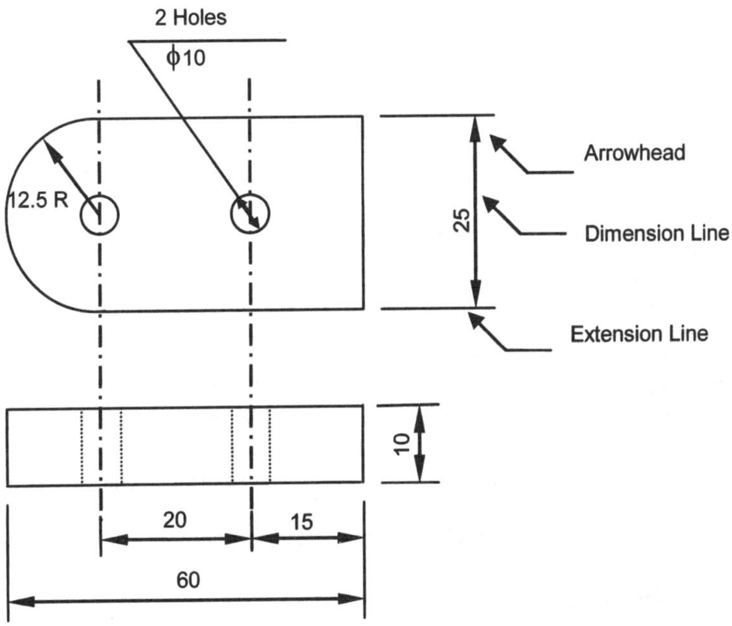

Fig. 1.20 Dimensioning terminology.

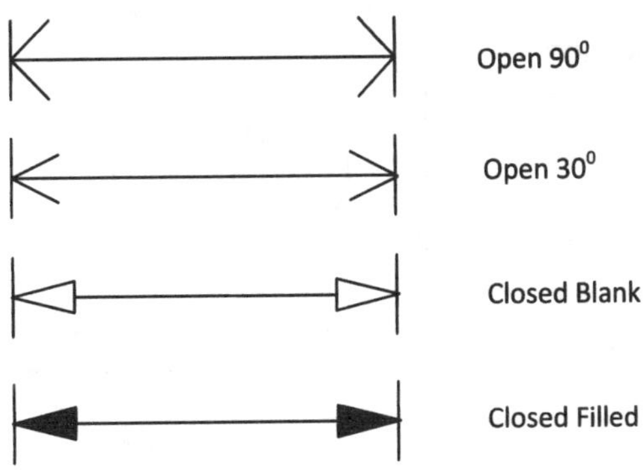

Fig. 1.21 Types of arrowheads.

1.5.2 System of Dimensioning

Dimensions are placed in such a manner that it is clear and readable. There are two recommended systems for placing dimensions on a drawing as follows:

(i) **Aligned System:** In aligned system, all the dimensions are so placed that they may be read easily from the bottom or right hand edge of the drawing paper or sheet. Fig 1.22(a) shows aligned system of placing dimensions on a drawing.

(ii) **Unidirectional System:** In this system all the dimensions are so placed that a reader may read them from the bottom of the drawing sheet as shown in Fig. 1.22(b). Such system is useful for large drawings.

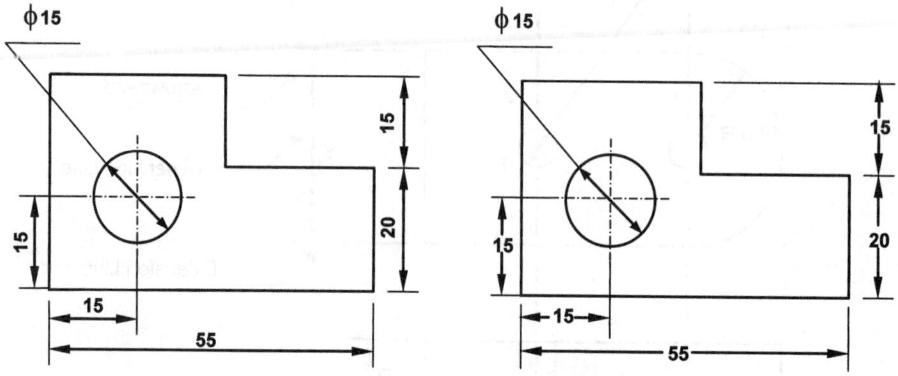

(a) Aligned System **(b)** Unidirectional System

Fig. 1.22 Systems of dimensioning.

1.5.3 Types of Dimensions

In engineering drawing, two types of dimensions are used e.g., Functional (size) dimension and Location dimension. The dimensions which are used to locate various sizes of the object such as length, breadth, diameter, radius etc., are known as size dimensions. The location dimensions are used to locate position of various constructional detail

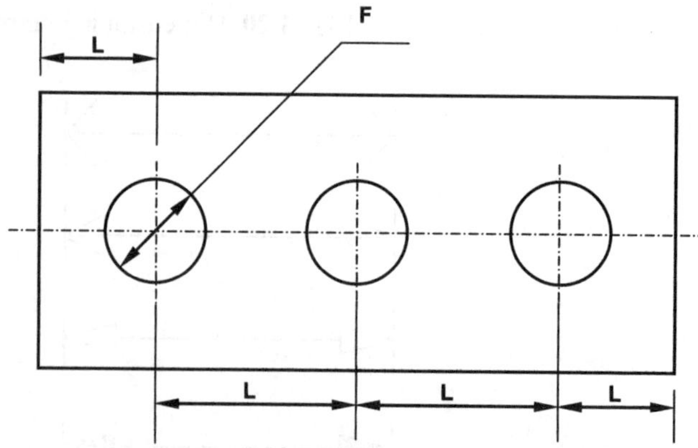

Fig. 1.23 Types of dimensions.

within the object. The functional and location dimensions are represented by letters 'F' and 'L' respectively as shown in Fig. 1.23.

1.5.4 General Rules of Dimensioning

Dimensions should be clear, readable and easy to understand.

1. All dimensions should be placed outside the view. Marking of dimensions inside the view is permitted only when it is not possible to place them outside the view.
2. Dimension lines should be drawn at least 8 mm away from the visible outlines of the view.
3. Dimensional value must be placed near the middle of dimension line. If it is not possible to do so then, it can be placed above the extended portion of the dimension line. Arrowheads are usually drawn within the limit of dimension line. If distance between two extension lines is not sufficient then dimensions may be placed as shown in Fig.1.24.

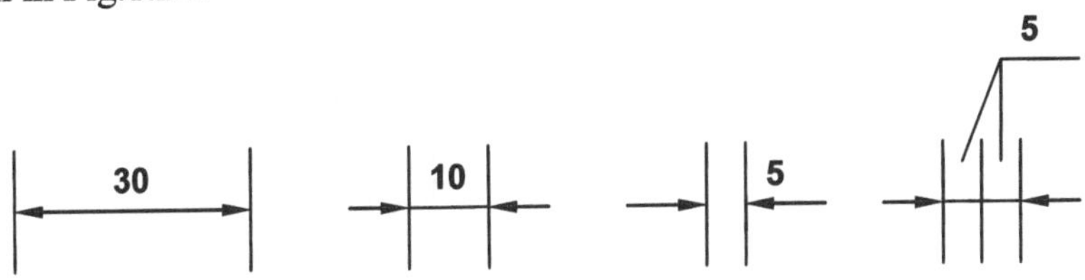

Fig. 1.24

4. Dimension of a particular feature should not be repeated not only in the same view but also in another view of the same object.
5. As far as possible every dimension should be marked on a drawing but none of dimension should be repeated.
6. Two dimension lines should not cross each other or dimension line should not intersect any other line of the drawing.
7. An outline of an object should not be used as a dimension line.
8. Centre line of any object should not be used as an extension line for placing dimensions.
9. As far as possible, aligned system of dimensioning should be used.

Method of placing dimensions to common features is presented in Fig. 1.25.

Object/ Feature	Illustration
Diameter of Circle	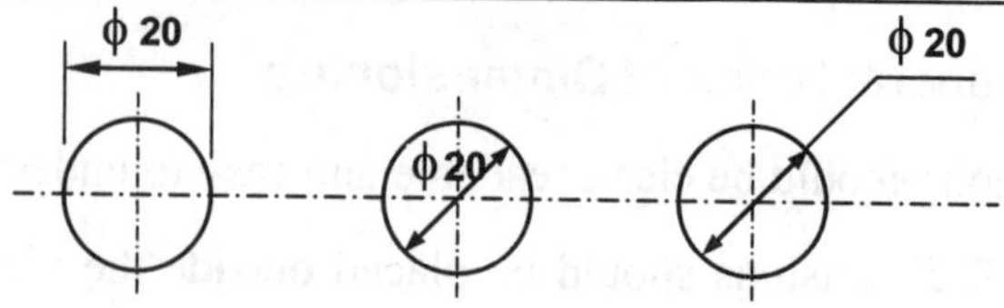
Angle	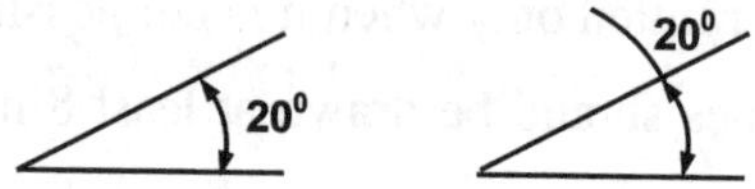
Spherical Object	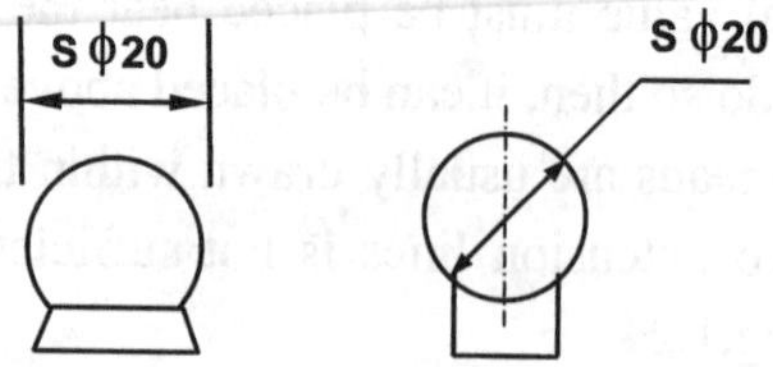
Square Object	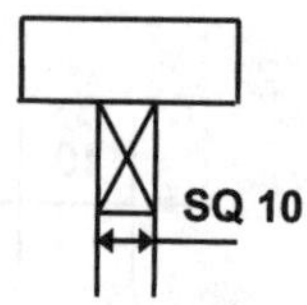
Chamfer	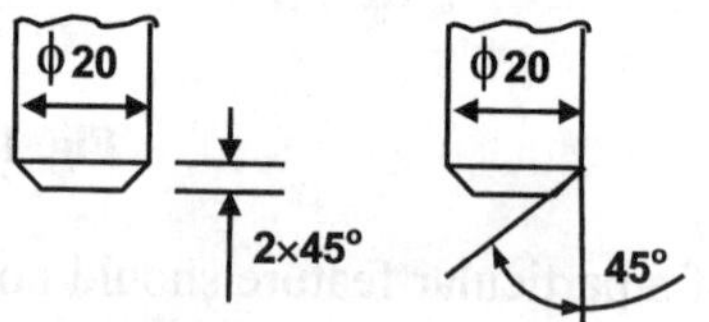
Counter Shunk	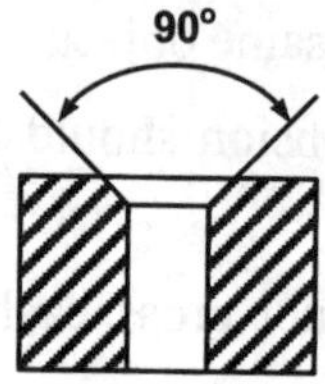
Taper (D-d)/L	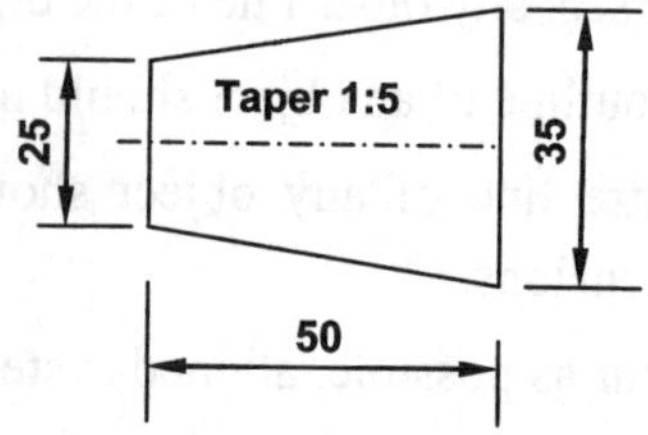

Fig. 1.25

1.6 General Preparation before Commencing Engineering Drawing

1. Arrange all drawing instruments required, clean them and check for their accuracy.
2. Design of drawing table and chair should facilitate comfortable sitting and proper movement of body. A bed arrangement will lead to early tiredness which may affect accuracy of work.
3. Natural air circulation and sufficient illumination in drawing hall or room must be ensured.
4. Keep the required instruments and accessories near the drawing table.
5. Fix the required size of drawing sheet on a drawing board with the help of drawing clips and clamp the mini drafter on the board in such a manner that its scale can move covering maximum area of drawing sheet.
6. Sharp the pencils of required grades to prepare their leads ready for use.
7. Draw margin lines and title block. The title block is placed in the lower right hand corner of drawing sheet. As per BIS, the recommended size of title block is 185×65 mm. A title block must contain following information.
   (a) Title of drawing
   (b) Drawing number
   (c) Name of firm
   (d) Scale
   (e) Symbol indicating method of projection
   (f) Date, name of person, checked by, approved by etc.

A typical layout of drawing sheet and a title block suitable for engineering students for class room practice, are presented in Fig. 1.26 and 1.27 respectively.

1.7 Geometrical Constructions

1.7.1 Basic Constructions

Engineering drawing of various objects requires basic knowledge of some geometrical constructions. Students are advised to understand the principles of geometrical constructions because it is the prerequisite for solving various problems of engineering drawing. Basic constructions very often required during drawing work are explained with the help of following examples.

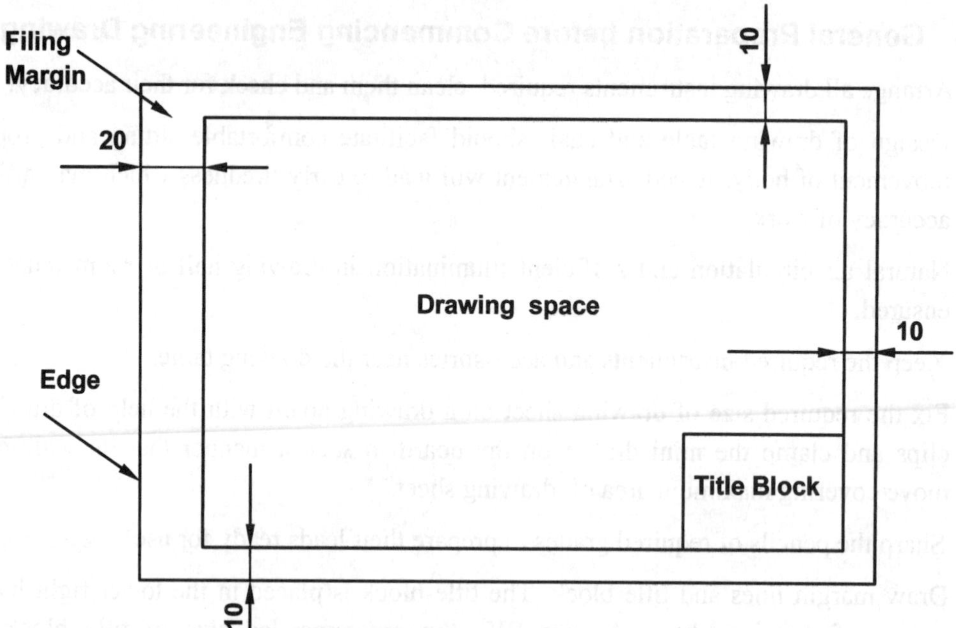

Fig. 1.26 Layout of drawing sheet.

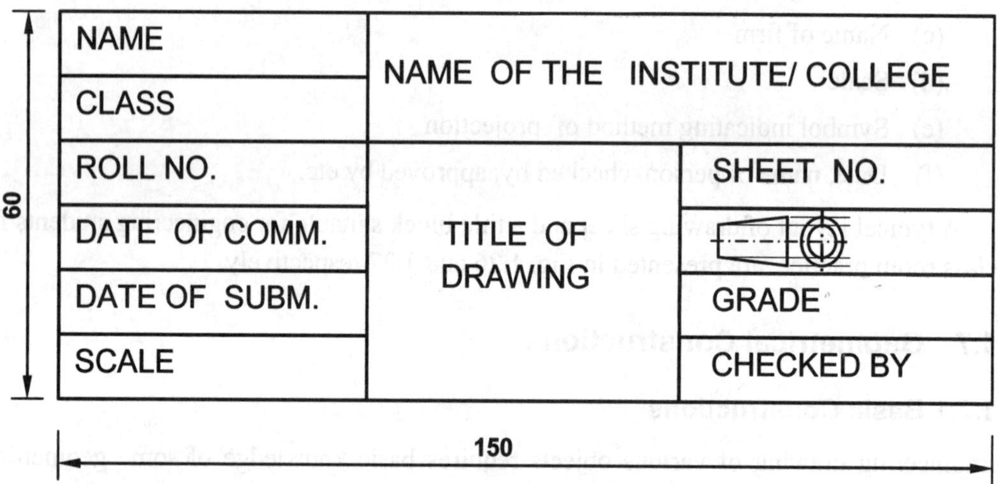

Fig. 1.27 Title block.

Problem 1.1 (Fig.1.28): *Draw a perpendicular bisector of a 60 mm long straight line.*

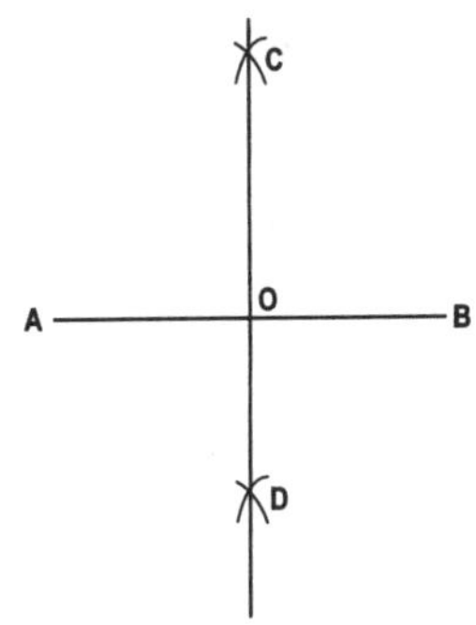

Fig. 1.28

Construction (Fig. 1.28):

1. First of all draw a straight line AB, 60 mm long.
2. Set compass to a length more than half the length of AB.
3. Keep the needle leg of the compass at point A and draw two arcs above and below line AB.
4. Now keeping the same length in compass, keep its needle leg at B. Draw two arcs above and below the line AB to intersect previous arcs at points C and D.
5. Draw a line passing through points C and D, which is the required perpendicular bisector of AB. The point O divides the line AB in two equal parts.

Problem 1.2 (Fig. 1.29): *Divide a straight line AB 70 mm long into six equal parts.*

Construction (Fig.1.29):

1. Draw a 70 mm long straight line AB.
2. Draw an another line AC inclined to AB with any suitable angle.
3. Mark six division points as 1', 2', 3', ... 6' on AC with the help of a divider or a compass. Join last division point (6′) to end point B.
4. Draw lines through 5', 4', 3', 2' and 1' parallel to line B6' meeting the line AB at 5,4,3,2 and 1 respectively.
5. Thus AB is divided into six equal parts.

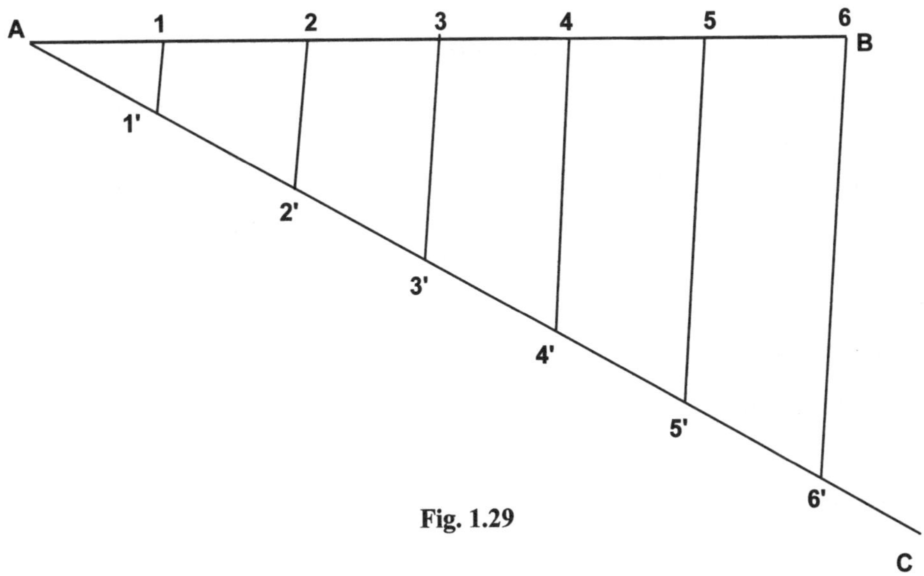

Fig. 1.29

Problem 1.3 (Fig.1.30): *Bisect a given angle.*

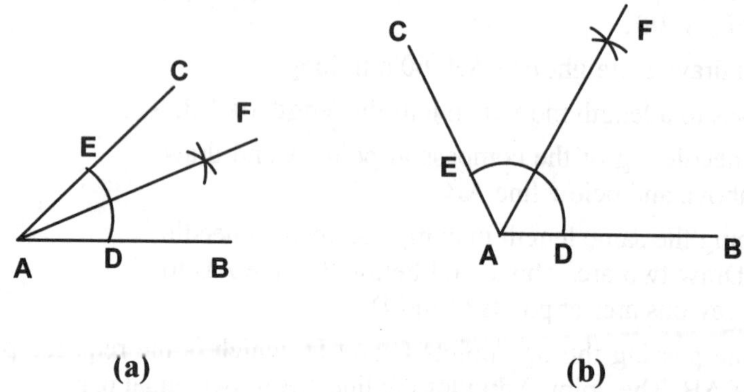

Fig. 1.30

Construction (Fig.1.30):

Let given angle ∠BAC is required to be bisected.

1. With A as centre and any convenient radius, draw an arc cutting AB and AC at points D and E respectively as shown in Fig.1.30(a) and (b).
2. With D as centre and any convenient radius, draw an arc within the included region.
3. With E as radius and same radius (as taken in step 2) draw an arc intersecting the previous arc at F. Draw a line joining A and F. AF is the required angle bisector of ∠BAC.

1.7.2 Construction of Regular Polygon

A polygon having equal sides and equal interior angles is termed as regular polygon. The knowledge of drawing regular polygons is required for solving various problems in engineering drawing. Several methods are available for drawing different types of regular polygons. Students are advised to use the method which produces an accurate construction in a faster way. Such methods can be understood well with the help of following examples.

Problem 1.4 (Fig. 1.31): *Draw an equilateral triangle having 50 mm long sides when*

(a) *one of the sides is horizontal*

(b) *one of the sides is vertical*

(c) *one of the sides is inclined at 45° with the horizontal.*

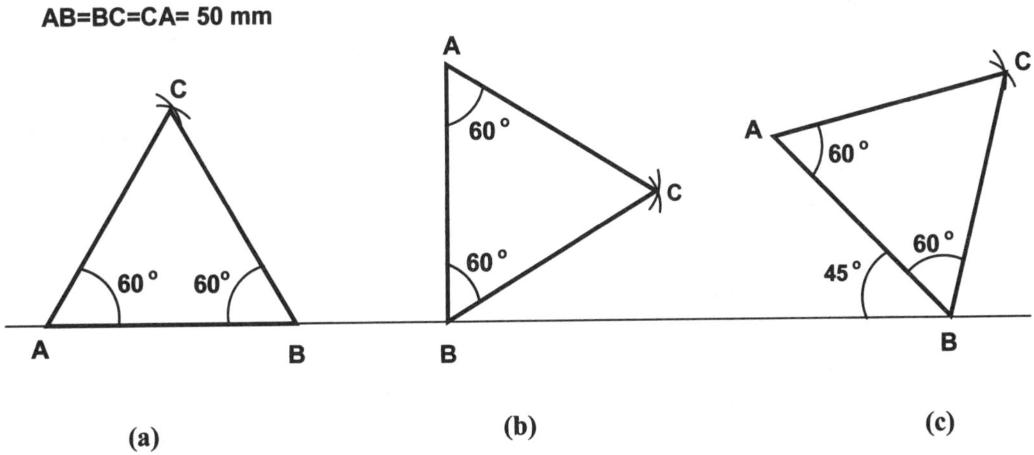

Fig. 1.31

Construction (Fig.1.31):

1. Draw a horizontal straight line AB, 50 mm long as shown in Fig. 1.31(a).
2. With A as centre and radius equal AB, draw an arc above line AB.
3. With B as centre and radius equal to AB, draw another arc to intersect previous arc at C.
4. Join A to C and B to C. The triangle ABC is required equilateral triangle.
5. Similarly draw the triangles for other positions of side AB as shown in Figs. 1.31 (b) & (c).

Problem 1.5 (Fig.1.32): *Construct a square with 45 mm long sides if*

(a) one of the sides is horizontal

(b) one of the sides is inclined 30° to horizontal.

(c) two sides are equally inclined (45°) with horizontal

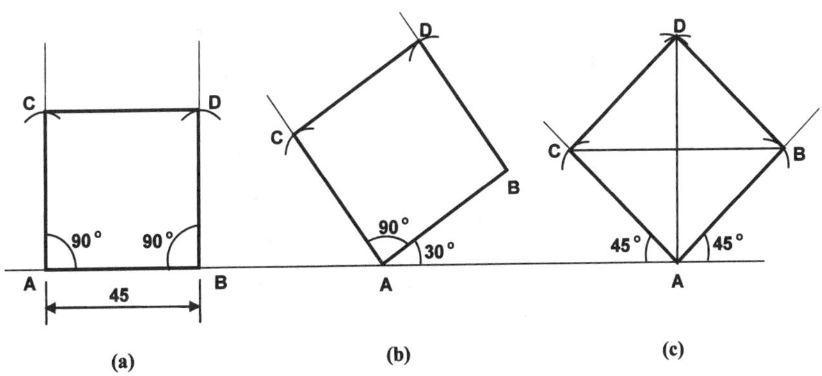

Fig. 1.32

Construction (Fig.1.32):

1. Draw a horizontal straight line AB, 45 mm long as shown in Fig.1.32(a).
2. Draw perpendicular lines at A and B.
3. With A as centre and radius equal to AB, draw an arc to intersect the perpendicular at D.
4. Similarly, obtain point C on another perpendicular.
5. Join C to D. Thus, ABCD is the required square.
6. Similarly draw the squares for other positions of side AB as shown in Figs.1.32 (b) & (c).

Problem 1.6 (Fig. 1.33): *Construct a square having its diagonals 60 mm long.*

Construction (Fig.1.33):

1. Draw a horizontal line AC, 60 mm long as diagonal of square.
2. Draw a vertical line BD, 60 mm long as perpendicular bisector of AC such that AC also bisect BD.
3. Join A, B, C and D to obtain required square.

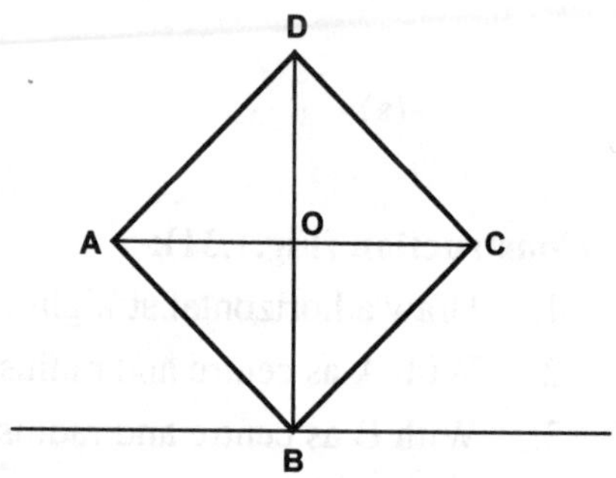

Fig. 1.33

Problem 1.7 (Fig.1.34): *Draw a regular pentagon with 50 mm sides.*

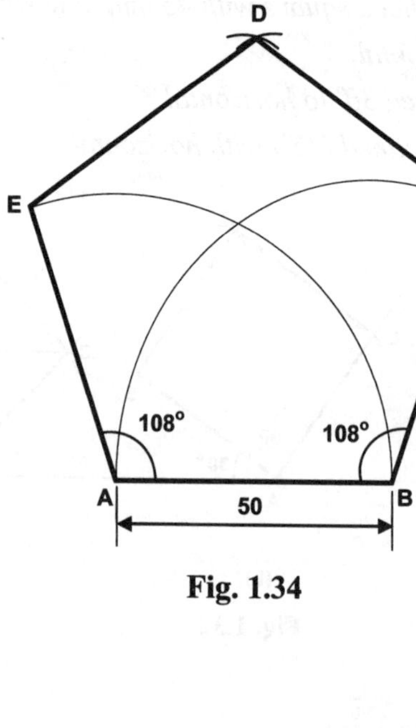

Fig. 1.34

Construction (Fig.1.34):

1. Draw a line AB 50 mm long.
2. Draw lines BC and AE 50 mm long and inclined at 108^0 with AB.
   With E and C as centres and radius equal to 50 mm, draw arcs intersecting at D.
3. Join points E and C to D. ABCDE is the required pentagon.

Problem 1.8 (Fig. 1.35): *Draw a regular hexagon with 50 mm sides.*

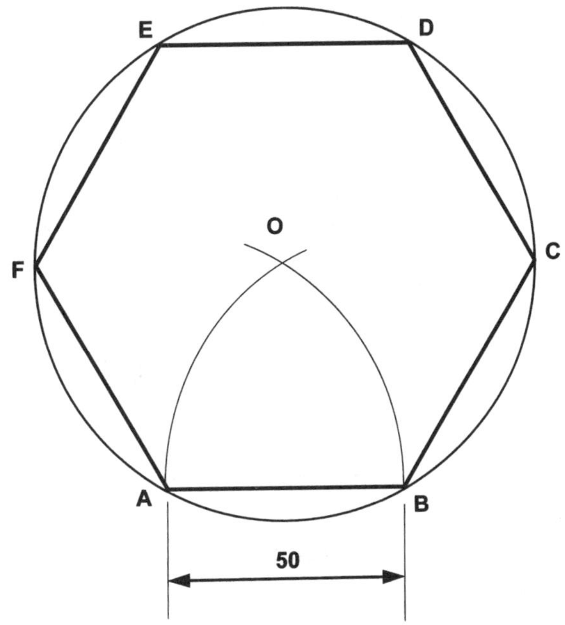

Fig. 1.35

Construction (Fig.1.35):

1. Draw line AB equal to 50 mm as one of the sides of hexagon.
2. With A and B as centres and radius equal to 50 mm, draw two arcs to intersect each other at O.
3. With O as centre and radius equal to 50 mm, draw a circle passing through A and B.
4. Using compass with radius equal to 50 mm draw arcs cutting the circle at C, D, E and F.
5. Join all the points with straight lines. ABCDEF is the required hexagon.

1.7.3 General Methods for Regular Polygons

General method can be used for drawing regular polygons with any number of sides. The construction of a regular hexagon using this method is explained with the help of following problem.

Problem 1.9 (Fig. 1.36): *Draw a regular hexagon with 45 mm sides.*

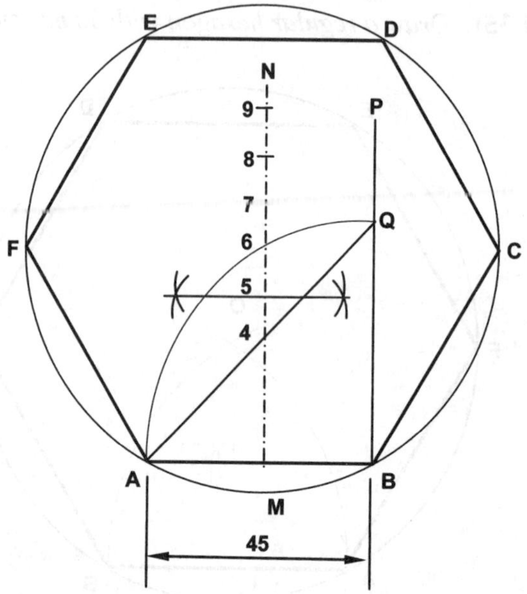

Fig. 1.36

Construction (Fig.1.36):

1. Draw a line AB of given side (45 mm). Erect a perpendicular BP at point B.
2. With B as centre and radius equal to AB draw an arc to intersect the perpendicular at Q. Join A to Q and draw MN as a perpendicular bisector of AB.
3. Mark point 4 at the intersection of MN and line AQ and mark point 6 at the intersection of MN and arc AQ. Bisect the line segment 4-6 and mark point 5 between 4 and 6. (For drawing polygons with 7,8,9..etc. sides, mark points 7, 8, 9... etc., on line MN such that 45 = 56 = 67 = 78.. etc.)
4. For drawing a regular hexagon, take 6 as centre and radius equal to A6, draw a circle passing through A and B. Taking a radius equal to AB divide the circle into six equal parts. Mark the divisions as C, D, E and F. Join all the division points to get a regular hexagon ABCDEF.

Note: *[Using the appropriate division point (4, 5, 6, 7 etc.,) on the line MN, a polygon of required sides can be constructed following the procedure explained in step 4].*

Exercises

1. What do you understand by Engineering Drawing? Why it is known as universal language of engineers? Explain.
2. Write the names of major drawing instruments with their uses?
3. How would you draw horizontal, vertical, inclined and parallel lines with the help of mini drafter? Explain with the help of suitable diagrams.
4. Explain the use of French curves in engineering drawing?
5. Draw different types of lines used in engineering practice.
6. Write the following sentence in freehand using single stroke vertical capital letters of 10 mm height.

   "Drawing is the Universal Language of Engineers"
7. Write freehand, in single stroke inclined capital letters of 10 mm height the following sentence.

   "Knowledge of Engineering Drawing is Essential For Engineers"
8. Write freehand, in single stroke vertical lower case letters of 5 mm height the following sentences.

   "Work is worship".
9. Write freehand, in single stroke inclined lower case letters of 5 mm height the following sentences.

   "Importance of man machine and materials in manufacturing".
10. What is the importance of dimensioning? Explain in brief.
11. What are the aligned system and unidirectional system of dimensioning? Explain in brief.
12. With the help of neat diagram, explain dimension line, extension line and leader line?
13. Draw and explain different types of arrow head used for dimension lines?
14. Write general rules of dimensioning?
15. Illustrate the method of dimensioning the following :

    (i) Diameter of Circle (ii) An angle (iii) Spherical objects (iv) Square objects (v) Holes (vi) Chamfers (vii) Countersunk hole and (viii) Taper.
16. Draw a perpendicular bisector of a 70 mm long straight line.

17. Draw a line AB, 80 mm long and divide it into nine equal parts.
18. Draw an equilateral triangle having 40 mm long sides when
    (i) one of the sides is horizontal (ii) one of the sides is vertical (iii) one of the sides is inclined at 45° with the horizontal.
19. Construct a square with 45 mm long sides if
    (i) one of the sides is horizontal (ii) one of the sides is inclined 30° to horizontal (iii) Two sides are equally inclined (45°) with horizontal
20. Draw a regular pentagon with 45 mm sides.
21. Draw a regular hexagon with 55 mm sides using any two methods.

CHAPTER 2

Scales

2.1 Introduction

Any object can be represented on drawing sheet/paper using its reduced or enlarged dimension if it is not possible to draw with its actual size. If the size of drawing sheet permits, drawing can be made with actual size. Such drawing is known as full size drawing. Very large objects such as buildings, bridges, heavy machineries, map of towns etc., cannot be represented on drawing with their actual sizes. For such objects, it is required to reduce the size of objects in some standard or given proportion for preparing their drawing. When the size of an object is very small such as tiny components of watch etc., then drawing is prepared with its enlarged dimensions. The proportion by which the dimensions of object is either reduced or enlarged in drawing, is termed as scale.

In other words, scale is a device which is graduated with different divisions representing actual distances according to some standard or given proportion. Scales can be classified as follows:

(a) **Full Size Scale:** When an object is drawn with its full size on drawing, the scale used is termed as full size scale. For such scale, "1:1" is indicated on the stick/scale.

(b) **Reducing Scale:** When the actual length of the object is reduced on drawing, the scale used is termed as reducing scale. The standard proportions for such scale are 1:2, 1:10, 1:15, 1:20, 1:50, 1:100 and so on.

(c) **Enlarging Scale:** When the object is drawn with the enlarged dimensions, the scale used is termed as enlarging scale. The standard proportion for enlarging scale may be selected as 2:1, 5:1, 10:1 etc.

For complete description of an object on drawing, the 'scale' used for drawing must be mentioned. The scale can be indicated on drawing sheet, in the following three ways:

(i) **As engineering scale:** In the scale, the relation between dimension on drawing and the corresponding actual dimension of the object is mentioned numerically

For example: 1 mm = 1 mm for full size scale

1 mm = 10 m for reducing scale

10 mm = 2 mm for enlarging scale

(ii) **As graphical scale:** Graphical scale is drawn on the drawing sheet itself, which can be used directly to set off distances or convert drawing length into actual length.

For example: Scale shown on the map of India/world.

(iii) **As ratio of length on drawing to actual length:** In this method, ratio of length of object on drawing to the actual length of object is represented as 1/100, 1/500 etc. The above ratio is also known as representative fraction (R.F), which is explained in detail in the following section.

2.2 Representative Fraction (R.F)

Representative fraction is defined as the ratio of length of object on drawing to the actual length of the object.

$$\text{R.F.} = \frac{\text{Length of object on drawing}}{\text{Actual length of object}}$$

For example, if 1 cm long line on drawing represents 10 meters as actual length, then

$$\text{R.F.} = \frac{1\,\text{cm}}{10\,\text{m}} = \frac{1\,\text{cm}}{10\times 100\,\text{cm}} = \frac{1}{1000}$$

Similarly, when 20 mm long line represents 0.5 mm actual length, then

$$\text{R.F.} = \frac{20\,\text{mm}}{0.5\,\text{mm}} = \frac{40}{1}$$

2.3 Units of Measurements

Table 2.1 and 2.2 represent units for linear measurement in Metric and British system respectively.

Table 2.1 Metric system - linear measurement

10 millimetres (mm) = 1 centimeter (cm)
10 centimetres (cm) = 1 decimeter (dm)
10 decimeters (dm) = 1 metre (m)
10 meters (m) = 1 decametre (da m)
10 decametre (da m) = 1 hectometre (hm)
10 hectometers (hm) = 1 kilometre (km)

Table 2.2 British system - linear measurement

12 inches = 1 feet
3 feet = 1 yard
220 yards = 1 furlong
8 furlongs = 1 mile

Some useful conversions are shown in Table 2.3

Table 2.3 Conversion factors

1 inch = 2.54 cm
1 mile = 1.609 km
1 hectare = 10,000 m^2
1 acre (ac) = 4840 $yard^2$

2.4 Types of Scale

In engineering practice, following types of scales are commonly used.

(a) Plain scale (b) Diagonal scale (c) Scale of chords

(d) Comparative scale (e) Vernier scale

2.5 Plain Scale

The plain scale consists of a line suitably divided into desired equal parts (primary divisions). The first part of which is further divided into smaller units as secondary divisions. These divisions show sub units of primary division or unit.

For example: If primary divisions represent centimetres then secondary divisions may be millimetres.

The plain scale is used to represent either two consecutive units or units and its sub units.

2.5.1 Information Required for Construction of a Plain Scale

For construction of scales on drawing sheet, following data is required.

(i) The representative fraction (R.F)

(ii) Maximum length/distance which can be measured or read

(iii) Minimum length/distance to be shown

(iv) Primary unit and its sub unit to be represented

2.5.2 Construction of a Plain Scale (Step-by-Step Construction)

The method of construction of a plain scale can be well understood with the help of following example.

The step-by-step construction has been presented for one problem only for easy understanding. Subsequent problems have been explained with steps written in text.

Problem 2.1 (Fig. 2.1): *Construct a plain scale of R.F.* $= \frac{1}{4}$ *to show decimetres and centimetres and long enough to measure up to 6 decimetre. Mark a distance of 3.8 decimetre on it.*

Given data: R.F. $= \frac{1}{4}$

Maximum length = 6 decimetre

Units: decimetres and centimetres

Distance to be marked = 3.8 d.m.

Determine length of scale (LOS) or L_s as follows:

$$\text{L.O.S} = \text{R.F.} \times \text{Maximum length to be measured}$$
$$= \frac{1}{4} \times 6 \times 10 = 15 \text{ cm}$$

Construction (Fig. 2.1):

Construct the scale as per following steps:

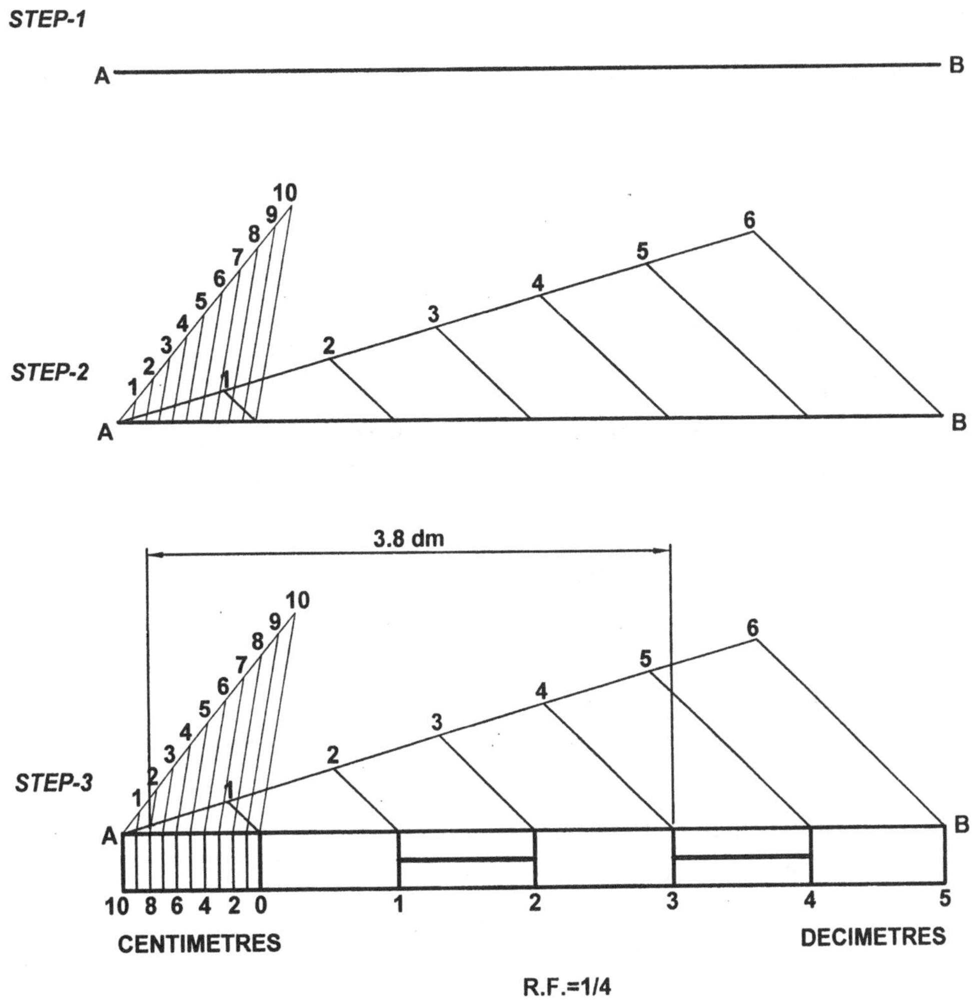

Fig. 2.1

Step 1: Draw a line AB, 15 cm long to represent 6 decimetres length.

Step 2: Divide the line AB into 6 equal parts. (Since 15 cm long line represents actual length of 6 dm, so 1/6th of the above line will represent 1 dm and hence '6' parts are required).

Step 3: Complete the rectangle with length equal to AB and width equal to 10 mm (Width may be taken 3-10 mm). Draw vertical lines from all division points of line AB. Mark '0' at first division and 1, 2, 3... at subsequent divisions. In order to distinguish the divisions clearly, draw thick and dark horizontal lines in the centre of alternate divisions. Write primary unit (dm) on the right hand side and secondary unit (cm) on the left hand side and R.F. in the middle of the scale. Show a distance of 3.8 dm as shown in Fig. 2.1.

Problem 2.2 (Fig. 2.2): *Construct a scale of 1:60 to show meters and decimetres and long enough to measure up to 6 metres.*

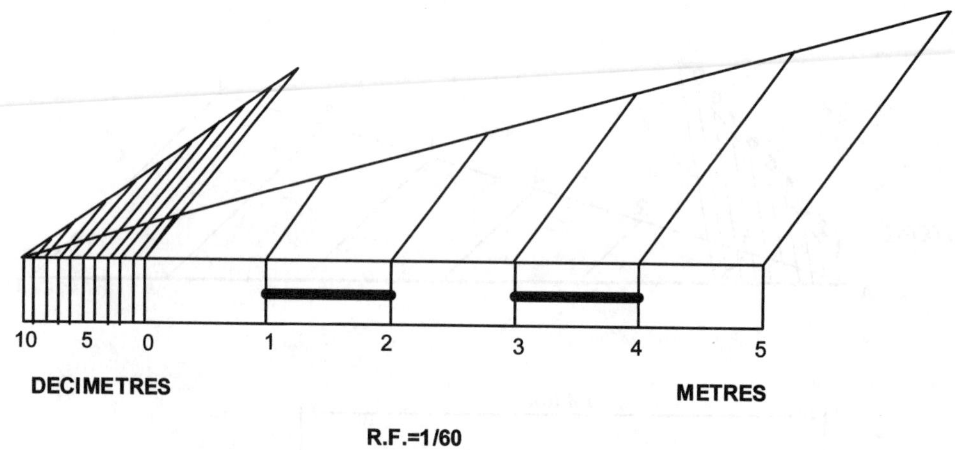

Fig. 2.2

Construction (Fig. 2.2):

(i) Given R.F. = $\frac{1}{60}$

(ii) Calculate length of scale as follows:

$$\text{LOS} = \text{R.F.} \times \text{Max. distance}$$

$$= \frac{1}{60} \times 6 \times 100 = 10 \text{ cm}$$

(iii) Draw a line of 10 cm length to represent 6 meters. Divide it into six equal parts.

(iv) Divide first part into 10 equal parts. Each division represents 1 dm.

(v) Complete the rectangle with some convenient width.

(vi) Mark '0' at first primary division and 1, 2, 3.. on subsequent divisions.

(vii) Complete the construction as shown in the figure.

Problem 2.3 (Fig.2.3): *On a map of a city, a distance of 36 km between two localities is shown by a line of 45 cm length. Calculate its R.F. and construct a plain scale to read kilometres and hectometres. Show a distance of 9.3 km on it.*

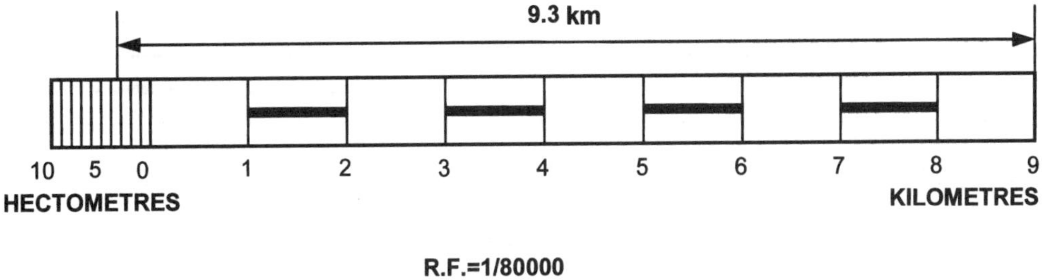

Fig. 2.3

Construction (Fig.2.3):

1. Find the R.F as below:

$$\text{R.F.} = \frac{\text{Length on drawing}}{\text{Actual length}}$$

$$\text{R.F.} = \frac{45\,\text{cm}}{36\,\text{km}} = \frac{45\,\text{cm}}{36\times1000\times100\,\text{cm}} = \frac{1}{80{,}000}$$

2. Find the length of scale as below:

$$\text{LOS} = \text{R.F.} \times \text{Max. Distance}$$

$$= \frac{1}{80{,}000}\times 10\,\text{km}(\text{assumed})$$

$$= \frac{1}{80{,}000}\times10\times1000\times100\,\text{cm} = 12.5\,\text{cm}$$

3. Draw a 12.5 cm long line to represent 10 km and divide it into 10 equal parts each representing one kilometre.
4. Complete the rectangle with length 12.5 cm and width equal to 10 mm.
5. Divide the first division into 10 equal parts each representing 1 hectometre.
6. Write primary unit (kilometres) secondary unit (Hectometres) and R.F.
7. Mark a distance of 9.3 km on the scale.

Problem 2.4 (Fig. 2.4): *A circular racing field of 1 km^2 is represented on a map by a circular area of 1 cm^2. Construct a plain scale to show kilometres and long enough to measure up to 15 km. Show a distance of 13 km on it.*

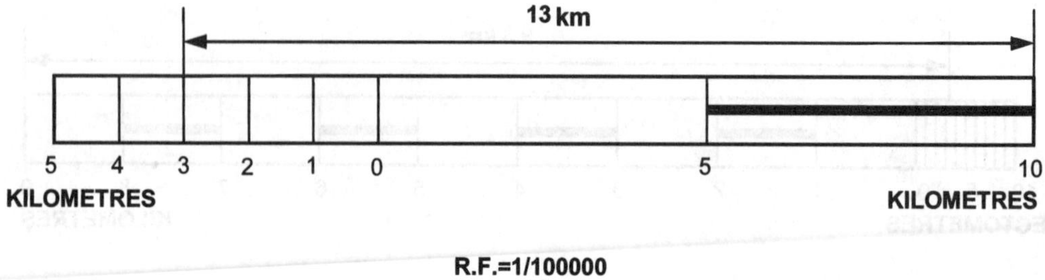

Fig. 2.4

Construction (Fig. 2.4):

1. Find R.F. of the scale as follows

$$\text{R.F.} = \sqrt{\frac{1\ \text{cm}^2}{1\ \text{km}^2}} = \frac{1\ \text{cm}}{1\ \text{km}} = \frac{1}{1 \times 1000 \times 100}$$

2. Calculate length of scale as below:

$$\text{LOS} = \text{R.F.} \times \text{Max. Distance}$$

$$= \frac{1}{100000} \times 15 \times 1000 \times 100 = 15\ \text{cm}$$

3. Draw a line 15 cm long and divide it into 3 equal parts.
4. Divide the first (left) division into 10 equal parts.
5. Complete the rectangle. Write primary unit, secondary unit and R.F. as shown in figure.
6. Show a distance of 13 km on it.

Problem 2.5 (Fig. 2.5): *On a map a rectangular plot of 100 sq. km^2 is represented by a rectangular area of 4 cm^2. Draw a scale long enough to measure up to 50 km. Mark a distance of 36 km on it.*

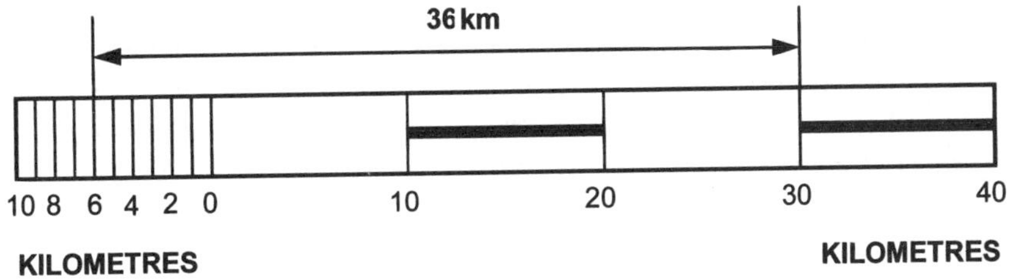

R.F.=1/500000

Fig. 2.5

Construction (Fig. 2.5):

1. Determine R.F. of scale as below:

$$\text{R.F.} = \sqrt{\frac{4\ \text{cm}^2}{100\ \text{km}^2}} = \frac{2\ \text{cm}}{10\ \text{km}} = \frac{2}{10 \times 1000 \times 100}$$

$$= \frac{1}{500000}$$

2. Find the length of scale as follows:

$$\text{LOS} = \text{R.F.} \times \text{Max. Length}$$

$$= \frac{1}{500000} \times 50 \times 1000 \times 100 = 10\ \text{cm}$$

3. Draw a line of 10 cm length to represent 50 km and divide it into 5 equal parts. Each part will represent 10 km.
4. Divide first division into 10 equal parts each representing 1 km.
5. Complete the rectangle. Write the primary unit, secondary unit and R.F. as shown in the figure.
6. Mark a distance of 36 km on it.

Problem 2.6 (Fig. 2.6): *A cube of 5 cm sides represents a water tank of 8000 m^3 volume. Determine the R.F. and construct a scale to show metres and long enough to measure up to 60 meters. Show a distance of 43 metres on it.*

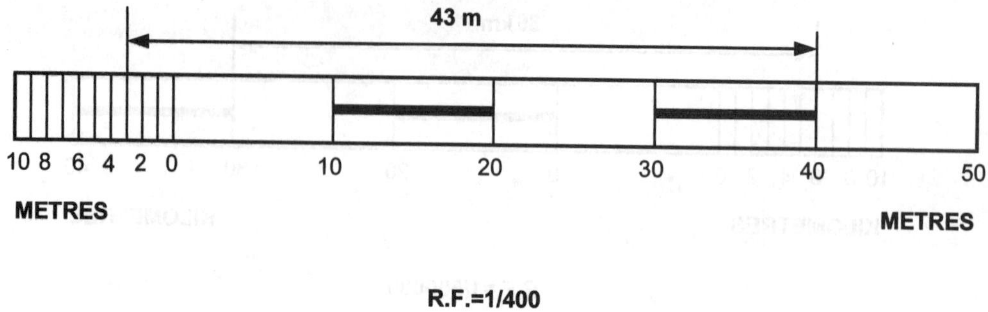

Fig. 2.6

Construction (Fig. 2.6):

1. Determine R.F. of scale as below:

$$\text{R.F.} = \sqrt[3]{\frac{\text{Volume of cube}}{\text{Volume of tank}}} = \sqrt[3]{\frac{125\ \text{cm}^3}{8000\ \text{m}^3}}$$

$$\text{R.F.} = \frac{5\ \text{cm}}{20\ \text{m}} = \frac{5}{20 \times 100} = \frac{1}{400}$$

2. Find the length of scale as below:

$$\text{LOS} = \text{R.F.} \times \text{Max. Length}$$

$$= \frac{1}{400} \times 60 \times 100 = 15\ \text{cm}$$

3. Draw a line of 15 cm length and divide it into 6 equal parts, so that each division represents 10 meters.
4. Divide the first division into 10 equal parts each representing 1 metre.
5. Complete the rectangle and write primary unit, secondary unit and R.F.
6. Mark a distance of 43 metre on it.

Problem 2.7 (Fig.2.7): *Construct a scale of R.F. 1:12 to read feet and inches and long enough to measure up to 6 feet. Indicate a distance of 4 feet 9 inches on it.*

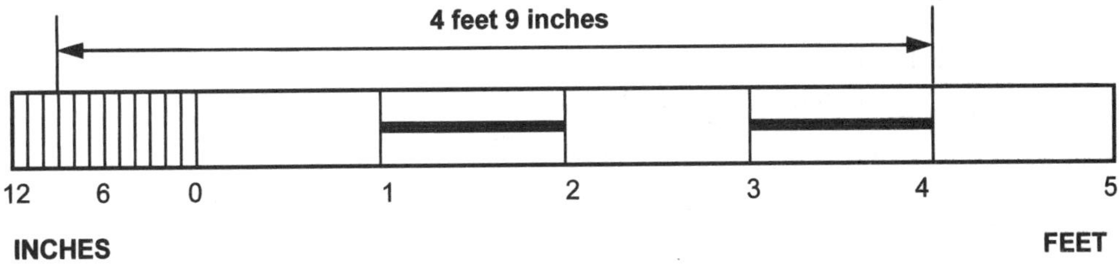

R.F.=1/12

Fig. 2.7

Construction (Fig.2.7):

1. Given R.F. = $\frac{1}{12}$
2. Find the length of scale as follows:

$$\begin{aligned} \text{LOS} &= \text{R.F.} \times \text{Max. Length} \\ &= \frac{1}{12} \times 6 \times 12 = 6 \text{ inches} \\ &= 6 \times 2.54 \text{ cm} = 15.24 \text{ cm} \end{aligned}$$

3. Draw a line 15.24 cm long. Divide it into 6 equal parts so that each part represents 1 foot.
4. Divide the first division into 12 equal parts each part representing 1 inch.
5. Complete the rectangle and mark the division points across the width of rectangle. Write FEET and INCHES to the right side and left side respectively and R.F in the middle of scale.
6. Mark a distance of 4 feet from the right side and 9 inches from left side.

Problem 2.8 (Fig.2.8): *On a map a distance of 2 yards is represented by 1 inch. Find the R.F. and construct a plain scale to show yards and feet and long enough to measure up to 10 yards. Show a distance of 5 yards and 2 feet on it.*

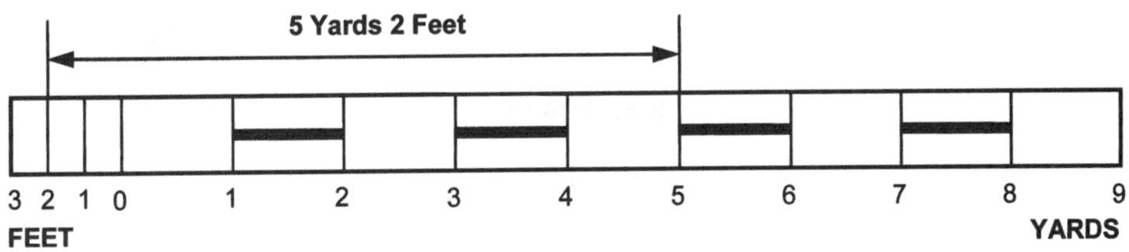

R.F.=1/72

Fig. 2.8

Construction (Fig. 2.8):

1. Find the R.F. of the scale as follows:

$$\text{R.F.} = \frac{1 \text{ inch}}{2 \text{ yards}} = \frac{1 \text{ inch}}{2 \times 3 \times 12 \text{ inch}} = \frac{1}{72}$$

2. Calculate length of scale as follows:

$$\text{LOS} = \text{R.F.} \times \text{Max. Distance}$$

$$= \frac{1}{72} \times 10 \times 3 \times 12 = 5 \text{ inches}$$

$$= 5 \times 2.54 = 12.7 \text{ cm}$$

3. Draw a line 12.7 cm long to represent 10 yards. Divide this line into 10 equal parts each representing 1 yard.
4. Divide first division into 3 equal parts, each representing 1 foot.
5. Complete the rectangle. Draw vertical divisions across the width of rectangle. Write primary units, secondary unit and R.F. as shown in Fig. 2.8. Mark a distance of 5 yards and 2 feet as shown in the above figure.

Problem 2.9 (Fig.2.9): *Construct a scale of R.F =* $1/80000$ *to show miles and furlongs and long enough to measure up to 5 miles. Show a distance of 3 miles and 4 furlongs on it.*

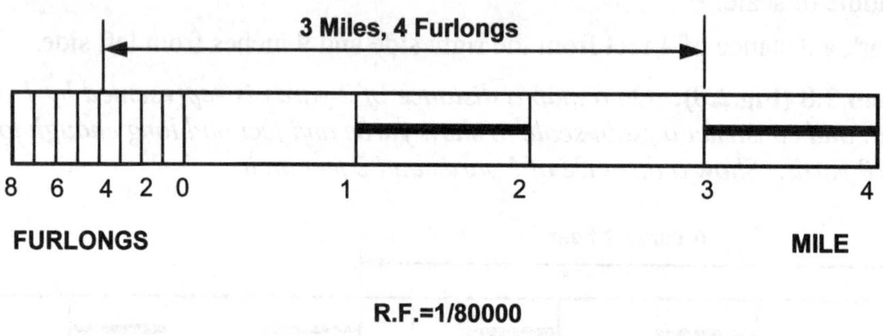

Fig. 2.9

Construction (Fig.2.9):

1. Given R.F. = $\frac{1}{80000}$

2. Calculate length of scale as follows:

$$LOS = R.F. \times Max.\ Distance$$

$$= \frac{1}{80,000} \times 5 \text{ miles} = \frac{1}{16000} \text{ miles}$$

$$\frac{1}{16000} \times 8 \times 220 \times 3 \times 12 = 3.96 \text{ inches } (= 10.05 \text{ cm})$$

Since,

1 mile = 8 furlong; 1 furlong = 220 yard; 1 yard = 3 feet; 1 feet = 12 inches

3. Draw a line 10.05 cm long to represent 5 miles. Divide it into 5 equal parts, each part representing 1 mile.
4. Divide first division into 8 equal parts each representing 1 furlong
5. Complete the rectangle and draw vertical division lines along the width of the rectangle.
6. Write primary unit, secondary unit and R.F. and show a distance of 3 miles 4 furlongs on the scale as shown in the Fig. 2.9.

Problem 2.10 (Fig. 2.10): *The distance between two cities is 120 km. A train covers this distance in 4 hours. If R.F. of the scale is* $1/200000$ *construct a scale to measure the distance covered by the train in a single minute. Show the distance covered by the train in 36 minutes.*

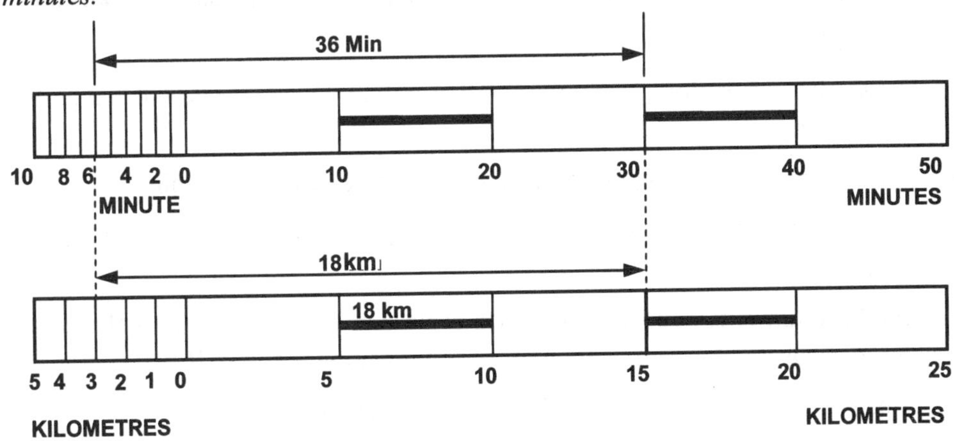

R.F.=1/200000

Fig. 2.10

Construction (Fig. 2.10):

1. Given R.F. = $\frac{1}{200000}$
2. Find the length of scale as below:

But before finding length of scale, it is required to find relation among L.O.S, Maximum time and corresponding maximum distance which can be obtained as below.

(a) From R.F. = $\frac{1}{200000}$

We know that, 1 cm represents 200000 cm

or 1 cm reps. 2 km

or 1 km is reptd. by $\frac{1}{2}$ cm

From speed of the train,

i.e., Speed = $\frac{120}{4}$ = 30 km/hr

or 30 km is covered in 1 hr = 60 min

$\therefore$ 1 km is covered in $\frac{60}{30}$ = 2 min

From (2.1) and (2.2), we get

$\frac{1}{2}$ cm reps. 1 km and 2 min.

or 1 cm reps. 2 km and 4 min.

Assuming L.O.S as 15 cm, we can say that 15 cm represents 30 km and also 60 minutes.

3. Draw a line 15 cm long representing 60 min time. Divide it into 6 equal parts each representing 10 min.
4. Divide the first division into 10 equal parts each representing 1 min.
5. Mark "MINUTES" on primary and secondary divisions. This is the required time scale.
6. Similarly draw its equivalent distance scale just below the time scale. Show the distance covered by train in 36 minutes as shown in the Fig. 2.10.

Note: A pair of two such scales is also known as comparative scale.

Problem 2.11 (Fig. 2.11): *A train is running at a constant speed of 40 km/hr. Construct a plain scale to read up to a kilometre. The scale should be long enough to measure up to 50 km. The R.F of the scale is* $\frac{1}{250000}$. *Indicate the time required by the train to cover a distance of 26 km.*

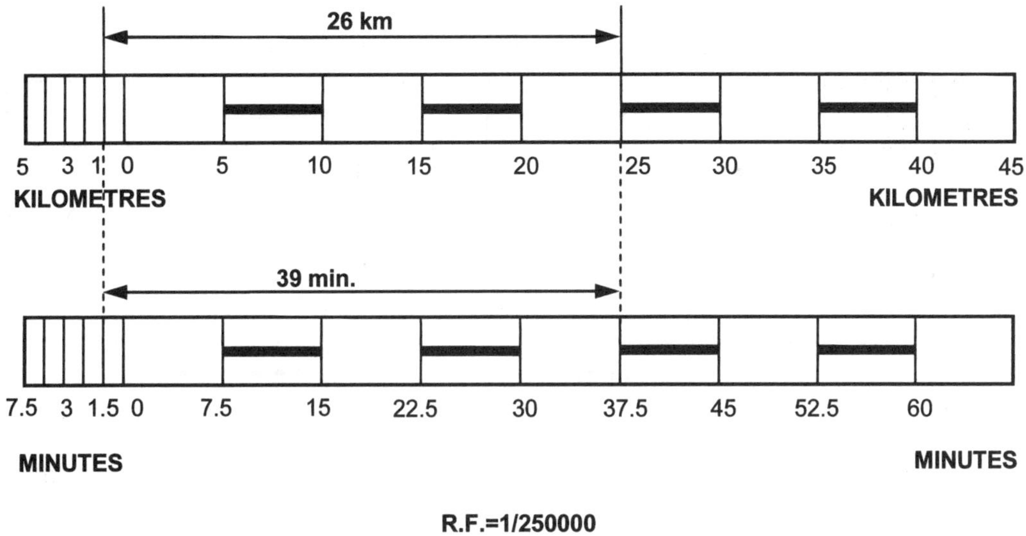

Fig. 2.11

Construction (Fig. 2.11):

1. Given R.F. = $\dfrac{1}{250000}$
2. Calculate length of scale as follows:

$$\text{LOS} = \text{R.F.} \times \text{Max. Distance}$$
$$= \frac{1}{250000} \times 50 \times 1000 \times 100 = 20 \text{ cm}$$

3. Draw a line 20 cm long representing a distance of 50 km.
4. Divide this line into 10 equal parts, each representing 5 km. Divide the first division into 5 equal parts so that each will represent 1 km.
5. Similarly draw its equivalent time scale just below the distance scale. Indicate the time required by train to cover a distance of 26 km as shown in Fig. 2.11.

Problem 2.12 (Fig. 2.12): *The distance between two cities is 180 km. A passenger train covers this distance in 6 hours. Construct a plain scale to measure time up to single minute. The R.F. of the scale is* $\frac{1}{200000}$. *Indicate the distance covered by the train in 38 minutes.*

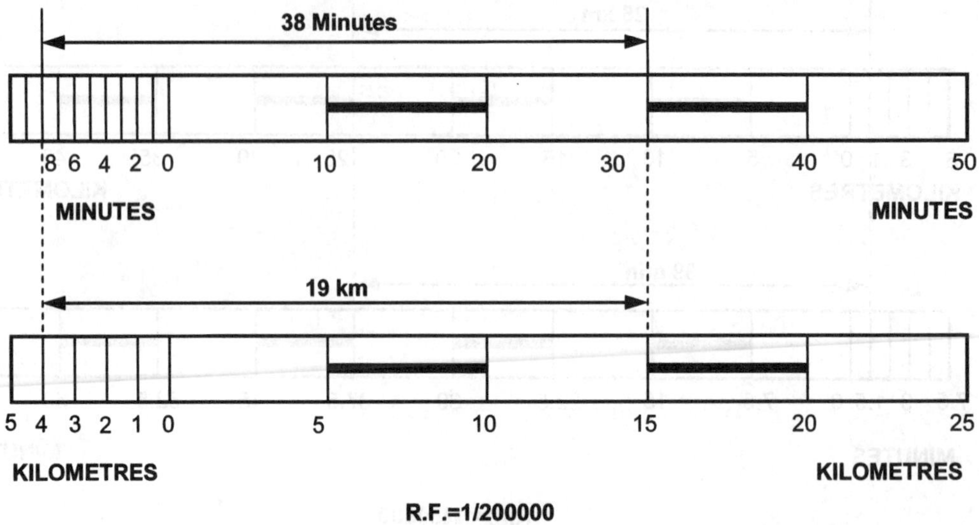

Fig. 2.12

Construction (Fig. 2.12):

1. Given, R.F. = $\frac{1}{200000}$)
2. Since, maximum distance/maximum time is not mentioned in the problem, assume the length of scale as 15 cm.
3. Find the relationship among L.O.S., maximum length and maximum time as follows:

   From R.F. ⇒ 1 cm reps. 20,0000 cm (1 cm represents 20,0000 cm)

   or 1 cm reps. 2 km

   ∴ 15 cm reps. 30 km

   From speed ⇒ 180 km / 6 hours

   i.e., 30 km/hour or 30 km/60min

   From (2.3) and (2.4) we can say, that

   15 cm represents a distance = 30 km (or 60 min time)

4. Draw a line 15 cm long to represent 60 minutes. Divide it into 6 equal parts, each representing 10 min.
5. Divide first division again into 10 equal parts, each representing 1 minute.
6. Complete the rectangle, write the primary unit, secondary unit and R.F. below the scale. This is the required time scale.

7. Similarly draw a distance scale just below the time scale. Indicate 38 minutes on the time scale. Indicate 38 minutes on the time scale and its corresponding distance on the distance scale as shown in Fig. 2.12.

2.6 Diagonal Scale

Diagonal scale is used to show three units for example metre, decimetre and centimetre. It is very useful to represent very small distance such as 0.1 mm etc. A very short line can be further divided using the principle of diagonal division as discussed below.

2.6.1 Principle of Diagonal Division

Let a short line AB is required to be divided into any number of equal parts say 10. Then it can be divided using the method of diagonal divisions which is explained as below.

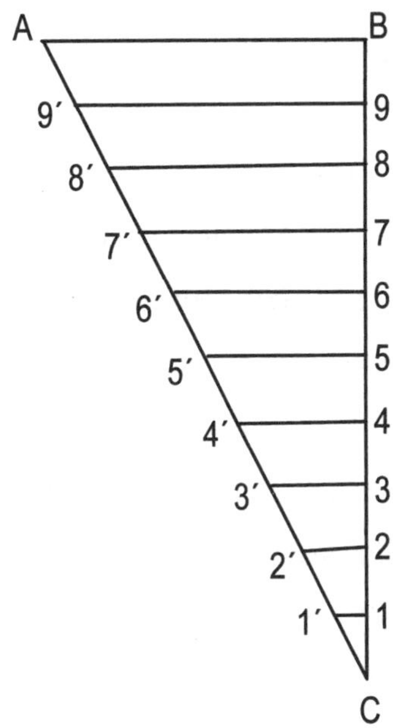

Fig. 2.13 Method of diagonal divisions.

1. Draw a line AB of given length as shown in Fig. 2.13
2. At B draw a perpendicular line BC of any length.
3. Divide BC into 10 equal parts.

4. Mark divisions as 9, 8…etc.
5. Join A to C.
6. Draw lines parallel to AB through points 9, 8, 7….etc.
7. Mark 1', 2', 3' at the intersection of these parallel lines and inclined line AC.

From Fig. 2.13, it is clear that, triangles 1'1C, 2'2C, 3'3C etc., are similar to Δ ABC.

Therefore, $\frac{AB}{BC} = \frac{1'1}{1C} = \frac{2'2}{2C} = ...$

As, $1C = \frac{1}{10}$ BC therefore, $11' = \frac{1}{10}$ AB, $22' = \frac{2}{10}$ AB and so on. In this way the line AB can be divided into 10 divisions.

2.6.2 Construction of Diagonal Scale

1. Find the R.F. of the scale.
2. Calculate length of scale as follows

$$LOS = R.F. \times \text{Maximum distance}$$

3. Draw a rectangle having its length equal to LOS and width as 5 cm.
4. Mark the main divisions and sub divisions as usual.
5. Divide left vertical side of the rectangle into required parts.
6. Draw a diagonal lines through divisions points of sub units.
7. Write main unit, sub units, and third unit and mention R.F.

Problem 2.13 (Fig. 2.14): *Construct a diagonal scale of R.F. 1:50 to read meters, decimetres and centimetres and long enough to measure up to 6 metres. Mark a distance of 4.38 metres on it.*

Construction (Fig. 2.14):

***Step* 1:** Find the R.F of the scale. Here, given R.F. = $\frac{1}{50}$

***Step* 2:** Calculate length of scale as

$$L_s = \frac{1}{50} \times 6 \times 100 = 12 \text{ cm}$$

***Step* 3:** Draw a rectangle of length AB = 12 cm and width AD of any convenient size.

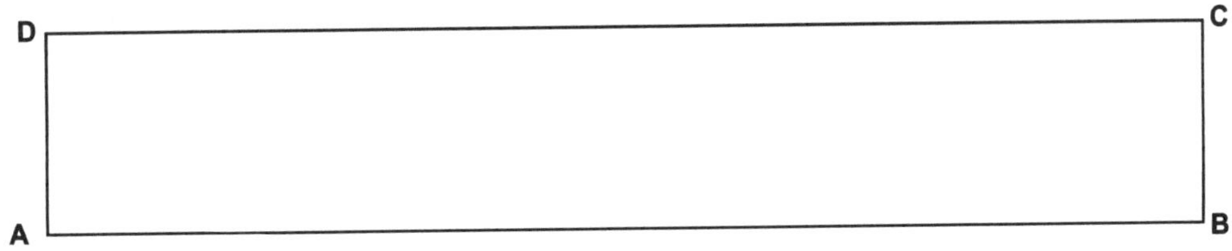

Fig. 2.14(a)

***Step* 4:** Divide the length of rectangle ABCD into six equal parts. (Since maximum length to be represented is 6 metres).

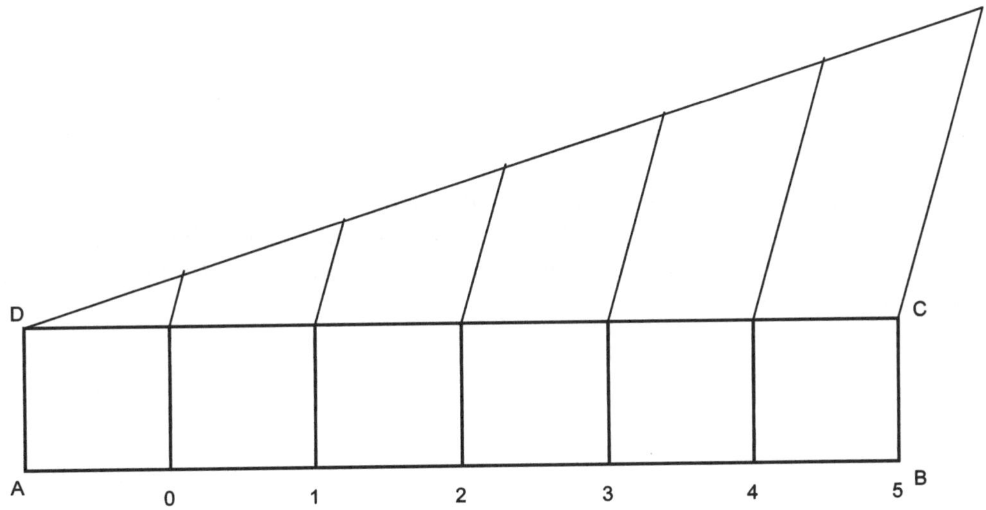

Fig. 2.14(b)

***Step* 5:** Since, subunit (decimetre) is $\frac{1}{10}^{th}$ of main unit (metre) and centimetre is also $\frac{1}{10}^{th}$ of a decimetre. Therefore, divide first division into 10 equal parts (as secondary divisions) and mark 10 division points on the upper edge of first division. Divide AD into 10 equal parts (as tertiary divisions).

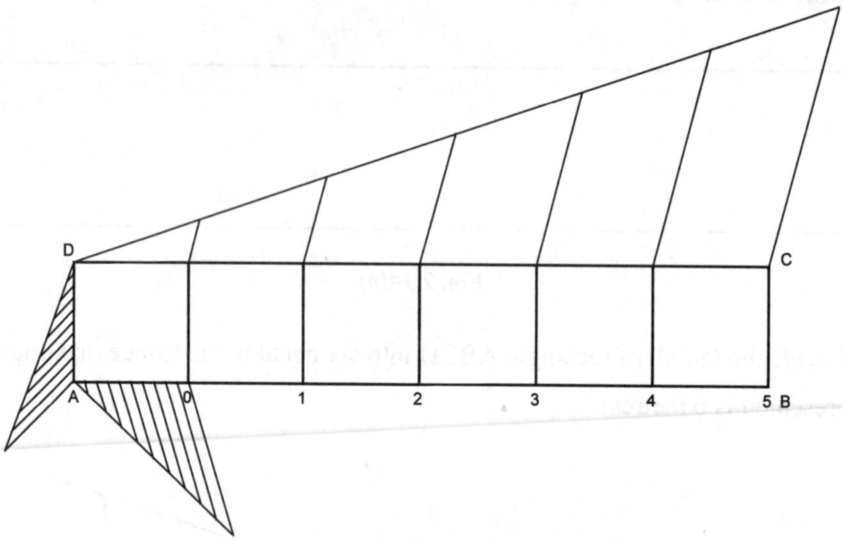

Fig. 2.14(c)

***Step* 6:** Draw diagonal lines through secondary division points. Also draw horizontal lines through tertiary division points. Complete the scale by darkening outlines and division lines. Keep all construction lines thin and light. Mark a distance of 4.8 metres on it as shown in the Fig. 2.14(d).

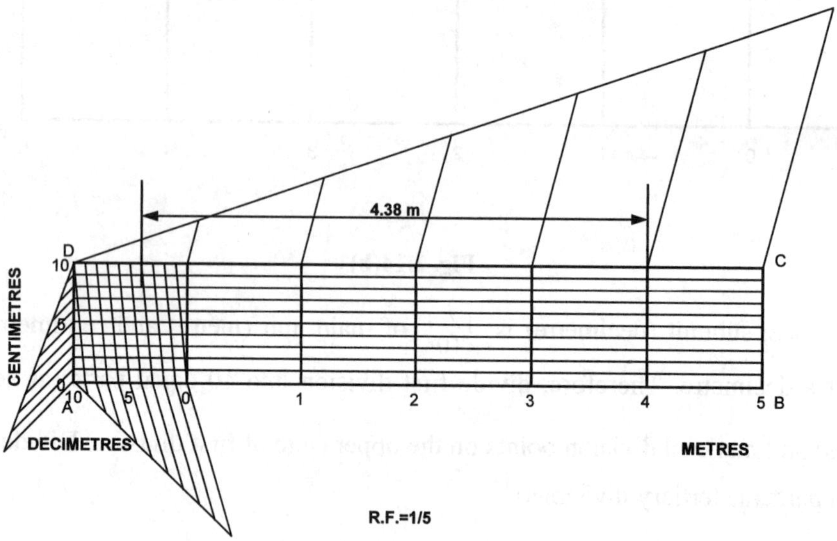

Fig. 2.14(d)

Problem 2.14 (Fig. 2.15): *Construct a diagonal scale of R.F.* $= \dfrac{1}{4000}$ *to measure single metre and long enough to measure up to 600 metres. Indicate a distance of 378 metres on it.*

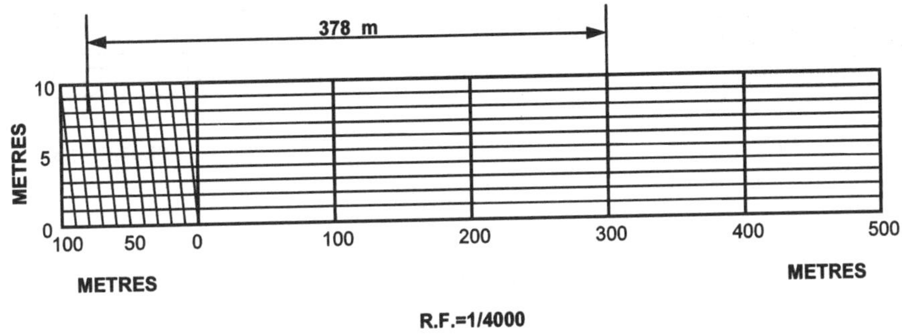

Fig. 2.15

Construction (Fig. 2.15):

1. Find the R.F. of the scale. Here R.F. $= \dfrac{1}{4000}$ (given)
2. Calculate length of scale as below.

   LOS = R.F. × Maximum distance

   $$= \frac{1}{4000} \times 600 \times 100 = 15 \text{ cm}$$

3. Draw a rectangle of length 15 cm and width 5 cm (or any convenient length).
4. Divide the length of rectangle into six equal parts. Mark the division points each representing 100 metres.
5. Divide the first division (left of zero) into 10 equal parts each representing 10 metres.
6. Divide the width of rectangle into 10 equal parts. Each represents one metre.
7. Complete the construction following the steps as mentioned in problem 2.13. Mark a distance of 378 metres on it.

Problem 2.15 (Fig. 2.16): *The distance between two locations by road is 4 kilometres. It is represented on a map by 1 centimetre long line. Determine the R.F. and construct a diagonal scale to show kilometres and hectometres. Show a distance of 56.7 km on it.*

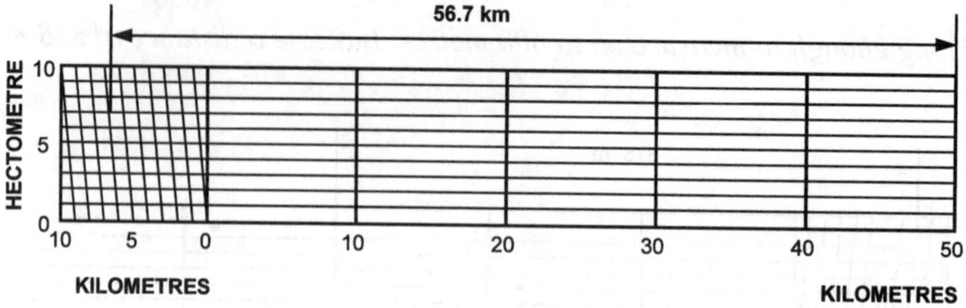

Fig. 2.16

Construction (Fig. 2.16):

1. Find the R.F. of the scale as follows:

$$\text{R.F.} = \frac{1\text{ cm}}{4\text{ km}} = \frac{1\text{ cm}}{4 \times 1000 \times 100\text{ cm}} = \frac{1}{400000}$$

2. Calculate length of scale as below:

   Since, maximum distance is not given, one can assume as 15 cm.

   Since, $\text{R.F.} = \frac{1}{400000}$

   Which shows, 1 cm represents 400000 cm

   or 1 cm represents 4 km

   15 cm will represent, $4 \times 15 = 60$ km

3. Draw a rectangle of length 15 cm and width 50 mm
4. Divide the length of rectangle into six equal parts each representing 10 km.
5. Divide the first division again into 10 equal parts such that each secondary division represents 1 kilometre.
6. For representing $1/10^{\text{th}}$ of a kilometre i.e., hectometre, divide the width of rectangle into 10 equal parts.
7. Complete the construction following steps as explained in problem 2.13.

Problem 2.16 (Fig. 2.17): *On a map 1 cm long line represents 40 metres. Find the R.F. and construct a suitable scale to measure metres, tens of metre and hundreds of metre. Mark a distance of 436 metres on the scale.*

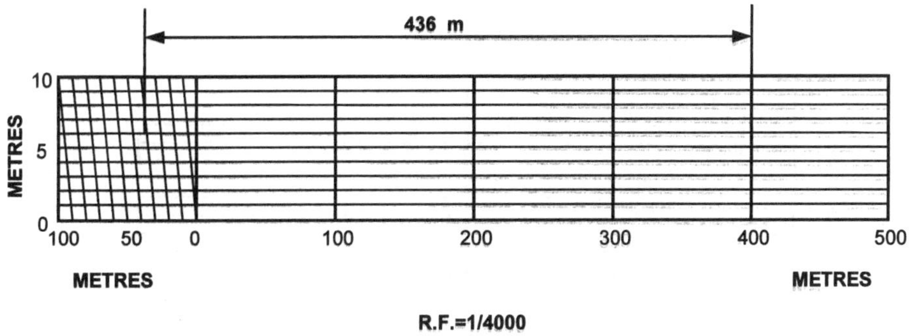

Fig. 2.17

Construction (Fig. 2.17):

1. Find the R.F. of the scale as below.

$$\text{R.F.} = \frac{1\,\text{cm}}{40\,\text{m}} = \frac{1\,\text{cm}}{40 \times 100\,\text{cm}} = \frac{1}{4000}$$

2. Find the length of scale as below.

   Since, maximum distance to be measured on the scale is not given, one can assume the length of scale as 15 cm

   so, $\because$ 1 cm represents 40 meters

   $\therefore$ 15 cm will represent = 15 × 40 = 600 metres

3. Draw a rectangle having length equal to 15 cm and width as 5 cm. Divide its length into six equal parts each representing 100 metres.
4. Divide the first division into 10 equal parts so that each division represents 10 metres. Draw diagonal line through each division.
5. Divide the width of rectangle into 10 equal parts.
6. Complete the construction and mark a distance of 436 metres on the scale as shown in Fig. 2.17.

Problem 2.17 (Fig. 2.18): *On a map, a rectangular plot of area 40,000 m^2 is represented by a similar rectangle of 25 cm^2. Find the R.F. and construct a diagonal scale to read up to one metre and long enough to measure up to 400 metres. Mark a distance of 225 metres on it.*

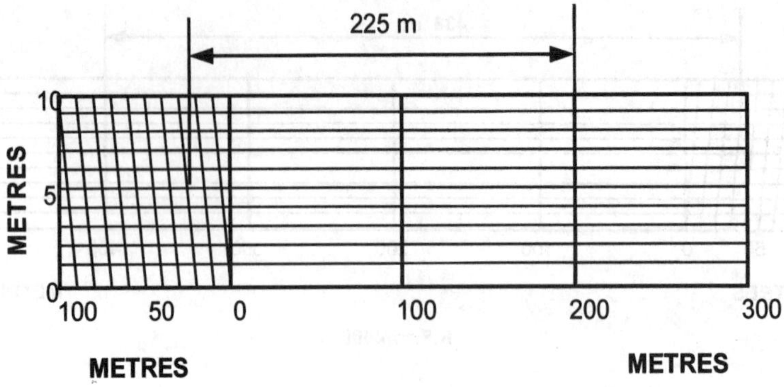

Fig. 2.18

Construction (Fig. 2.18):

1. Find the R.F. of the scale as below

$$\text{R.F.} = \sqrt{\frac{25\ \text{cm}^2}{40{,}000\ \text{m}^2}} = \frac{5\ \text{cm}}{200\ \text{m}} = \frac{1}{4000}$$

2. Find the length of scale as below

   Length of scale = R.F. × Maximum distance

$$\text{LOS} = \frac{1}{4000} \times 400 \times 100 = 10\ \text{cm}$$

3. **Draw a rectangle of length 10 cm and width 5 cm. Divide length of the rectangle into four equal parts, each representing 100 metres.**

4. **Divide first primary division into ten equal parts each representing 10 metre and draw diagonal lines through these points.**

5. Divide the width of rectangle into ten equal parts.

6. Complete the construction and mark a distance of 225 metres as shown in Fig. 2.18

Problem 2.18 (Fig. 2.19): *Construct a diagonal scale of R.F. = 1/36 showing yards, feet and inches. The scale should be long enough to measure up to 4 yards. Show a distance of 3 yards 1 foot and 7 inches.*

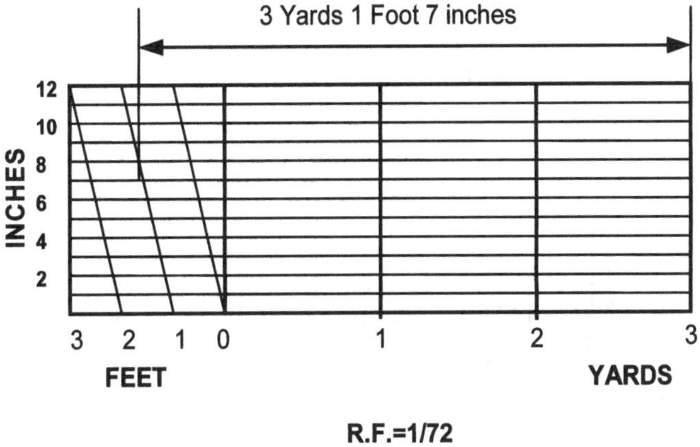

Fig. 2.19

Construction (Fig. 2.19):

1. Find the length of scale as below:

$$\text{LOS} = \text{R.F.} \times \text{Maximum distance}$$
$$= \frac{1}{36} \times 5 \times 3 \times 12 = 5 \text{ inches } (12.7 \text{ cm})$$

2. Draw a rectangle of length 12.7 cm and width 5 cm. Divide its length into four equal parts each representing 1 yard.
3. Divide first main division into three equal parts each representing 1 foot.
4. Divide the width of rectangle into 12 equal parts.
5. Complete the construction as shown in Fig. 2.19.

Problem 2.19 (Fig. 2.20): *Construct a scale to measure 5 km, 1/8 of km and 1/40 of km in which 1 km is represented by 4 cm. Mark a distance of 2.775 km on it.*

Construction (Fig. 2.20):

1. Find R.F. of the scale as follows:

$$\text{R.F.} = \frac{4 \text{ cm}}{1 \text{ km}} = \frac{4 \text{ cm}}{1 \times 1000 \times 100 \text{ cm}} = \frac{1}{25000}$$

2. Find the length of scale as below:

Since maximum distance is not given in the problem, it may be assumed.

Let the maximum distance to be measured is 4 km. Then,

$$\text{LOS} = \text{R.F.} \times \text{Maximum distance}$$

$$= \frac{1}{25000} = 4 \times 1000 \times 100 = 16 \text{ cm}$$

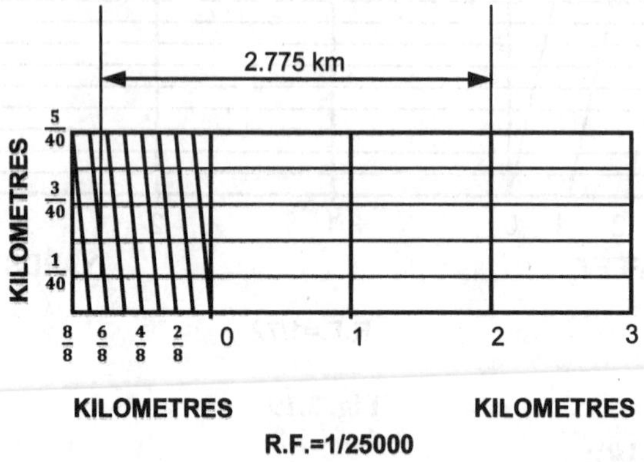

Fig. 2.20

3. Draw a rectangle of length 16 cm and width equal to 5 cm. Divide its length into 4 equal parts each representing 1 km.
4. Divide first division into 8 equal parts each representing 1/8 of a km.
5. Divide width of the rectangle into 5 equal parts to represent 1/40 of a kilometre.
6. Complete the construction and mark a distance if 2.775 km on its as shown in Fig. 2.20.

Since, $\frac{1}{8}$ km = 0.125 km or $\frac{1}{40}$ km = 0.025 km

$\therefore$ 2.775 km = 2 + 6 × 0.125 + 1 × 0.025

So to mark the distance of 2.775 km, take 2nd division of primary units, 6th division on secondary and 1st division on tertiary units.

2.7 Scale of Chords

The scale of chords is used to measure or set the required angle. It is useful when protractor is not available. The construction of the scale of chords is explained with the help of following example.

Problem 2.20 (Fig. 2.21): *Construct a scale of chords showing 5° divisions. Set off the angle of 35° and 135° with the help of it.*

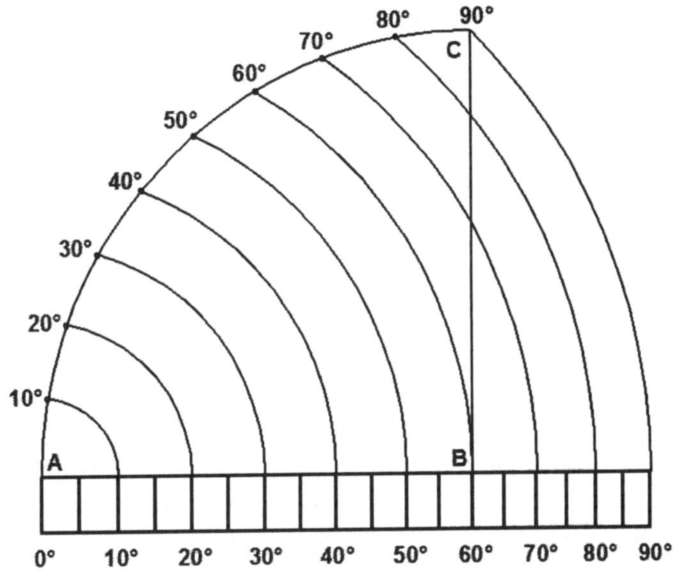

Fig. 2.21(a)

Construction (Fig. 2.21(a) and (b)):

1. Draw a line AB of any convenient length.
2. With B as centre and radius equal to AB draw an arc.
3. Draw a perpendicular line at B cutting the arc at C.
4. Divide the arc AC into nine equal parts. Mark division point as 10°, 20°…etc.
5. With centre A and radius equal to A-10° turn down the division to the line AB. Similarly turn down all the divisions to the line AB. Extend the line AB beyond B.
6. Complete the construction as shown in the Fig 2.21(a).
7. To set-off a required angle (say 35°) draw a straight line OX of any convenient length as shown in Fig. 2.21(b). With O as centre and radius equal to AB draw an arc PR. Now with radius equal to 0-35 and P as centre draw an arc cutting PR at Q. Join O to Q.

   $\angle POQ = 35^o$

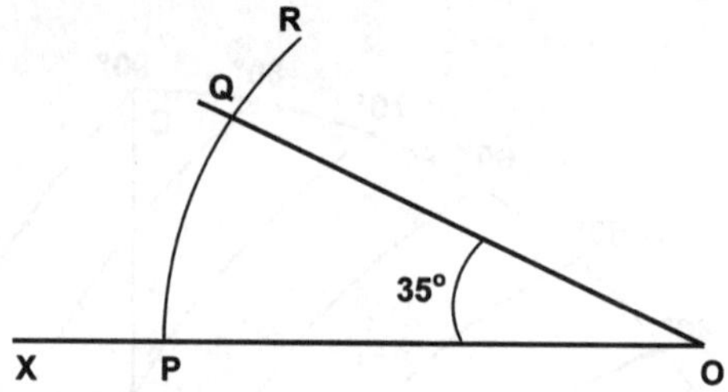

Fig. 2.21(b)

8. To construct an angle of 135° (Fig. 2.21(c)), draw a straight line OX of any convenient length. With O as centre and radius equal to AB draw an arc PS. Since the required angle is greater than 90° then it will be drawn in two parts (say 75 + 60 = 135°). Now with P as centre and radius equal to 0-75 (Fig. 2.21(a)), draw an arc cutting PS at Q. Now take Q as centre and radius equal to 0-60, cut arc on arc PS at R. Join O to R. Thus, angle POS is the required angle of 135°.

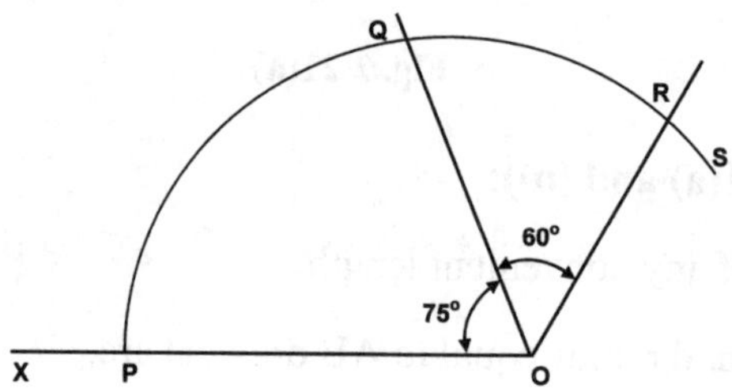

Fig. 2.21(c)

Problem 2.21 (Fig. 2.22): *Measure the given angle ($\angle$ MON) using scale of chords.*

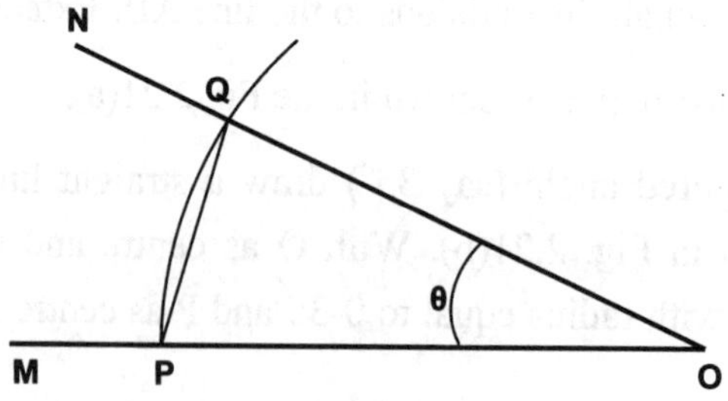

Fig. 2.22

Construction (Fig. 2.22):

1. Take O as centre and radius equal to AB (From Fig. 2.21) draw an arc PQ.
2. Join P to Q. Now with the help of divider or compass take distance equal to PQ and measure on horizontal scale, which will give an angle of 30°.

Exercises

1. On a map 1 cm long line represents 60 cm. Find the R.F and construct a scale to show metres and decimetres and long enough to measure up to 6 metres. Show a distance of 4 metres and 69 decimetres on it.
2. Construct a plain scale to show metres and decimetres when 3 centimetres represents 2 metres and long enough to measure up to 5 metres. Show a distance of 4.5 metres on it.
3. The distance between two cities is 160 kilometres and it is covered by a passenger train in 4 hours. Draw a plain scale to measure the time up to single minute. The R.F of the scale is $\frac{1}{200000}$. Show the distance covered by the train is 37 minutes.
4. Construct a scale of 1:50 showing metres and decimetres and long enough to measure up to 8 metres. Show a distance of 6.5 metres on it.
5. An area of 225 cm^2 on a map represents an area of 81 m^2 on a field. Find the R.F. and draw a scale to show metres and decimetres and long enough to measure upto 9 metres. Show a distance of 8 m 5 dm on it.
6. A cube of 8 cm^3 represents a cubical tank of 1000 m^3 volume. Find the R.F. and construct a scale to show meters and long enough to measure up to 60 metres. Mark a distance of 53 metres on it.
7. On a map 3 cm long line represents 1 metre. Find the R.F. and construct a diagonal scale to show metres, decimetres and centimetres and long enough to measure up to 5 metres. Mark a distance of 2.72 metres on it.
8. Construct a diagonal scale of R.F. = $\frac{1}{5000}$ to show metres and long enough to measure up to 600 metres. Show a distance of 536 metres on it.
9. An area of 144 cm^2 represents an area of 36 km^2 on the field. Find the R.F. and construct a diagonal scale to show kilometres, hectometres and decametres and long enough to measure up to 8 kilometres. Show a distance of 6 kilometres, 5 hectometres and 4 decametres on the scale.
10. Construct a diagonal scale to measure 0.01 and 0.1 of a kilometre and long enough to measure up to 6 kilometres. The R.F. of scale is $\frac{1}{40}$. Mark a distances of 5.67 kilometres on the scale.

11. On a building plan 1 centimetre line represents 3 metre distance. Find the R.F. and construct a scale to measure up to 50 metres. Indicate a distance of 38.7 metres on it.

12. On a map 2 inch long line represents 1.25 yards. Find the R.F. and construct a diagonal scale to show yard, feet and inches. The scale is long enough to measure up to 6 yards. Show a distance of 3 yards 3 feet and 3 inches on it.

13. A room of 125 m^3 volume is shown by a cube of 2 cm side. Find the R.F. and construct a diagonal scale to show metres and decimetres. The scale should be long enough to measure up to 50 metres.

14. Construct a scale of chords showing 10° disvisions with any suitable dimensions. Set-off angles of 45° and 120° with the help of it

15. Construct a scale of chords showing 5° divisions with any suitable dimensions. Draw a triangle ABC with $\angle A = 40^o$ $\angle B = 65^o$ and $\angle C = 75^o$, using the scale of chords.

CHAPTER 3

Conic Sections

3.1 Introduction

Many objects which we use in our daily life, consist of curved profiles. In order to prepare drawings of such objects, the knowledge of various curves such as ellipse, parabola, hyperbola etc., is essential.

In this chapter, definitions, method of construction and applications of conic sections have been discussed with the help of suitable problems. A few problems have also been discussed with detailed construction steps for easy understanding.

3.2 The Cone

A cone is formed when a right angle triangle is rotated about its altitude. Altitude of the triangle becomes the height of cone. Various terms with reference to a cone, are illustrated in Figs. 3.1(a) and (b).

3.3 Conic Sections and their Applications

When a right circular cone is cut by a section plane inclined to its axis in different positions, sections obtained are known as conic sections as explained below.

(i) When a cutting plane is perpendicular to the axis of the cone, the section obtained is known as circle as shown in Fig. 3.2(a) and (b).

Practical applications: Circle or circular objects are used in variety of applications such as disc, rings, cover plates etc.

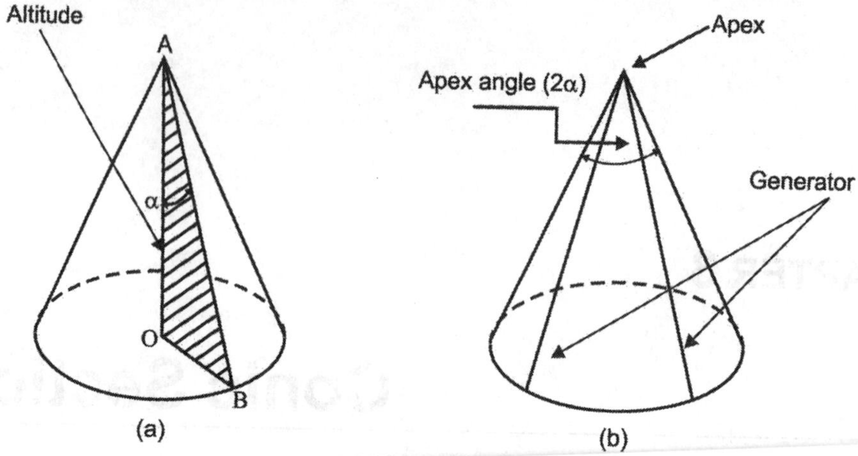

Fig. 3.1 Formation of cone.

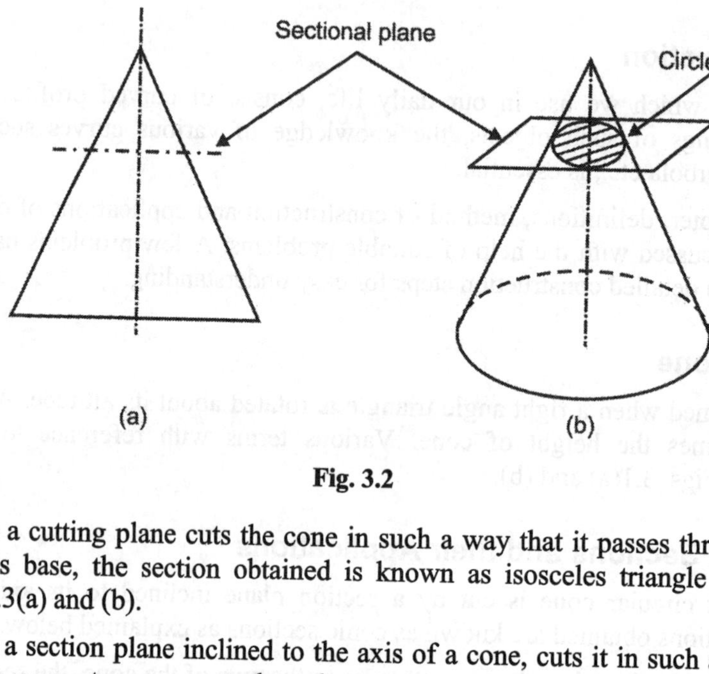

Fig. 3.2

(ii) When a cutting plane cuts the cone in such a way that it passes through its apex and its base, the section obtained is known as isosceles triangle as shown in Fig. 3.3(a) and (b).

(iii) When a section plane inclined to the axis of a cone, cuts it in such a way that its all the generators are cut, then the section obtained is known as ellipse. The position of cutting plane and section obtained are shown in Fig. 3.4(a) & (b).

Practical applications: Elliptical curves are used in bridges, dams, man holes of boilers, stuffing box etc.

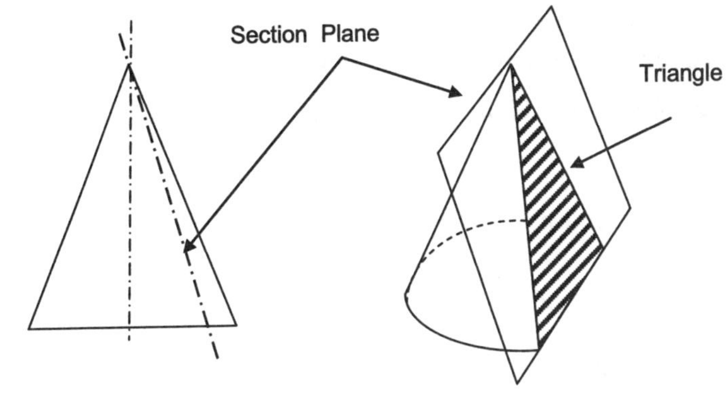

(a) (b)

Fig. 3.3

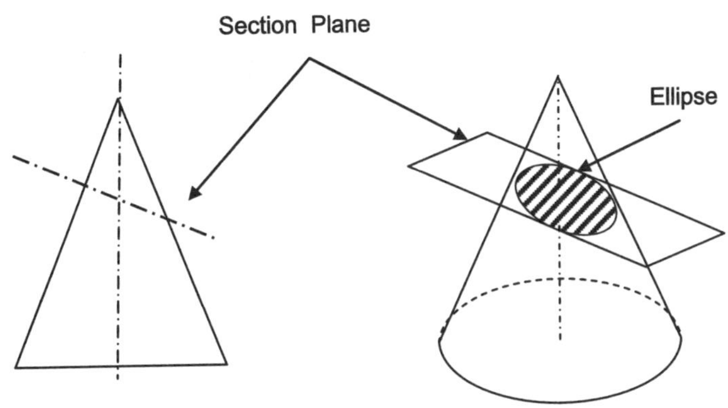

(a) (b)

Fig. 3.4

(iv) When a section plane is inclined to the axis of cone but parallel to one of its generators, the section obtained is known as parabola as shown in Fig. 3.5(a) and (b).

Practical applications: Parabolic curves are used in solar photovoltaic concentrator, sound and light reflectors, machine tools, buildings, bridges, arches, path of the object thrown from the earth at some angle (projectile's path) etc.

(v) When the section plane cuts both the parts of double cone, the section obtained is known as hyperbola. In this case, the angle which a cutting plane makes with the axis, must be less than half the apex angle as shown in Fig. 3.6 (a) and (b).

When the section plane is parallel to the axis of the cone, then the section obtained is known as rectangular hyperbola.

Practical applications: Hyperbolic curves are used in cooling towers water channel etc.

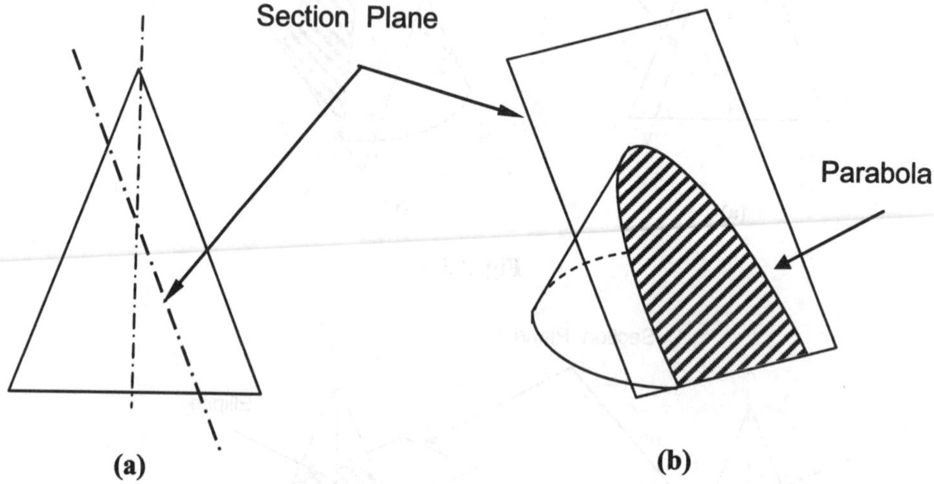

Fig. 3.5

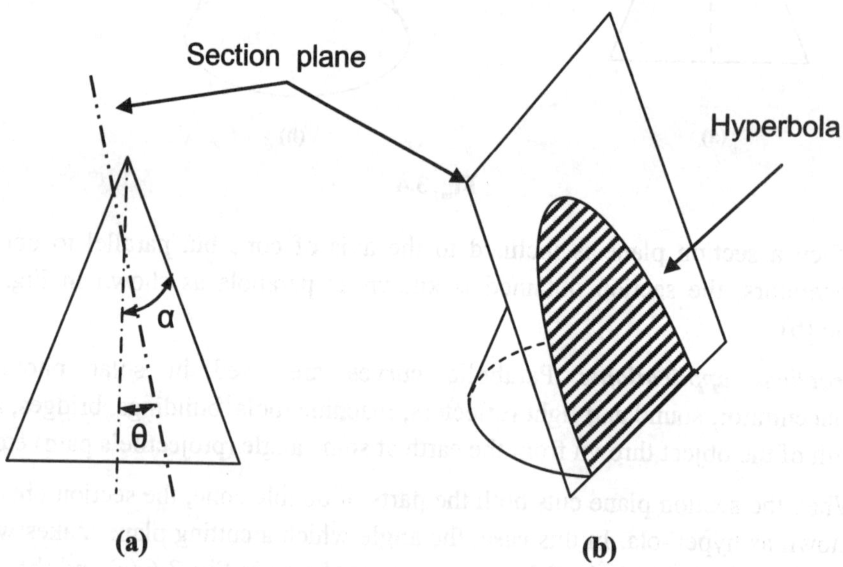

Fig. 3.6

3.4 General Definition of a Conic

The conic is defined as a locus of a points which moves in a plane in such a way that the ratio of its distance from a fixed points (focus) and a fixed straight line (directrix) is always constant. The ratio is called eccentricity (e).

$$e = \frac{\text{Distance of a point from the focus}}{\text{Distance of the same point from the directrix}}$$

$e < 1$ for ellipse; $e = 1$ for parabola; $e > 1$ for hyperbola

3.5 Ellipse

The ellipse can be defined as the locus of a point moving in a plane in such a way that the ratio of its distance from a fixed point (focus) and from a fixed straight line (directrix) is constant and always less than one. Fig. 3.7 shows different elements of an ellipse.

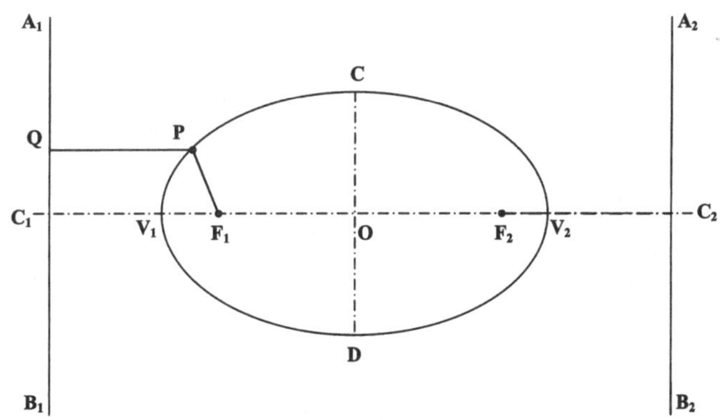

Fig. 3.7

In Fig. 3.7,

A_1B_1 and A_2B_2 – Directrices

F_1 and F_2 – Foci

V_1 and V_2 – Vertices

$C_1 C_2$ – Principal axis

$V_1 V_2$ – Major axis

$$\text{Eccentricity, } e = \frac{\text{Shortest distance of a point from focus}}{\text{Shortest distance of the point from directrix}}$$

$$= \frac{PF_1}{PQ} = \text{constant}$$

Ellipse can also be described mathematically as follows:

$$\frac{x^2}{a^2}+\frac{y^2}{b^2}=1$$

3.5.1 Construction of Ellipse

There are several methods for the construction of an ellipse. Selection of the appropriate method depends on the given input data for solving the problems. Table 3.1 presents different methods and the input data required for construction.

Table 3.1 Methods of construction of ellipse.

S.No.	Method	Input data required
1.	General method (Eccentricity method)	(i) Distance of focus from directrix (ii) Eccentricity (e)
2.	Arcs of circles method (Intersecting arcs method)	(i) Major axis (ii) Minor axis
3.	Concentric circles method	(i) Major axis (ii) Minor axis
4.	Oblong method (i) Rectangle method	(i) Major axis (ii) Minor axis
	(ii) Parallelogram method	(i) Sides of parallelogram (ii) Included angles between them

3.5.2 Construction of Ellipse by General Method (Eccentricity Method)

Construction of ellipse by this method is explained with the help of following example.

Problem 3.1 (Fig.3.8): *Draw an ellipse when the distance of the focus from its directrix is 60 mm and eccentricity is 2/3. Also draw a tangent and a normal at a point on the curve whose shortest distance from the directrix is 75 mm.*

Construction (Fig.3.8):

***Step* 1:** Draw a directrix DD' and principal axis CC' perpendicular to DD'. Mark focus F on the principal axis, 60 mm away from the directrix. Divide CF into five equal parts mark V on second division point left of the focus such that $\frac{VF}{VC}=\frac{2}{3}=e$.

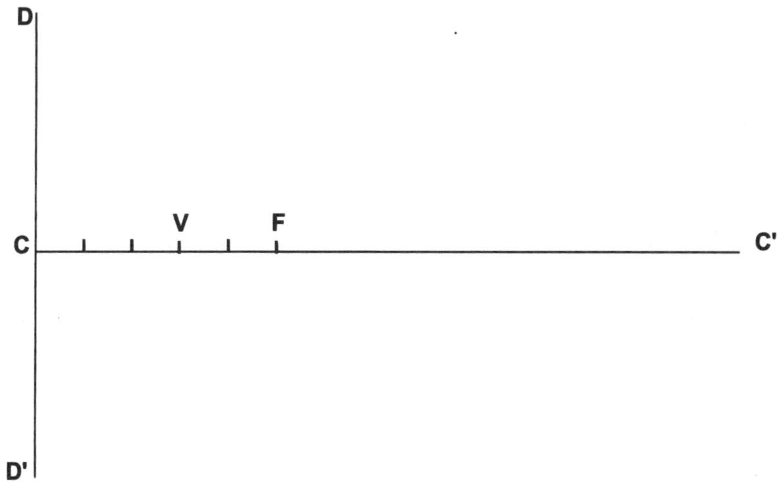

Fig. 3.8(a)

Step 2: Draw a vertical line at point V. With centre V and radius equal to VF, draw an arc intersecting the vertical line at V'. Draw an inclined line passing through C and V' of any convenient length.

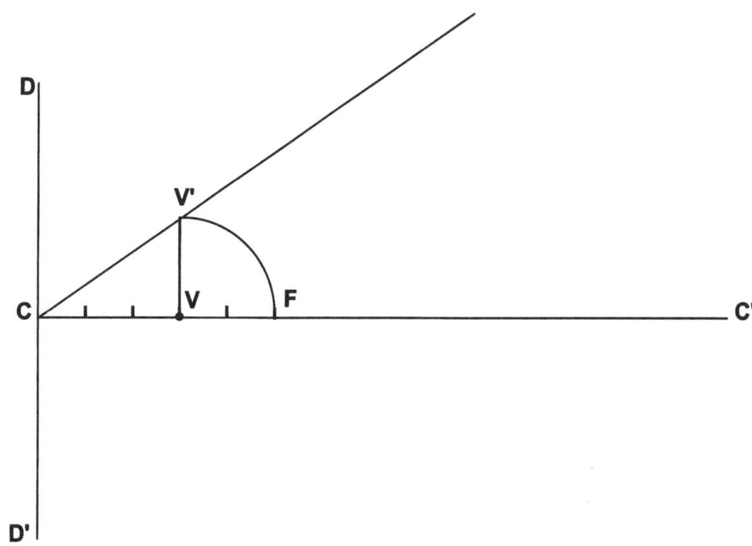

Fig. 3.8(b)

Step 3: Mark some points 1, 2, 3...etc., (not necessary at equal distances) on the axis. Draw vertical lines through these points to meet the inclined line at 1',2',3'.... etc.

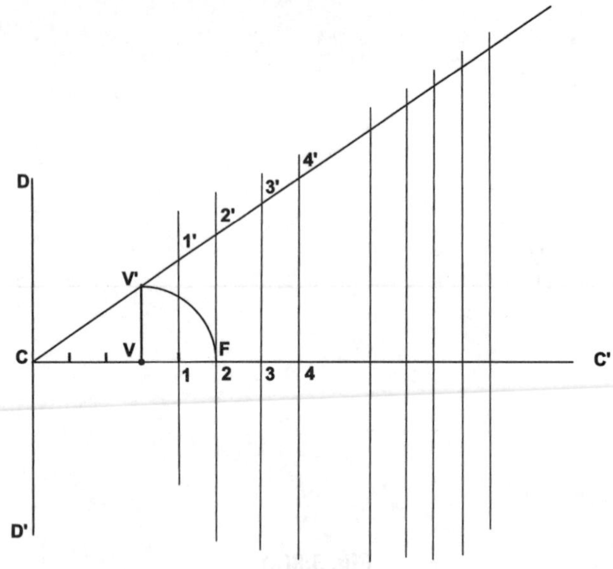

Fig. 3.8(c)

***Step* 4:** With centre F and radius equal to 1 – 1', draw arcs on perpendicular line through 1 to obtain two points P_1 and P_1'. Similarly, with centre F and radius equal to 2 – 2', 3 – 3' ..etc., draw arcs on respective lines to obtain P_2, P_2', P_3 and P_3'.. etc.

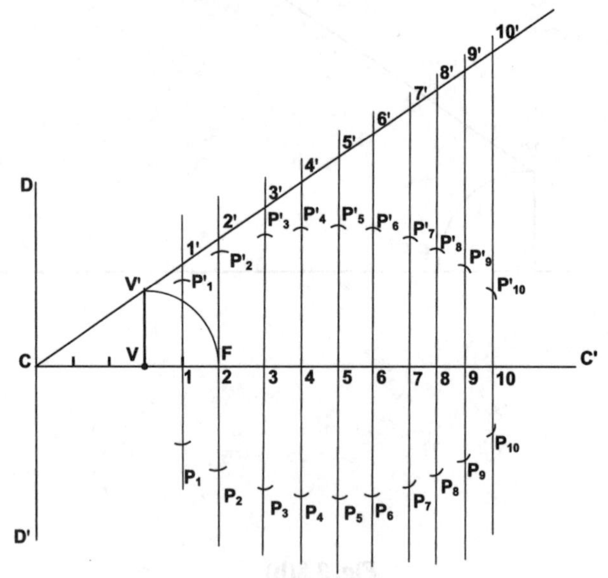

Fig. 3.8(d)

***Step* 5:** Join P_1, P_2, P_3 ... etc., to get a smooth closed curve which is the required ellipse.

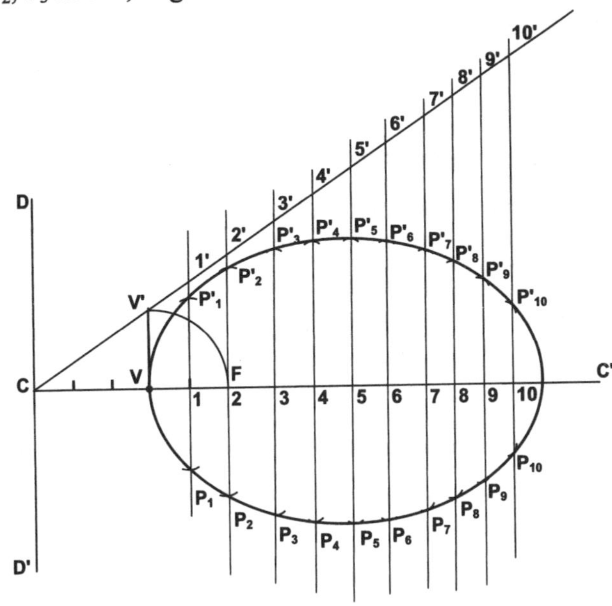

Fig. 3.8(e)

***Step* 6:** (To draw tangent and normal)

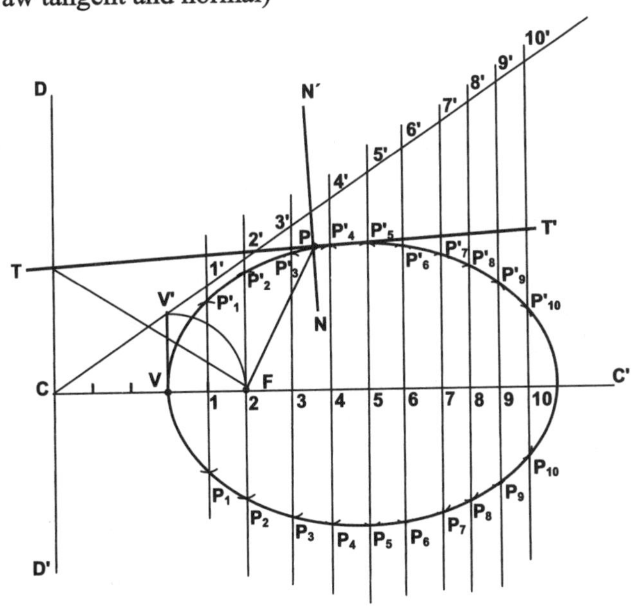

Fig. 3.8(f)

Mark a point P on the curve at a distance of 75 mm from the directrix. Join P to F. Draw a line perpendicular to PF to intersect the directrix at T. Join T to P and extend it to T'. TT' is the required tangent. Draw another line at P perpendicular to TT', which is the required normal (NN').

3.5.3 Construction of Ellipse by Arcs of Circle Method

This method can be used when (i) the length of major and minor axes are given (ii) length of only major axis and distance between foci are given. The procedure of construction is explained below with the help of following problem.

Problem 3.2 (Fig.3.9): *Construct an ellipse using arcs of circles method if the major axis is 120 mm long and foci are 90 mm apart.*

Construction (Fig.3.9):

Step 1: Draw a line AB, 120 mm long as major axis. Mark foci F_1 and F_2 on it so that $F_1F_2 = 90$ mm. Mark points 1, 2, 3....etc., between F_1 and F_2 (not necessarily at equal distances).

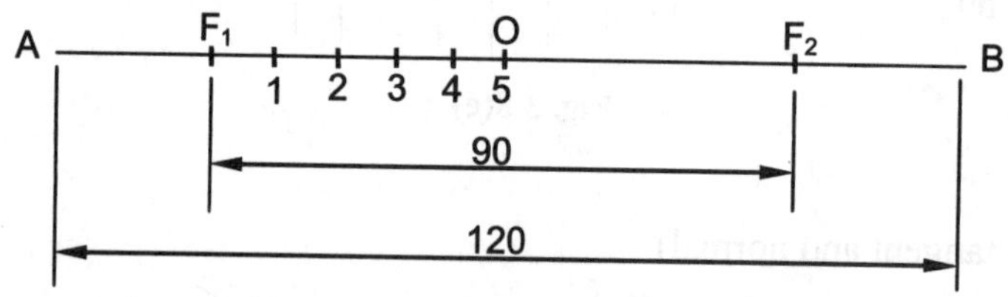

Fig. 3.9(a)

Step 2:

(i) With F_1 and F_2 as centres and A1 as radius, draw arcs on both sides of the major axis.

(ii) With F_1 and F_2 as centres and B1 as radius, draw arcs intersecting previous arcs (drawn in step 1).

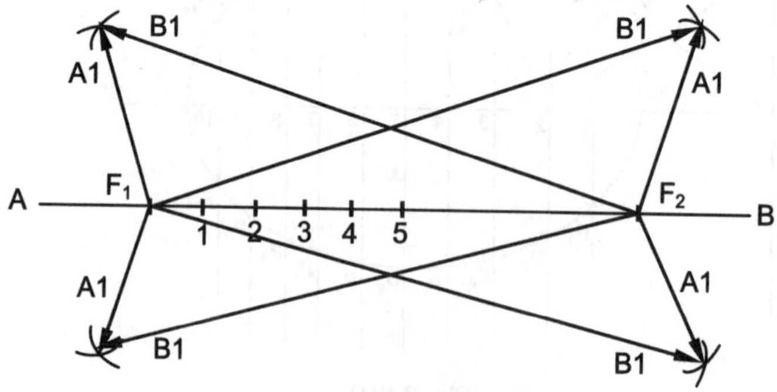

Fig. 3.9(b)

***Step* 3:** Repeat step 2 with radius equal to A2 and B2, A3 and B3...etc., to obtain more points. Mark them as P_1, P_2...etc., and P'_1, P'_2...etc.

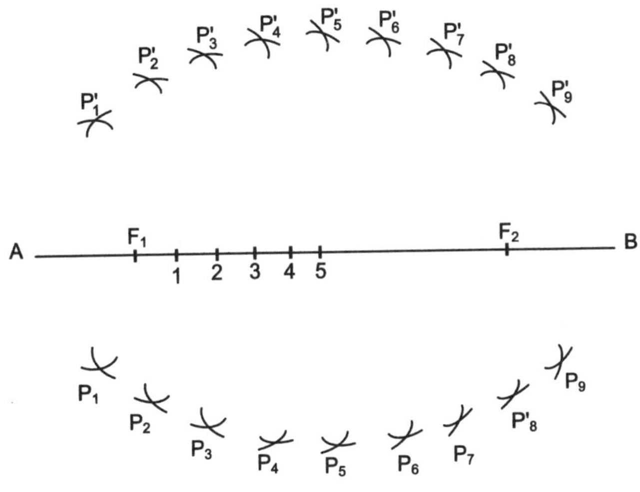

Fig. 3.9(c)

***Step* 4:** Join all the points to obtain a smooth and closed curve, which is the required ellipse.

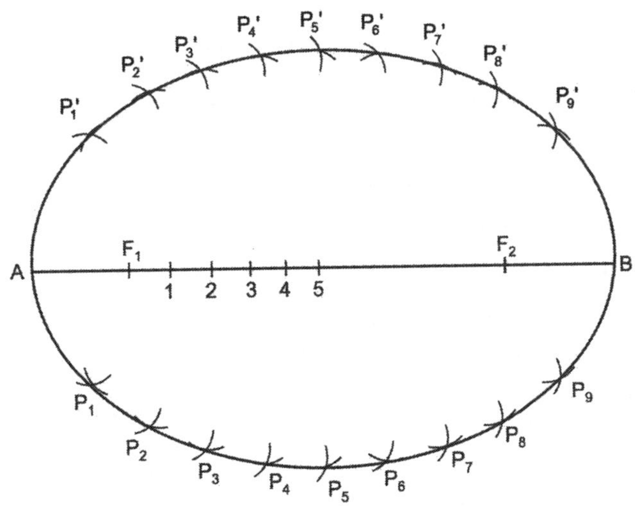

Fig. 3.9(d)

Problem 3.3 (Fig. 3.10): *Draw an ellipse when the length of major and minor axes are 90 mm and 60 mm respectively. Draw a tangent and a normal at a point Q, 70 mm from the centre.*

Construction (Fig. 3.10):

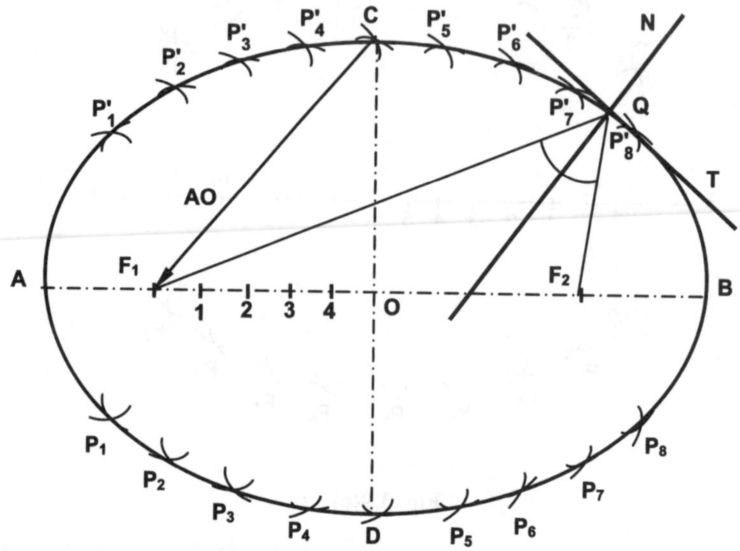

Fig. 3.10

1. Draw major axis AB and minor axis CD equal to 90 mm and 60 mm respectively.
2. With C as centre and radius equal to AO, draw arcs on the major axis to locate foci F_1 and F_2.
3. Draw the ellipse following the steps explained in problem 3.2.

   To draw tangent and normal

4. Mark a point Q on the curve 70 mm away from the centre.
5. Join Q to F_1 and F_2. Bisect the angle F_1 Q F_2, which is the required normal. Draw another line perpendicular to the normal to obtain the tangent.

3.5.4 Construction of Ellipse by Concentric Circles Method

This method is used to construct an ellipse when length of major and minor axes are given. The method of construction is explained with the help of following problem.

Problem 3.4 (Fig. 3.11): *Construct an ellipse having major and minor axes 100 mm and 60 mm long respectively using concentric circles method.*

Construction (Fig. 3.11):

Step 1: Draw major axis AB and minor axis CD, 100 mm and 60 mm long respectively. With O as a centre and radius equal to OA and OC, draw two concentric circles, as shown in Fig. 3.11(a).

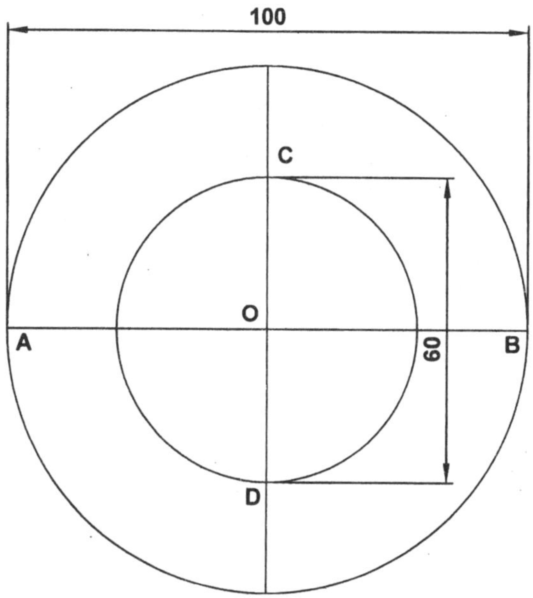

Fig. 3.11(a)

Step 2: Divide the circles into twelve equal parts. Mark division points on smaller circle as 1, 2, ...etc., and on bigger circle as 1',2',3'...etc. Draw twelve radial lines joining O to 1',2',... etc.

Step 3: Draw a horizontal line through 1 and a vertical line through 1' such that these lines meet each other at right angle. Mark the point of intersection as P_1. Repeat the above step with points 2 and 2′, 3 and 3′.... etc., to obtain points P_2, P_3, P_4 .... P_{12}.

Step 4: Join all the points, P_1, P_2 ... etc., by a smooth curve to get the required ellipse as shown in Fig. 3.11 (d).

AB = 100 mm

CD = 60 mm

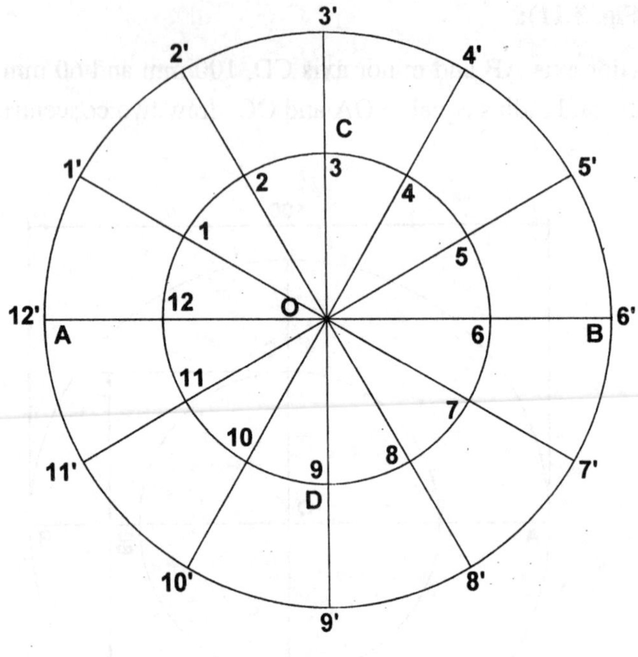

Fig. 3.11(b)

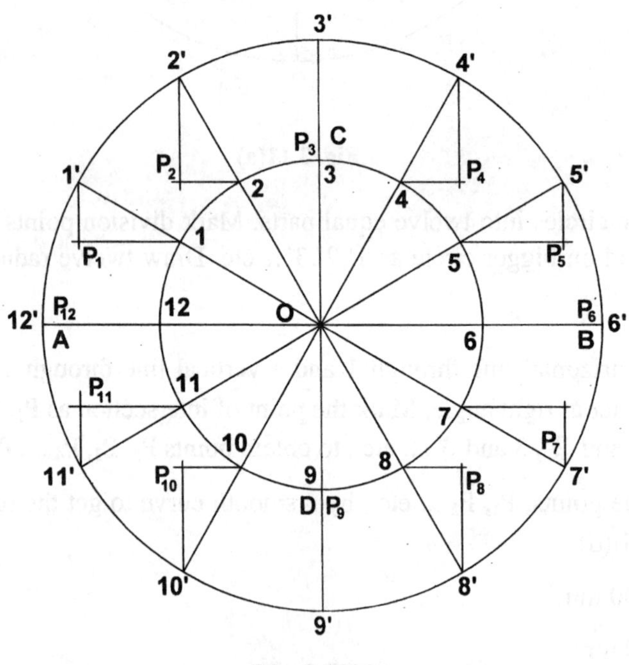

Fig. 3.11(c)

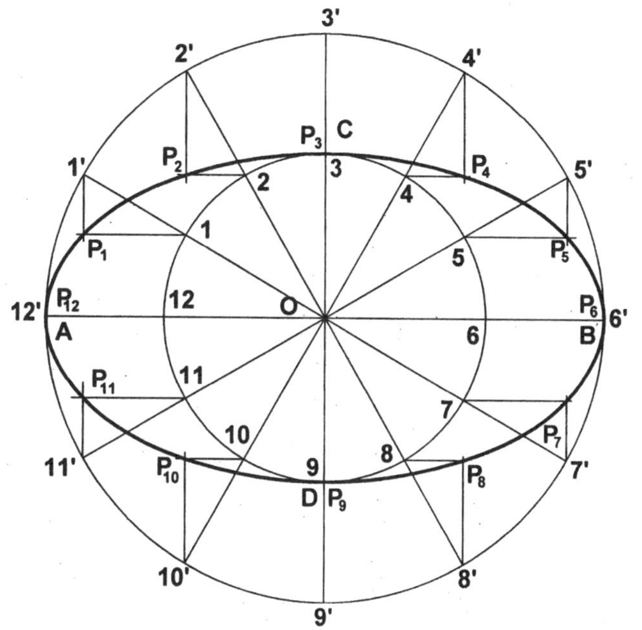

Fig. 3.11(d)

3.5.5 Construction of Ellipse by Oblong (Rectangle) Method

This method can be used to draw an ellipse when the length of major and minor axes are given. The detailed procedure is explained in problem 3.5.

Problem 3.5: *The major and minor axes of an ellipse are 120 mm and 80 mm respectively. Draw the ellipse using oblong method.*

Construction (Fig. 3.12)

***Step* 1:** Draw major axis AB = 120 mm and minor axis, CD = 80 mm. Complete the rectangle EFGH.

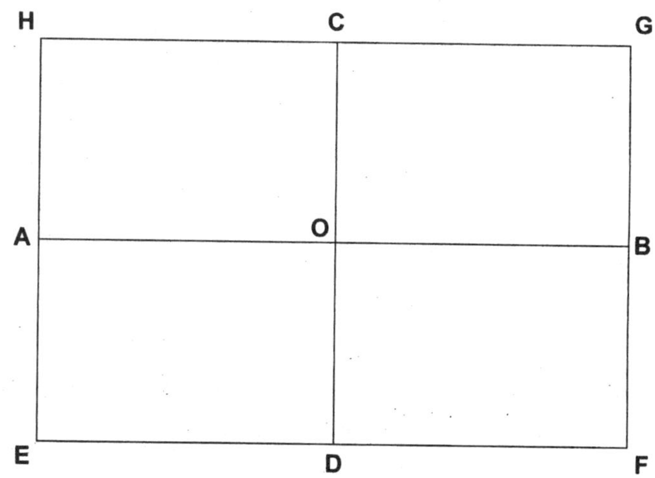

Fig. 3.12(a)

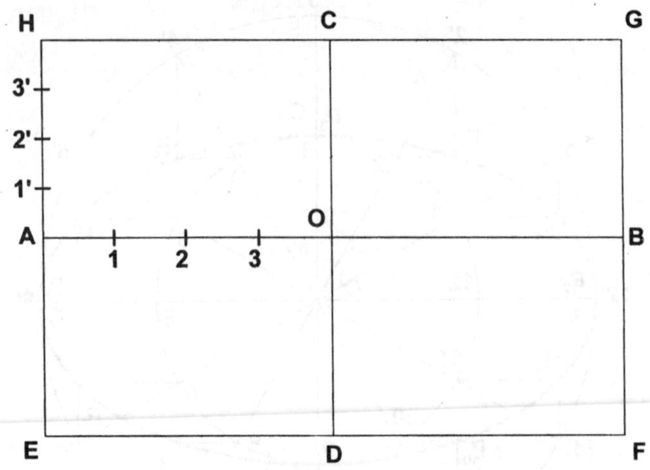

Fig. 3.12(b)

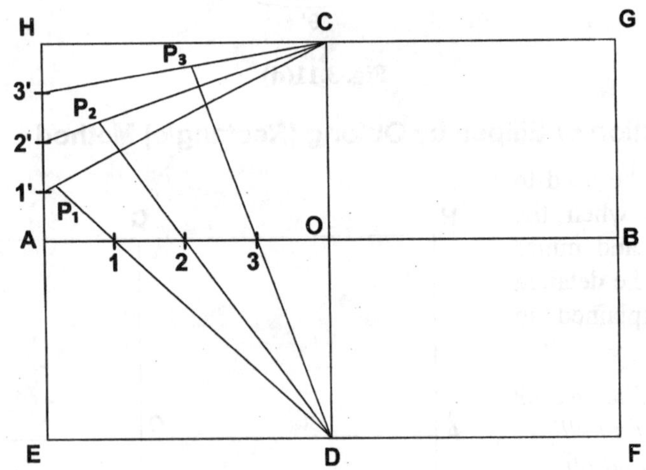

Fig. 3.12(c)

***Step* 2:** Divide AO and AH into same number of equal parts (say four). Mark divisions as 1, 2, 3 on AO and 1′ 2′, and 3′ on AH.

***Step* 3:** Draw lines joining C to 1', 2' and 3'. From point D, draw lines through 1, 2 and 3 intersecting the lines C1', C2' and C3' at P_1, P_2 and P_3 respectively.

***Step* 4:** Draw horizontal lines from P_1, P_2 and P_3 to the right of the minor axis CD. Mark P_4, P_5 and P_6 on these lines considering the symmetry of the figure about CD.

***Step* 5:** Draw vertical lines through P_1, P_2, P_3, P_4, P_5 and P_6 and mark the remaining points on these lines considering symmetry of the figure about AB. Join all the points (P_1.... P_{12}) by a smooth curve, which is the required ellipse.

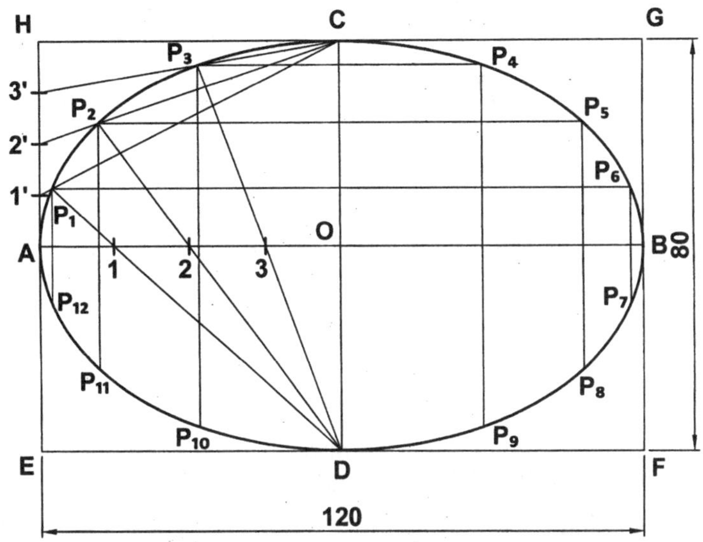

Fig. 3.12(d)

Important Note for Students

Construction procedures for problems 3.1, 3.2, 3.4 and 3.5 have been explained in detail using a number of steps for easy understanding. But from examination point of view, students are advised to draw as per final step only (i.e., not detailed steps).

Problem 3.6 (Fig. 3.13): *Inscribe an ellipse in a parallelogram whose sides are 120 mm and 80 mm and included angle between them is 70°.*

Construction (Fig. 3.13):

1. Draw adjacent sides of the parallelogram as below.

   PQ = 120 mm, PS = 80 mm and ∠SPQ = 70° and complete the parallelogram PQRS.

2. Mark A, B, C and D as mid points of PS, QR, SR and PQ respectively.
3. Draw major axis, AB and minor axis CD.

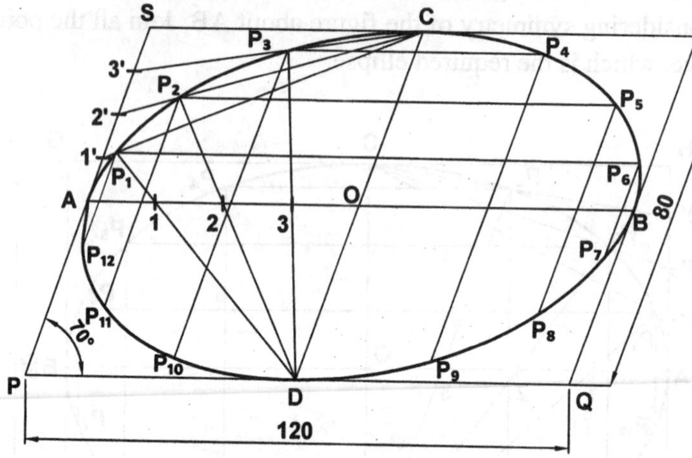

Fig. 3.13

4. Divide AO and AS into same number of equal parts (say four parts). Mark division points as 1, 2 and 3 on AO and 1', 2' and 3' on AS.
5. Draw lines joining C to 1′, 2′ and 3′.
6. From D, draw lines passing through 1, 2 and 3 intersecting the lines C1', C2' and C3' at P_1, P_2 and P_3 respectively.
7. Other points in the remaining three quadrants can be obtained as explained in Problem 3.5.
8. Join P_1, P_2, P_3 ... etc., by smooth curve to get the required ellipse.

3.6 Parabola

A Parabola can be defined as the locus of a point moving in a plane in such a way that the ratio of its distance from a fixed point (focus) and from a fixed straight line (directrix) is constant and is always equal to unity.

A Parabola can also be described mathematically as follows:

$$y^2 = 4\,ax$$

or

$$x^2 = 4\,ay$$

It can be constructed using the following methods.

(a) General method (eccentricity method)

(b) Offset method

(c) Oblong method (rectangle and parallelogram methods)

(d) Tangent method

3.6.1 Construction of Parabola by General Method

This method requires distance of focus from the directrix and eccentricity (e). The procedure for construction of a parabola by this method is explained with the help of following problem.

Problem 3.7 (Fig. 3.14): *Draw a Parabola when the distance of the focus from the directrix is 50 mm and eccentricity is equal to one. Also draw a tangent and a normal at a point on the curve 45 mm from the directrix.*

Construction (Fig. 3.14):

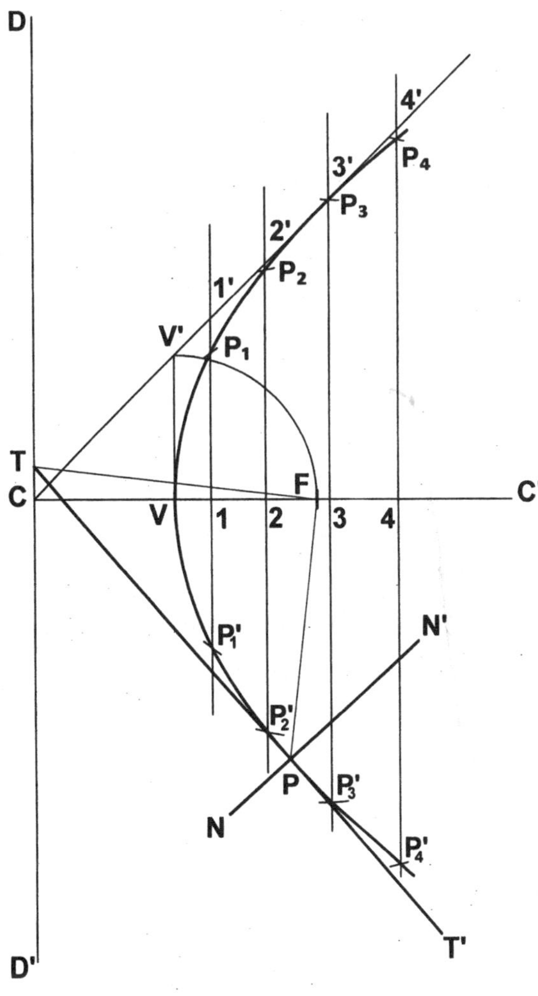

Fig. 3.14

1. Draw a directrix DD' and a principal axis CC' as shown in Fig. 3.14.
2. Mark focus, F on CC' at a distance of 50 mm from directrix and divide CF into two equal parts. Mark V as a mid point of CF.
3. Draw a vertical line through point V. With V as centre and radius equal to VF, draw an arc intersecting vertical line at V'. Mark some points 1, 2, 3, .. etc., to the right of V on the axis. Draw a number of vertical lines through these points intersecting the inclined line at 1',2',3'.. etc.
4. With F as centre and radius equal to 1−1', draw arcs cutting vertical line through 1 at P_1 and P_1'. Similarly with F as centre and radius equal to 2-2', 3-3' etc., obtain P_2, P_3... etc. Join P_1, P_2 ...etc., by a smooth curve.
5. Mark a point P on the curve at a distance of 45 mm from the directrix. Join P to F. Draw a line perpendicular to PF to intersect the directrix at T. Join T to P and extend it to T'. TT' is the required tangent. Draw another line at P perpendicular to TT', which is the required normal (NN').

3.6.2 Construction of Parabola by Offset Method

Consider following equation of parabola

$$x^2 = 4ay$$

$$\Rightarrow \quad x^2 \propto y$$

Offset method is based on above equation, in which offset of a point varies as square of its distance from starting point.

Refer Fig. 3.15 where, AB and EF are the given base and axis respectively of a parabola. The offset can be calculated as follows:

$$1P_1 = \left(\frac{1}{4}\right)^2 DF$$

$$2P_2 = \left(\frac{1}{2}\right)^2 DF$$

$$3P_3 = \left(\frac{3}{4}\right)^2 DF$$

Points P_1', P_2' and P_3' can be obtained from the symmetry of the curve.

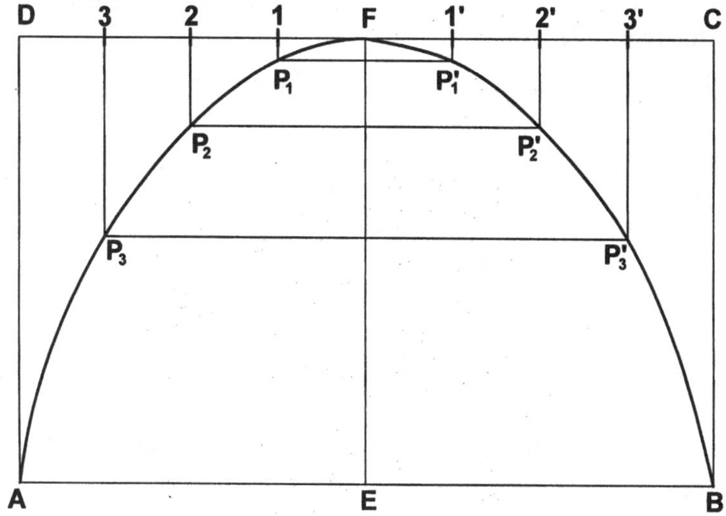

Fig. 3.15

Problem 3.8 (Fig. 3.16): *Construct a parabola using offset method, if its base and axis are 128 mm and 80 mm long respectively.*

Construction (Fig. 3.16):

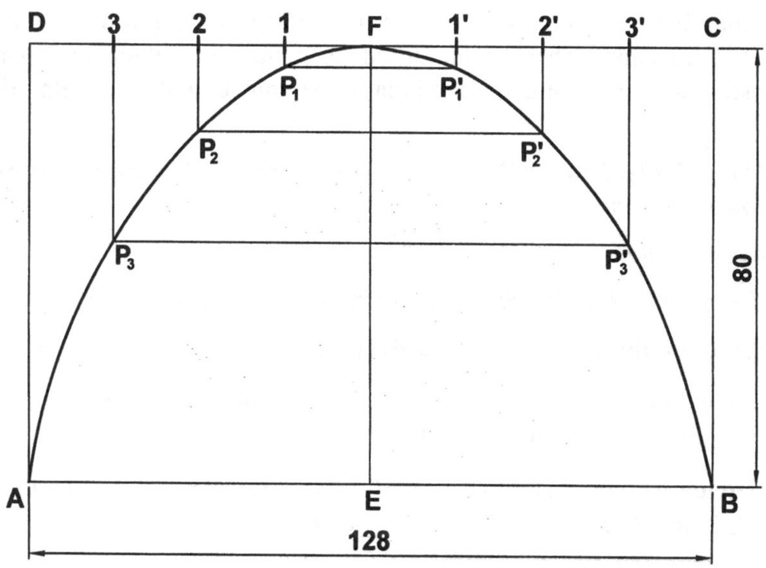

Fig. 3.16

1. Draw line AB = 120 mm and EF = 80 mm as base and axis respectively and complete the rectangle ABCD.
2. Divide DF and FC into same number of equal parts (say four)
3. Calculate the offsets as follows:

$$1P_1 = \left(\frac{1}{4}\right)^2 DF = \left(\frac{1}{4}\right)^2 \times 64 = 4 \text{ mm}$$

$$2P_2 = \left(\frac{1}{2}\right)^2 DF = \left(\frac{1}{2}\right)^2 \times 64 = 16 \text{ mm}$$

$$3P_3 = \left(\frac{3}{4}\right)^2 DF = \left(\frac{3}{4}\right)^2 \times 64 = 36 \text{ mm}$$

4. Draw vertical lines through points 1, 2, and 3 of length $1P_1$, $2P_2$ and $3P_3$ respectively.
5. Join P_1, P_2, P_3 and A with a smooth curve.
6. Draw horizontal lines through P_1, P_2 and P_3 and vertical lines through 1', 2' and 3'.
7. Mark P_1', P_2' and P_3' at the point of intersection of horizontal and vertical lines.
8. Join F, P_1', P_2' P_3' and B by a smooth curve.

3.6.3 Construction of Parabola by Oblong Method

This method may be used to construct a parabola when the length of base and axis are given. This method can further be classified as rectangle method and parallelogram method. Construction procedure of parabola is explained with the help of following problem.

Problem 3.9 (Fig. 3.17): *Draw a Parabola having its base 120 mm and axis 80 mm long, using the following methods.*

(a) Rectangle method

(b) Parallelogram method when the angle between base and axis is 70°.

(a) Construction (by rectangle method) (Fig. 3.17):

1. Draw lines AB (base) and EF (axis), 120 mm and 80 mm long respectively. Complete the rectangle ABCD.
2. Divide AE and AD into four parts. Mark divisions as 1, 2 and 3 on AE and 1', 2' and 3' on AD.

3. Draw lines joining F to 1', 2' and 3'. Draw perpendicular lines through points 1, 2 and 3 intersecting previous lines at P_1, P_2 and P_3 respectively.
4. Join A, P_1, P_2, P_3 and F with a smooth curve.

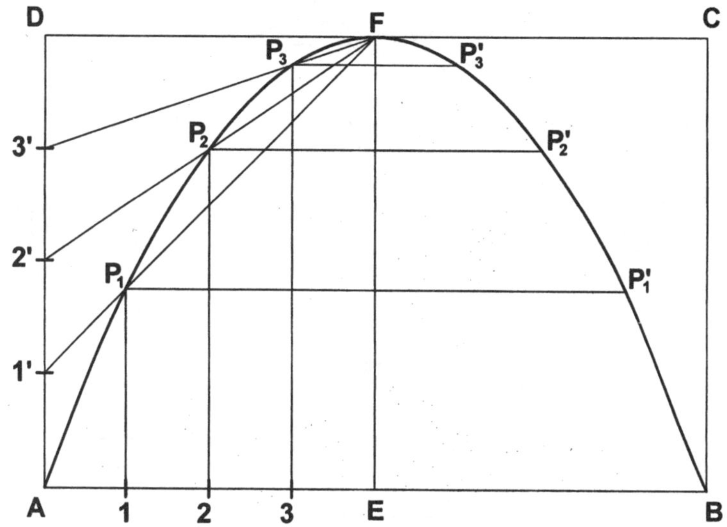

Fig. 3.17

5. Draw horizontal lines from P_1, P_2 and P_3. Considering symmetry of the curve about EF, obtain other points P_1', P_2' and P_3'.
6. Join F, P_3', P_2', P_1' and B with a smooth curve to get right part of the parabola.

(b) Construction (by Parallelogram method) (Fig. 3.18):

1. Draw a line AB, 120 mm long as base of parabola.
2. Draw the axis, EF at the mid point of AB and inclined at 70° to it.
3. With the base AB. Construct a parallelogram ABCD.
4. Divide AE and AD into four equal parts and mark divisions as 1, 2 and 3 on AE and 1', 2' and 3' on AD.
5. Draw straight lines joining F to 1', 2' and 3'. Through points 1, 2 and 3 draw lines parallel to EF intersecting the lines F_1', F_2' and F_3' at P_1, P_2 and P_3 respectively.
6. Join A, P_1, P_2, P_3 and F to obtain a smooth curve.

7. Obtain points P_1', P_2' and P_3' using the procedure explained in the part (a) of this problem.

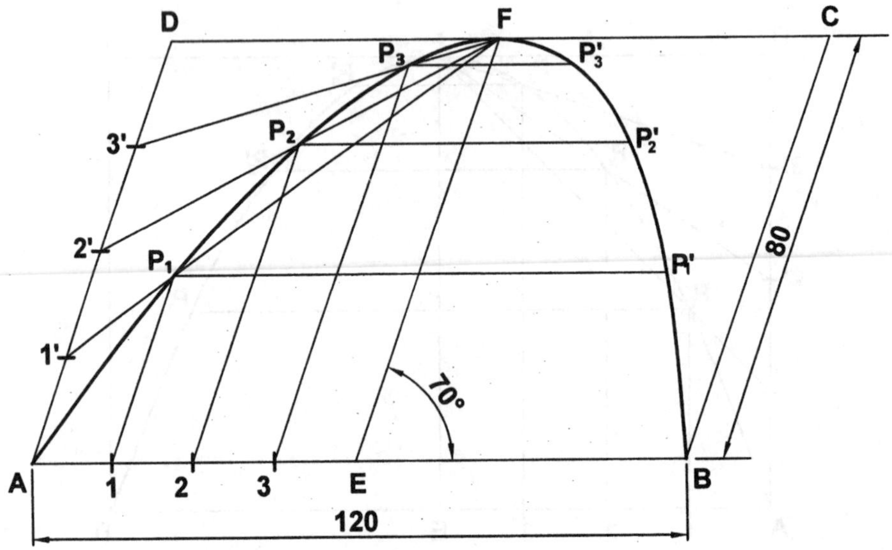

Fig. 3.18

3.6.4 Construction of Parabola by Tangent Method

This method is used to draw a parabola when the length of its base and the axis are given. The construction procedure is explained with the help of following problem.

Problem 3.10 (Fig. 3.19): *Draw a parabola having its base 120 mm and axis 50 mm long using Tangent method.*

Construction (Fig. 3.19):

1. Draw a line AB, 120 mm long as base of the parabola.
2. Erect a 50 mm long perpendicular axis, CD at the mid point of AB.
3. Extend CD to O such that CD = DO.
4. Join A and B to O.
5. Divide OA and OB into same number of equal parts (say eight parts).
6. Mark division points as 1, 2, 3....etc., as shown in Fig. 3.19.

7. Join 1 to 1, 2 to 2....etc. Draw a smooth curve tangential to lines 1-1, 2-2,....etc., to obtain the required parabola.

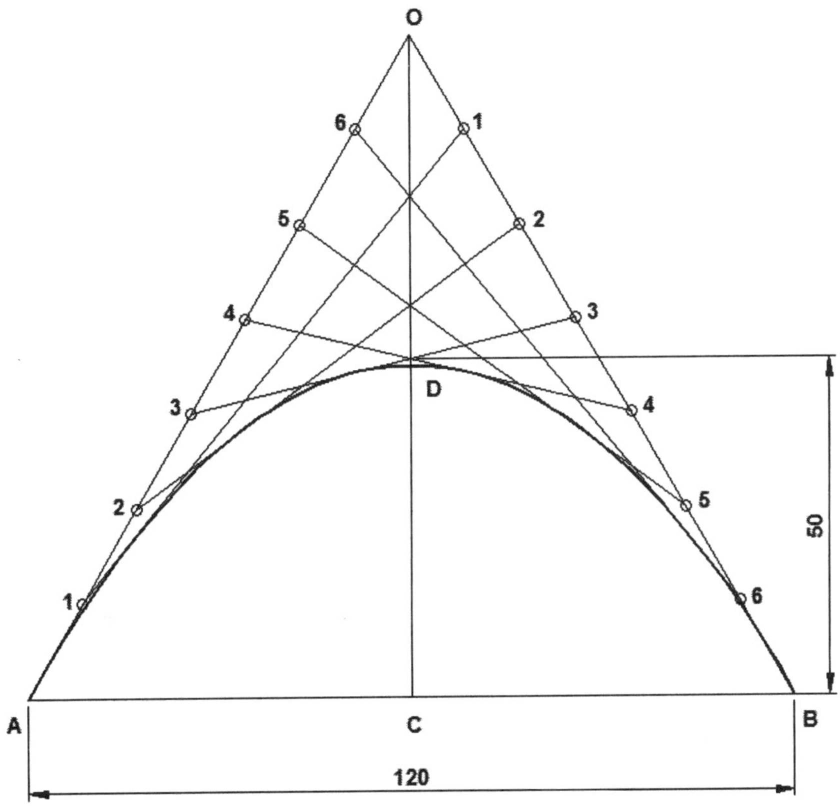

Fig. 3.19

Problem 3.11 (Fig. 3.20): *A shot is discharged from the ground at an inclination of 45° with horizontal. It returns to the ground at a distance of 100 mm away from the point of discharge. Draw the path traced out by the shot.*

Construction (Fig. 3.20):

1. Draw an Isosceles triangle ABO having its base AB, 100 mm long and $\angle OAB = \angle OBA = 45°$.
2. Mark C as mid point of AB and join O to C.
3. Draw the parabola using tangent method as explained in problem 3.10.

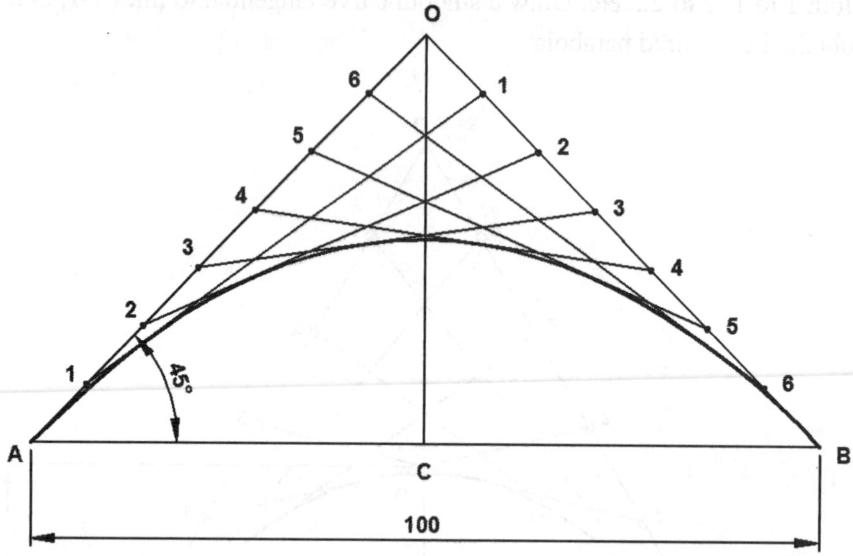

Fig. 3.20

3.7 Hyperbola

A hyperbola can be defined as locus of a point moving in a plane in such a way that the ratio of its distance from fixed point and from a fixed straight line is constant and is always greater than one i.e., $e > 1$.

Mathematically, a hyperbola can be described by,

$$\frac{x^2}{a^2} - \frac{y^2}{b^2} = 1$$

3.7.1 Construction of Hyperbola by General Method (Eccentricity Method)

This method requires distance of focus from the directrix and eccentrity(e). The procedure of construction of the hyperbola using this method is explained with the help of following problem.

Problem 3.12 (Fig. 3.21): *Draw a parabola when the distance between directrix and focus is 50 mm and eccentricity is 3/2. Also draw a tangent and a normal at a distance of 45 mm from directrix.*

OR

Draw a locus of a point moving in a plane in such a way that the ratio of its distance from a fixed point and from a fixed line is 3/2. The fixed point is 50 mm away from the fixed straight line. Also draw a tangent and normal at a point on the curve 45 mm away from the directrix.

Construction (Fig. 3.21):

1. Draw a directrix DD' and a horizontal principal axis CC'.
2. Mark focus F on CC' at a distance of 50 mm from the directrix. Divide CF into five equal parts.
3. Mark vertex V, on third division point from the focus, so that,

$$\frac{VF}{VC} = \frac{3}{2}$$

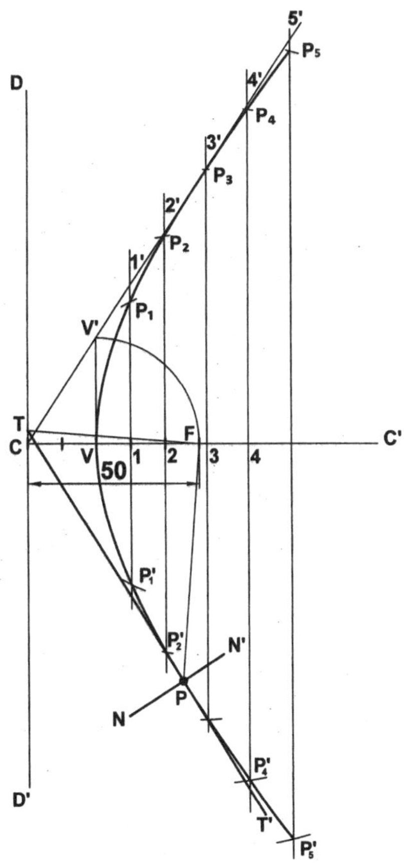

Fig. 3.21

4. Erect a perpendicular at V. With V as centre and radius equal to VF, draw an arc intersecting the perpendicular line at V'.

5. Join C to V' and extend the line.
6. Mark few points 1, 2, 3 etc., on the axis. Through these points, draw lines perpendicular to the axis and intersecting the line through V' at $1', 2', 3'$ ...etc.
7. With F as centre and radius equal to $11'$ draw two arcs intersecting the line through 1 at P_1 and P_1'.
8. Similarly with F as centre and radius equal to $22', 33', 44'$ etc., obtain P_2, P_2', P_3, P_3' etc.
9. Join P_1, P_1', P_2, P_2' ... etc., by a smooth curve.
10. Draw tangent and normal at a given point using the steps explained in problem 3.1.

3.7.2 Construction of Hyperbola by Intersecting Arcs Method

Hyperbola can also be defined as a plane curve generated by a point P moving in such a way that the difference of its distances from two fixed points is constant and is equal to transverse axis. (refer Fig. 3.22)

F_1 and F_2 are Foci

V_1 and V_2 are vertices

As per above definitions,

$PF_1 - PF_2$ = constant = $V_1 V_2$ other terms used with a hyperbola are illustrated in Fig. 3.23, as follows.

$X_1 Y_1$ and $X_2 Y_2$ are asymptotes $D_1 D_1'$ and $D_2 D_2'$ are directrices and CC' is conjugate axis.

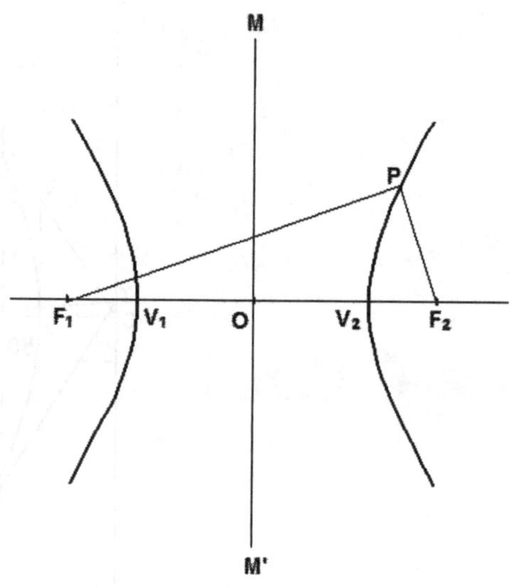

Fig. 3.22

Intersecting arcs method can be used to draw hyperbola or two branches of a hyperbola when distance between foci and the distance between vertices are given. Following problem illustrates the method of its construction and determination of asymptotes, directrices and conjugate axis of a given hyperbola.

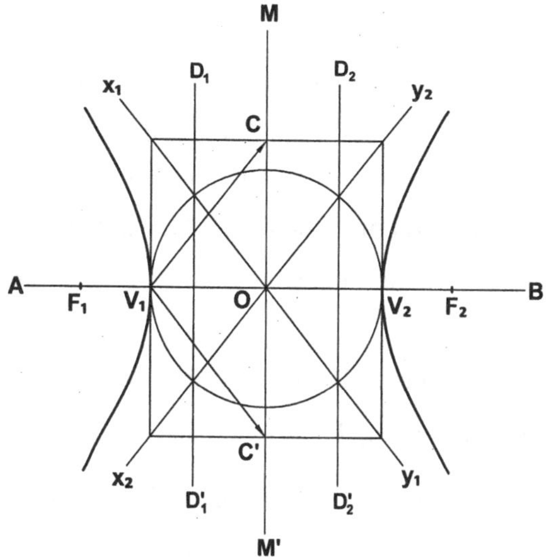

Fig. 3.23

Problem 3.13 (Fig. 3.24): *Construct a hyperbola (two branches) if the distance between its foci is 80 mm and distance between its vertices is 50 mm. Also draw a tangent and a normal at a point 35 mm from one of the foci.*

OR

Draw the locus of a point P moving in such a way that the difference of its distances from two fixed points, is always constant and is equal to 50 mm. Two fixed points are 80 mm apart. Also draw a tangent and a normal at a point 30 mm from one of its foci.

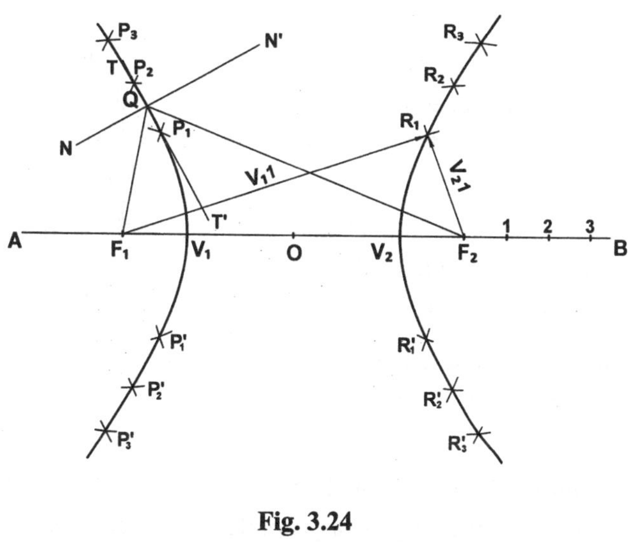

Fig. 3.24

Construction (Fig. 3.24):

1. Draw a horizontal principal axis AB and mark a point O on it.
2. Mark foci F_1 and F_2 and vertices V_1 and V_2 on the axis such that distance between F_1 and F_2 is 80 mm and distance between V_1 and V_2 is 50 mm.
3. Mark points 1, 2, 3... etc., on one side of AB (say to the right of O).
4. With F_1 and F_2 as centres and radius equal to $V_2 1$, draw arcs above and below the AB.
5. Now with the same centres (F_1 and F_2) and radius equal to $V_1 1$, draw arcs intersecting previous arcs at P_1, P_1', R_1 and R_1'.
6. Repeat above steps with points 2, 3, 4... etc., and obtain P_2, P_2'... and R_2, R_2' ...etc.
7. Join all the points (P_1, P_2, ..., R_1, R_2,... ,P_1',P_2'...., R_1',R_2'....etc.,) by a smooth curves to obtain the required hyperbola.
8. Mark a point Q on the curve 30 mm from the focus (F_1). Join F_1 and F_2 to Q. Bisect $\angle F_1QF_2$. The angle bisecting line is the required tangent TT'. Draw a line NN' through Q perpendicular to TT', which is the required normal.

Problem 3.14 (Fig. 3.25): *Locate a symptotes and directrixes of a hyperbola drawn in problem 3.13.*

Construction (Fig. 3.25):

1. Through point O, draw a line MM' perpendicular to AB.
2. With centre V_1 and radius equal to OF_1, draw two arcs to intersect MM' at C and C'.
3. Draw horizontal lines through C and C' and vertical lines through V_1 and V_2 to obtain a rectangle PQRS.
4. Draw diagonals of rectangle PQRS and extend them, which are required a symptotes.

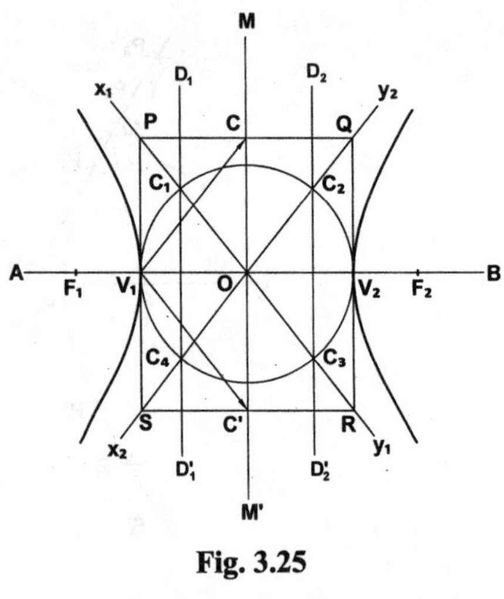

Fig. 3.25

5. With O as centre and radius equal to OV_1, draw a circle intersecting the asymptotes at C_1, C_2, C_3 and C_4.

6. Draw a line passing through C_1 and C_4. Mark it as D_1D_1', which is one of the directrices.
7. Similarly, draw a line passing through C_2 and C_3. Mark it as D_2D_2', which is another directrix.

3.7.3 Rectangular Hyperbola by Asymptotes Method

Rectangular hyperbola is a curve traced out by a point moving in a plane in such a way that the product of its distances from two fixed lines at right angles to each other is constant. The two fixed lines are known as asymptotes.

The asymptote method is used to draw rectangular hyperbola when position of a point P with respect to asymptotes is given. When the angle between two asymptotes is 90°, the method is referred as orthogonal asymptotes method. On the other hand, if the angle between asymptotes is less than 90°, the method is referred as oblique asymptotes method.

The procedure of drawing a rectangular hyperbola is explained with the help of following problem.

Problem 3.15 (Fig. 3.26): *A point P is 30 mm and 60 mm from two straight lines, which are at right angle to each other. Draw a rectangular hyperbola through point P.*

Construction (Fig. 3.26):

1. Draw a symptotes OA and OB at right angle to each other.
2. Mark point P such that it is 60 mm from OA and 30 mm from OB.
3. Through P, draw lines parallel to OA and OB.
4. Mark some points 1, 2, 3… etc., on PD.
5. Join O to 1, 2, 3 …etc., by straight lines intersecting EF at 1',2',3'... etc.
6. From 1, 2…etc., draw vertical lines and from 1',2',3'.... etc., draw horizontal lines. Mark P_1, P_2..etc., at the intersection of horizontal and vertical lines.
7. Take two points on the left of P (say 5 and 6). From O draw lines passing through 5 and 6 intersecting EF at 5' and 6'.
8. Draw horizontal lines through 5' and 6' vertical lines through 5 and 6, intersecting the horizontal lines at P_5 and P_6.
9. Join all the points (P_1 .... P_6) by a smooth curve, which is the required rectangular hyperbola.

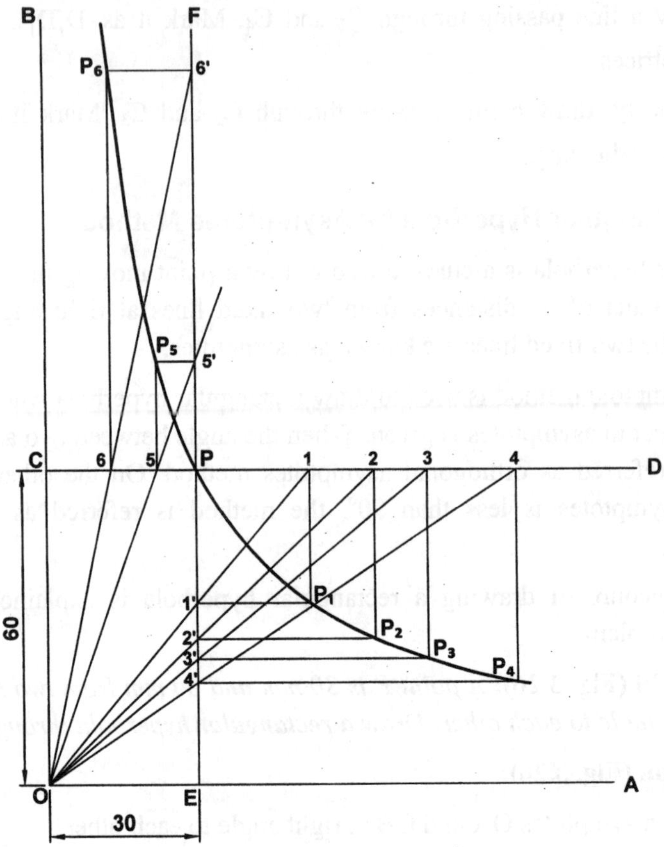

Fig. 3.26

Problem 3.16 (Fig. 3.27): *Two straight lines OA and OB having an angle of 75° between them. A point P is 60 mm from OA and 40 mm from OB. Draw a hyperbola through point P using oblique asymptotes method.*

Construction (Fig. 3.27):

1. Draw asymptotes OA and OB such that $\angle BOA = 75^\circ$
2. Draw a line EF parallel to OB and 40 mm away from it. Draw another line CD parallel to OA and 60 mm away from it. Mark point P at the point of Intersection of EF and CD.
3. Mark points 1, 2, 3 and 4 on PD and 5 and 6 on PC.
4. From O, draw lines passing through 1, 2, 3...etc., and intersecting EF at 1',2',3'... etc.

5. From 1',2',3'... etc., draw line parallel to CD and from 1, 2, 3...etc., draw lines parallel to EF.
6. Mark P_1, P_2, P_3...etc., at the intersecting of these lines as shown in Fig. 3.27.
7. Join all the points (P_1 to P_6) by a smooth curve.

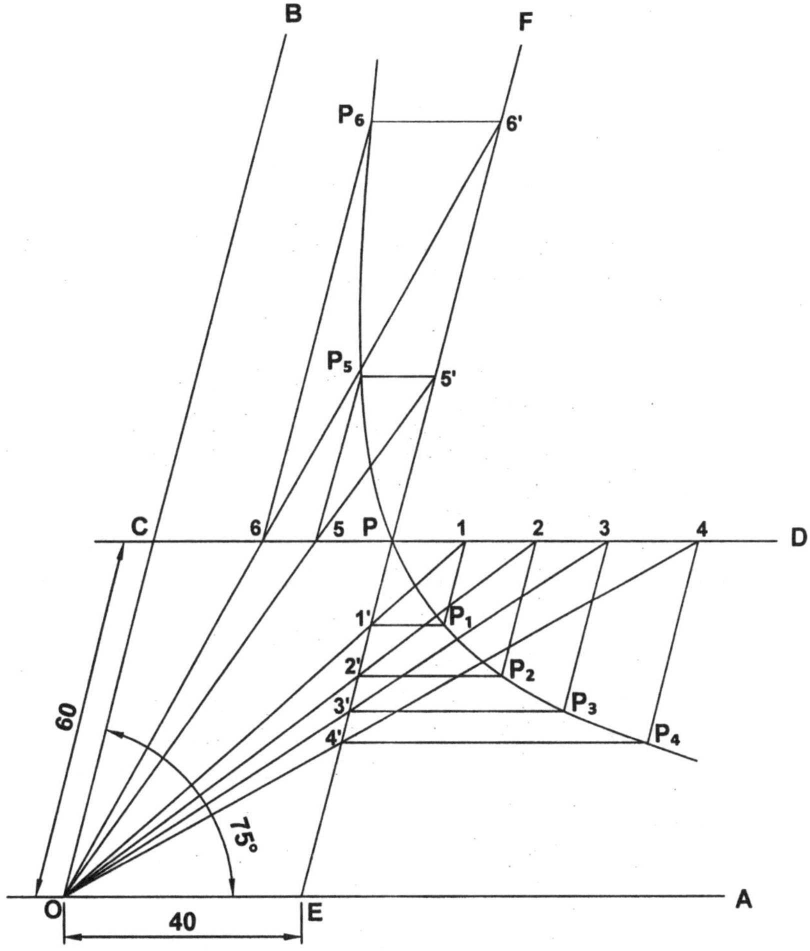

Fig. 3.27

3.8 Additional Problems on Conics

Problem 3.17 (Fig. 3.28): *The major and minor axis of an ellipse are 120 mm 80 mm long respectively. Find the foci and construct an ellipse using intersecting arcs method. Also draw a tangent and a normal at a point 35 mm below the major axis.*

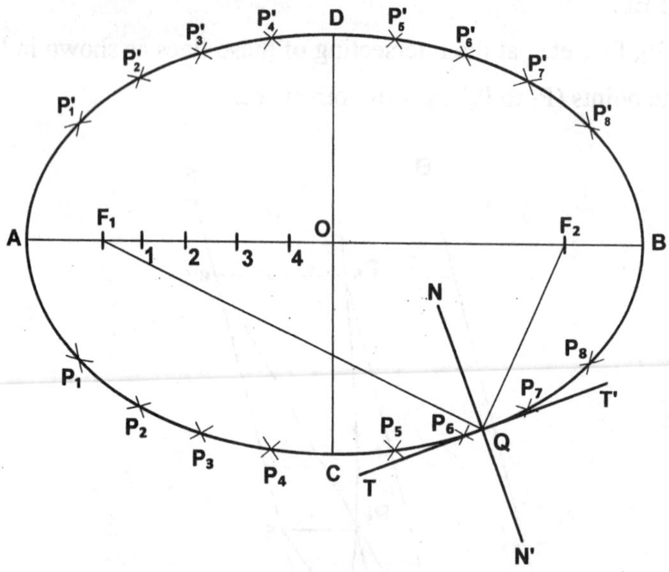

Fig. 3.28

Construction (Fig. 3.28):

1. Draw major axis AB and minor axis CD equal to 120 mm and 80 mm respectively.
2. With C or D as centre and radius equal to AO, draw arcs on major axis to locate F_1 and F_2.
3. Complete the construction following the steps explained in problem 3.2.

Problem 3.18 (Fig. 3.29): *Draw an ellipse, one half of it by concentric circle method and other half by oblong method when its major axis is 120 mm and foci are 20 mm away from its ends. Also draw a tangent and a normal at a point 35 mm above the major axis.*

Construction (Fig. 3.29):

1. Draw major axis AB, 120 mm long. Mark F_1 and F_2, 20 mm from end A and B respectively.
2. Mark O as mid point of AB. Draw a line C'D' perpendicular to AB and passing through O.
3. With F_1 as centre and radius equal to OA, draw two arcs cutting C'D' at C and D. (now CD represents minor axis.)
4. With O as centre and radius equal to OC and OA, draw two concentric semi circles.
5. Divide the semi circles into six equal parts and mark the division points as 1, 2, 3.... etc., and 1', 2', 3'... etc.

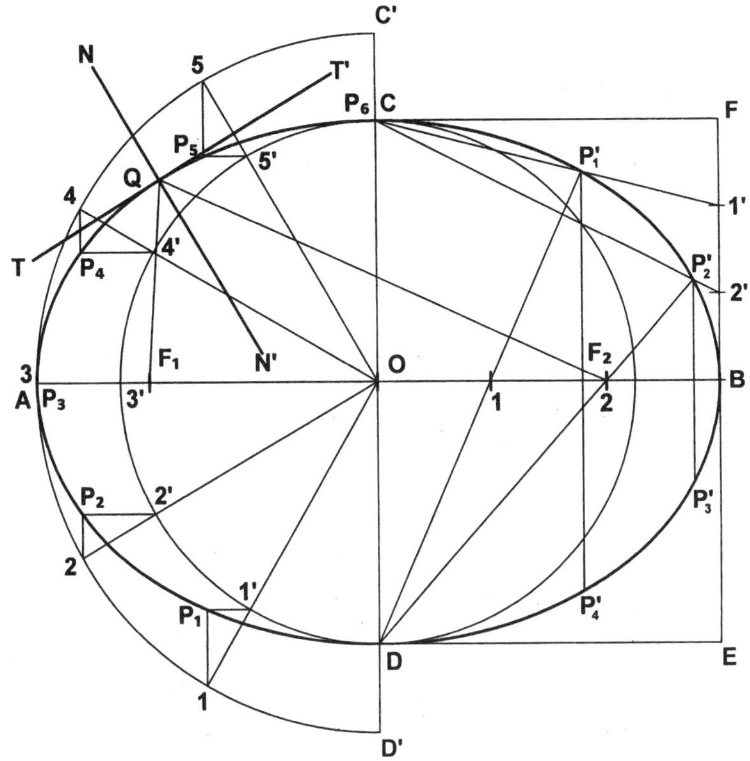

Fig. 3.29

6. Draw horizontal lines through 1', 2', 3'... etc., and vertical lines through 1,2,3 .. etc.
7. Mark P_1, P_2, P_3 ... etc., at the intersection of horizontal and vertical lines.
8. Join P_1, P_2, P_3 .... etc., by a smooth curve to obtain the left part of ellipse.
9. To draw the right part of the ellipse using oblong method, complete the rectangle CDEF.
10. Divide OB and BF into four equal parts. From D, draw lines passing through 1, 2, 3... etc.,
11. From C, draw lines passing through 1', 2', 3'... etc. Mark P_1' and P_2', at the intersection of these lines. Obtain points P_3' and P_4' considering symmetry of the curve about OB.
12. Join all the points (P_1', P_2', .. etc.,) by a smooth curve to obtain right part of the ellipse.

Problem 3.19 (Fig. 3.30): *A ball is thrown up in the air. It reaches a maximum height of 140 metres and travels a horizontal distance of 120 metres. Draw the path of the ball assuming it to be parabolic.*

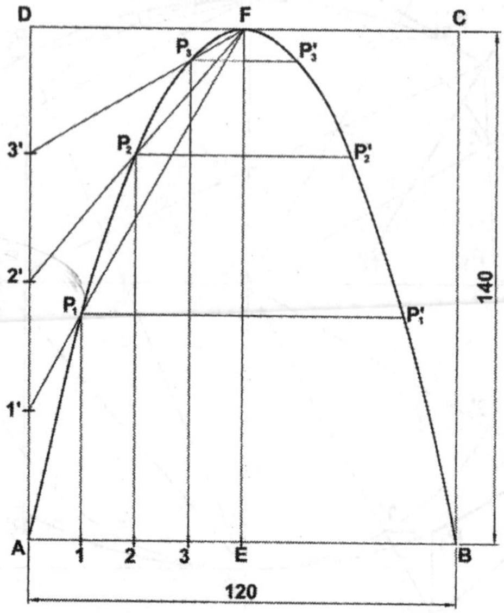

Fig. 3.30

Construction (Fig. 3.30):

1. Draw base AB and altitude EF using suitable scale.
2. Complete rectangle ABCD.
3. Construct a parabola using oblong method as explained in Problem 3.9.

Problem 3.20 (Fig. 3.31): *A stone is thrown up in the sky from a building 5 metres high. In its highest flight, the stone just crosses 12 metres high tree. Trace the path of the projectile if the horizontal distance between building and tree is 4 metres. Find the distance of the point from the building where the stone strikes on the ground.*

Construction (Fig. 3.31):

1. Draw a ground line GG'G" and mark G and G', 8 m apart using suitable scale.
2. Draw vertical lines at G and G' to represent 12 metres height and complete the rectangles GSRG' and PQRS.
3. Draw a parabola inside the rectangle PQRS using oblong method as explained in Problem 3.9.
5. Extend line PQ and mark points 8,9 ... etc., such as 6 7=7 8=8 9.
6. Mark 8', 9'.. etc., on QG' such that 6' 7'= 7' 8'=8' 9'.

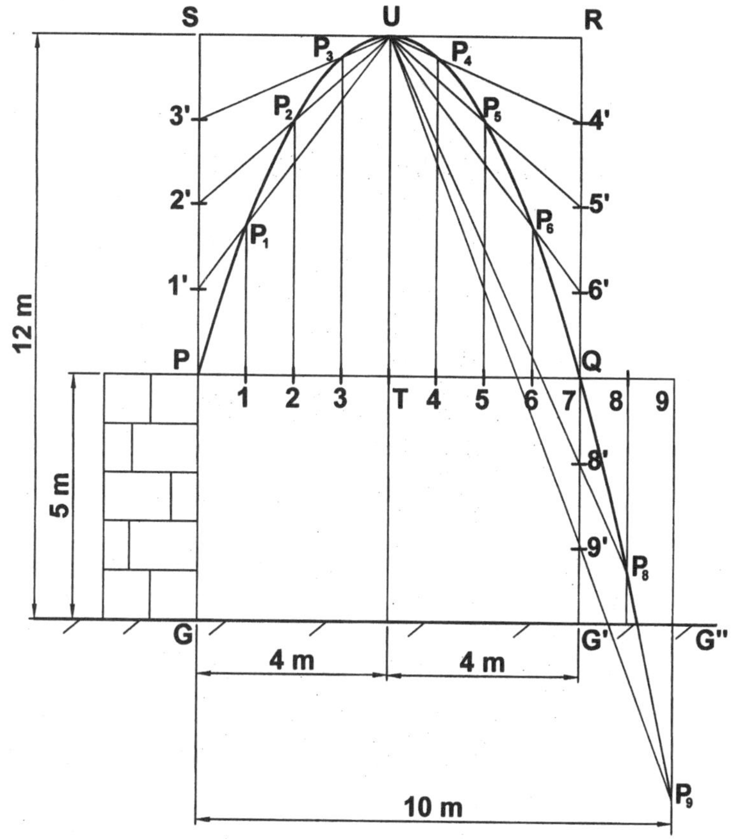

Fig. 3.31

7. From U, draw lines passing through 8' and 9'. Also draw vertical lines through 8 and 9 intersecting the lines through 8' and 9' at P_8 and P_9 respectively.
8. Extend the parabola by joining Q, P_8 and P_9.
9. Locate the point X on the ground level where the stone will strike and measure its horizontal distance from the building.

Problem 3.21 (Fig. 3.32): *A fountain jet discharges water from the ground level at an inclination of 50° to the ground. The jet travels a horizontal distance of 8 metres from the point of discharge and falls on the ground. Trace the path of the jet. Name the curve.*

Construction (Fig. 3.32):

The path of the fountain jet will be parabolic, which can be drawn using tangent method as discussed below.

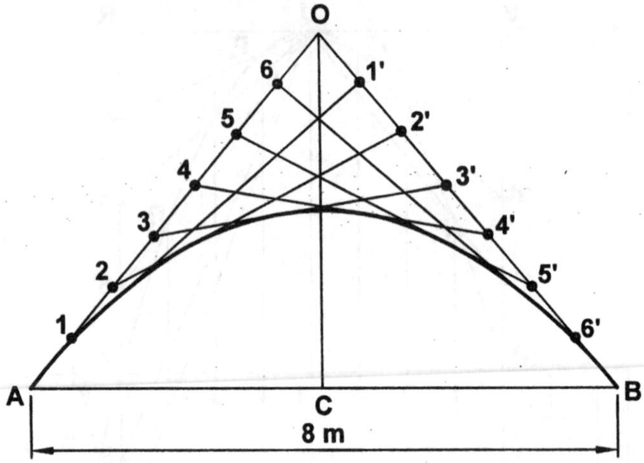

Fig. 3.32

1. Draw base AB representing 8 metres length using suitable scale.
2. At the ends of AB, draw two lines OA and OB inclined at 50° to AB such that OAB is an isosceles triangle.
3. At the mid point of the AB, mark C and join it to O.
4. Divide OA and OB into same number of equal divisions (say seven).
5. Mark division points 1, 2, 3...etc., on OA and 1', 2', ... etc., on OB as shown in Fig. 3.32.
6. Join 1 to 1', 2 to 2', 3 to 3'... etc.
7. Draw a smooth curve tangential to these lines which is the required parabola.

Problem 3.22 (Fig. 3.33): *Trace the locus of a point, such that the difference between the distances of the point from two fixed points 80 mm apart is constant and is equal to 60 mm. Name the curve.*

Construction (Fig. 3.33):

From the definition, it is clear that the curve is hyperbola and it is drawn as discussed below.

1. Draw axis AB of any convenient length.
2. Mark F_1 and F_2 on AB such that F_1F_2=80 mm.

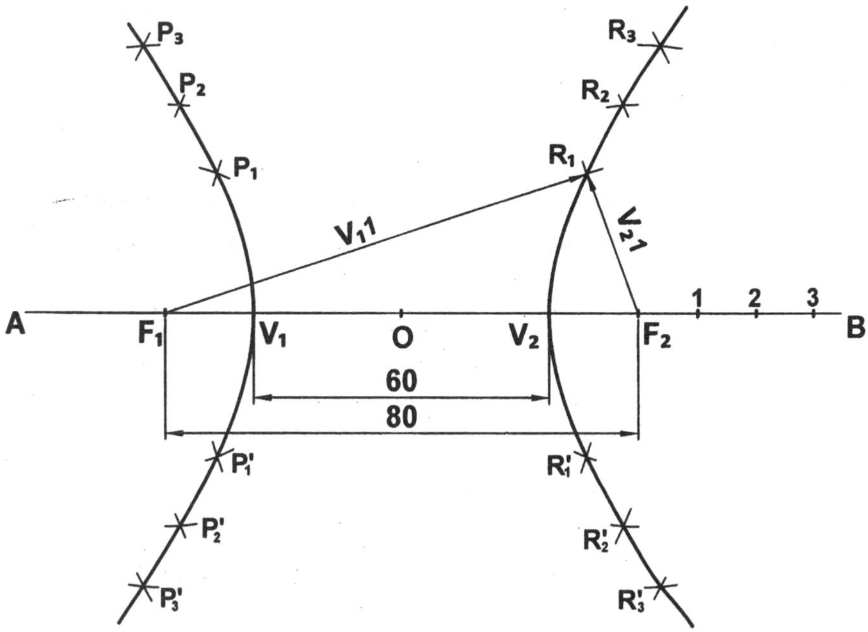

Fig. 3.33

3. Mark O as a mid point of F_1F_2.
4. Mark V_1 and V_2 such that V_1V_2=60 mm. Mark points 1, 2, 3...etc., on the AB.
5. With F_1 and F_2 as centres and radius equal to $V_1 1$ and $V_2 1$ respectively, draw the arcs on both sides of the axis to obtain R_1 and R_1'.
6. With F_1 and F_2 as centres and radius equal to $V_2 1$ and $V_1 1$ respectively, draw the arcs on both sides of the axis to obtain P_1 and P_1'.
7. Repeat above steps for points 2, 3, 4... etc., to obtain remaining points.
8. Join P_1, P_2,... and P_1', P_2', ... by a smooth curve to obtain first branch of hyperbola.
9. Join R_1, R_2, ... and R_1', R_2', ... by a smooth curve to obtain other branch of hyperbola.

Exercises

1. Draw an ellipse when the distance of the focus from its directricx is 50 mm and eccentricity is 2/3. Also draw a tangent and a normal at a point on the curve 65 mm from the directrix.
2. Major axis of an ellipse is 100 mm long and its foci are 80 mm apart. Draw the ellipse using arcs of circles method.

3. Draw an ellipse using arcs of circles method if its major axis and minor axis are 110 mm and 70 mm long respectively.

4. Two fixed points A and B are 100 mm apart. Trace the complete path of a point P moving in such a way that the sum of its distances from A and B is always constant and equal to 120 mm. Name the curve.

5. Directrices of an ellipse are 160 mm apart and its vertices are 100 mm apart. Construct an ellipse, one half it by concentric circles method and other half by oblong method.

6. Inscribe an ellipse in a parallelogram having sides 140 mm and 90 mm long and included angle is 60°.

7. A plot of land is in the shape of a rectangle 25m × 20m. Inscribe an elliptical flower bed in it.

8. A fixed point is 75 mm from a fixed straight line. Draw the locus of a point P moving in such a way that its distance from the fixed straight line is equal to its distance from the fixed point. Name the curve.

9. A ball is thrown up in the air reaches a maximum height of 50 metres and travels a horizontal distance of 70 metres. Trace the path of the ball assuming it to be parabolic.

10. A base and axis of a parabola are 120 mm and 60 mm respectively. Draw the parabola using offset method.

11. A shot is discharged from a gun which is kept inclined at 40° from the ground. It returns to the ground at a point 110 metres from the point of the discharge. Trace the path of the shot.

12. The vertex of a hyperbola is 60 mm from its focus. If eccentricity is 3/2, draw the curve. Also draw a tangent and a normal at a point on the curve 70 mm from the directrix.

13. A point P is 40 mm and 20 mm away from two fixed mutually perpendicular straight lines OX and OY respectively. Draw a hyperbola through the given point P.

14. Draw two brances of a hyperbola when its transverse axis is 60 mm long and foci are 80 mm apart. Locate its directrix and determine the eccentricity.

15. A rectangular hyperbola has its foci 70 mm apart. Locate its vertices and directrices graphically and draw two branches of a hyperbola.

16. The asymptotes of a hyperbola are inclined at 70° to each other. A point P on the curve is 25 mm and 40 mm away from the asymptotes. Draw the hyperbola through point P.

CHAPTER 4

Engineering Curves

4.1 Introduction

Apart from conics, there are another curves, which are also used widely in engineering practices such as roulettes (cycloidal curves), involutes, spirals etc.

In this chapter, we shall discuss method of construction and applications of roulettes, involutes of circle, and polygons, archemedian and logarithmic spirals.

4.2 Cycloid

Cycloid is a curve generated by a point on the circumference of a circle which rolls without slipping along a straight line. The circle which rolls on a straight lines is called as generating circle and the fixed straight line is called directing line.

Applications: Cycloid is used in the design of gear tooth profile.

The method of construction of a cycloid is explained with the help of following problem.

Problem 4.1 (Fig. 4.1): *Draw a cycloid if diameter of rolling circle is 50 mm. Also draw a tangent and normal at a point on the curve 45 mm above the base line.*

OR

A circle of 50 mm diameter rolls along a straight line without slipping. Trace the locus of the point P on the circumference of the circle. Name the curve. Also draw a tangent and a normal at a point on the curve 45 mm above the base line.

Construction (Fig. 4.1):

Step 1: Draw a circle of diameter 50 mm and a line PA tangential to it equal to circumference of the circle (πD).

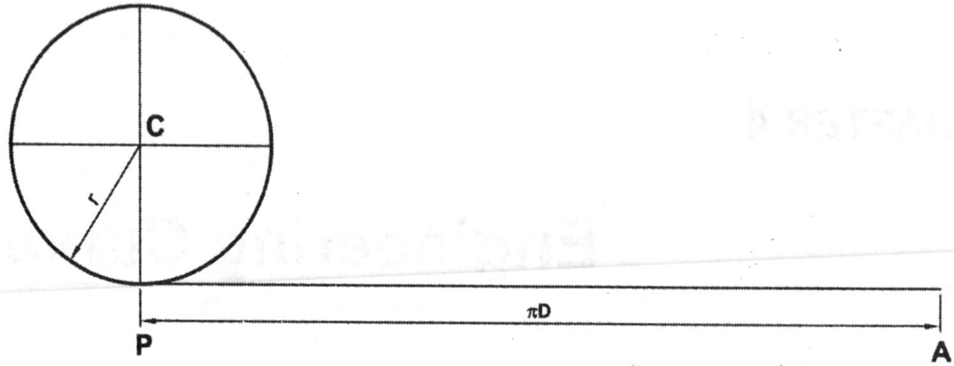

Fig. 4.1(a)

Step 2: Divide the circle and base line PA into same number of equal parts (say 12). Mark the division point as 1, 2, 3...etc., on PA and 1', 2', 3'... etc., on the circumference of the circle. From C, draw a line CB parallel to PA. At points 1, 2, 3.. etc., draw lines perpendicular to PA and intersecting CB at C_1, C_2, C_3 etc.

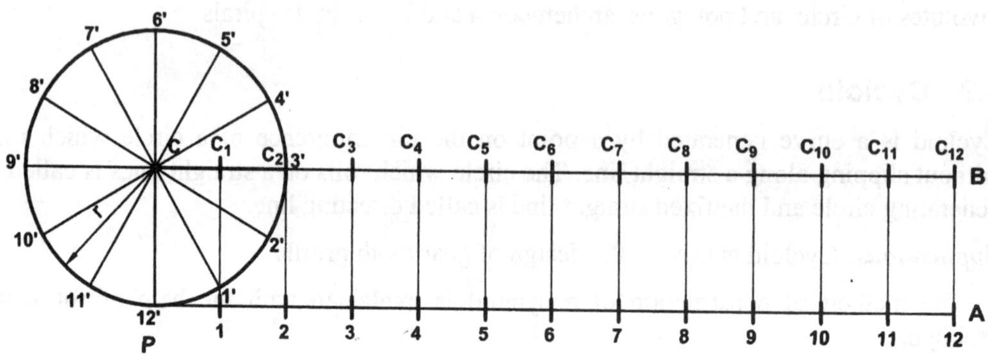

Fig. 4.1(b)

Step 3: Through 1', 2', 3'...etc., draw lines parallel to PA. With centre C_1 and radius equal to r (25 mm), draw an arc cutting the line through 1' at P_1. Similarly with C_2, C_3 ... etc., as centres and radius equal to r (25 mm), draw arcs on the lines through 2', 3'... etc., respectively intersecting at P_2, P_3 ... etc., respectively. Join P_1, P_2 ... etc., by a smooth curve, which is the required cycloid.

Step 4: (Tangent and normal at a given point on the curve).

Mark a point Q on the curve, 45 mm above the base line. With Q as centre and radius equal to r, draw an arc cutting on the line through C at M. Through M, draw a line perpendicular to PA intersecting at N. Draw a line through N and Q, which is the required normal. Through Q, draw a line perpendicular to NN' . This is the required tangent TT'.

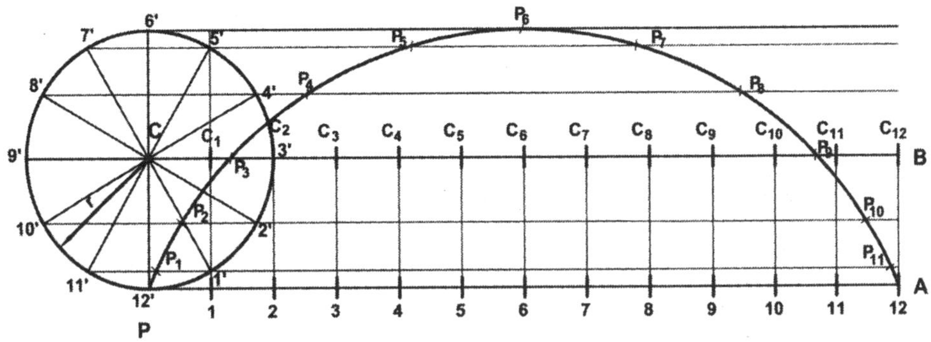

Fig. 4.1(c)

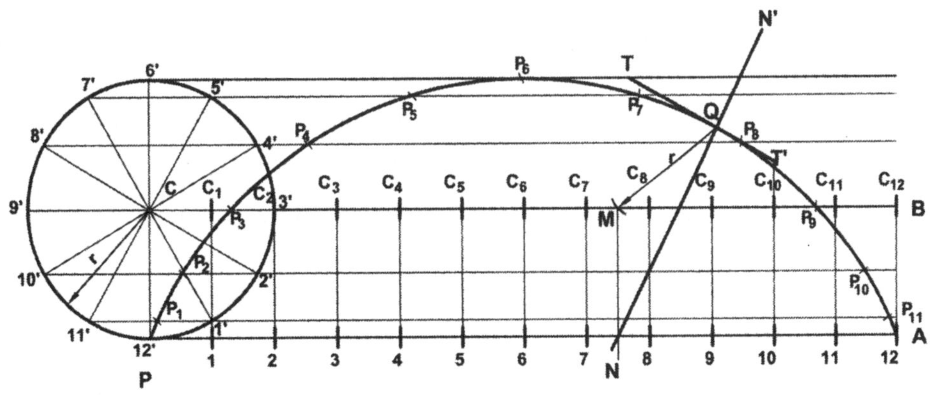

Fig. 4.1(d)

4.3 Epicycloid

It is defined as a curve generated by a point on the circumference of a circle which roles without slipping on another circle outside it.

Applications: Epicycloid is used in mechanisms, geer tooth profile, rotary pumps, blower etc.

The method of construction can be understood with the help of following example.

Problem 4.2 (Fig. 4.2): *Draw an epicycloid if the diameters of directing and rolling circles are 150 mm and 50 mm respectively. Also draw a tangent and normal at a point 120 mm away from the centre of directing circle.*

OR

A circle of 50 mm diameter rolls along the outside of another circle of 150 mm diameter without slipping. Trace the locus of a point on the circumference of the smaller circle for one complete revolution. Name the curve and draw a tangent and a normal at a point 120 mm from the centre of the directing circle.

Construction (Fig. 4.2):

1. Calculate subtending angle θ as below:

$$\theta = \frac{r}{R} \times 360 = \frac{25}{75} \times 360^\circ = 120^\circ$$

2. Mark any point O. With O as centre and radius equal to 75 mm, draw an arc PA subtending an angle of 120° i.e., $\angle POQ = 120^\circ$.
3. Extend line OP such that CP = 25 mm
4. With C as centre and radius equal to 25 mm, draw a generating circle and divide it into 12 equal parts. Mark division points as 1, 2, 3... etc., (in anticlockwise direction).
5. Divide arc PA, into 12 equal parts and mark divisions as 1', 2', 3'... etc. Draw radial lines through 1', 2', 3'... etc., and meeting at O.
6. With O as centre and radius equal to O1, O2... etc., draw a number of arcs through 1, 2, 3, ... etc.
7. Extend the radial line through 1', 2'... etc., to intersect central arc CB at C_1, C_2 ... etc., respectively.
8. With C_1 as centre and radius equal to 25 mm, draw an arc intersecting the arc through 1 at P_1.
9. Similarly, with centres C_2, C_3, C_4 ....etc., and radius equal to 25 mm, draw arcs intersecting the corresponding arcs to obtain P_2, P_3, P_4 ... etc.
10. Draw a smooth curve through points P_1, P_2 ... etc. which is the required epicycloid.

To draw Tangent and Normal

11. Mark a point P' on the curve, 120 mm from the centre of directing circle.
12. With P' as centre and radius equal to r (25 mm) draw an arc cutting the centre arc at C'. Draw line joining O and C' intersecting base arc at N.
13. Draw a line passing through N and P' which is the required normal (NN').
14. Draw a line TT' through P' perpendicular to the normal which is required tangent.

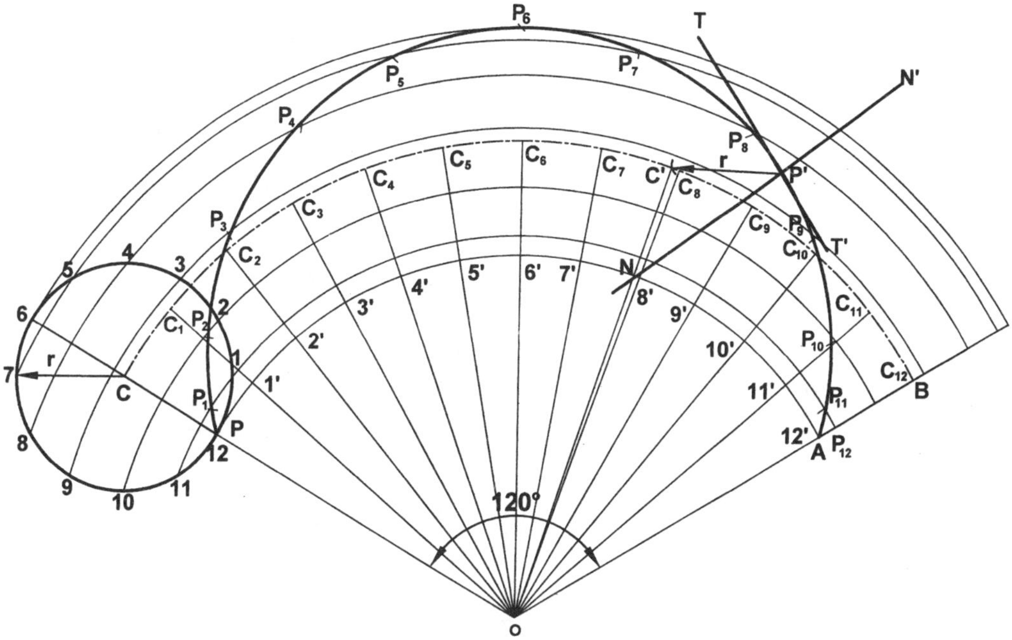

Fig. 4.2

4.4 Hypocycloid

It is defined as a curve generated by a point on the circumference of a circle, which rolls without slipping on another circle inside it. The rolling circle is called generating circle and fixed circle is called directing circle.

Applications: The hypocycloid curve is used in metal cutting tools, various mechanisms, rotary pumps, blowers etc.

Problem 4.3 (Fig. 4.3): *Draw a hypocycloid if the diameter of directing circle and rolling circle are 150 mm and 50 mm respectively. Also draw a tangent and a normal at point 25 mm from the centre of directing circle.*

OR

A circle of 50 mm diameter rolls along the inside of another circle of 150 mm diameter without slipping. Trace the locus of a point on the circumference of smaller circle for one complete revolution. Name the curve. Also draw a tangent and a normal at a point 25 mm from the centre of directing circle.

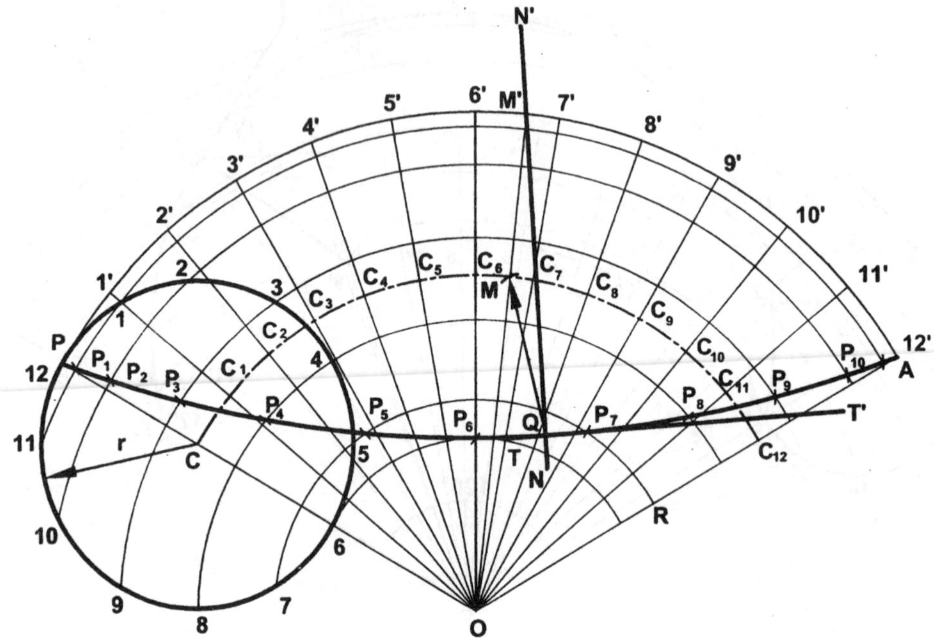

Fig. 4.3

Construction (Fig. 4.3):

1. Calculate the angle (θ) subtended by base arc as below:

$$\theta = \frac{r}{R} \times 360 = \frac{25}{75} \times 360 = 120^\circ$$

2. Mark a point O. With O as centre and radius equal to 75 mm, draw an arc PA such that $\angle POA = 120^\circ$.
3. Mark a point C on OP such that PC = 25 mm.
4. With C as centre and radius equal to 25 mm draw circle (generating circle). Divide it into 12 equal parts and mark the divisions as 1, 2, 3....etc.
5. Also divide the base arc PA into 12 equal parts and mark the divisions as 1', 2', 3' ... etc.
6. Draw radial lines through 1', 2', ... etc., meeting at O and intersecting the centre arc at C_1, C_2, C_3... etc.
7. With C_1, C_2, C_3... etc., as centres and radius equal to r (25 mm) draw the arcs cutting the arcs through 1, 2, 3 ... etc., at P_1, P_2, P_3 ... etc., respectively.

8. Draw a smooth curve passing through P_1, P_2, P_3 ...etc. This curve is the required hypocycloid.

To draw tangent and normal

9. Mark a point Q on the curve at a distance of 25 mm from the centre of directing circle.
10. With Q as centre and radius equal to r (= 25 mm), draw an arc intersecting centre arc at M.
11. Join O to M and extend the line to intersect the base arc at M'.
12. Draw a line passing through M' and Q. The line NN' is the required normal.
13. Draw a tangent (TT') passing through Q and perpendicular to NN'.

4.5 Involute

It is defined as a curve traced out by an end of string/thread which is unwound from a circle or a polygon keeping the thread tight.

Applications: The involute curve is used in the design of tooth profile of large gears. It is also used as cam profile very commonly.

Problem 4.4 (Fig. 4.4): *Draw an involute of a circle of 40 mm diameter. Also draw a tangent and a normal at a point 90 mm from the centre of the circle.*

Construction (Fig. 4.4):

1. Draw a circle of 40 mm diameter. Mark a point P on its circumference. Draw a tangential line PA equal to πD.
2. Divide the circle and the line PA into same number of equal parts (say 12).
3. Mark the division points on the circle as 1, 2, 3.. etc.
4. Mark the divisions on line PA as 1', 2', 3'.... etc.
5. Draw tangents at point 1, 2, 3, ...etc.
6. With 1 as centre and radius equal to P1', draw an arc intersecting the tangent through 1 at P_1.
7. Similarly, with 2, 3, 4... etc., as centres and radii equal to P2', P3', P4' etc., draw arcs intersecting tangents at P_2, P_3, P_4 ... etc., respectively.
8. Draw a smooth curve through points P_1, P_2, P_3 ... etc. This is the required involute.

To draw a tangent and a normal

9. Mark a point Q on the curve, 90 mm from the centre of circle.
10. Join O to Q and mark M as a mid-point of OQ.
11. With M as centre and radius equal to OM, draw a semi circle intersecting the circle at N.
12. Draw a line passing through N and Q. This is the required normal (NN'). Draw a line (tangent) TT' perpendicular to normal and passing through Q.

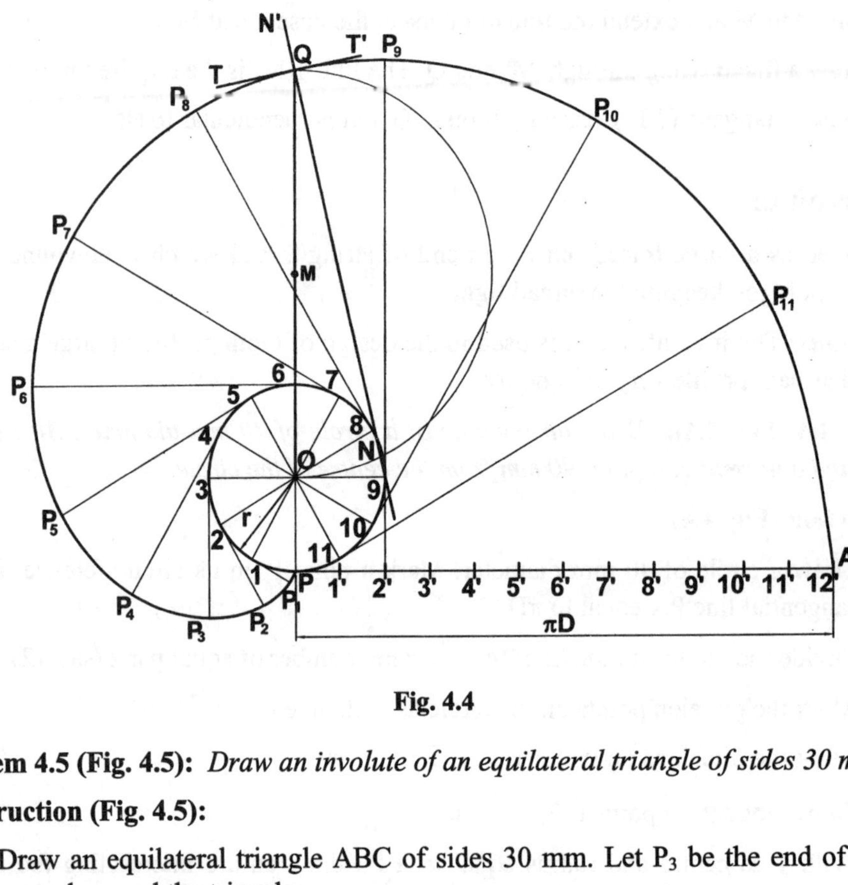

Fig. 4.4

Problem 4.5 (Fig. 4.5): *Draw an involute of an equilateral triangle of sides 30 mm.*

Construction (Fig. 4.5):

1. Draw an equilateral triangle ABC of sides 30 mm. Let P_3 be the end of a thread wound around the triangle.
2. Assuming the thread to be unwound in clockwise direction, extend the sides of the triangle beyond the corner points.
3. With A as centre and radius equal to AC, draw an arc to intersect extended line AB at P_1.

4. With B as centre and radius equal to twice the side of triangle (i.e., AC+ AB), draw an arc intersecting extended BC line at P_2.
5. With C as centre and radius equal to three times the side (i.e., AB + BC + CA), draw an arc cutting extended AC line at P_3.
6. The curve thus obtained is the required involute.

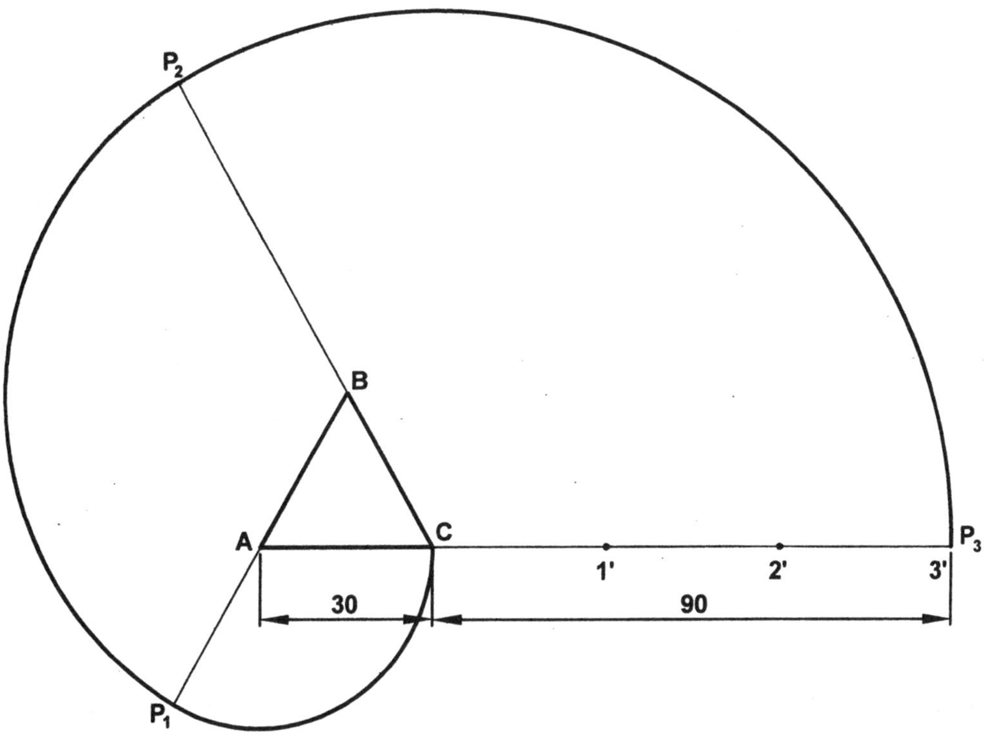

Fig. 4.5

Problem 4.6 (Fig. 4.6): *Draw an involute of a square of side equal to 30 mm.*

Construction (Fig. 4.6):

1. Draw a square ABCD of side equal to 30 mm
2. Produce the sides AB, BC, CD and AD.
3. With A, B, C and D as centres and radii equal to 30 mm, 60 mm, 90 mm and 120 mm respectively draw arcs intersecting the corresponding produced lines at P_1, P_2, P_3 and P_4 respectively.

4. The curve obtained through P_1, P_2, P_3 and P_4 is the required involute.

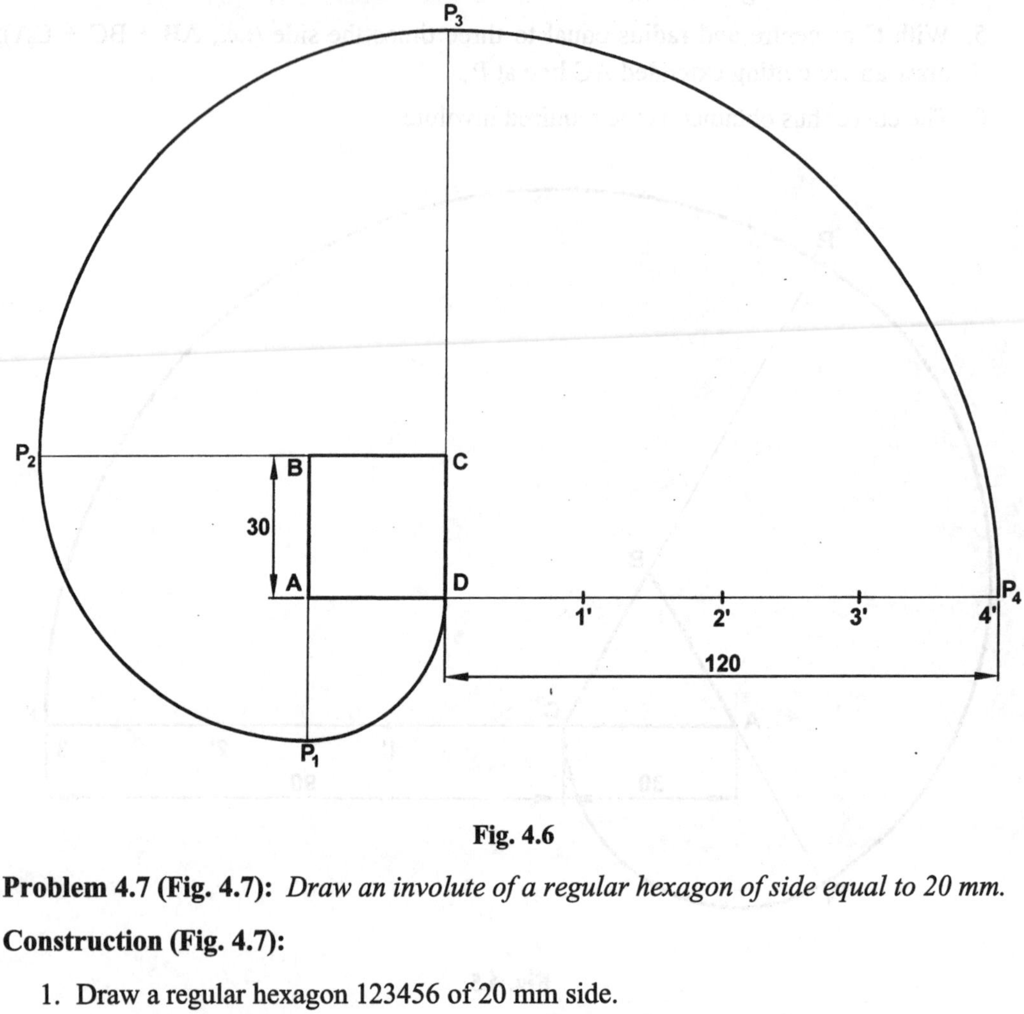

Fig. 4.6

Problem 4.7 (Fig. 4.7): *Draw an involute of a regular hexagon of side equal to 20 mm.*

Construction (Fig. 4.7):

1. Draw a regular hexagon 123456 of 20 mm side.
2. Produce sides 12, 23, 34, 45 and 56 as shown in Fig. 4.7.
3. With centres 1, 2, 3 etc., and radii equal to 20 mm, 40 mm, 60 mm etc., draw arcs intersecting the corresponding extended lines at P_1, P_2, ...P_6.
4. Thus smooth curve passing through P_1, P_2, P_3, P_4, P_5 and P_6 is the required involute.

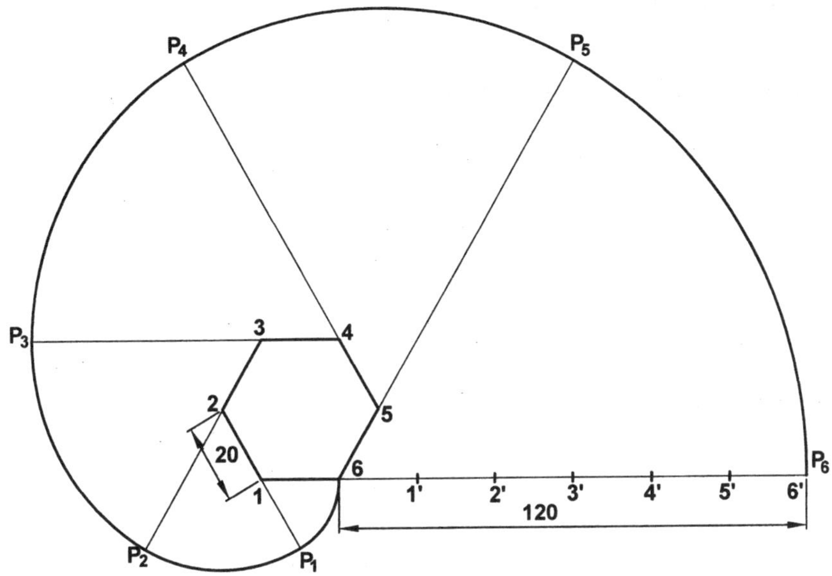

Fig. 4.7

4.6 Spirals

When a line rotates in a plane about one of its ends and at the same time a point moves along the line continuously towards or away from the fixed point, the curve traced out by the moving point is known as spiral. Various terms used with reference to the spiral are explained as follows: (Refer Fig.4.8)

(i) **Pole:** The fixed point of a line about which it rotates is called as pole (Point O).

(ii) **Radius Vector:** The line joining any point on the curve with the pole is known as radius vector (say line OP_{13}).

(iii) **Vectorial Angle:** The angle between the radius vector line and the initial position of the line is known as vectorial angle (say $\angle POP_1$)

(iv) **Convolution:** Each complete revolution of a point around the pole is termed as a convolution. The spiral may be drawn for any number of convolutions.

In engineering practice, following spirals are used commonly.

4.6.1 Archemedian Spiral

It is a curve traced out by a point moving along a straight line towards and away from the pole with uniform angular velocity.

The archemedian spiral is used in the design of teeth profile of helical gears, cam profile etc.

Problem 4.8 (Fig. 4.8): *Draw an archemedian spiral for 1½ convolutions. The shortest and greatest radii are 18 mm and 90 mm respectively. Also draw a tangent and a normal at a point 55 mm from the pole.*

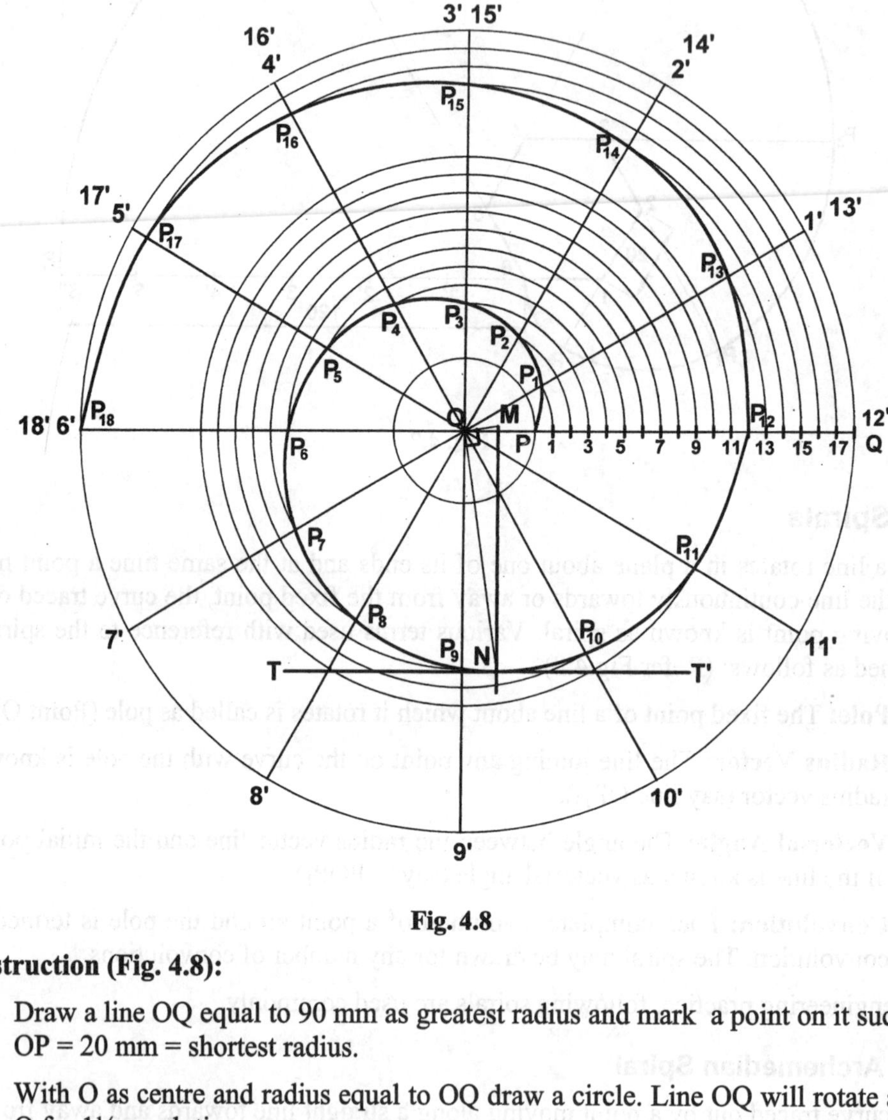

Fig. 4.8

Construction (Fig. 4.8):

1. Draw a line OQ equal to 90 mm as greatest radius and mark a point on it such that, OP = 20 mm = shortest radius.
2. With O as centre and radius equal to OQ draw a circle. Line OQ will rotate around O for 1½ revolutions.
3. Divide PQ into 18 equal parts and mark divisions as 1, 2, 3...etc., (Since, one revolution is divided into twelve equal parts).

4. Divide the circle into twelve equal parts, and mark and divisions as 1', 2', 3'...etc.
5. With O as centre and radius equal to O1, draw an arc intersecting radial line O1' at P_1.
6. Similarly, with O as centre and radii equal to O2, O3 ... etc., draw arcs intersecting the line O2', O3', ... etc., at P_2, P_3 etc., respectively.
7. Draw a smooth curve passing through P_1, P_2... P_{18}. This is the required archemedian spiral.

To draw tangent and normal

8. Mark a point N on the curve at a distance of 55 mm from the pole. Join N to O by a straight line.
9. Draw a line OM perpendicular to ON such that,

$$OM = \frac{OP_3 - OP}{1.57} = 8.9 \text{ mm } [\text{since angle between } OP_3 \text{ and } OP \text{ is } 90^\circ = \frac{\pi}{2} = 1.57]$$

10. Draw a line joining M and N which is required normal.
11. Draw another line passing through N and perpendicular to MN. This is the required tangent.

Problem 4.9 (Fig. 4.9): *A point moves along a bar at an uniform speed. The bar rotates about its one end O at an uniform speed. Name and construct the path of a point P starting from a position 20 mm away and moving upto 65 mm away from the fixed end of bar during its one revolution. Also draw a tangent and a normal at a point 40 mm away from the pole O.*

Construction (Fig. 4.9):

Since, the bar rotates about O at an uniform speed, the path traced out by a point moving along it, will be an archemedian spiral. Following steps are taken to construct the curve.

1. Draw a line OA of length equal to 65 mm. Mark a point P, 20 mm away from O.
2. With O as centre and radius equal to OA, draw a circle and divide it into twelve equal parts. Mark the divisions as 1', 2', 3'.. etc.
3. Divide line PA into twelve equal parts and mark the divisions as 1, 2, 3 ...etc.
4. Join O to 1', 2', 3'... etc.
5. With O as centre and radius equal to O1, draw an arc intersecting the radial line O1' at P_1.
6. Similarly with O as centre and radii equal to O2, O3 ... etc., draw arcs intersecting the radial lines O2', O3'... etc. at P_2, P_3...etc. respectively.
7. Draw a smooth curve through P_1, P_2, P_3...P_{12}. This is required archemedian spiral.
8. Draw a tangent and a normal at a point 40 mm away from O as explained in problem 4.8.

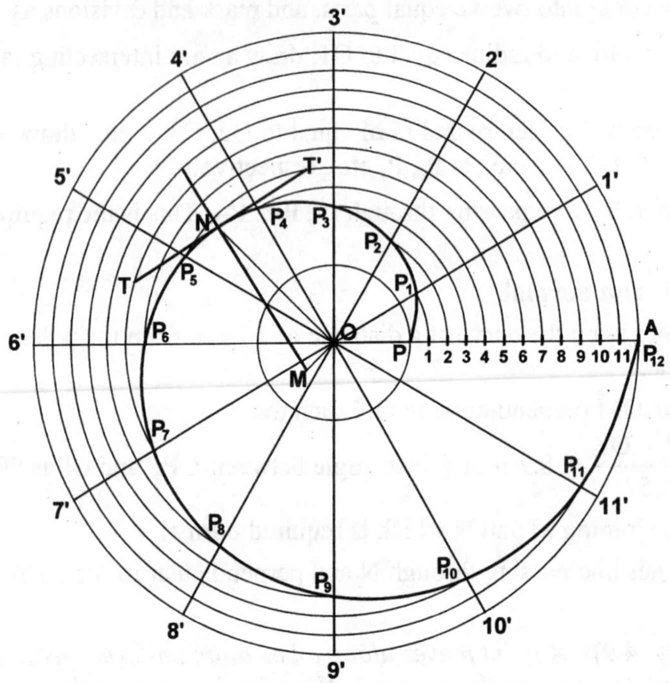

Fig. 4.9

Problem 4.10 (Fig. 4.10): *A link OA, 75 mm long rotates about O in anticlockwise direction. A point P on the link, 20 mm away from O, moves and reaches the end A, while the link has rotated through 2/5 of a revolution. Trace the path of point P assuming movement of link and point P to be uniform.*

Construction (Fig. 4.10):

1. Draw a line OA, 75 mm long representing the link OA.
2. Mark a point P on it, 20 mm away from O.
3. With O as centre and radius, equal to OA, draw an arc AA' such that $\angle AOA' = 144^\circ$ (Since $\frac{2}{5}$ of a revolution $= \frac{2}{5} \times 360^\circ = 144^\circ$).
4. Divide line PA and arc AA' into same number of equal parts (say twelve).
5. With O as centre and radius equal to O1, draw an arc intersecting radius line O1' at P_1.
6. Similarly, obtain P_2, P_3....etc., and join them by a smooth curve.

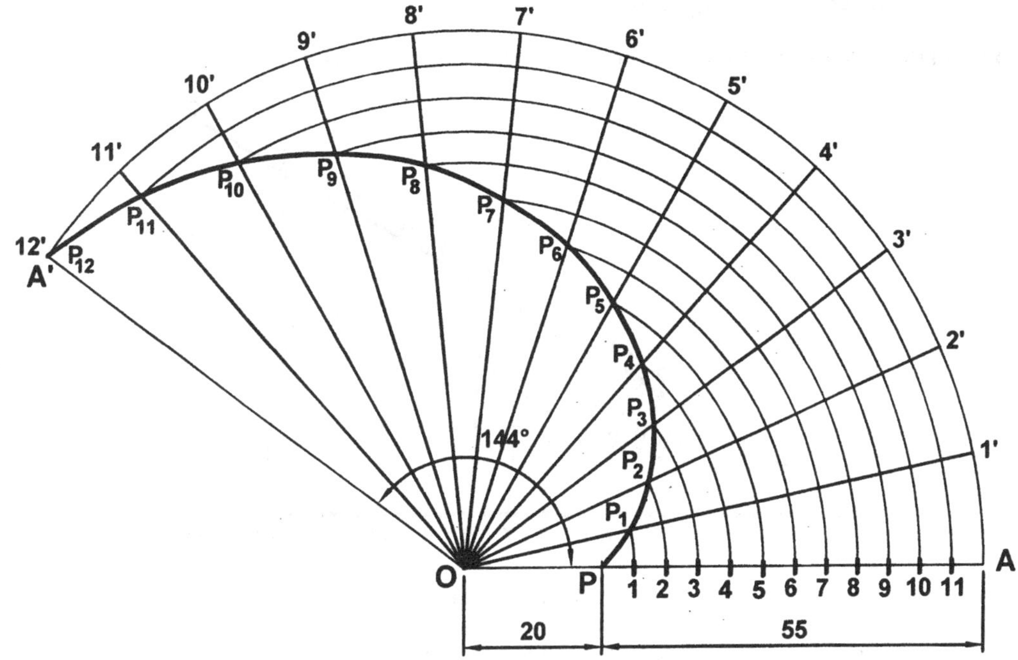

Fig. 4.10

4.6.2 Logarithmic Spiral

It is a curve traced out by a point moving along a rotating line such that the ratio of the lengths of consecutive radius vectors enclosing equal angles is always constant.

In other words the values of vectorial angles are in arithmetic progression and corresponding lengths of radius vectors are in geometrical progression.

The method of construction of a logarithmic spiral is explained with the help of following problem.

Problem 4.11 ***(Fig. 4.11)*****:** *Draw a logarithmic spiral of one convolution, if the length of shortest radius is 20 mm and the ratio of lengths of radius vectors enclosing an angle of 30° is 10/9. Also draw a tangent and a normal at a point on the curve 55 mm from the pole.*

Construction (Fig. 4.11(a) and (b)):

To determine lengths of radius vectors (Refer Fig. 4.11(a))

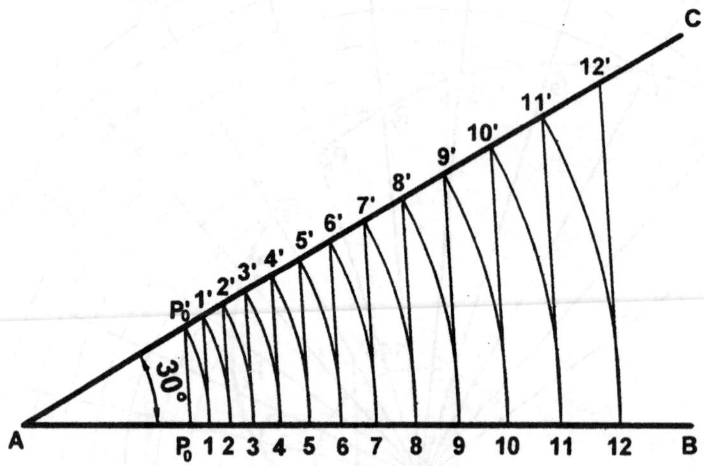

Fig. 4.11 (a)

1. Draw lines AB and AC such that the angle between them is 30°.
2. Mark a point P_o on AB such that $AP_o = 20$ mm.
3. Mark a point P_o' on AC such that,

$$\frac{AP_o'}{AP_o} = \frac{10}{9}$$

4. Join P_o and P_o'.
5. With A as centre and radius equal to AP_o', draw an arc intersecting AB at 1.
6. Through 1 draw a line parallel to P_oP_o' intersecting AC at 1'.
7. Similarly obtained points 2, 3, 4 ... on AB.

To draw logarithmic spiral (Refer Fig. 4.11(b))

8. Draw a line OA of any length and mark on it P_o such that $OP_o = 20$ mm (shortest radius)
9. Draw a number of radial lines through O such that the angle between adjacent radius vectors is 30°.
10. Mark points P_1, P_2, P_3 ... etc., on consecutive radius vectors such that $OP_1 = A1$, $OP_2 = A2$ ... etc.
11. Draw a smooth curve through P_1, P_2, P_3 ... etc., which is the required logarithmic spiral.

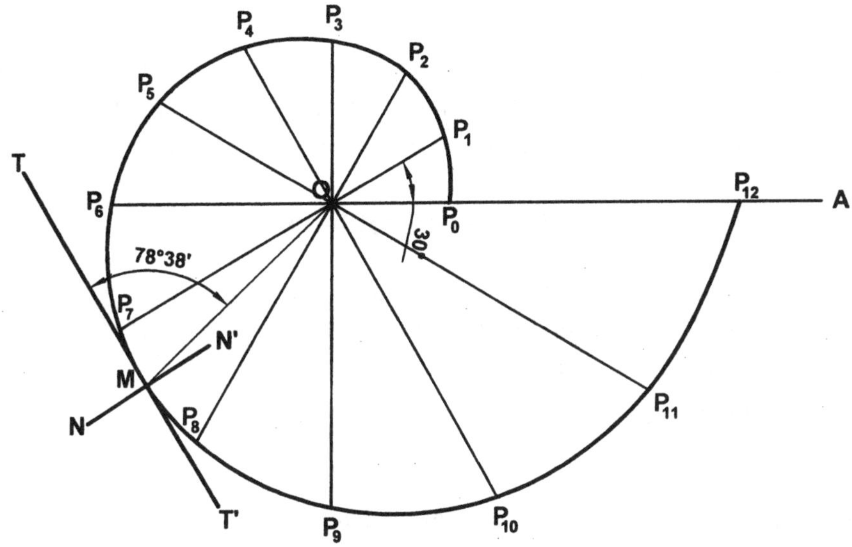

Fig. 4.11(b)

To draw a tangent and normal

Consider following equations of logarithmic spiral,

$$r = a^{\theta}$$

where, r = radius vector

θ = vectorial angle

a = constant

Hence, $\log r = \theta \log a$

or $\log a = \frac{1}{\theta} \log r$

$$\log a = \frac{6}{\pi} \log \frac{10}{9}$$

If α is the angle between the tangent to the curve and the line joining pole and given point, then it can be calculated as below:

$$\tan \alpha = \frac{\log_e}{\log_a} = \frac{\log 2.718}{\frac{6}{\pi}\log\frac{10}{9}}$$

or $\alpha = 78°38'$

12. Mark a point M on the curve 55 mm away from the pole. Join O to M. Draw a line TT' inclined at $\alpha = 78°38'$ to OM, which is the required tangent. Draw another line NN' perpendicular to TT'. The line NN' is the required normal.

Additional Problems

Problem 4.12 (Fig. 4.12): *A circle of 40 mm diameter rolls on a horizontal line for half revolution and then on a vertical line upward for remaining half revolution. Draw the curve traced out by a point on the circumference of the circle.*

Construction (Fig. 4.12):

1. Draw a rolling circle having diameter equal to 40 mm.
2. Mark a point P on the circumference of the circle.
3. Draw a line PA tangent to the circle equal to $\pi D/2$. Divide the circle into 12 equal parts. Mark the divisions as 1, 2, 3... etc.
4. Divide the line PA into six equal parts. Mark the divisions as 1', 2' ... etc.
5. Draw vertical lines through 1', 2', 3' .... 6' intersecting CB at $C_1, C_2, C_3 \ldots C_6$.
6. With C_1, as centre and radius equal to r (20 mm), draw an arc intersecting the line through 1 and P_1.
7. Similarly with $C_2, C_3 \ldots C_6$ as centres and radius equal to 20 mm, draw arcs intersecting lines through 2, 3...6 at $P_2, P_3 \ldots P_6$ respectively.
8. Through A and C_6, draw a vertical line. With C_6 as centre and radius equal to 20 mm draw a circle and divide it into twelve equal parts. Mark divisions on its circumference as shown in Fig. 4.12. (This circle represents its position after half revolution).
9. Draw a vertical base line BC, tangential to the circle. Divide it into six equal parts, and mark the divisions as 7', 8' ... 12'.
10. Draw horizontal lines through 7', 8' ... etc., intersecting vertical centre line at C_7, $C_8, C_9 \ldots$ etc.
11. With $C_7, C_8, C_9 \ldots$ etc., as centres and radius equal to 20 mm, draw arcs intersecting lines through 7, 8, 9 .. etc., at $P_7, P_8, P_9 \ldots$ etc.
12. Draw a smooth curve through $P_1, P_2 \ldots P_{12}$.

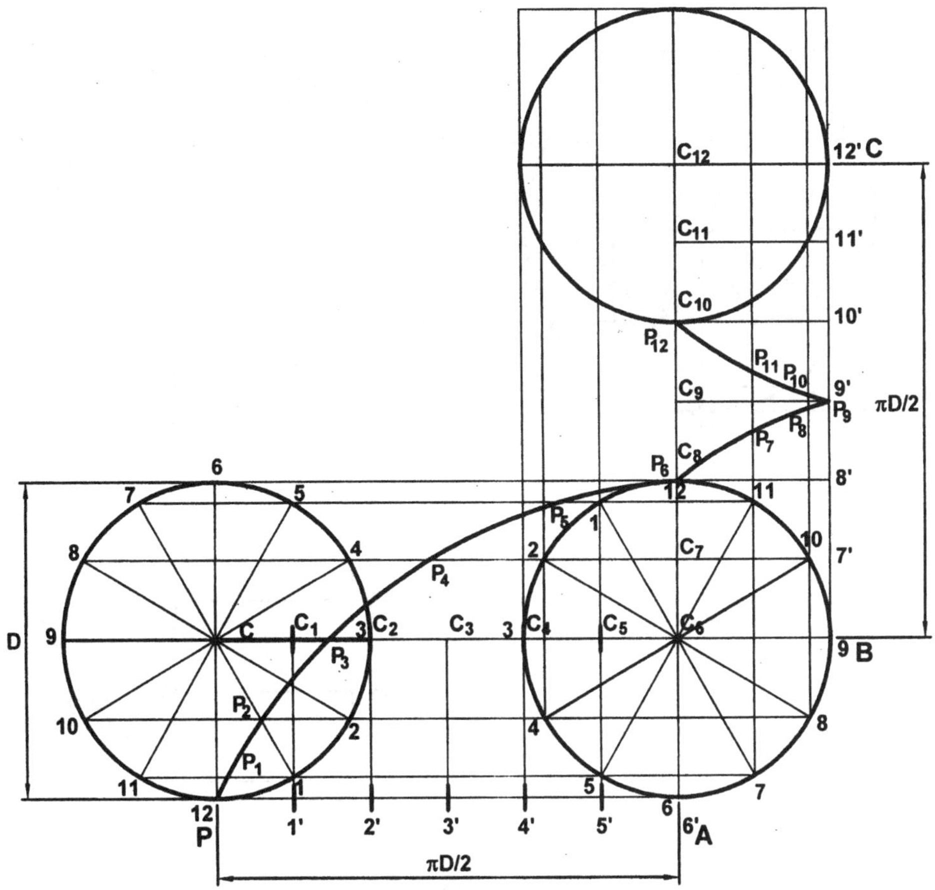

Fig. 4.12

Problem 4.13 (Fig. 4.13): *A circle of 40 mm diameter rolls on a horizontal line for a half revolution and then on a vertical line downwards for another half revolution. Draw the curve traced out by a point P on the circumference of the circle. Assume that the horizontal and the vertical lines constitute a corner.*

Construction (Fig. 4.13):

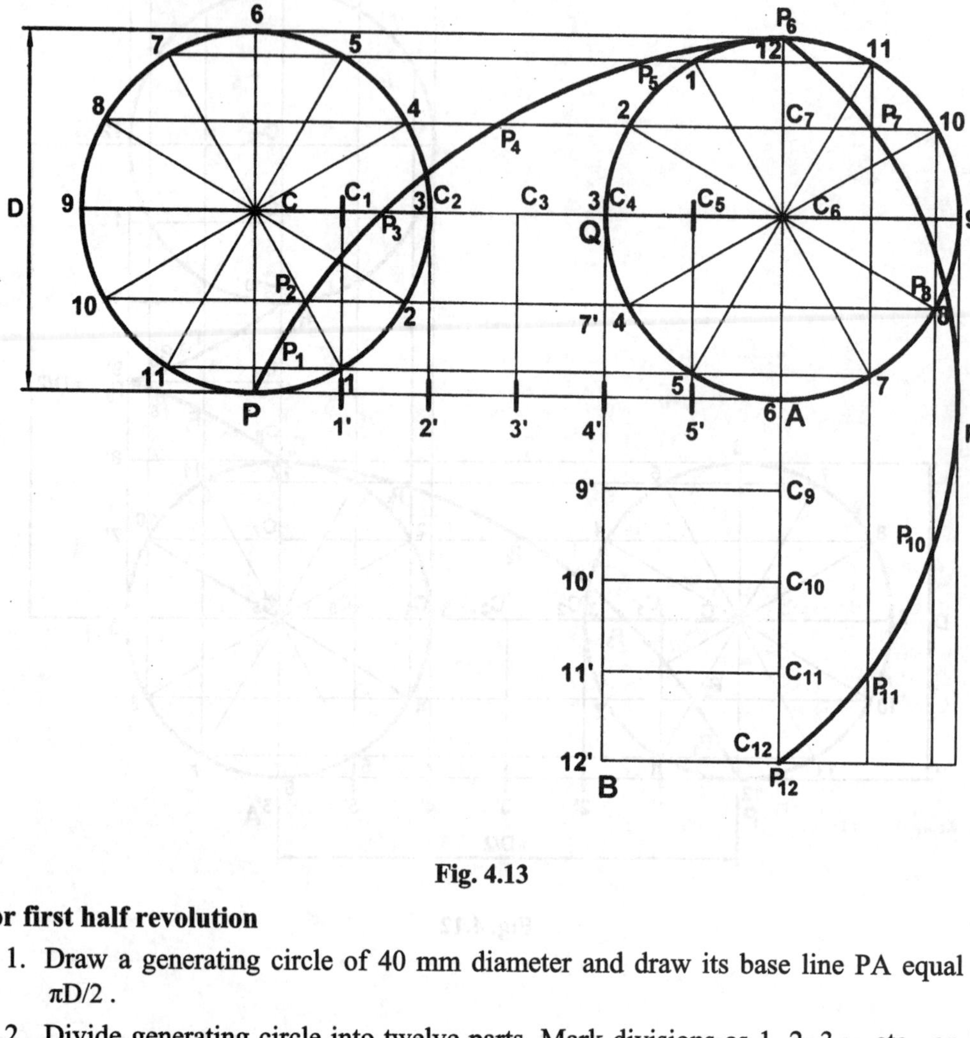

Fig. 4.13

For first half revolution

1. Draw a generating circle of 40 mm diameter and draw its base line PA equal to $\pi D/2$.
2. Divide generating circle into twelve parts. Mark divisions as 1, 2, 3 ... etc., on its circumference.
3. Divide line PA into six equal parts and mark divisions as 1', 2', 3'...6'.
4. Draw vertical lines through 1', 2'... etc., to intersect centre line at C_1, C_2, C_3 ... etc., respectively.
5. With C_1, C_2, ... etc., as centres and radius equal to 20 mm draw arcs intersecting lines through 1, 2, 3 ... etc., at P_1, P_2, P_3 ... etc., respectively.
6. Join P_1, P_2, ... P_6 by a smooth curve.

For second half revolution

7. With C_6 as centre and radius equal to 40 mm, draw a circle representing the its position after half revolution.
8. Draw a vertical base line QB equal to $\pi D/2$ and divide it into six equal parts. Mark divisions as $7', 8' \ldots 12'$.
9. Draw horizontal lines through $7', 8' \ldots$ etc., to obtain centres C_7, C_8, C_9 ... etc., respectively.
10. With C_7, C_8, C_9 ... etc., as centres and radius equal to 20 mm, draw arcs intersecting lines through 7, 8, 9 ... etc., at P_7, P_8, P_9 ... etc., respectively.
11. Draw a smooth curve through P_7, P_8, P_9 ... etc.

Problem 4.14 (Fig. 4.14): *A circle of 50 mm diameter rolls without slipping inside another circle of diameter 100 mm. Trace the locus of point on the circumference of smaller circle. Name the curve.*

Construction (Fig. 4.14):

In this problem, diameter of directing circle is double the diameter of rolling circle i.e.,

$$R = 50 \text{ mm}$$

$$r = 25 \text{ mm}$$

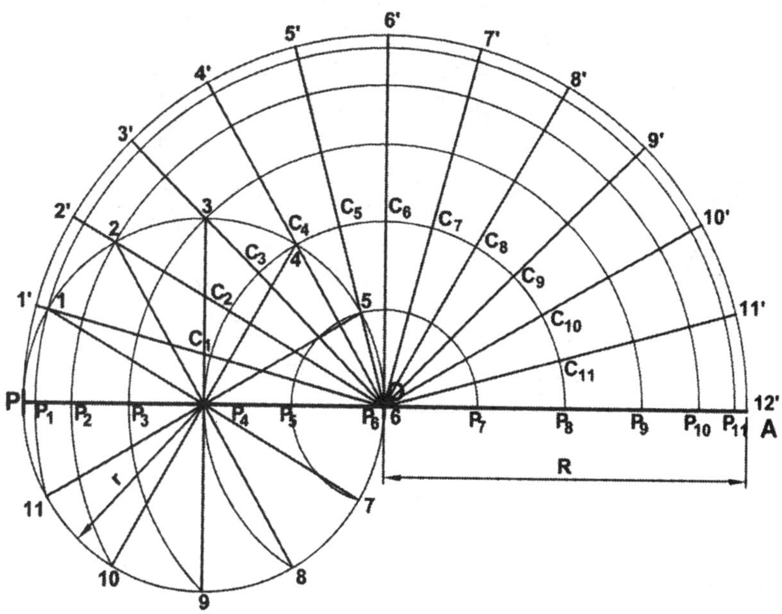

Fig. 4.14

Since, smaller circle rolls without slipping inside the another circle, the curve traced by point P, will be a hypocycloid. For such hypocycloid,

$$\theta = \frac{r}{R} \times 360 = \frac{25}{50} \times 360 = 180^{o}$$

To construct the hypocycloid using given data, follow the steps of construction as discussed in problem 4.3. It can be noted that, in this case, point on the circumference of rolling circle generates a straight line.

Problem 4.15 (Fig. 4.15): *Draw an epicyclod if the diameter of generating and directing circles are 40 mm.*

Construction (Fig. 4.15):

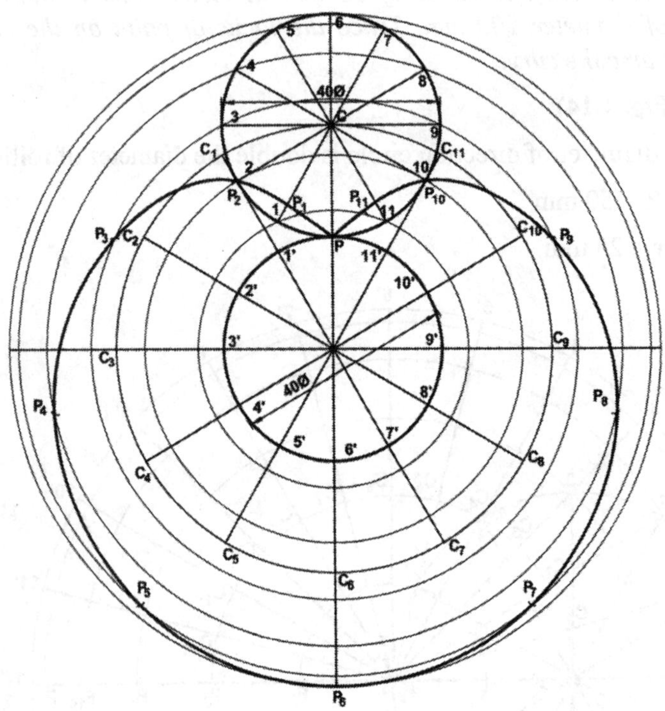

Fig. 4.15

Given, R = 20 mm

r = 20 mm

$$\therefore \quad \theta = \frac{r}{R} \times 360 = \frac{20}{20} \times 360 = 360^{o}$$

To construct, the epicycloids with the given data, follow the steps of construction as explained in problem 4.2. The curve traced by point P is in the shape of heart, therefore it is also called as cardiod.

Exercises

1. A circle of 50 mm diameter rolls along a straight line without slipping. Draw the curve traced out by a point P on the circumference of the circle for one complete revolution. Also draw a tangent and a normal at point on the curve 35 mm from the base line.
2. A circle of 46 mm diameter rolls on a horizontal line for a half revolution and then on a vertical line downwards for another half revolution. Draw the curve traced out by a point P on the circumference of the circle. Assume that the horizontal and vertical lines constitute a corner.
3. A circle of 50 mm diameter rolls on the circumference of another circle of 175 mm diameter and outside it. Draw the curve traced out by a point on the rolling circle for its one complete revolution. Also draw a tangent and a normal at any point on the curve.
4. A circle of 50 mm diameter rolls on another circle of 75 mm diameter with internal contact. Draw a curve traced out by a point P on the circumference of rolling circle for one complete revolution. Also draw a tangent and a normal at any point on the curve.
5. A cycle wheel of 50 cm diameter rolls over a culvert of 175 cm diameter. Draw the path traced out by a point on the circumference of the wheel for one complete revolution. Assume suitable scale.
6. A circle of 40 mm diameter rolls on the outside of a base circle of the same diameter. Draw the curve traced out by a point on the rolling circle for one complete revolution of it.
7. A circle of 50 mm diameter rolls without slipping on another circle of diameter 100 mm outside it. Draw the path traced out by a point on the circumference of rolling circle for one complete revolution.
8. Draw an epicycloid when the radii of rolling circle and directing circle are 25 mm and 75 mm respectively. Also draw a tangent and a normal at a point 110 mm from the centre of directing circle.
9. Draw the involute of a circle of 40 mm diameter. Also draw a tangent and a normal at a point 95 mm form the centre of the circle.
10. A line rolls over a square of 30 mm side without slipping. Draw the curve traced out by a point on the line. Name the curve.
11. Draw an involute of an equilateral triangle of 45 mm side.

12. Draw an involute of a hexagon of 30 mm side.
13. Draw an involute of a regular pentagon of 25 mm side.
14. An inelastic string attached to the circumference of a circle of 50 mm diameter, is wound completely. Draw the curve traced out by another end of string if it is unwound keeping the string tight.
15. Construct an archemedian spiral of 1½ convolutions given the greatest diameter 120 mm and shortest diameter of 30 mm.
16. Construct an archemedian spiral of two convolutions given greatest and shortest radii as 84 mm and 12 mm respectively.
17. A point moves along a bar at an uniform speed. The bar rotates about its one end O at an uniform speed. Name and construct the path a point P starting from a position 20 mm away and moving up to 60 mm away from the fixed end of bar during its one revolution. Draw tangent at a point 45 mm away from O.
18. In a logarithmic spiral, the shortest radius is 40 mm. The length of adjacent radius vectors enclosing 30° are in the ratio 9:8. Construct the spiral for one revolution. Also draw a tangent and a normal at any point on the curve.
19. Draw a logarithimic spiral of one convolution given the shortest radius 20 mm and the ratio of the length of adjacent radii enclosing 30° as 10:9.
20. A point P moves towards another point O, 60 mm from it and reaches it while moving around it. Draw the curve traced out by the point P. Also draw a tangent and a normal at any point on the curve.
21. A link OA, 80 mm long rotates about O in anticlockwise direction. A point Q on the link, 20 mm away from O, moves and reaches the end A, while the link has rotated through 2/5 of a revolution. Assuming the movements of the link and the point to be uniform, trace the path of point Q.

CHAPTER 5

Orthographic Projections

5.1 Projection

Image or representation of any object on a plane or screen is known as projection of that object. Projection helps to describe objects (machine parts, building etc.,) graphically on a paper which is very important for engineers, architectures, manufacturers etc. For example, projections of a building are presented as front view (elevation) and top view (plan) which are very useful for construction and planning. Various views of a machine parts are useful for complete understanding and to present their technical description.

5.2 Types of Projection

In engineering practice, two main types of projections are used commonly i.e., pictorial projection and orthographic projection.

(a) **Pictorial Projection:** When all three dimensions such as length, breadth and height of an object are shown in one view, such projection is termed as pictorial projection. The pictorial projections are classified as isometric projection, oblique projection and perspective projection.

(b) **Orthographic Projection:** In an orthographic projection, the object is represented either by two or three views such as front view, top view and side view which are two dimensional figures. Orthographic projection is discussed in detail in the following section.

5.3 Orthographic Projection

The projection of any object, obtained when all the projectors are parallel to each other but are perpendicular to the plane of projection, is called as orthographic projection. To obtain the orthographic projection of any object, following three items (Fig. 5.1) are required.

(a) The object
(b) The observer
(c) The plane of projection (horizontal or vertical)

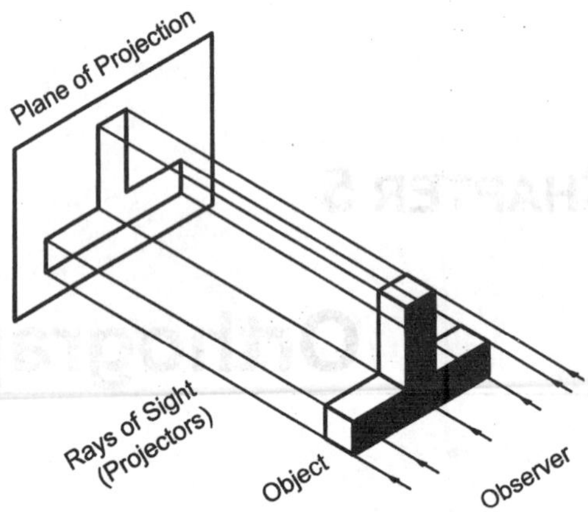

Fig. 5.1

The object is placed between the plane and observer. Imagine that the rays of sight are passing through various points on the contour of the object and are meeting at different points on the reference plane. Joining all points in correct sequence, one will get a figure which is called the projection of the object. Since single view obtained in such a way is not sufficient to describe the object completely therefore, another views such as top view and side view are also required for this purpose.

5.3.1 Important Terminology

(a) **Plane of Projection or Reference Plane:** To obtain orthographic projections of any object, two planes (horizontal and vertical) are employed which are called as reference planes or principal planes of projections.

(b) **Vertical Plane (V.P):** The reference plane, which is vertical is called vertical plane (V.P). It is also known as frontal plane. Front view is obtained on the V.P.

(c) **Horizontal Plane (H.P):** The reference plane, which is horizontal, is called horizontal plane (H.P). Top view or plan is obtained on the H.P.

(d) **Auxiliary Plane (A.P):** The plane which is inclined at any angle with the principal plane is termed as auxiliary plane (A.P).

(e) **Profile Plane or Auxiliary Vertical Plane:** The plane which is perpendicular to both the principal planes, is called auxiliary vertical plane (AVP) or profile plane (P.P).

(f) **Reference Line:** The line of intersection of two principal planes is known as reference line or ground line. It is usually denoted by xy line.

(g) **Front View or Elevation:** If the object is viewed from the front, then the projection obtained on vertical plane (V.P) is known as the front view or elevation.

(h) **Top View or Plan:** If the object is viewed from the top, then the projection obtained on the H.P is known as the top view or plan.

(i) **Side View:** If the object is viewed from its side (left or right), then the projection obtained on an auxiliary vertical plane (AVP) is known as side view.

5.3.2 Four Quadrants

When the horizontal and vertical planes are assumed to be extended beyond the reference line, four quadrants are formed as shown in Fig. 5.2. These quadrants are named as first, second, third and fourth quadrants taking in anticlock-wise direction.

In order to obtain the projections of any object, the object is first assumed to be situated in one of the four quadrants. Then horizontal and vertical projectors are drawn from the given object meeting vertical and horizontal planes respectively giving front view on the V.P whereas top view on the H.P. To bring both the views on a plane sheet of paper, H.P is rotated in clockwise direction.

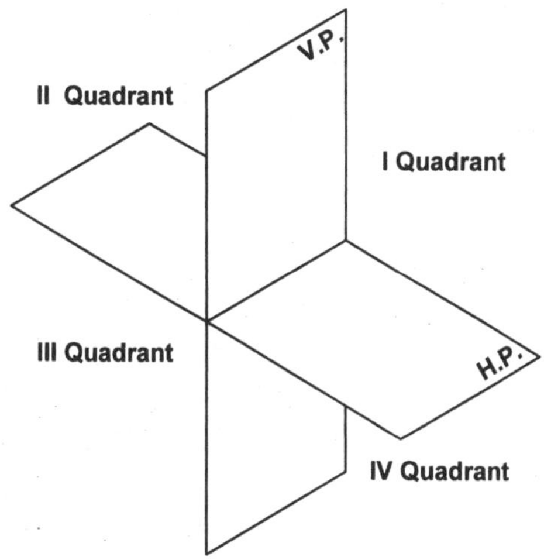

Fig. 5.2 Four quadrants.

The position of an object placed in different quadrants can be described as follows.

1. First quadrant – Above H.P. and in front of V.P.
2. Second quadrant – Above H.P. and behind V.P.
3. Third quadrant – Below H.P. and behind V.P.
4. Fourth quadrant – Below H.P. and in front of V.P.

If the object is situated in first quadrant, then method of obtaining projections is known as first angle projection method.

Similarly, for second, third and fourth quadrant, the methods are called as second angle, third angle and fourth angle projection methods respectively. Second and fourth angle projection methods are not used in engineering practice because of difficulty arises in handling front view and top view which may be overlapped. First angle and third angle projection methods are discussed in the following sections.

5.4 First Angle Projection Method

In this method, the object is assumed to be situated in first quadrant i.e., above the H.P and in front of the V.P. The object remains between the reference plane and the observer as shown in Fig. 5.3(a).

The front view appears in V.P and the top view on the H.P. When horizontal plane (H.P) is rotated in clockwise direction, top view comes below the front view as illustrated in Fig. 5.3(b).

Consider an auxiliary vertical plane (A.V.P) perpendicular to both H.P and the V.P. Looking from left, the side view is obtained on the A.V.P. If the A.V.P is brought in the same plane, then left side view appears towards right of the front view. Similarly the right side view will appears to the left of the front view.

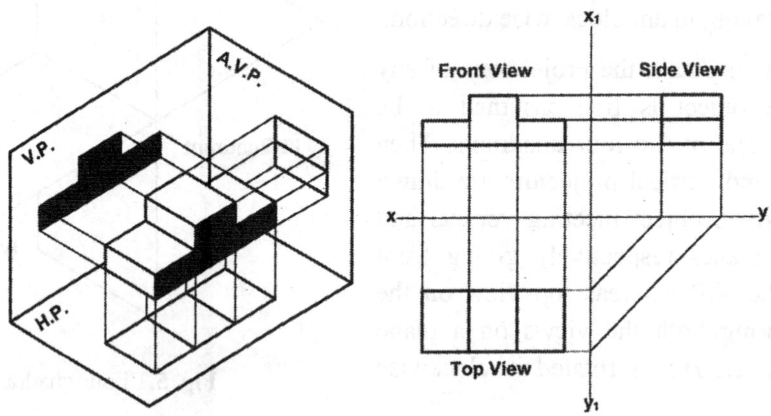

(a) (b)

Fig. 5.3

5.5 Third Angle Projection

In third angle projection method, the object is assumed to be situated in third quadrant i.e., below H.P and behind the V.P. Both the reference planes are assumed to be transparent. The plane is between the object and the observer as shown in Fig. 5.4(a). Front view appears on the V.P and top view on the H.P, since both V.P and H.P are assumed to be transparent. When the horizontal plane (H.P) is rotated in clockwise direction, the top view comes above the front view as shown in Fig. 5.4(b).

As side view is obtained on A.V.P and when A.V.P is brought in the same plane as that of front view, then right side view appears to the right and left side view appears to the left of the front view. The salient features of first and third angle projection methods are presented in Table 5.1.

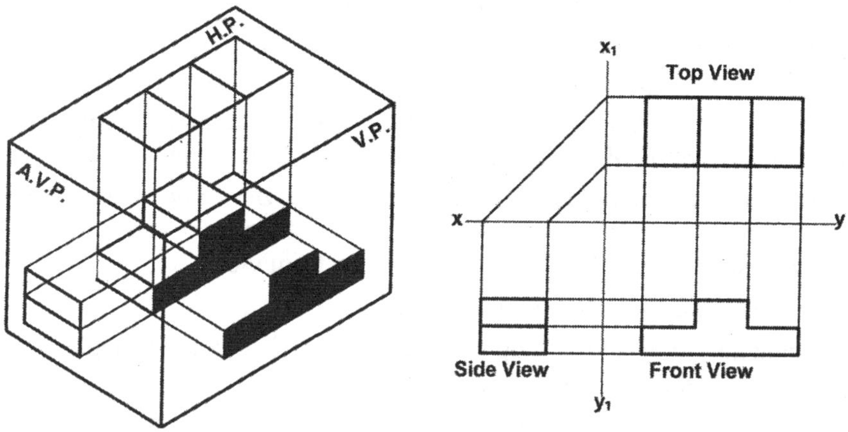

(a) (b)

Fig. 5.4

Table 5.1 Salient features of first angle and third angle projection methods.

S.No.	First angle projection	Third angle projection
1.	The object is assumed to be situated in first quadrant.	The object is assumed to be situated in third quadrant.
2.	The object is kept between observer and plane of projection.	The plane of projection lies between observer and object.
3.	The plane of projection is assumed to be non-transparent.	The plane of projection is assumed to be transparent.
4.	Front view is drawn above the reference line whereas top view is drawn below the reference line.	Top view is drawn above the reference line and front view is drawn below the reference line.
5.	Left side view is drawn to the right and right side view is drawn to the left of the front view.	Left side view is drawn to the left and right side view is drawn to the right of the front view.
6.	B.I.S recommended Symbol 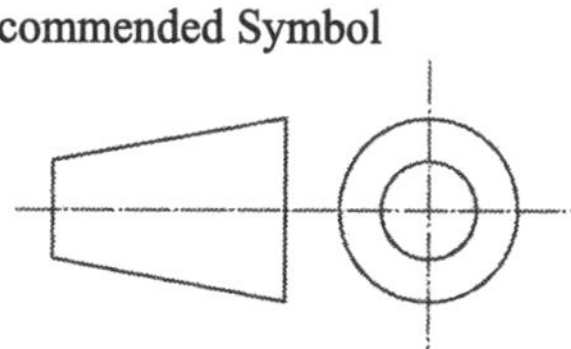	B.I.S recommended Symbol 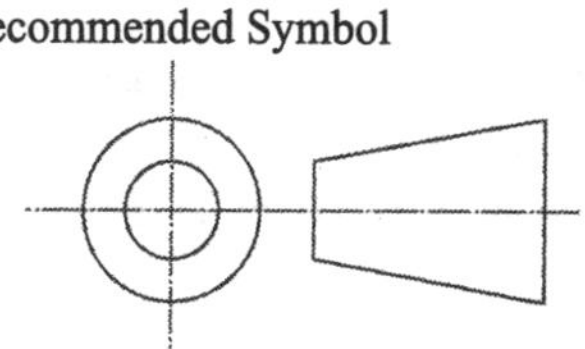

5.6 General Procedure for Obtaining Orthographic Projections

If the pictorial view of any object is given and it is required to draw orthographic projections (i.e., front, top and side view) from it, then following steps are helpful.

(i) Observe the pictorial view carefully for dimension and direction of each view. For example, direction of arrow head indicates front face of the object.

(ii) Reserve the required space for front, top and side view. For example, in first angle projection, left side view is drawn to the right of the front view. So space for side view must be reserved to the right side of the front view.

(iii) Draw reference lines (horizontal and vertical).

(iv) Imagine the front, side and top views are enclosed in three rectangles. Observe the length and breadth of these rectangles.

(v) Draw outlines of front view, top view and side view (as rectangles) using thin construction lines.

(vi) Begin your drawing preferably from circular part with its centre axis.

(vii) Complete the construction as per details given in pictorial view.

(viii) Erase the unwanted outlines of rectangles.

(ix) Draw invisible edges/holes by dotted lines.

(x) Provide all necessary dimensions to each view.

5.6.1 Things to Remember

(a) Front view and top view are aligned through vertical projectors.

(b) Front and side views are aligned through horizontal projectors.

(c) Surface parallel to reference plane will be seen with true size and shape in that plane.

(d) Surface perpendicular to reference plane will be seen as straight lines in the plane.

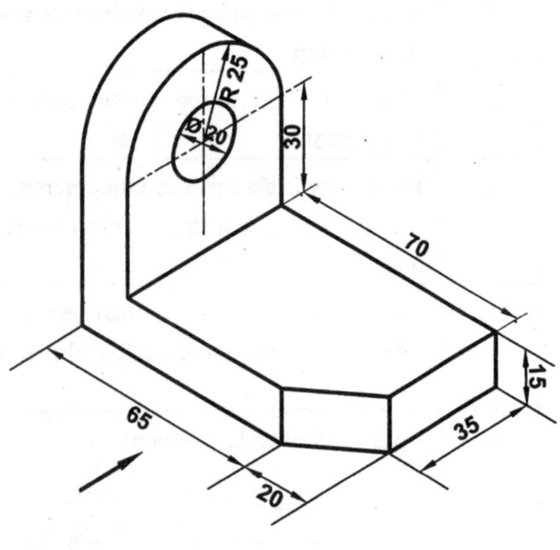

Fig. 5.5

Problem 5.1: *Pictorial view of a bearing block is shown in Fig. 5.5. using first angle projection method draw following views.*

(i) *Front view*

(ii) *Top view*

(iii) *Right hand side view*

Construction: Assuming the object in first quadrant, draw the required views using following three steps.

Step 1: Draw reference lines and rectangular boxes of 85 mm × 70 mm (for front view), 85 mm × 50 mm (for top view) and 50 mm × 70 mm for side view.

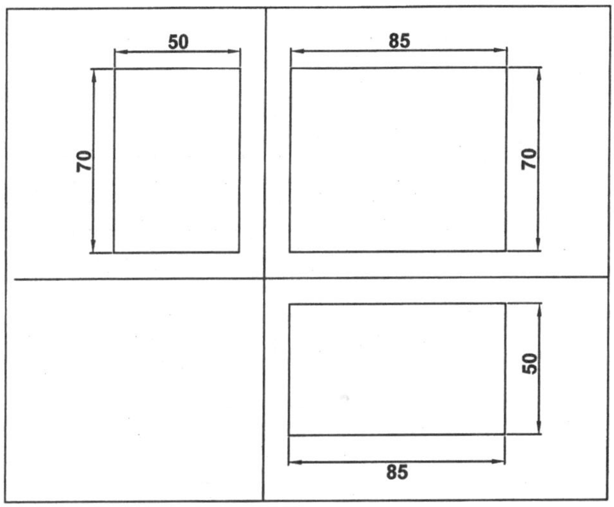

Fig. 5.6 (Step 1).

Step 2: Begin from centre line of the hole. Draw a circle representing hole of $\phi 20$ in side view. Draw internal parts as per details shown in pictorial view.

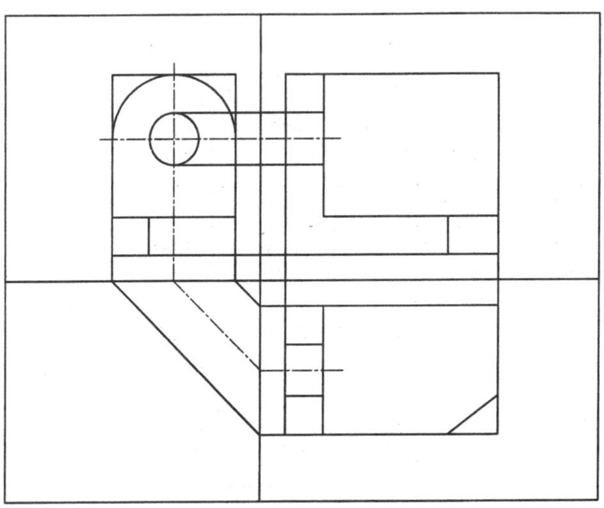

Fig. 5.7 (Step 2).

Step 3: Erase unwanted construction lines. Dark the final figures. Mark dimensions as shown in the Fig. 5.8.

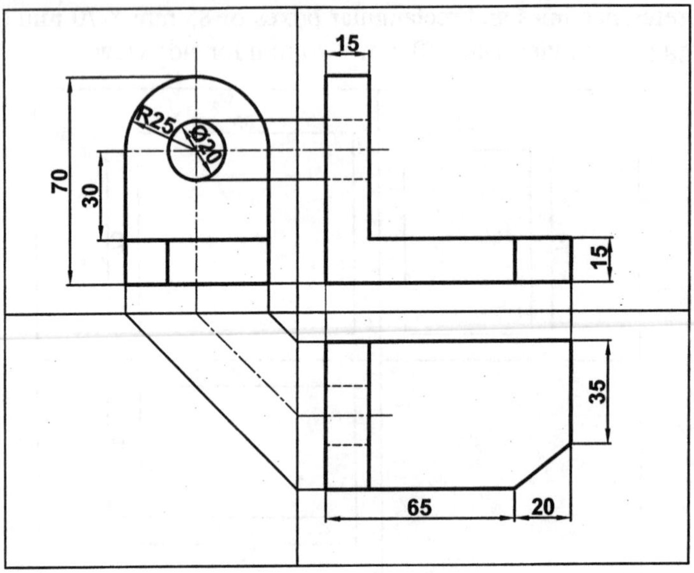

Fig. 5.8 (Step 3).

Problem 5.2: *The isometric view of an object is shown in Fig. 5.9, using first angle projections, draw its front view, top view and left side view.*

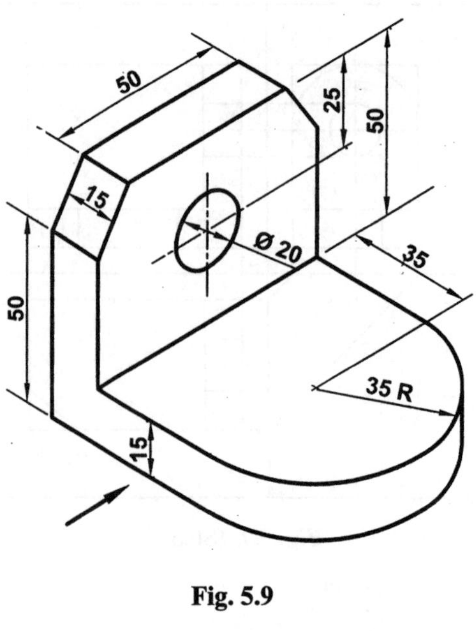

Fig. 5.9

Construction: (Refer Fig. 5.10)

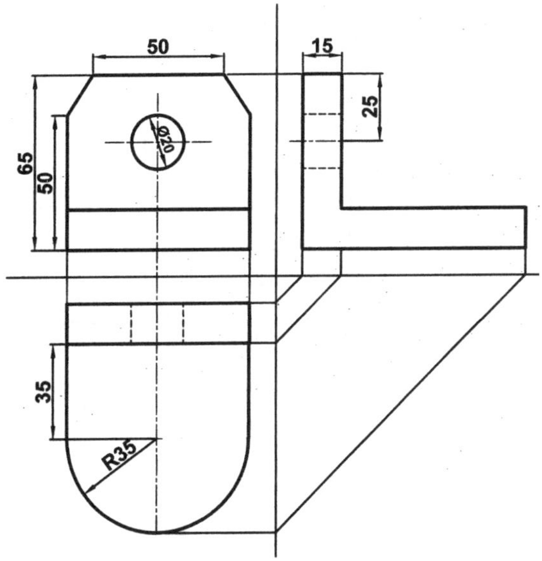

Fig. 5.10

Problem 5.3: *Draw the following views of the object shown in Fig. 5.11.*

(i) *Front view*

(ii) *Right side view*

(iii) *Top view*

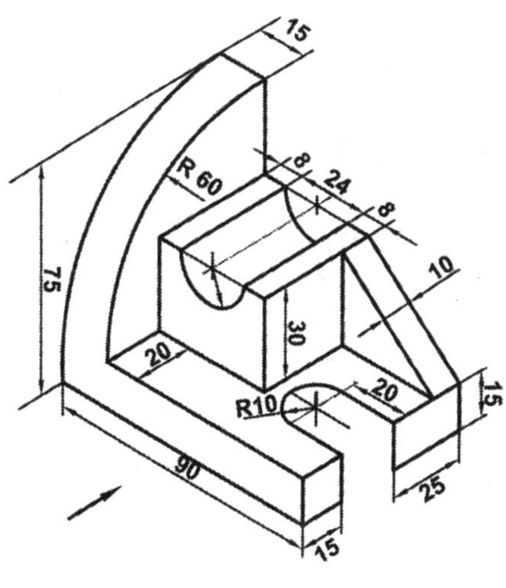

Fig. 5.11

Construction: (Refer Fig. 5.12)

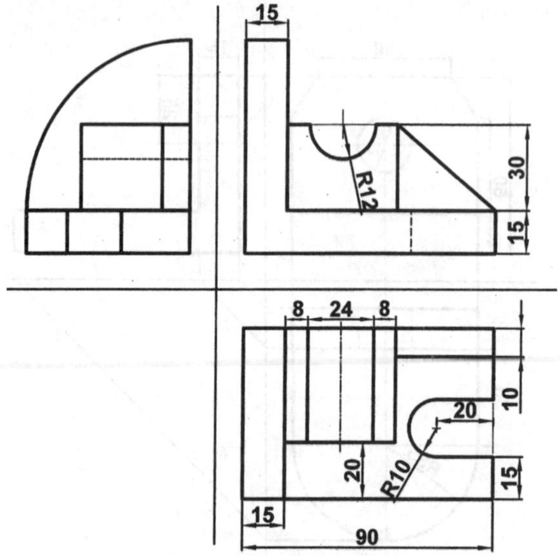

Fig. 5.12

Problem 5.4: *The pictorial view of an object is shown in Fig. 5.13. Using first angle projections draw the following views.*

(i) *Front view*

(ii) *Top view*

(iii) *Side view (right)*

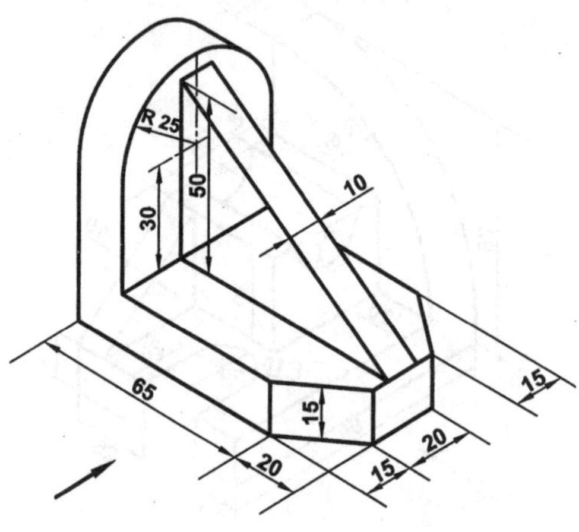

Fig. 5.13

Construction: (Refer Fig. 5.14)

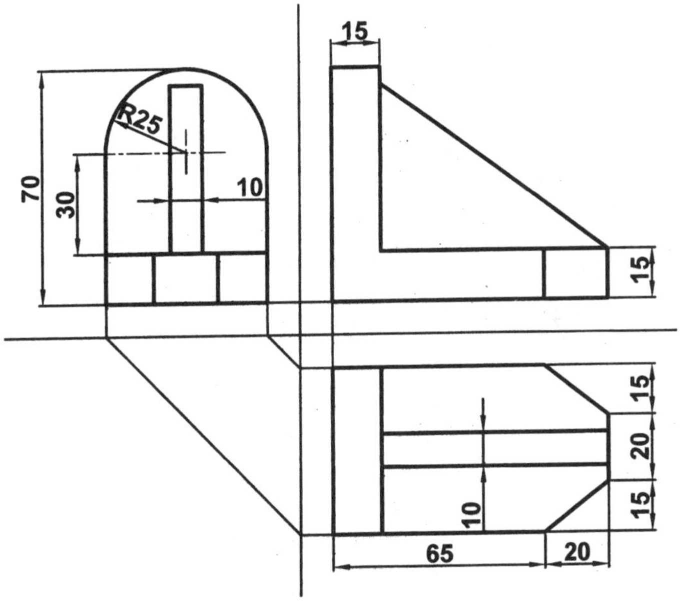

Fig. 5.14

Problem 5.5 : *Draw following three views of the bracket shown in Fig. 5.15.*
(i) Front view from the direction of arrow (ii) Top view and (iii) Right side view.

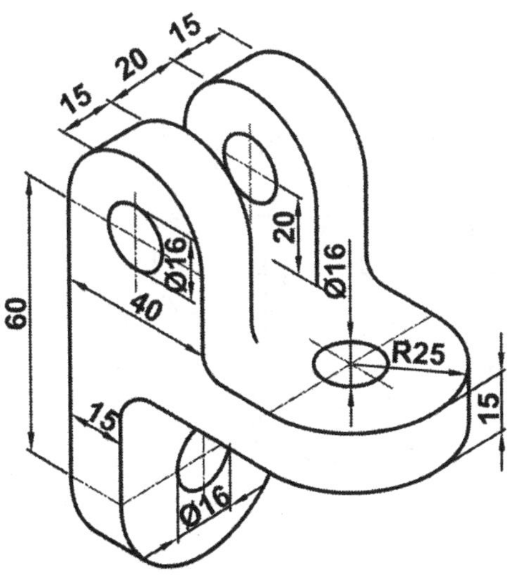

Fig. 5.15

Construction: Refer Fig. 5.16

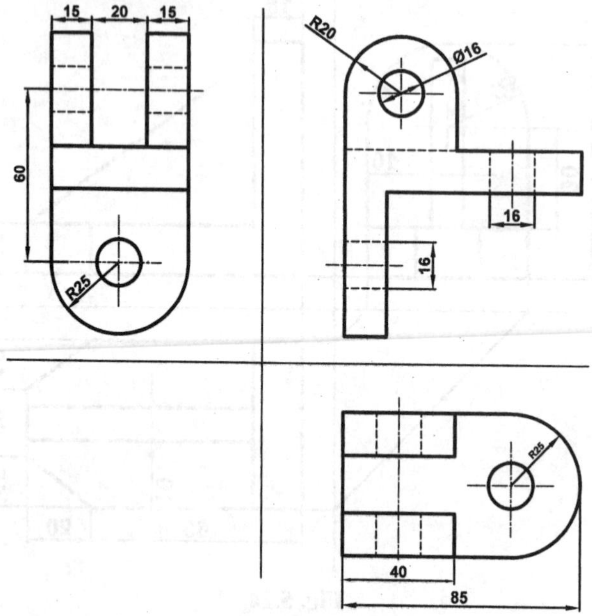

Fig. 5.16

Problem 5.6: *Draw front view and top view of a bearing shown in Fig. 5.17*

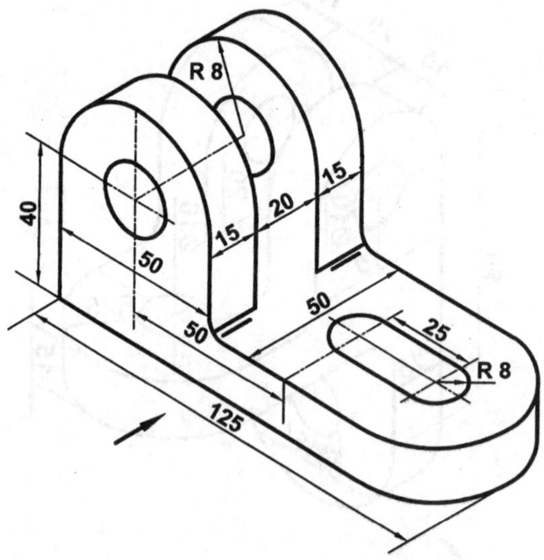

Fig. 5.17

Construction: Refer Fig. 5.18

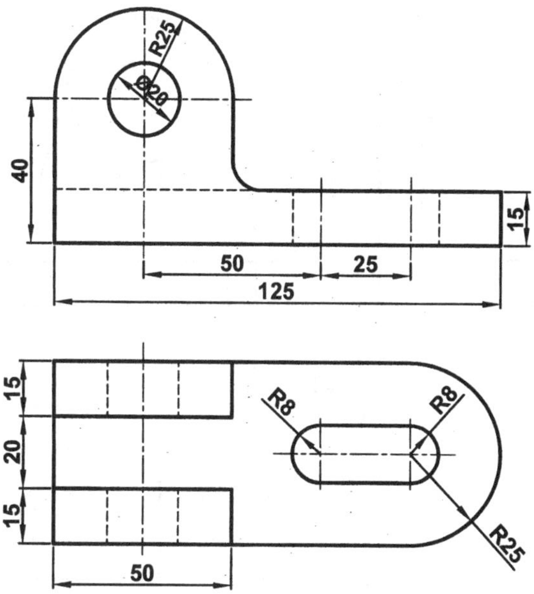

Fig. 5.18

Problem 5.7: *Draw the following view of the object shown in Fig. 5.19.*

(i) Front view from arrow (ii) Top view and (iii) Side view.

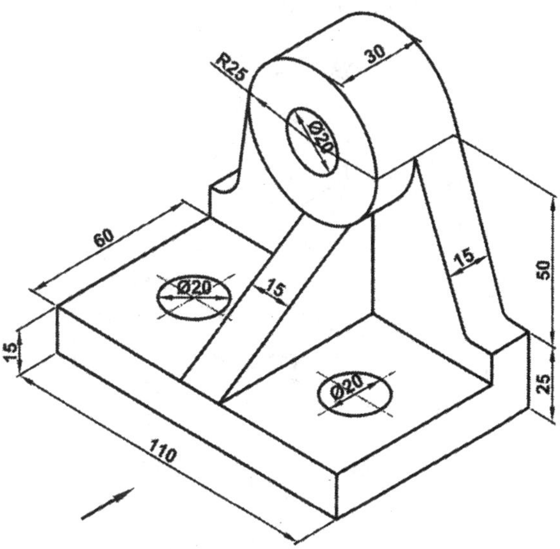

Fig. 5.19

Construction: Refer Fig. 5.20

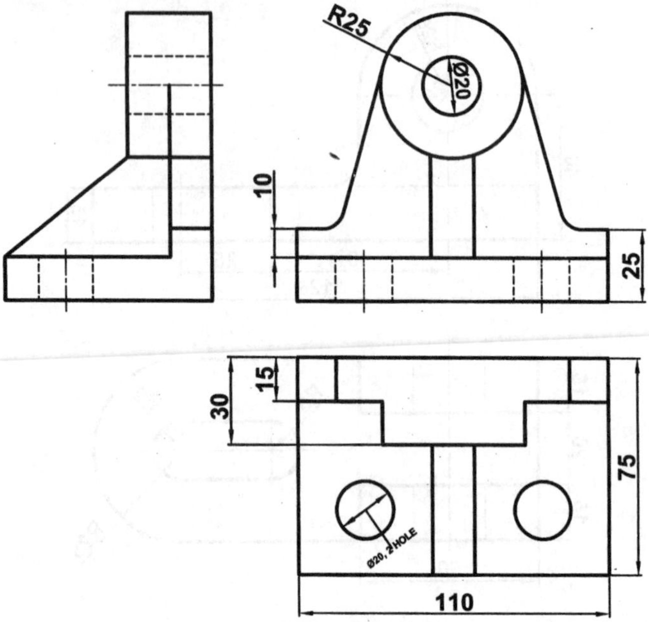

Fig. 5.20

Problem 5.8: *Draw the following views of the object shown in Fig. 5.21*
(i) Front view (ii) Top view and (iii) Left side view.

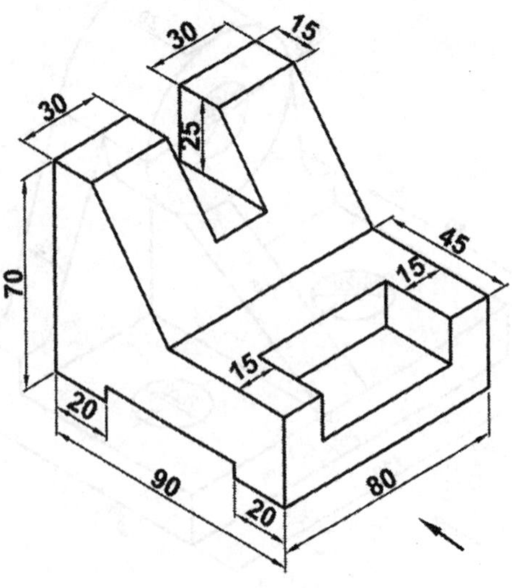

Fig. 5.21

Construction: Refer Fig. 5.22

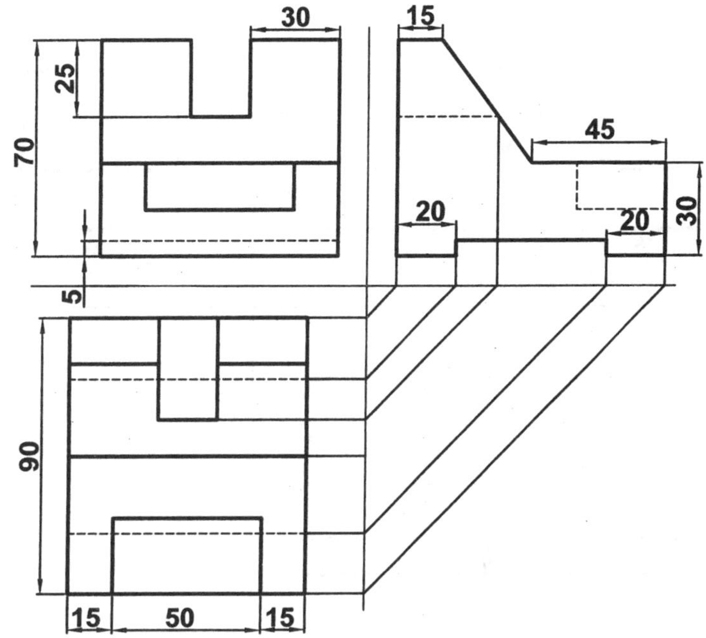

Fig. 5.22

Problem 5.9: *Draw front view, top view and left side view of a V-block shown in Fig. 5.23. using first angle projection method.*

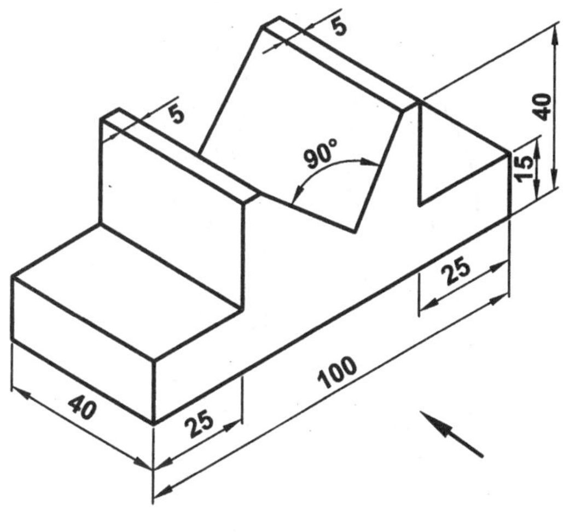

Fig. 5.23

Construction: Refer Fig. 5.24.

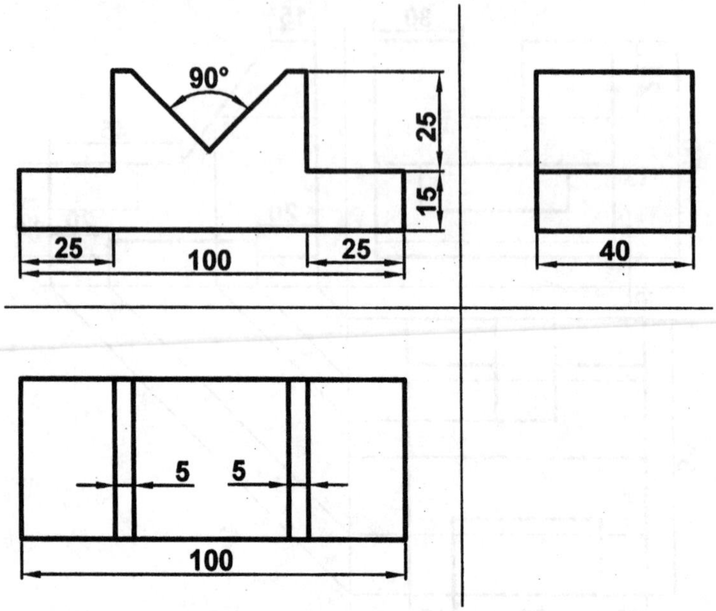

Fig. 5.24

Problem 5.10: *The pictorial view of an object is shown in Fig. 5.25. Using first angle projections method, draw front view, top view and left side view.*

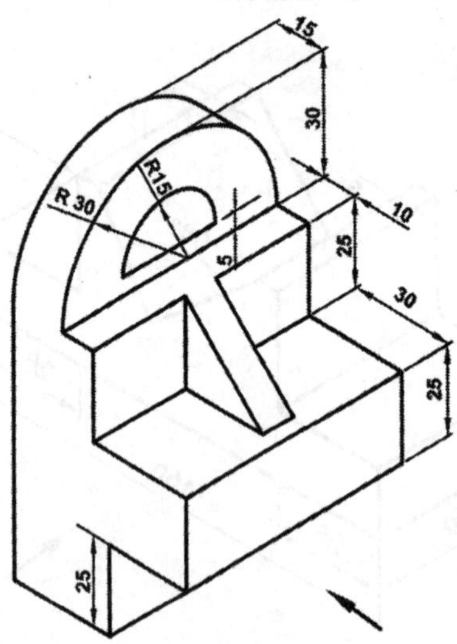

Fig. 5.25

Construction: Refer Fig. 5.26

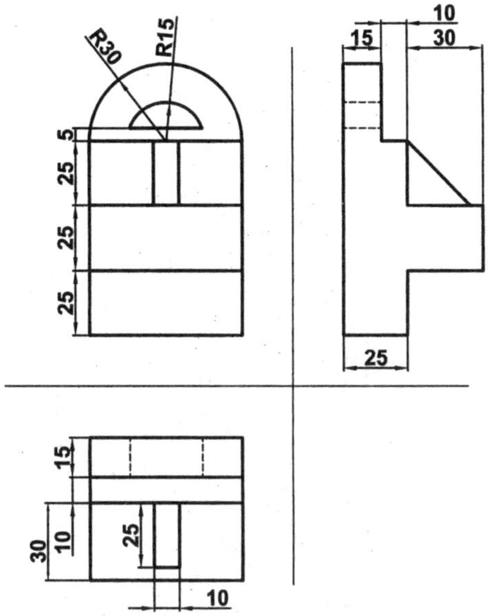

Fig. 5.26

Exercises

1. Draw front view, top view and side view of an object shown in Fig. 5.27 using first angle projection method.

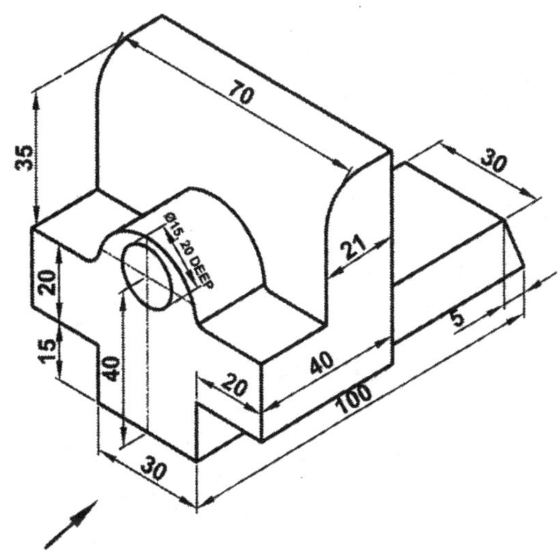

Fig. 5.27

2. The pictorial view of an object is shown in Fig. 5.28 using first angle projection, draw front view, top view and side view.

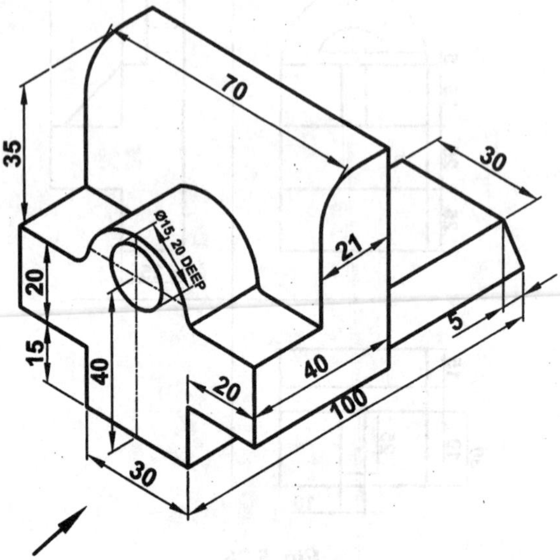

Fig. 5.28

3. The pictorial view of a block is given in Fig. 5.29 using first angle projection method, draw front view, top view and side view.

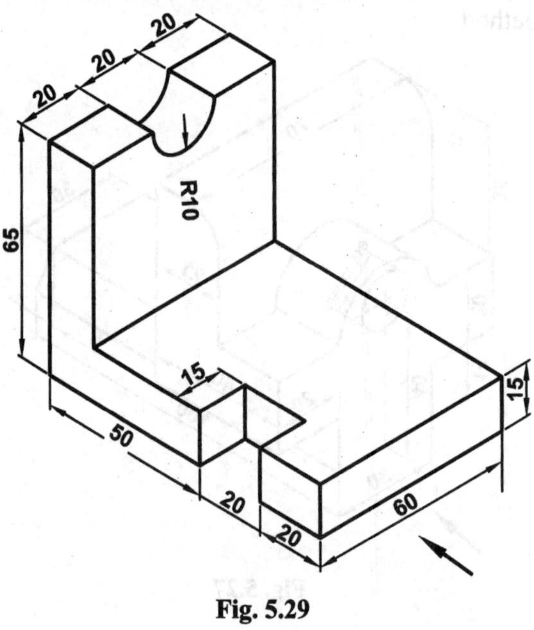

Fig. 5.29

4. The pictorial view of a block is shown in Fig. 5.30 using first angle projection method, draw front view, top view and side view

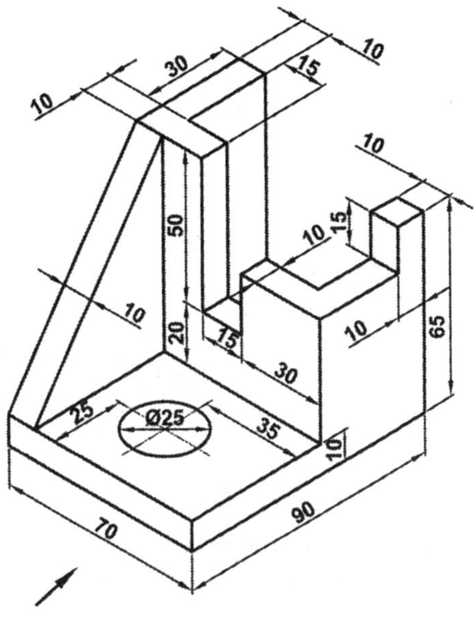

Fig. 5.30

5. Draw front view and top view of the object shown in Fig. 5.31 using first angle projection method.

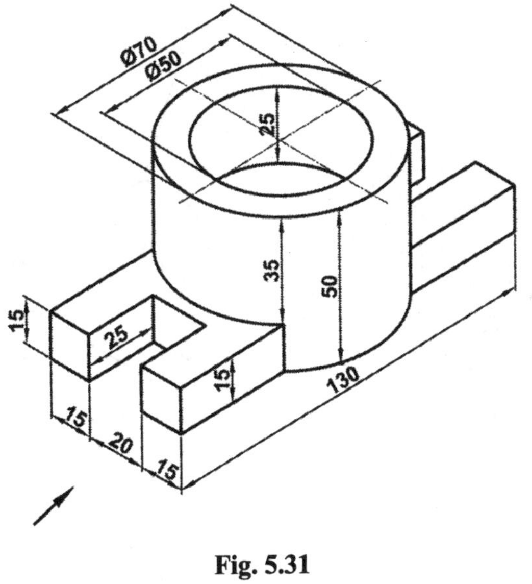

Fig. 5.31

6. Draw the following views of an object shown in Fig. 5.32, using first angle projection method.

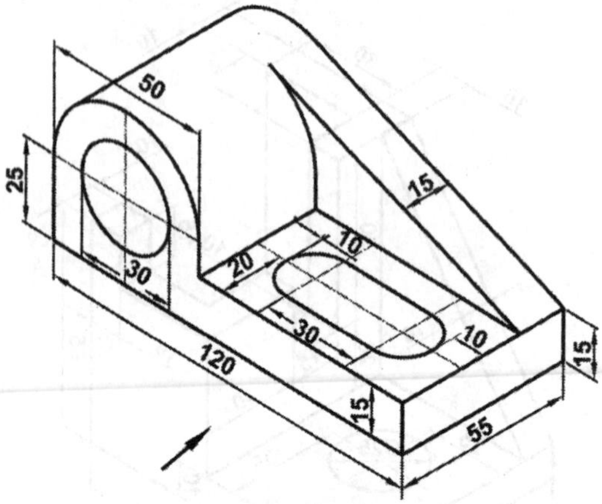

Fig. 5.32

7. The pictorial view of an object is shown in Fig. 5.33 using first angle projection method, Draw its front view, top view and side view.

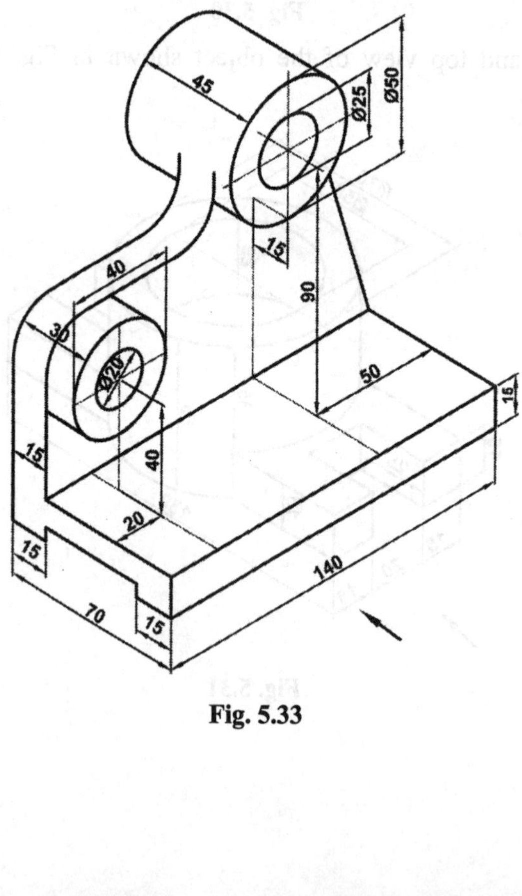

Fig. 5.33

8. Draw front view, top view and side view of the object shown in Fig. 5.34.

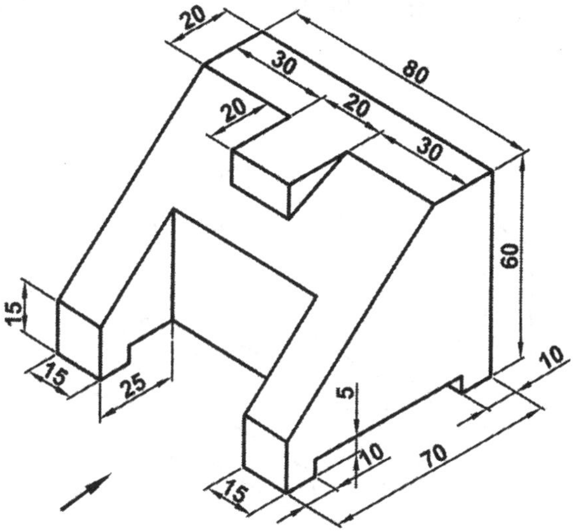

Fig. 5.34

9. The pictorial view of an object is shown in Fig. 5.35 using first angle projection method, Draw its front view from the direction of arrow, top view and right hand side view.

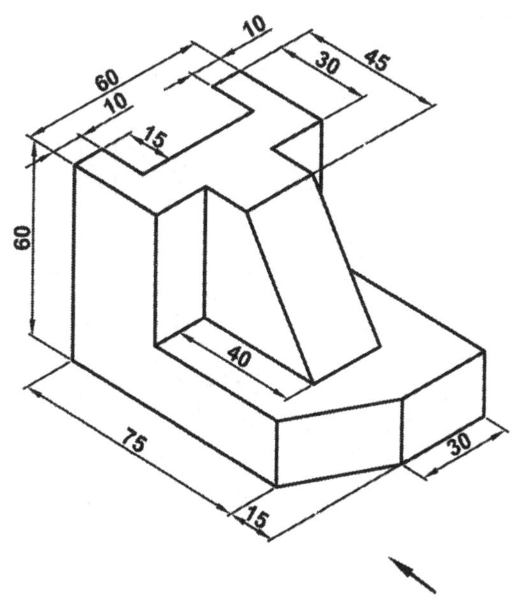

Fig. 5.35

10. The pictorial view of an object is shown in Fig. 5.36 using first angle projection method, draw the front view from the direction of arrow, top view and side view.

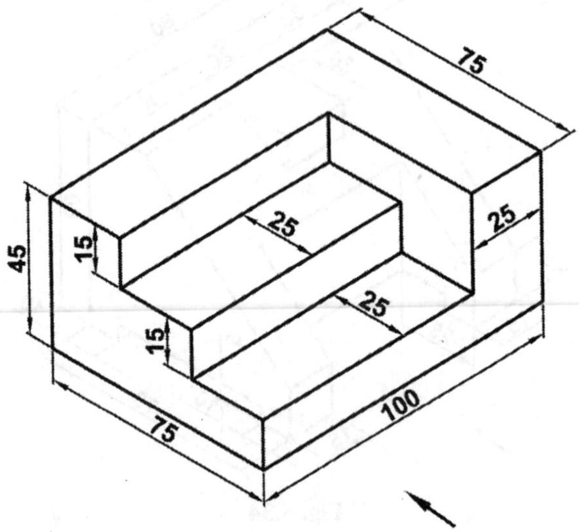

Fig. 5.36

11. Draw front view, top view and side view of a block shown in Fig. 5.37 using first angle projection method.

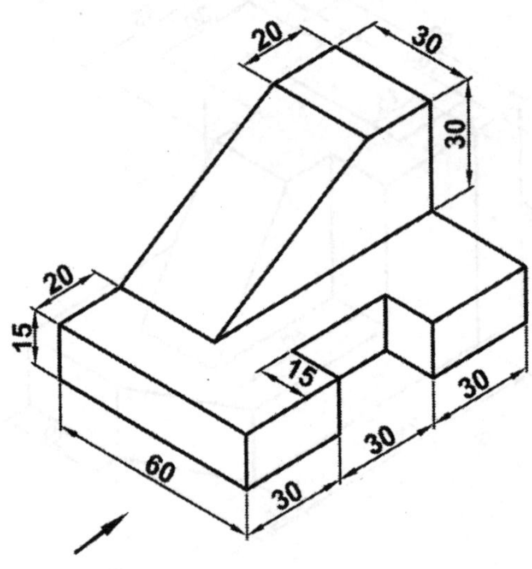

Fig. 5.37

12. The pictorial view of an object is shown in Fig. 5.38. Draw its front view, top view and side view using first angle projection method.

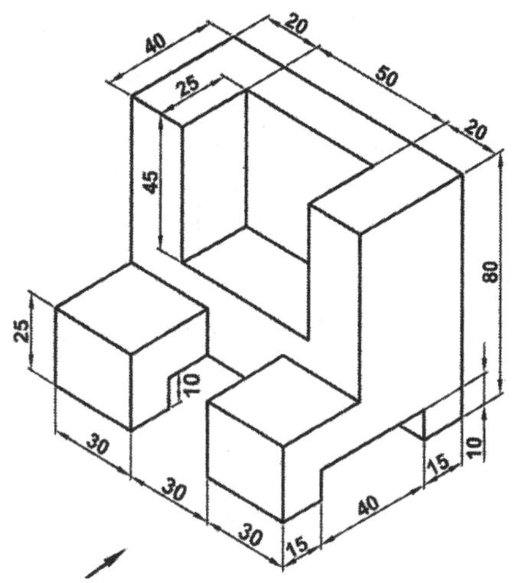

Fig. 5.38

13. Draw front view, top view and side view of a block shown in Fig. 5.39 using first angle projection method.

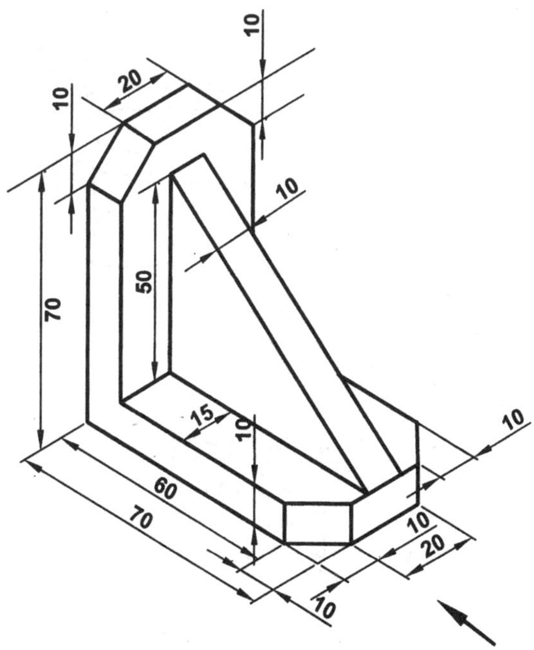

Fig. 5.39

14. The pictorial view of an object is shown in Fig. 5.40. Draw its front view, top view and side view using first angle projection method.

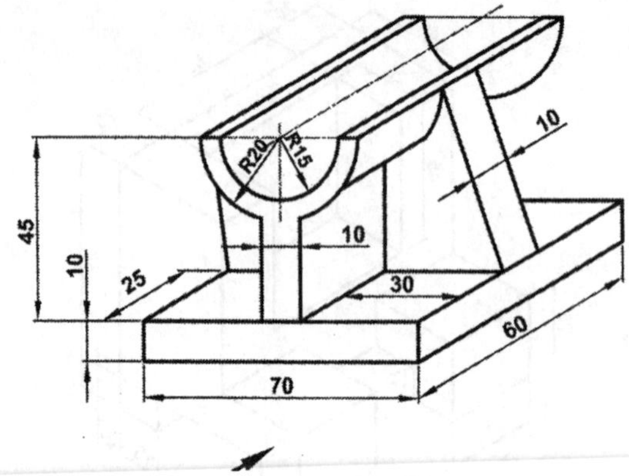

Fig. 5.40

15. Draw the following views of a block shown in Fig. 5.41.

    (i) Front view
    (ii) Top view and
    (iii) Left hand side view.

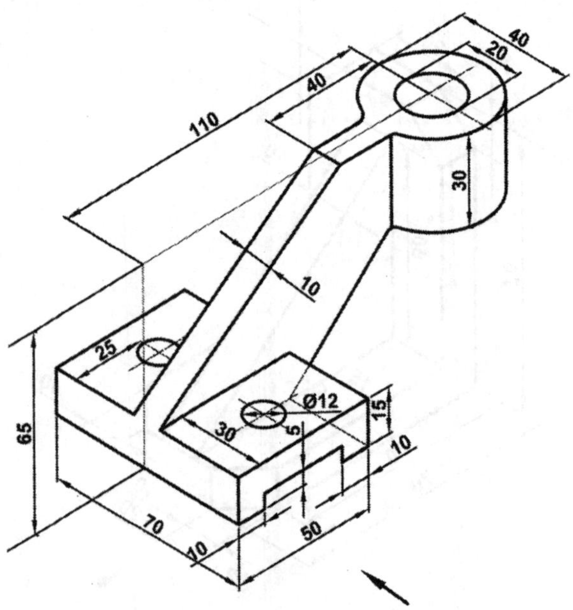

Fig. 5.41

CHAPTER 6

Projections of Points

6.1 Introduction

A point may be defined as smallest geometrical element having no dimensions. It is represented as dot. A point may be situated in any one of the four quadrants which are formed by two mutually perpendicular planes.

To obtain the projections of a point, first of all the given point is assumed to be situated in the given quadrant. Perpendicular projectors are then drawn from the point to meet the planes of projection. Now by rotating the horizontal plane in clockwise direction, both the views (i.e., front view and top view) are brought in the same plane. On the drawing, actual position of the point is represented by capital letter A, B, C, etc. The corresponding front view, top view and side view are designated by a', b', c',.., a, b, c,..and a", b", c"... respectively.

6.2 Point above H.P. and in front of V.P. (i.e., in first quadrant)

Problem 6.1 (Fig. 6.1): *A point P is situated 40 mm above the H.P. and 65 mm in front of the V.P. Draw its front view and top view.*

Graphical Representation (Fig. 6.1(a)):

Point P is situated in first quadrant as shown in Fig. 6.1(a), such that it is 40 mm above the H.P. and 65 mm in front of the V.P. Drawing horizontal and vertical projectors through P to meet V.P. and H.P. respectively, we will get front view 40 mm above the xy line and when horizontal plane is rotated clockwise, the top view will appear 65 mm below the xy line.

Construction (Fig. 6.1(b)):

(i) Draw a reference line xy.

(ii) At any suitable point on xy line, draw a perpendicular projector.

(iii) On the projector, mark front view p', 40 mm above xy line.

(iv) Mark the top view p, 65 mm below xy line.

(v) Mark the dimensions as shown in Fig.6.1(b).

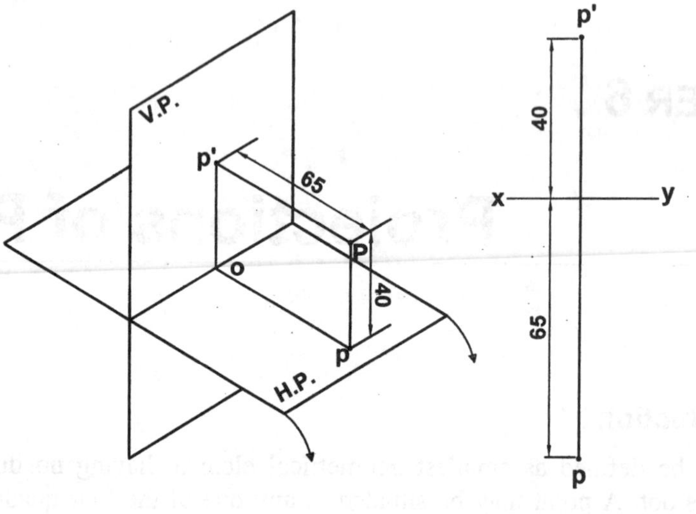

Fig. 6.1

6.3 Point above H.P. and behind V.P. (i.e., in second quadrant)

Problem 6.2 (Fig. 6.2): *A point A is 60 mm above the H.P. and 30 mm behind the V.P. Draw its projections.*

Graphical Representation (Fig. 6.2(a)):

Point A is situated in second quadrant as shown in Fig. 6.2(a), such that it is 60 mm above the H.P. and 30 mm behind the V.P. Drawing horizontal and vertical projectors through A to meet V.P. and H.P. respectively, we will get front view 60 mm above the xy line and when horizontal plane is rotated clockwise, the top view will appear 30 mm above the xy line.

Construction (Fig. 6.2(b)):

(i) Draw a reference line xy.

(ii) At any suitable point on xy line, draw a perpendicular projector.

(iii) On the projector, mark front view a', 60 mm above xy line.

(iv) Mark the top view a, 30 mm above xy line.

(v) Mark the dimensions as shown in Fig.6.2(b).

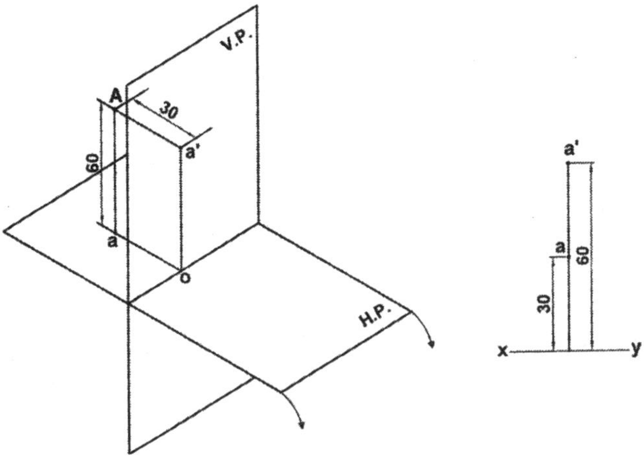

(a) (b)

Fig. 6.2

6.4 Point below H.P. and behind V.P. (i.e., in third quadrant)

Problem 6.3 (Fig. 6.3): *A point Q is 50 mm below the H.P. and 40 mm behind the V.P. Draw its front view and top view.*

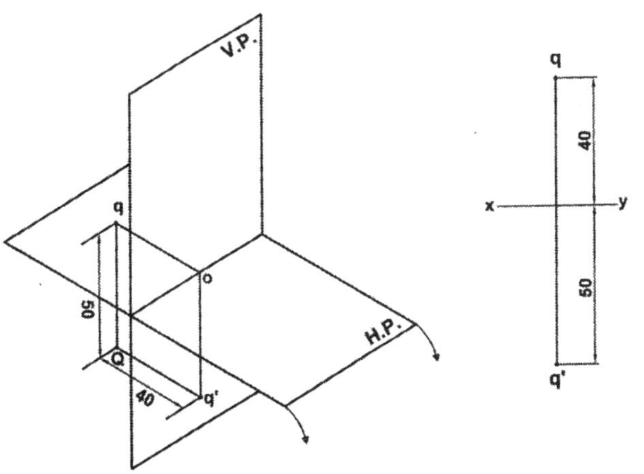

(a) (b)

Fig. 6.3

Graphical Representation (Fig. 6.3(a)):

Point Q is situated in third quadrant as shown in Fig. 6.3(a), such that it is 50 mm below the H.P. and 40 behind the V.P. Drawing horizontal and vertical projectors through Q to meet V.P. and H.P. respectively, we will get front view 50 mm below the xy line and when horizontal plane is rotated clockwise, the top view will appear 40 mm above the xy line.

Construction (Fig. 6.3(b)):

(i) Draw a reference line xy.

(ii) At any suitable point on xy line, draw a perpendicular projector.

(iii) On the projector, mark front view q', 50 mm below xy line.

(iv) Mark the top view q, 40 mm above xy line.

(v) Mark the dimensions as shown in Fig.6.3(b).

6.5 Point below H.P. and in front of V.P. (i.e., in fourth quadrant)

Problem 6.4 (Fig. 6.4): *A point P is situated 30 mm below the H.P. and 45 mm in front of the V.P. Draw its front view and top view.*

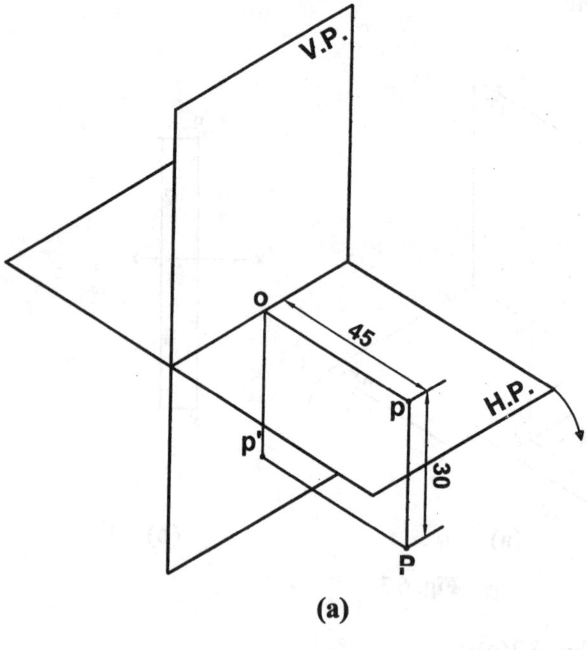

(a)

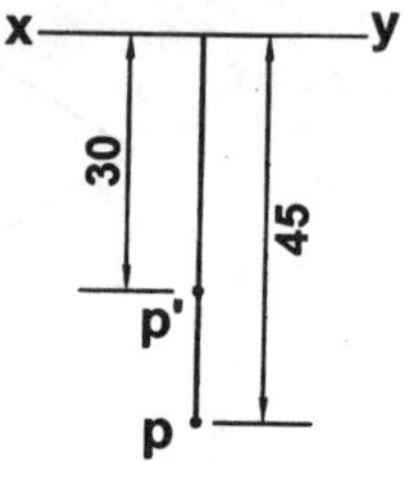

(b)

Fig. 6.4

Graphical Representation (Fig. 6.4(a)):

Point P is situated in fourth quadrant as shown in Fig. 6.4(a), such that it is 30 mm below the H.P. and 45 mm in front of the V.P. Drawing horizontal and vertical projectors through P to meet V.P. and H.P. respectively, we will get front view 30 mm below the xy line and when horizontal plane is rotated clockwise, the top view will appear 45 mm below the xy line.

Construction (Fig. 6.4(b)):

(i) Draw a reference line xy.

(ii) At any suitable point on xy line, draw a perpendicular projector.

(iii) On the projector, mark front view p', 30 mm below xy line.

(iv) Mark the top view p, 45 mm below xy line.

(v) Mark the dimensions as shown in Fig. 6.4(b).

6.6 Point on the H.P. and in front of the V.P.

Problem 6.5 (Fig. 6.5): *A point B is on the H.P. and 30 mm in front of the V.P. Draw its projections.*

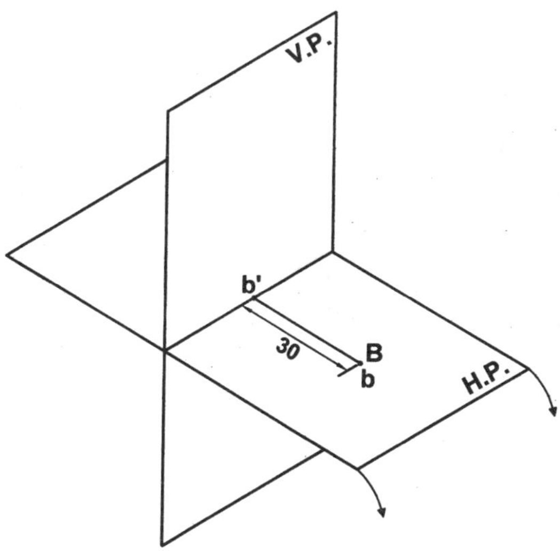

(a)

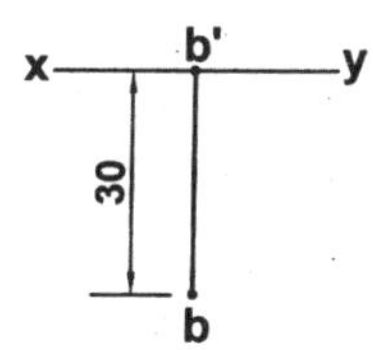

(b)

Fig. 6.5

Graphical Representation (Fig. 6.5(a)):

Point B is situated on H.P. as shown in Fig. 6.5(a), such that it is 30 mm in front of the V.P. Drawing horizontal projector through B to meet V.P, we will get front view on xy line and when horizontal plane is rotated clockwise , the top view will appear 30 mm below the xy line.

Construction (Fig. 6.5(b)):

(i) Draw a reference line xy.

(ii) At any suitable point on xy line, draw a perpendicular projector.

(iii) On the projector, mark front view b' on xy line.

(iv) Mark the top view b, 30 mm below xy line.

(v) Mark the dimensions as shown in Fig.6.5(b).

6.7 Point on the H.P. and behind the V.P.

Problem 6.6 (Fig. 6.6): *A point R is on the H.P. and 60 mm behind the V.P. Draw its projections.*

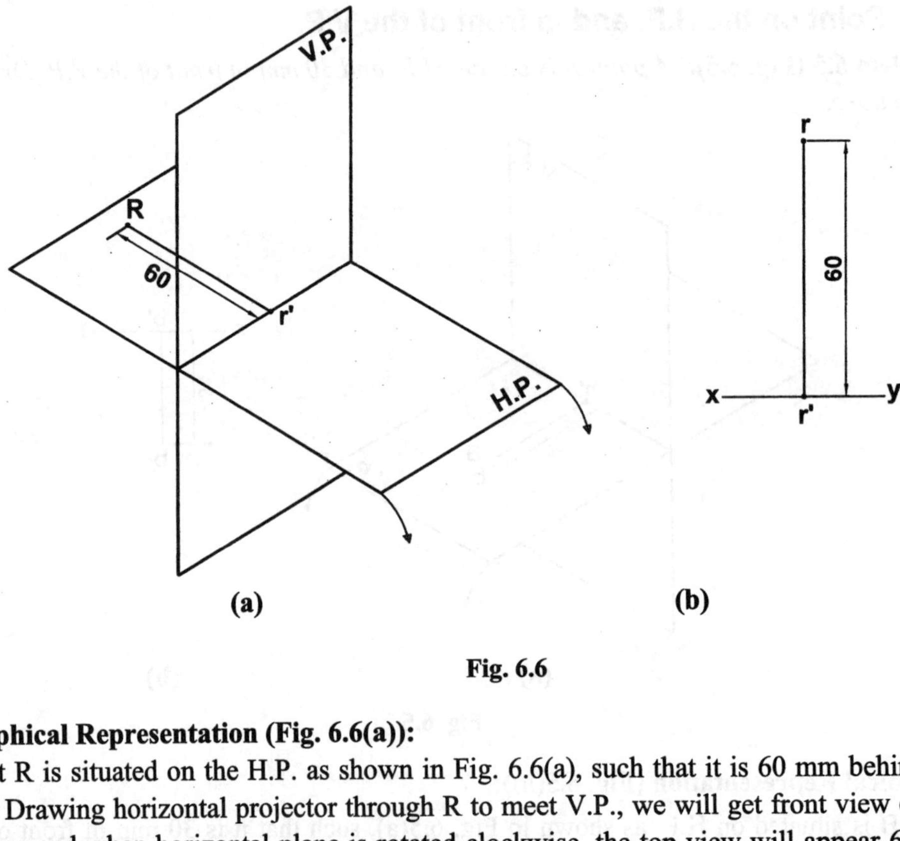

Fig. 6.6

Graphical Representation (Fig. 6.6(a)):

Point R is situated on the H.P. as shown in Fig. 6.6(a), such that it is 60 mm behind the V.P. Drawing horizontal projector through R to meet V.P., we will get front view on the xy line and when horizontal plane is rotated clockwise, the top view will appear 60 mm above the xy line.

Construction (Fig. 6.6(b)):

(i) Draw a reference line xy.

(ii) At any suitable point on xy line, draw a perpendicular projector.

(iii) On the projector, mark front view r' on the xy line.

(iv) Mark the top view r, 60 mm above xy line.

(v) Mark the dimensions as shown in Fig.6.6(b).

6.8 Point above H.P. and in the V.P.

Problem 6.7 (Fig. 6.7): *A point S is 70 mm above the H.P. and in the V.P. Draw its projections.*

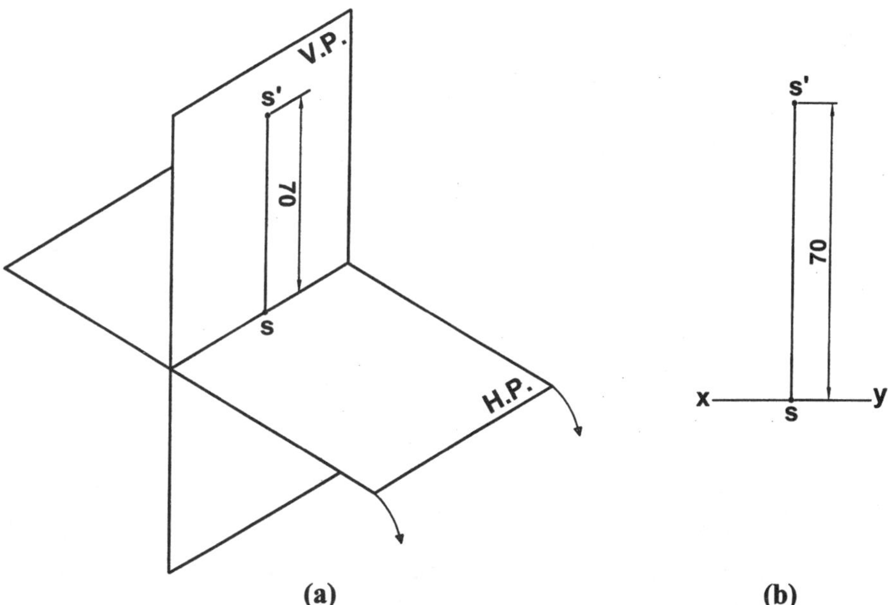

Fig. 6.7

Graphical Representation (Fig. 6.7(a)):

Point S is in V.P. as shown in Fig. 6.7(a), such that it is 70 mm above the H.P. Drawing vertical projector through S to meet H.P., we will get top view on the xy line. Front view will appear 70 mm above the xy line.

Construction (Fig. 6.7(b)):

(i) Draw a reference line xy.

(ii) At any suitable point on xy line, draw a perpendicular projector.

(iii) On the projector, mark front view s', 70 mm above xy line.

(iv) Mark the top view s, on the xy line.

(v) Mark the dimensions as shown in Fig. 6.7(b).

6.9 Point below H.P. and in the V.P.

Problem 6.8: *A point P is 65 mm below the H.P. and in the V.P. Draw its projections.*

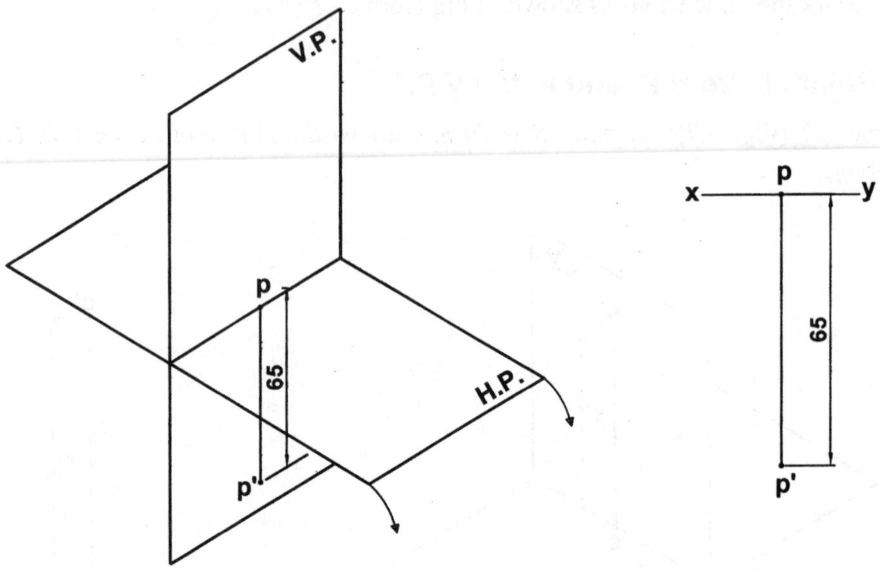

(a) (b)

Fig. 6.8

Graphical Representation (Fig. 6.8(a)):

Point P is in the V.P. as shown in Fig. 6.8(a), such that it is 65 mm below the H.P. Drawing vertical projector through P to meet H.P., we will get top view on the xy line and front view will appear 65 mm below the xy line.

Construction (Fig. 6.8(b)):

(i) Draw a reference line xy.

(ii) At any suitable point on xy line, draw a perpendicular projector.

(iii) On the projector, mark front view p', 65 mm below the xy line.

(iv) Mark the top view p, on the xy line.

(v) Mark the dimensions as shown in Fig.6.8(b).

6.10 Point in both H.P. and V.P.

Problem 6.9: *A point A is in both H.P. and V.P. Draw its projections.*

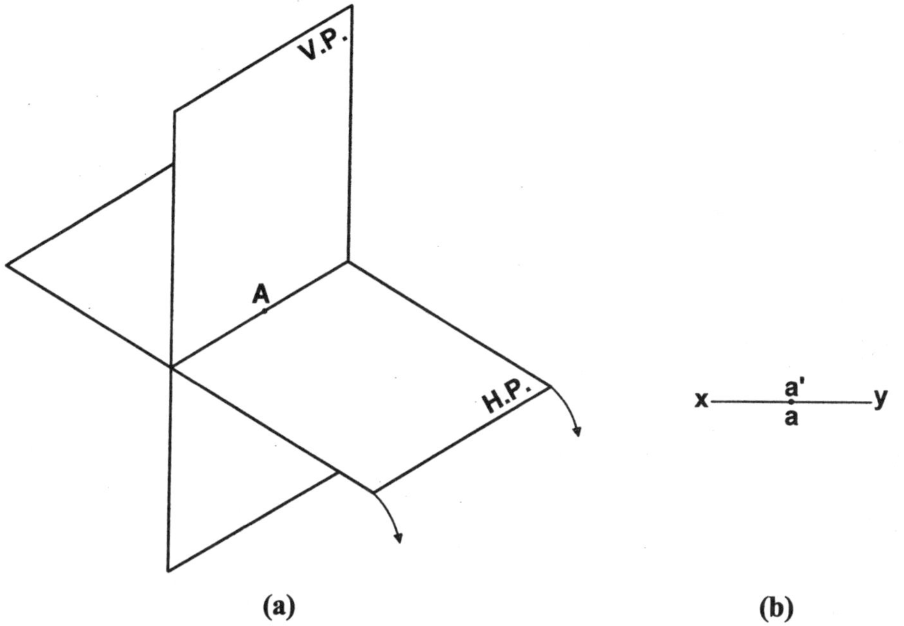

Fig. 6.9

Graphical Representation (Fig. 6.9(a)):

Point A is situated on both H.P. and V.P. as shown in Fig. 6.9(a). In this case we will get front view and top view coinciding each other on the xy line.

Construction (Fig. 6.9(b)):

(i) Draw a reference line xy.

(ii) On the xy line, mark front view a' and top view a, both coinciding at a point.

Problem 6.10: *Draw the projections of following points on the same ground line keeping the projectors 25 mm apart.*

(a) Point A is 40 mm above H.P. and 25 mm in front of the V.P.

(b) Point B is in the V.P. and 40 mm above the H.P.

(c) Point C is in the H.P. and 25 mm behind the V.P.

(d) Point D is 30 mm below the H.P. and 25 mm behind the V.P.

(e) Point E is 20 mm above the H.P. and 45 mm behind the V.P.

(f) Point F is 45 mm below the H.P. and 25 mm in front of the V.P.

(g) Point G is both in H.P. and V.P.

Solution:

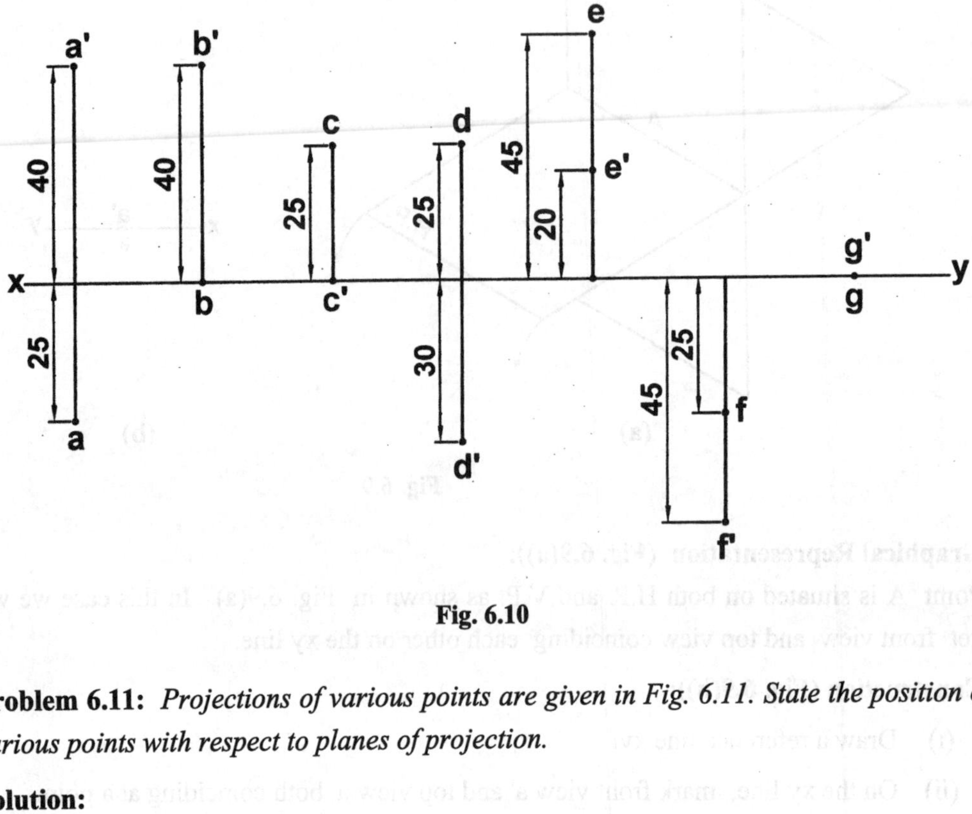

Fig. 6.10

Problem 6.11: *Projections of various points are given in Fig. 6.11. State the position of various points with respect to planes of projection.*

Solution:

(a) Point P is 25 mm above H.P. and 50 mm in front of the V.P.

(b) Point Q is 20 mm above H.P. and 40 mm behind the V.P.

(c) Point R is in the H.P. and 50 mm in front of the V.P.

(d) Point S is in the V.P. and 50 mm below the H.P.

(e) Point T is 30 mm below the H.P. and 40 mm behind the V.P.

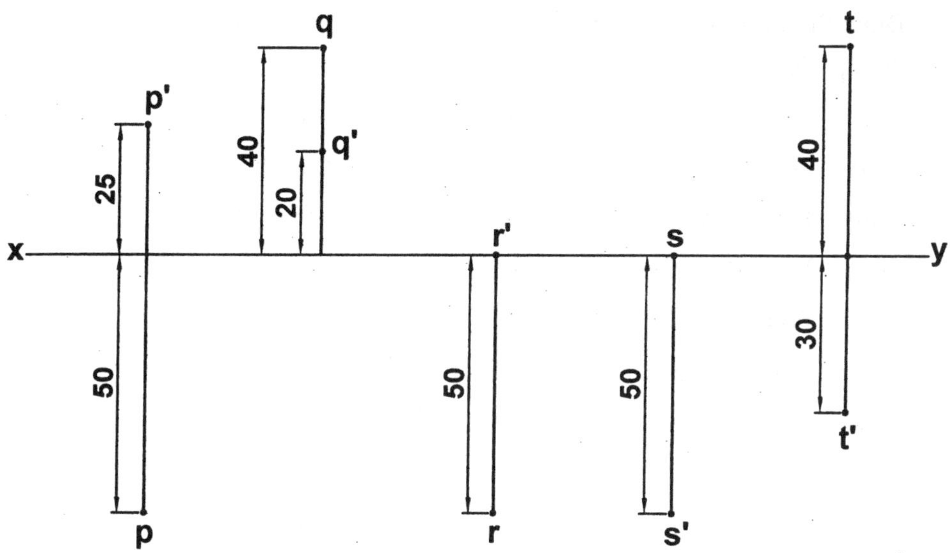

Fig. 6.11

Exercises

1. A point A is 30 mm above the H.P. and 40 mm in front of the V.P. Draw its projections.
2. A point P is 25 mm above the H.P. and 35 mm behind the V.P. Draw its projections.
3. A point B is 20 mm below H.P. and 35 mm behind the V.P. Draw its projections.
4. A point Q is 25 mm below the H.P. and 30 mm in front of V.P. Draw its projection.
5. A point P is 50 mm from both the principal planes. Draw its projections considering all the four quadrants.
6. A point A is 20 mm above the H.P. and 25 mm in front of V.P. Another point B is 25 mm behind the V.P and 45 mm below the H.P. Draw the projections of A and B keeping the projectors 50 mm apart.
7. A point P is 40 mm above the H.P. and 60 mm in front of V.P. Another point Q is in H.P. and 60 mm behind the V.P. The distance between their projectors is 90 mm. Draw the projections of the points. Also draw straight line joining their top and front views.

8. Draw the projections of the following points on the same reference line keeping their projectors 40 mm apart.
   (i) Point A is 30 mm above the H.P. and 50 mm in front of V.P.
   (ii) Point B is 20 mm above the H.P. and 45 mm behind the V.P.
   (iii) Point C is 30 mm below H.P. and 30 mm behind the V.P.
   (iv) Point D is 25 mm below the H.P. and 40 mm in front of V.P.
   (v) Point E is in the H.P. and 50 mm behind the V.P.
   (vi) Point F is in the V.P. and 45 mm below the H.P.
   (vii) Point G is in both the H.P. and V.P.

CHAPTER 7

Projections of Straight Lines

7.1 Introduction

A straight line can be defined as the shortest distance between any two points. The projections of straight line are drawn by joining the respective projections of its ends. For obtaining front view and top view of any given line, first of all line is assumed to be situated in a given quadrant or position, then the view which shows true length, is drawn first. Other view is obtained by projecting the first view.

7.2 Positions of Straight Lines

A line can have following positions with respect to two reference planes (i.e., Horizontal and Vertical planes).

1. Straight line parallel to both H.P. and V.P.
2. Straight line parallel to V.P. and perpendicular to H.P.
3. Straight line parallel to H.P. and perpendicular to V.P.
4. Straight line inclined to H.P. and parallel to V.P.
5. Straight line inclined to V.P. and parallel to H.P.
6. Straight line contained by one or both the planes.
7. Straight line inclined to both H.P. and V.P.
8. Straight line contained by a plane perpendicular to both the planes. (Profile Plane)

The procedures of obtaining projection of a straight line in different positions (as mentioned above) have been explained in the following sections.

7.3 Straight Line Parallel to both H.P. and V.P.

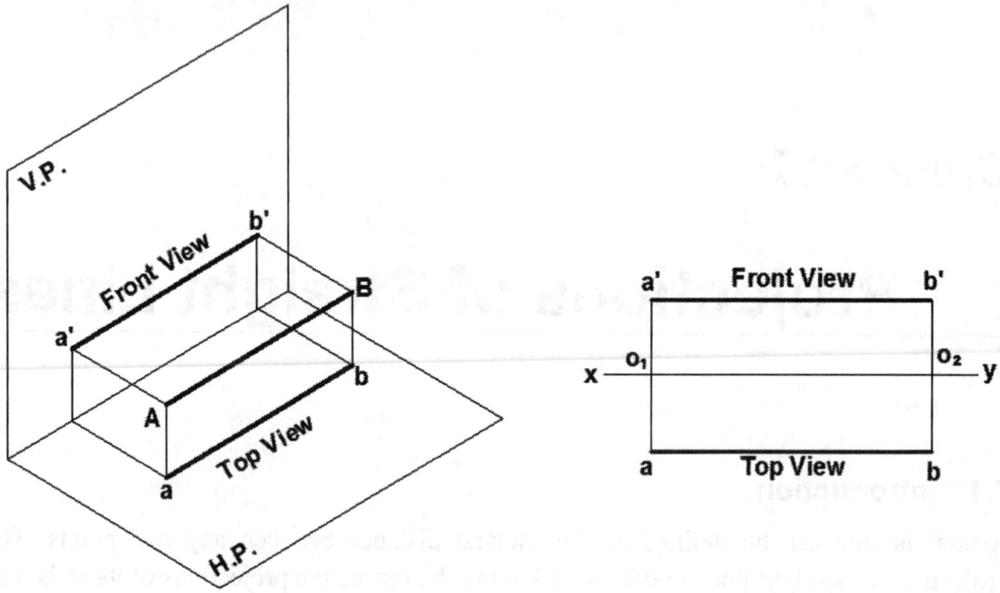

Fig. 7.1

Fig. 7.1 illustrates pictorial representation of a line AB parallel to both H.P. and V.P. The line is situated in first quadrant. In this case, both front view (a'b') and top view (ab) are parallel to line AB and their lengths are equal to true length of the line AB.

Problem 7.1 (Fig. 7.2): *A straight line AB, 55 mm long is parallel to both Horizontal Plane (H.P.) and Vertical Plane (V.P.). Its both ends A and B are 30 mm away from the H.P. and V.P. Draw its projections.*

Construction (Fig. 7.2):

1. First of all draw a reference line xy.
2. Mark a' and a as front view and top view respectively, 30 mm above and below the xy line on a vertical projector through o_1.These points represent the projections of end A of the line.
3. Mark point o_2 on xy line such that o_1o_2 = 55 mm and draw a vertical projector through point o_2.
4. On this vertical projector, mark points b' and b, 30 mm above and below the xy line, which represent projections of end B of the line.
5. Join a' to b' and a to b such that a'b' and ab represent front view and top view respectively of the line AB.

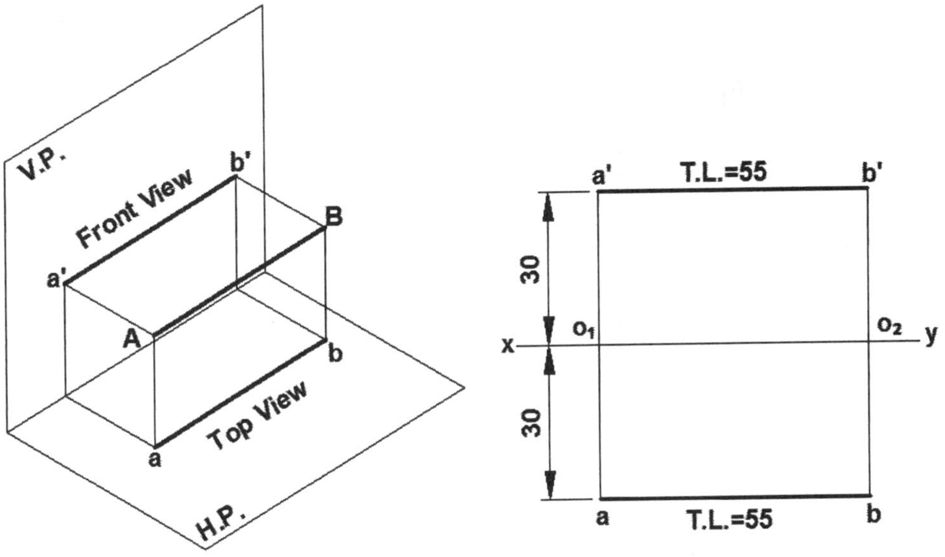

Fig. 7.2

7.4 Straight Line Parallel to V.P. and Perpendicular to H.P.

When a line is perpendicular to one plane, it will be parallel to another plane. Figure 7.3 shows a line PQ parallel to the V.P. and perpendicular to the H.P. Since, line is parallel to the V.P., its projection on V.P. (i.e., front view) represents the true length of the line. In such case, front view (with true length) is drawn first and top view is obtained by projecting the front view to the H.P. In this case p'q' is equal to the true length of the line and top view is a point.

Problem 7.2 (Figs. 7.3, 7.4): *A straight line AB, 50 mm long is parallel and 30 mm away from the V.P. and perpendicular to the H.P. If the end A is 20 mm above the H.P., Draw its projections.*

Construction (Figs. 7.3, 7.4):

1. Draw a reference line xy.
2. Draw a vertical projector at any convenient point say o on the xy line.
3. On this projector, mark point a', 20 mm above xy line and b' such that a' b'=50 mm. The line a' b' represents front view of the line AB.
4. On the vertical projector, mark a point 30 mm below the xy line which is the required top view.

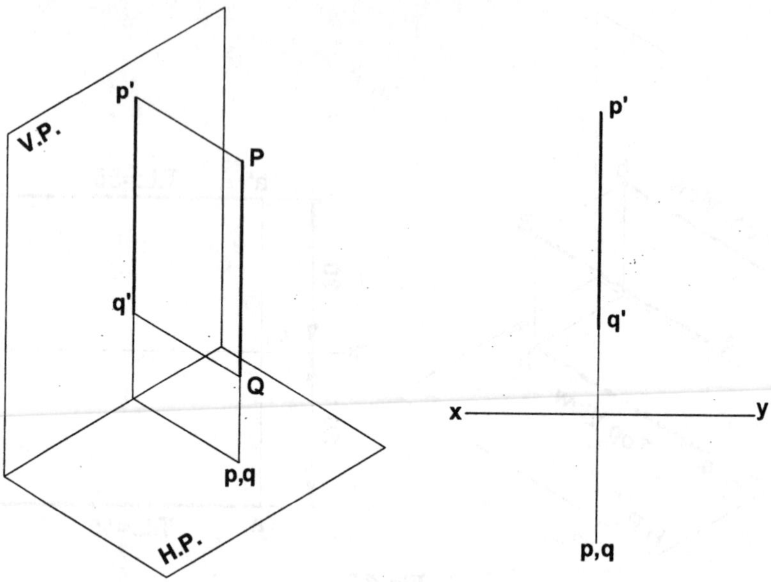

Fig. 7.3

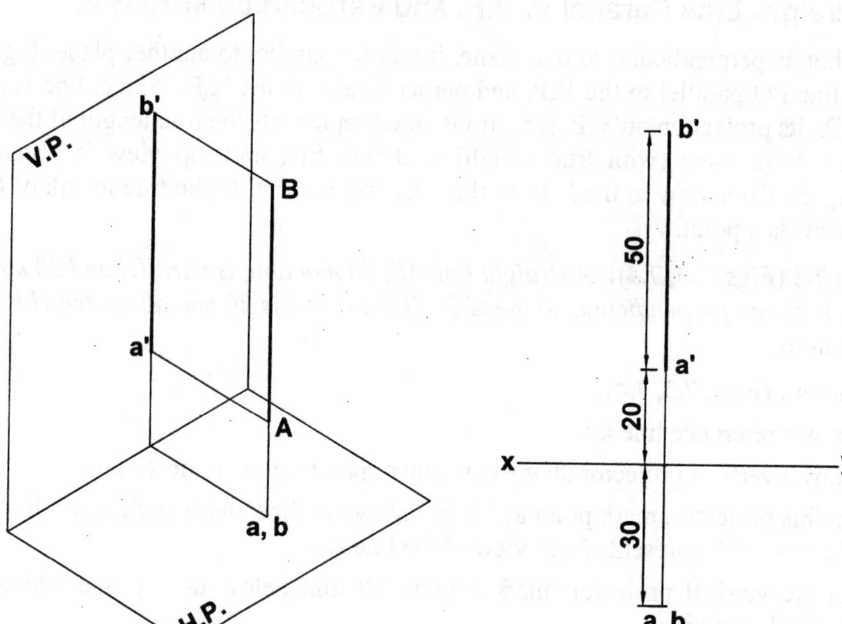

Fig. 7.4

7.5 Straight Line Parallel to H.P. and Perpendicular to V.P.

A line PQ, perpendicular to the V.P. and parallel to the H.P. is shown in Fig. 7.5. Since, the line is parallel to the H.P., its top view represents true length of the line PQ. The front view is a point.

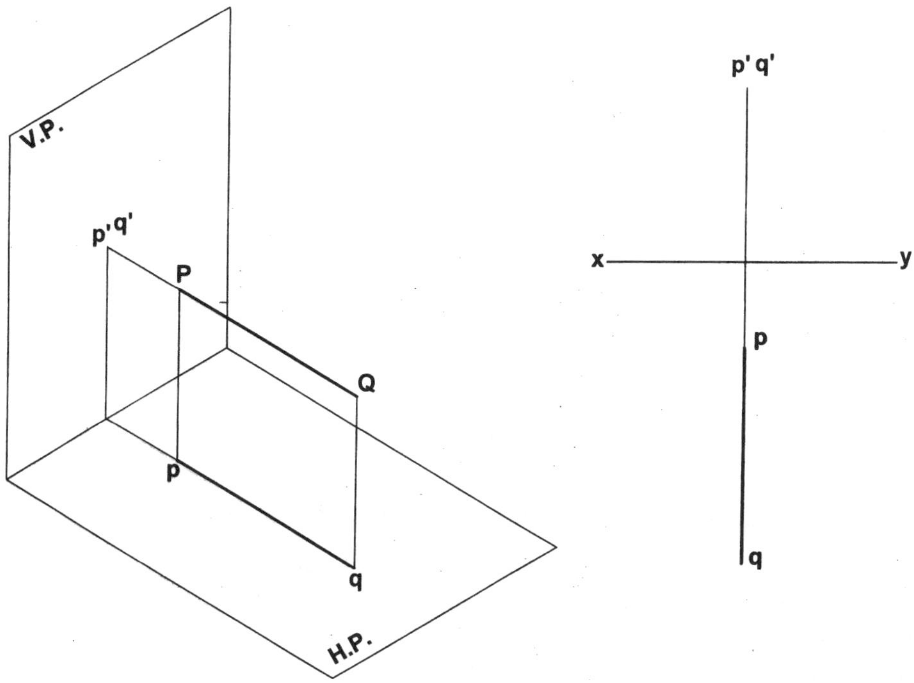

Fig. 7.5

Problem 7.3 (Fig. 7.6): *A 50 mm long line AB is parallel to the H.P. and perpendicular to the V.P. End A is 40 mm above H.P. and 20 mm in front of the V.P. Draw its projections.*

Construction (Fig. 7.6):

1. Draw a reference line xy.
2. Draw a vertical projector at any convenient point on the xy line.
3. Mark a point a', 40 mm above xy line and point a, 20 mm below the xy line.
4. Mark point b, 50 mm away from point a on the same projector.
5. Thus line ab shows the top view and line a'b' represents front view of line AB.

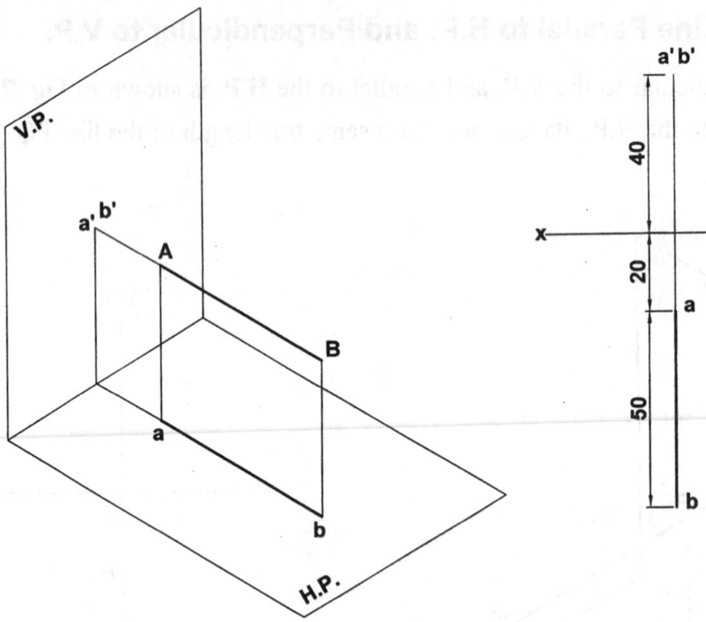

Fig. 7.6

7.6 Straight Line Inclined to H.P. and Parallel to V.P.

Consider a line AB inclined at θ with the H.P. and parallel to the V.P. as shown in Fig. 7.7. Since, the line is parallel to the V.P., its projection on the V.P. (i.e., front view) will have a length which is equal to the true length of the line AB. But its top view is shorter than true length and parallel to xy line.

Problem 7.4 (Fig. 7.8): *A line AB, 70 mm long has its end A is 20 mm above the H.P. and 30 mm in front of the V.P. The line is inclined at 45° to the H.P. and parallel to the V.P. Draw its projections.*

Construction (Fig. 7.8):

1. Draw a reference line xy and at any convenient point O on it, draw a vertical projector.
2. On this vertical projector, mark a', 20 mm above xy line and a, 30 mm below the xy line.
3. Through a', draw a line a' b', 70 mm long and inclined at 45° to the xy line.
4. Draw a vertical projector through b'.
5. Through point a, draw a line parallel to xy line which intersects the vertical projector at point b. Thus lines a' b' and ab are required front view and top view respectively.

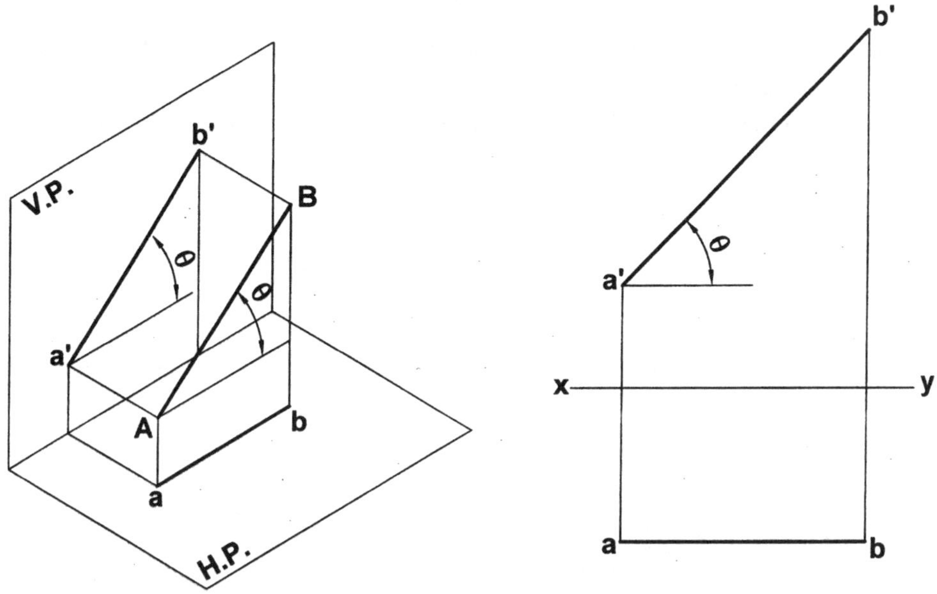

Fig. 7.7

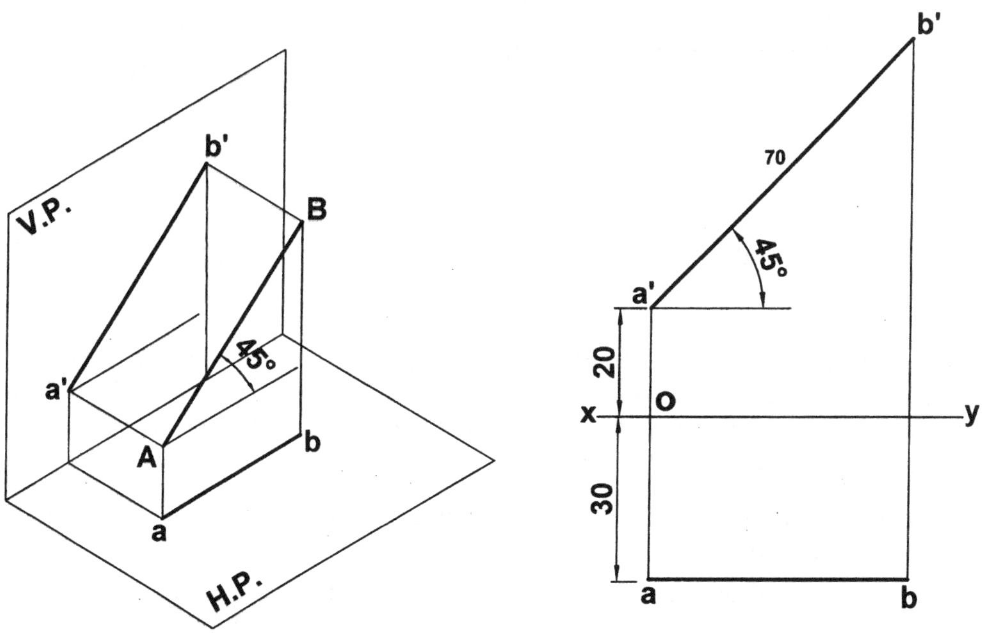

Fig. 7.8

7.7 Straight Line Inclined to V.P. and Parallel to H.P.

Consider a line AB inclined at ø with the V.P. and parallel to the H.P. as shown in Fig. 7.9. In this case, since the line is inclined to the V.P., its front view is shorter than its true length. The top view is a line equal to the true length of the line because the line is parallel to the H.P.

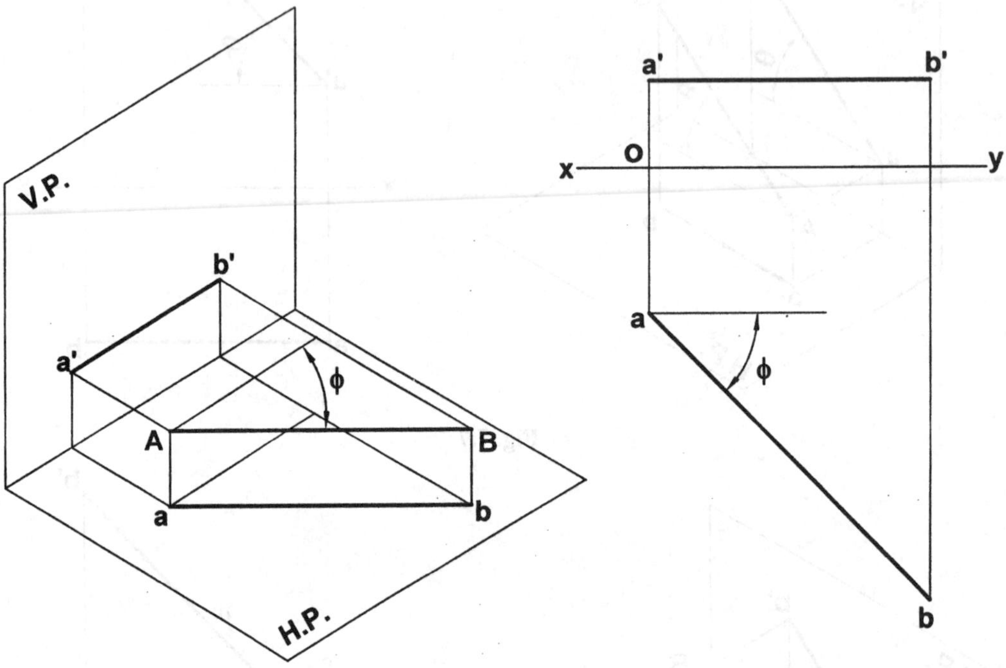

(a) (b)

Fig. 7.9

Problem 7.5 (Fig. 7.10): *The length of front view of a line AB parallel to H.P. and inclined at 45° to the V.P., is 50 mm. One end of the line is 15 mm above the H.P. and 25 mm in front of the V.P. Draw the projections of the line and determine its true length.*

Construction: (Fig. 7.10)

1. Draw a reference line xy and at any convenient point on the reference line, draw a vertical projector.
2. On this vertical projector, mark point a' 15 mm above the xy line and point a, 25 mm below the xy line to represent the projections of end A of the line.
3. Through a', draw a line a'b', 50 mm long and parallel to xy line.
4. Draw a vertical projector through b'.

5. Through point a, draw a line inclined a 45° to xy line, intersecting the vertical projector through b' at b.
6. Thus, line a'b' and ab represent front view and top view respectively of line AB. By measuring the length of line ab, true length of line AB can be obtained.

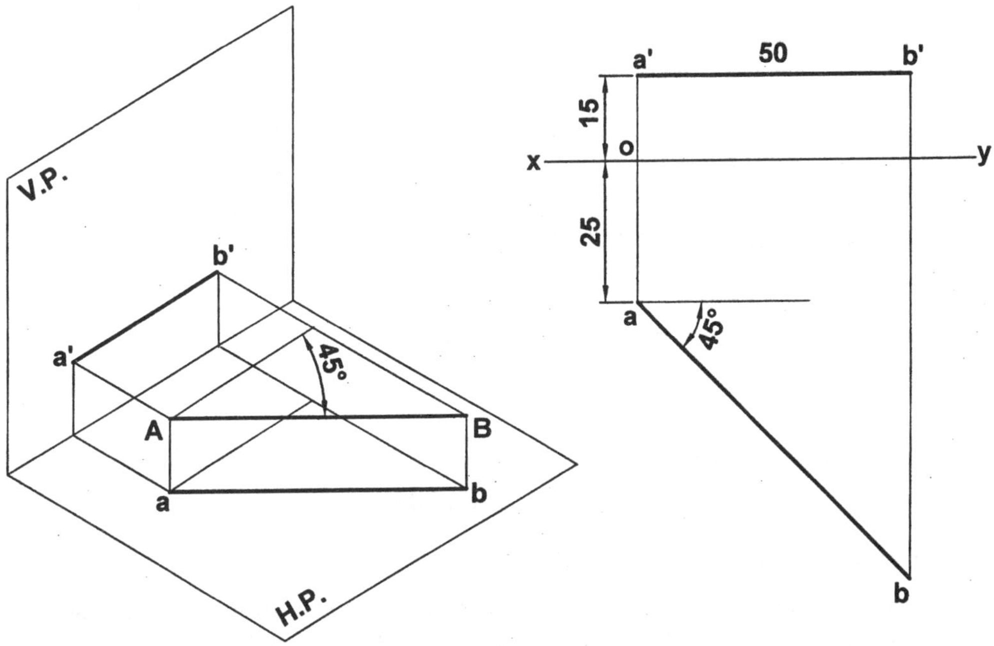

(a) (b)

Fig.7.10

7.8 Straight Line Contained by one or both the Planes

A straight line may be situated in one of the reference planes or both the reference planes. A straight line represents its true length in that plane to which the line is contained. If the line is contained by both the planes, its projections (i.e., front view and top view) will represent the true length and always coincide on the xy line. There could be following three cases.

Case 1: Line is Contained by the H.P.

Since the line PQ is contained by the H.P., its top view will represent its true length but length of the line in front view will be shorter and coincide with xy line as shown in Fig. 7.11.

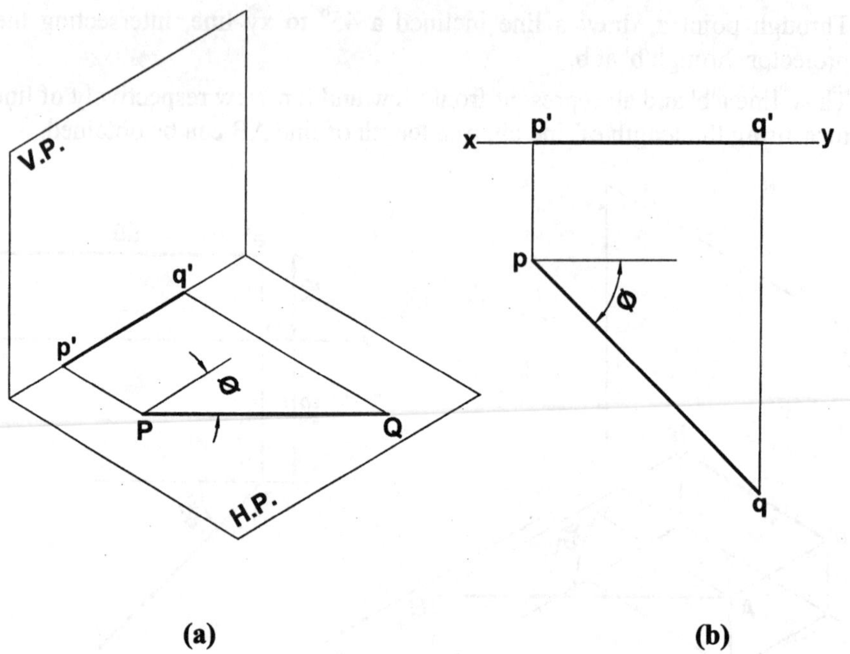

Fig.7.11

Case 2: Line is Contained by the V.P.

In this case, the line is contained by the V.P., therefore its front view represents true length. The length of line in top view is shorter than its true length and coincides with xy line as shown in Fig. 7.12.

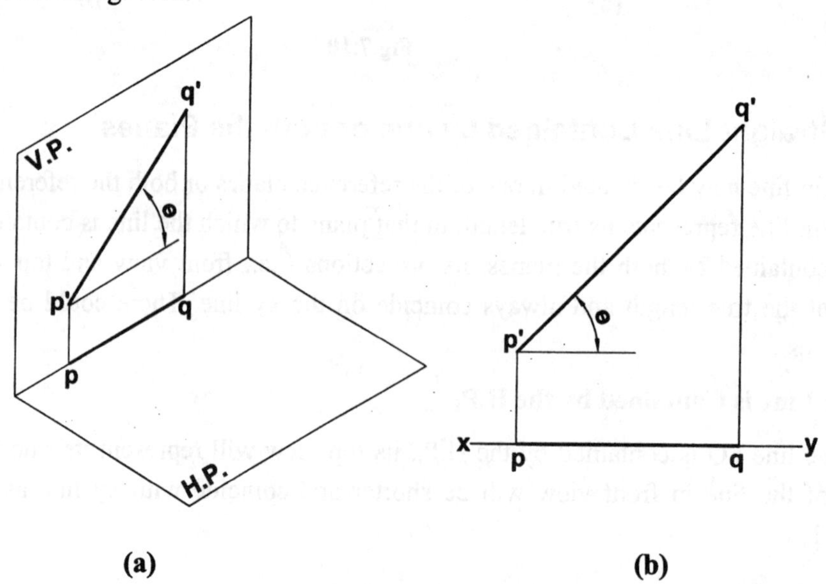

Fig.7.12

Case 3: Line is Contained by both the H.P. and the V.P.

A line PQ contained by both the H.P. and the V.P. is shown in Fig. 7.13. The front view and the top view coincide with xy line and their lengths are equal to true length of the line PQ.

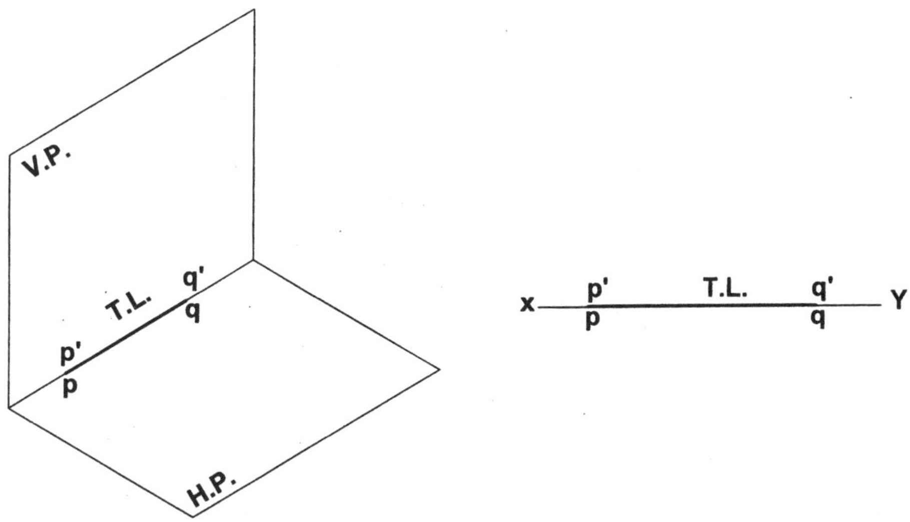

(a) (b)

Fig. 7.13

Problem 7.6 (Fig. 7.14): *A line AB, 65 mm long is contained by the H.P. and is inclined at 45° to the V.P. Its end A is in the H.P. and 25 mm in front of the V.P. Draw its projections.*

Construction (Fig. 7.14):

1. Draw a reference line xy.
2. Mark a point a' on xy line and a, 25 mm below xy line on a vertical projector through a'.
3. Through point a, draw a line ab, 65 mm long and inclined at 45° to the xy line. Line ab represents top view of the line AB.
4. Draw a vertical projector through b, intersecting the xy line at b'.
5. The line a'b' represents the front view.

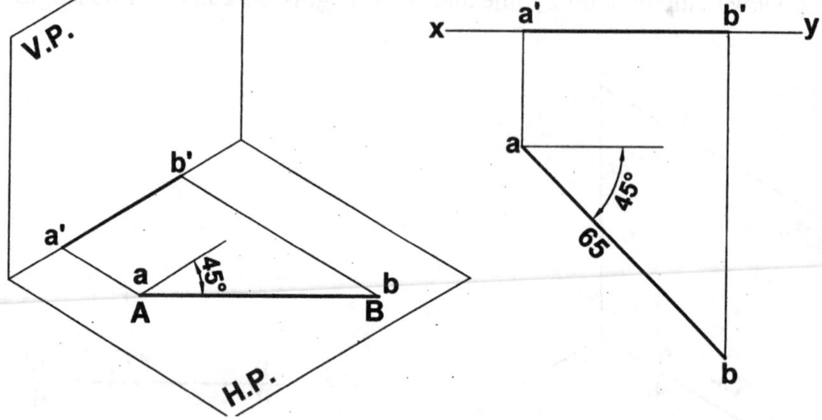

(a) (b)

Fig. 7.14

Problem 7.7 (Fig.7.15): *A line AB, 65 mm long is situated in the V.P. and is inclined at 30° to the H.P. The end A is 30 mm above the H.P. Draw its projections.*

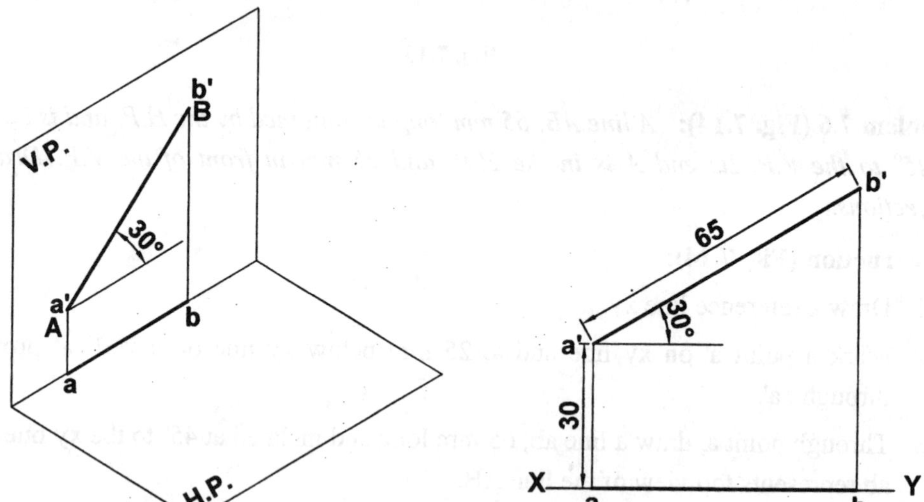

(a) (b)

Fig. 7.15

Construction (Fig. 7.15):

1. Draw a reference line xy.
2. Mark a point a on the xy line and a', 30 mm above xy line.
3. Through a', draw a line a'b', 65 mm long and inclined at 30° to the horizontal.
4. Drop a vertical projector through b' intersecting the xy line at b.
5. Thus line a'b' represents front view with true length and ab represents top view.

Problem 7.8 (Fig. 7.16): *Draw the projections of a 70 mm long line AB contained by both the H.P. and the V.P.*

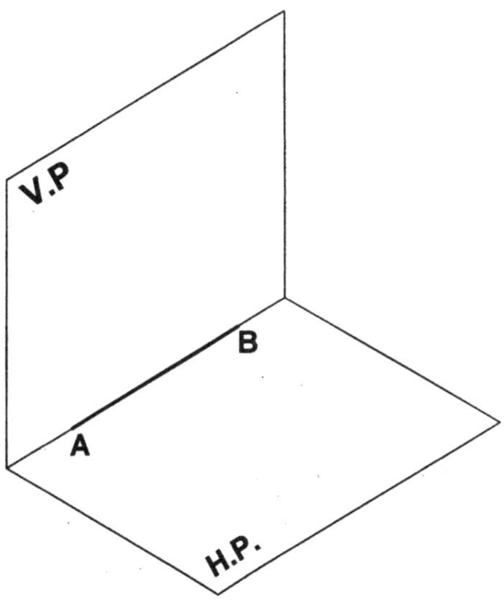

(a)

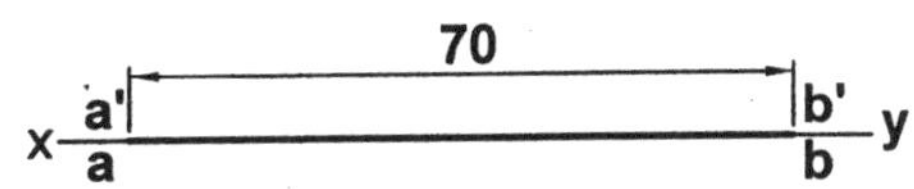

(b)

Fig. 7.16

Construction (Fig. 7.16):

1. Draw a reference line xy.
2. Mark a' and a coinciding each other on the xy line.
3. Similarly mark b' and b coinciding each other at a distance of 70 mm from point a and a'.
4. Thus, a'b' and ab represent front view and top view respectively which coincide each other on the xy line.

7.9 Projections of Straight Lines – at a Glance

Projections of lines in different positions are summarized in Table 7.1 as follows:

Table 7.1 Summary of Projections of Lines

S.No.	Position of Straight Line (AB)	Front View (FV)	Top View (TV)
1	Line parallel to both H.P. & V.P. 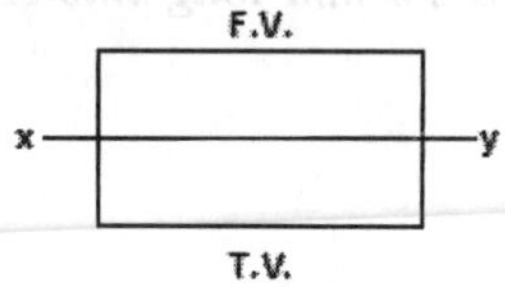 **Fig.7.17 (a)**	A line Parallel to xy line having True Length (T.L.)	A line Parallel to xy line having True Length (T.L.)
2	Line perpendicular to the H.P. 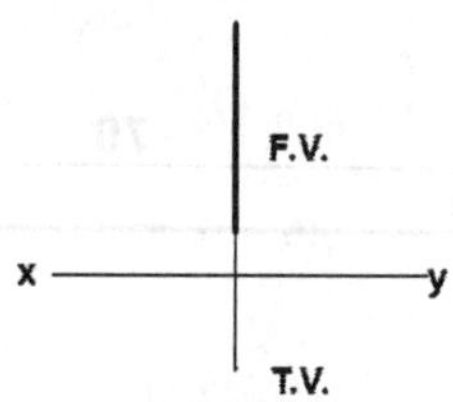 **Fig.7.17 (b)**	A line perpendicular to xy line having T.L.	A point
3	Line perpendicular to the V.P. 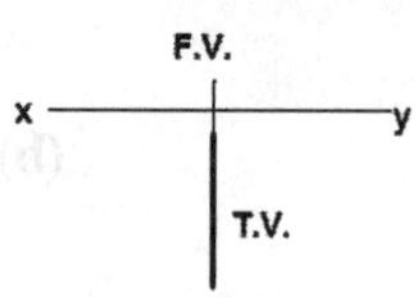 **Fig. 7.17 (c)**	A Point	A line perpendicular to xy line having T.L.
4	Line inclined to the H.P. & Parallel to the V.P. 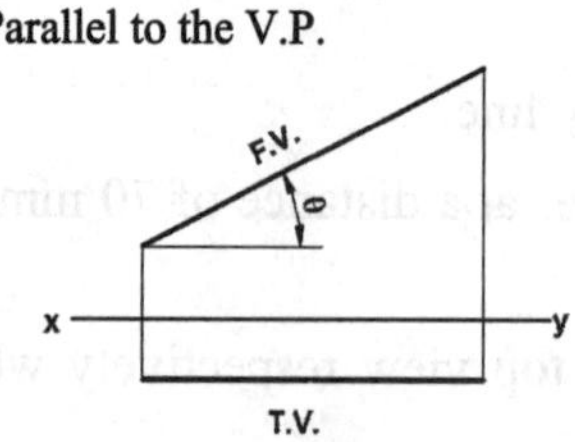 **Fig. 7.17 (d)**	A line inclined to xy line having T.L.	A line parallel to xy line & shorter than T.L.

Table 7.1 ***Contd...***

S.No.	Position of Straight Line (AB)	Front View (FV)	Top View (TV)
5	Line inclined to the V.P. & Parallel to the H.P. 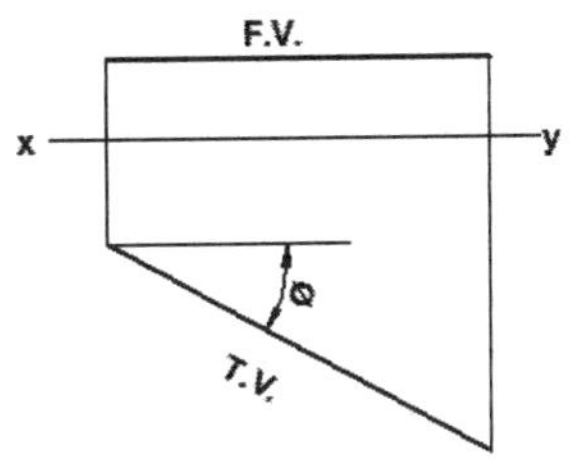 Fig. 7.17 (e)	A line parallel to xy line & shorter than T.L.	A line inclined to xy line having T.L.
6	Line contained by H.P. 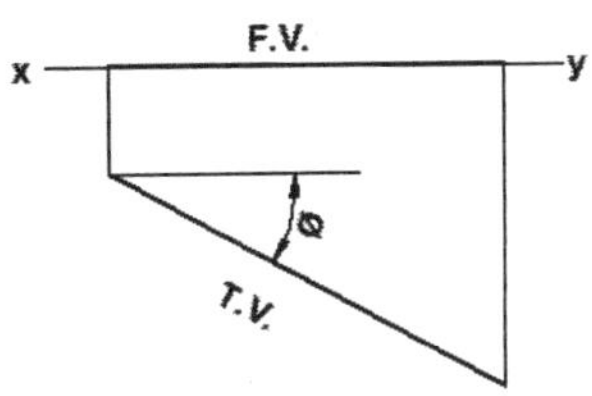 Fig. 7.17 (f)	A line coinciding with xy line & shorter than T.L.	A line inclined to xy line having T.L.
7	Line contained by V.P. 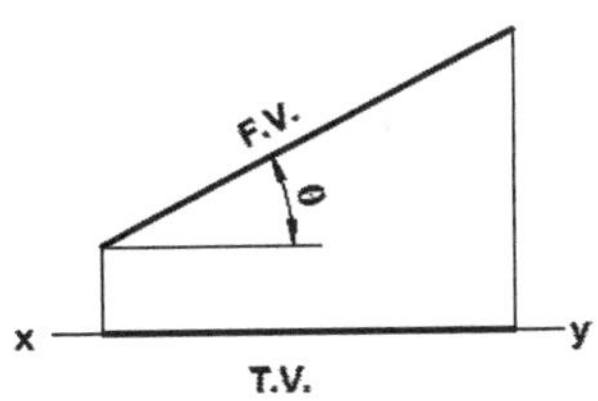 Fig. 7.17(g)	A line inclined with xy line	A line coinciding with xy line & shorter than T.L.
8	Line contained by H.P. and V.P. 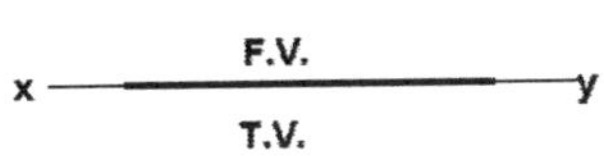 Fig. 7.17(h)	A line coinciding with xy line & having T.L.	A line coinciding with xy line & having T.L.

7.10 Things to Remember

1. A straight line represents its true length in that plane to which it is either parallel or contained. For Example: If a line is parallel to H.P., then its projection on the H.P. (i.e., top view) will represent true length.

2. The projection of a straight line on a plane to which it is inclined will have length shorter than its true length.
3. When a straight line is inclined to both the H.P. and the V.P., its projections are shorter than its true length but their angle of inclinations (apparent angles) with xy line are greater than the true inclination.
4. If the end position of a line is not given, then it is a common practice that one end is assumed to be on the xy line.
5. If the sum of inclination angles with the H.P. and the V.P. (i.e., θ & ø) is equal to 90° its front and top view will appear perpendicular to the xy line. The front view and top view will be contained by a plane perpendicular to H.P. and V.P. (Profile plane)

7.11 Straight Line Inclined to both H.P. and V.P.

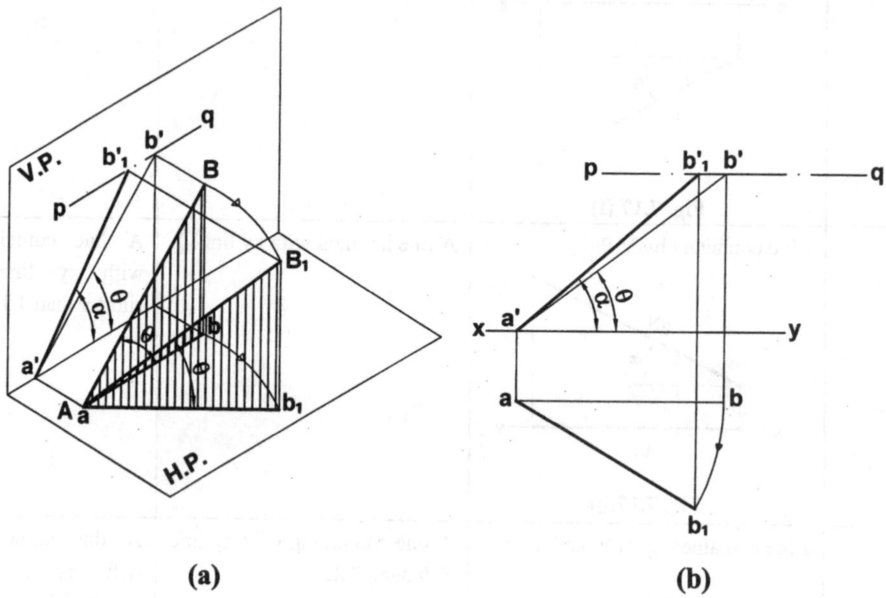

Fig. 7.18

Consider a case when a line AB, parallel to the V.P. and inclined at θ to the H.P. (in first quadrant) as shown in Fig. 7.18. The end A is in the H.P. The line AB can also be considered as hypotenuse of a right angle triangle ABb. In this position, a'b' and ab represent front view and top view of line AB respectively. Since, line is parallel to the V.P. its front view shows true length but top view is shorter than AB and parallel to xy line.

Now keeping the angle θ constant if the end B is rotated about end A to a position say B_1, then the line AB becomes inclined to V.P. also or we can say that the line is inclined to both the H.P. and the V.P. For this new position of line (say AB_1) we observe the following points:

1. In the front view b' will move to b_1' along a straight line pq parallel to xy line. This line pq is known as locus of end B in the front view. The new front view is shorter than AB and makes an angle of α with xy line.
2. In the top view, b will move to b_1 along an arc (having radius equal to ab and a as centre). The length in the top view (i.e., ab) remains constant for any position of the end B.

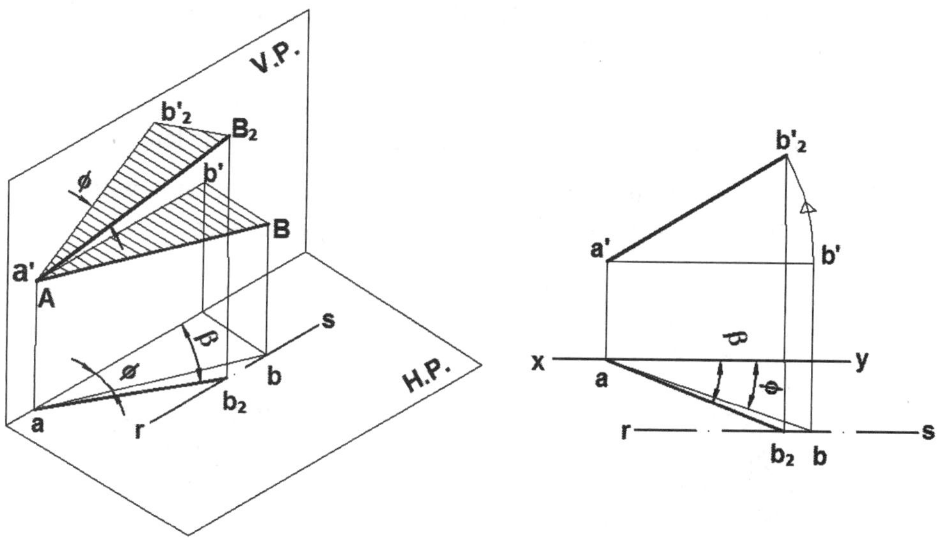

(a) (b)

Fig. 7.19

Consider another case (Fig. 7.19) when the line is parallel to the H.P. and inclined at ø with the V.P. and end A is in the V.P. The line AB can be considered as hypotenuse of a right angle triangle Ab'B making an angle of ø with the base. Front view and top view of line AB are a'b' and ab respectively.

Now, keeping end A fixed and angle ø constant if the end B is rotated about A to a new position say B_2, then the line becomes inclined to H.P. also or we can say that the line AB is inclined to both the planes. We observe the following points:

1. In the front view, b' moves to b_2' along an arc with centre a' and radius equal to a'b'. The length in front view remains constant irrespective of position of line.

2. In top view, b moves to b_2 along a line rs parallel to xy line. The line rs is also known as locus of end B in the top view. The new top view ab_2 is shorter than AB and makes an angle of β with xy line.

Hence, it can be concluded that when a line is inclined to both the H.P. (with θ) and the V.P. (with ø), its projections are shorter than its true length. The angle of inclinations of front view and top view (i.e., α and β) are greater than the true inclinations. The angles α and β are called apparent angles of inclination.

7.12 Projections of Straight Line Inclined to both the Planes when True Length, True Inclinations and Positions of End Points are given and $\theta + \phi \neq 90°$

From the previous section, it has been observed that when a line is inclined to both the H.P. and the V.P., neither front view nor top view represents the true length of a line.

Consider a line AB which is inclined at θ with the H.P. and at ϕ with the V.P. as shown in Fig. 7.20. To draw its projections following steps are taken:

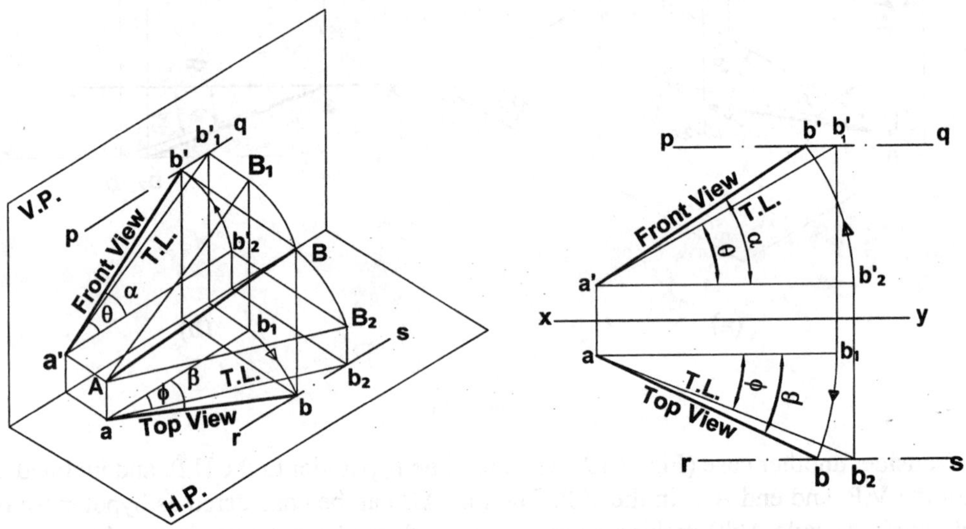

Fig. 7.20

Construction (Fig. 7.20):

1. Mark the projections of end point A (i.e., a' and a as per given position).

2. First of all, assume the line AB inclined at an angle of θ with the H.P. and parallel to the V.P. and draw its projections $a'b_1'$ and ab_1 as front view and top view respectively.
3. Now assume the line AB inclined at an angle of ϕ with the V.P. and parallel to the H.P. and draw its projections $a'b_2'$ and ab_2 as front view and top view respectively.
4. Through b_1', draw a locus (line) pq parallel to the xy line which represents locus of end B in front view.
5. Through b_2, draw a locus (line) rs parallel to xy line which represents locus of end B in top view.
6. With a' as centre and radius equal to $a'b_2'$ (length of front view), draw an arc intersecting the line pq at b'. Join a' and b' then a'b' represents the front view of line AB.
7. Similarly, with a as centre and radius equal to ab_1 (equal to length of top view), draw an arc to intersect rs at b. Join a to b then ab represents top view of the line AB.
8. Draw a vertical projector through b' which should pass through b. This can be used as a check also.

Problem 7.9 (Fig. 7.21): *A straight line AB, 65 mm long is inclined at 30° to the H.P. and 45° to the V.P. The end A is 20 mm above the H.P. and 15 mm in front of the V.P. Draw its projections.*

Construction(Step-by-step Solution):

Step 1 (Refer Fig. 7.21 (a))

Draw a xy line and mark the projections of end A (i.e., a', 20 mm above the xy line and a, 15 mm below the xy line). Assuming the line inclined at 30° to the H.P. and parallel to the V.P., draw its projections as follows:

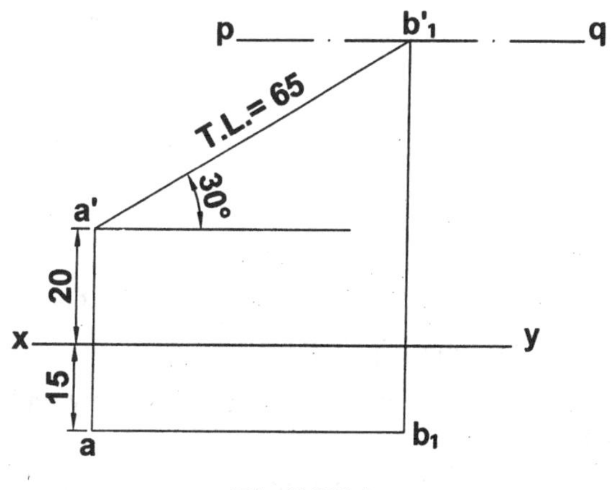

Fig. 7.21(a)

1. Through a', draw a line $a'b_1'$ equal to 65 mm and inclined at 30° to the xy line.
2. Through b_1', draw a horizontal locus line pq.
3. Draw a vertical projector through b_1'.
4. Through a, draw a horizontal line to intersect vertical projector through b_1' at b_1.

Here, $a'b_1'$ and ab_1 are front view and top view respectively.

Step 2 (Refer Fig. 7.21 (b))

Assuming the line inclined at 45° to the V.P. and parallel to the H.P., draw its projections as follows:

1. Through a draw a line ab_2 having length equal to 65 mm and inclined at 45° to the xy line.
2. Draw a horizontal line rs through b_2 to represent locus of end B in top view.
3. Draw a vertical upward projector through b_2.
4. Draw a horizontal projector through a' intersecting the vertical projector at b_2'.

   Hence, $a'b_2'$ and ab_2 are front view and top view for the assumed condition.

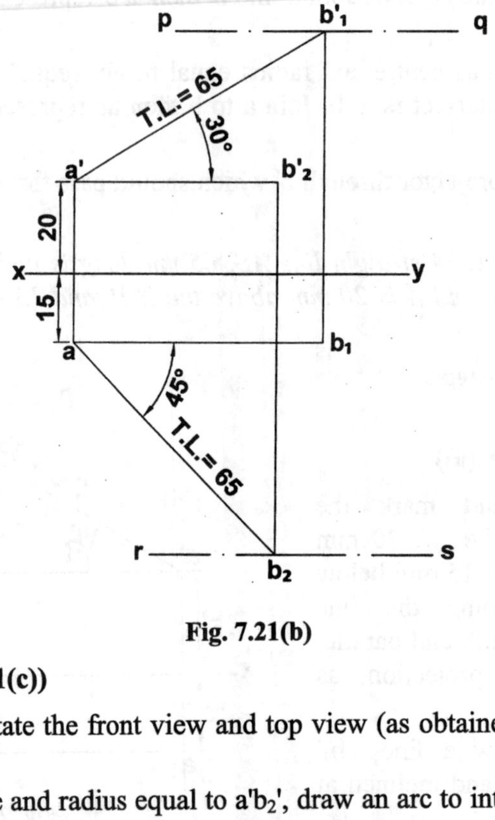

Fig. 7.21(b)

Step 3 (Refer Fig. 7.21(c))

In this step we will rotate the front view and top view (as obtained in previous steps) as follows:

1. With a' as centre and radius equal to $a'b_2'$, draw an arc to intersect the locus line pq at b'. Join a' to b'. The line a'b' is the required front view.
2. With a as centre and radius equal to ab_1, draw an arc to intersect the locus line rs at b.
3. Join a to b. The line ab is the final top view.

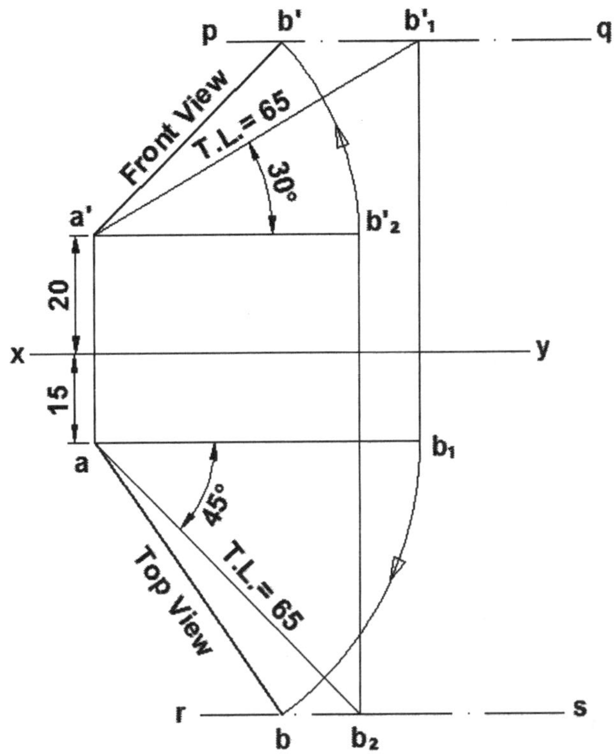

Fig. 7.21(c)

Problem 7.10 (Fig. 7.22): *A line AB 50 mm long has its end A in both the H.P. and the V.P. It is inclined at 30° to the H.P. and 45° to the V.P. Draw its projections.*

Construction (Fig. 7.22):

This problem is similar to the previous problem except the position of end A which is on both the H.P. and the V.P. The procedure for solving this problem will be same as that of problem 7.9.

1. Draw a xy line and mark the projections of end A. (i.e., a' and a coinciding points on xy line)
2. First of all assuming the line AB to be inclined at 30° to the H.P. and parallel to the V.P., draw $a'b_1'$ = 50 mm and inclined at 30° to the xy line as front view. Through b_1', draw the locus line pq parallel to xy line. Project b_1' on xy line so that ab_1 represents the top view in this assumed condition.

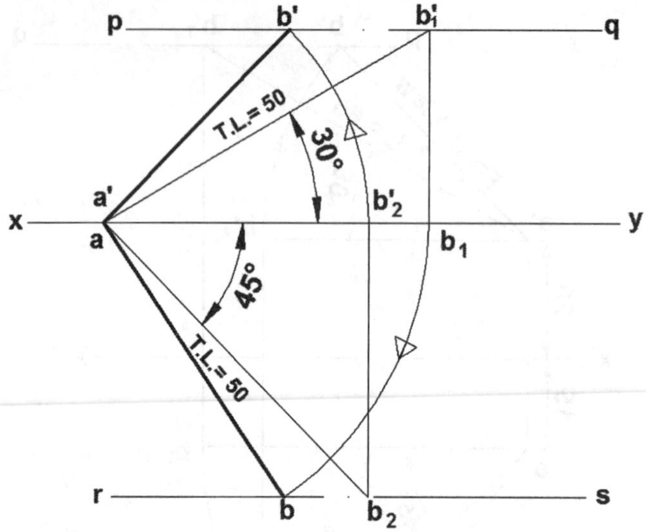

Fig. 7.22

3. Now assuming the line AB to be inclined at 45° to the V.P. and parallel to the H.P., draw ab_2, 50 mm long and inclined at 45° to the xy line. Through b_2, draw locus line rs parallel to xy line. Project b_2 upward on the xy line so that ab_2' represents its front view for the assumed condition.
4. With a' as centre and radius equal to $a'b_2'$, draw an arc to intersect locus line pq at b'. Draw a'b' as thick line to represent front view.
5. With a as centre and radius equal to ab_1, draw an arc to intersect locus line rs at b. Draw ab as thick line which is the required top view.

7.13 Projection of Straight Line Inclined to both the Planes when True Length, True Inclinations and Position of End Points are given and $\theta + \phi = 90°$

When a line is inclined to both the H.P. and the V.P. such that $\theta+\phi = 90°$, then such line is contained by a plane which is perpendicular to both the H.P. and the V.P. (also known as profile plane). The projection of such line can be drawn by following two methods.

Method I- By making the view parallel to xy line

Method II- By using profile plane

These methods are explained with the help of following examples.

Problem 7.11 (Fig. 7.23): *A line AB 90 mm long has its end A, 15 mm above the H.P. and 65 mm in front of the V.P. The line is inclined at 60° to the H.P. and 30° to the V.P. Draw its projections.*

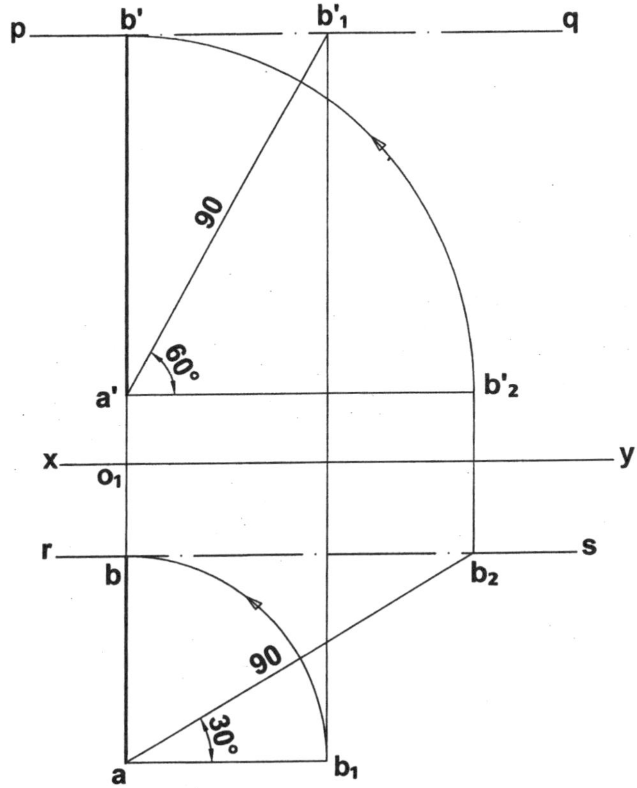

Fig. 7.23

Method-I (Fig. 7.23):

1. Draw a reference line xy.
2. Mark a point o_1 on the xy line and draw a perpendicular line through it.
3. On this line, mark a', 15 mm above and a, 65 mm below the xy line.
4. Draw a line $a'b_1'$, 90 mm long and inclined at 60° to the xy line.
5. Through b_1' draw a horizontal locus line.
6. Through b_1' draw a vertical projector and through a draw a horizontal projector. Mark b_1 at the intersection of these two lines.
7. Draw ab_2, 90 mm long and inclined at 30° to the horizontal.

8. Draw a horizontal line through a' and vertical line through b_2. Mark b_2' at the intersection of these lines.
9. Draw a horizontal locus line through b_2.
10. Now, with a as centre and radius equal to ab_1, draw an arc to intersect locus line through b_2 at b. Join a to b. Line ab is the required top view.
11. With a' as centre and radius equal to a'b_2', draw an arc to intersect horizontal locus line through b_1' at b'. Join a' to b', which is the required front view.

Method-II (Fig. 7.24)

In this method, side view of line is drawn first which is obtained on a profile plane. Front view and top view are obtained by projecting the side view horizontally and vertically downward. Construction steps are as follows:

1. Draw a xy line and a vertical reference line x_1y_1 perpendicular to xy line.
2. Mark a' 15 mm above xy line and a, 65 mm below the xy line.
3. Draw a horizontal line through a'. Also draw a horizontal line through a to meet x_1y_1 line at r. Now with o as centre and radius equal to 65 mm, draw an arc to meet xy line at a point r_1. Draw a vertical line through r_1 to meet horizontal line through a' at a". The point a" represents side view of end A.

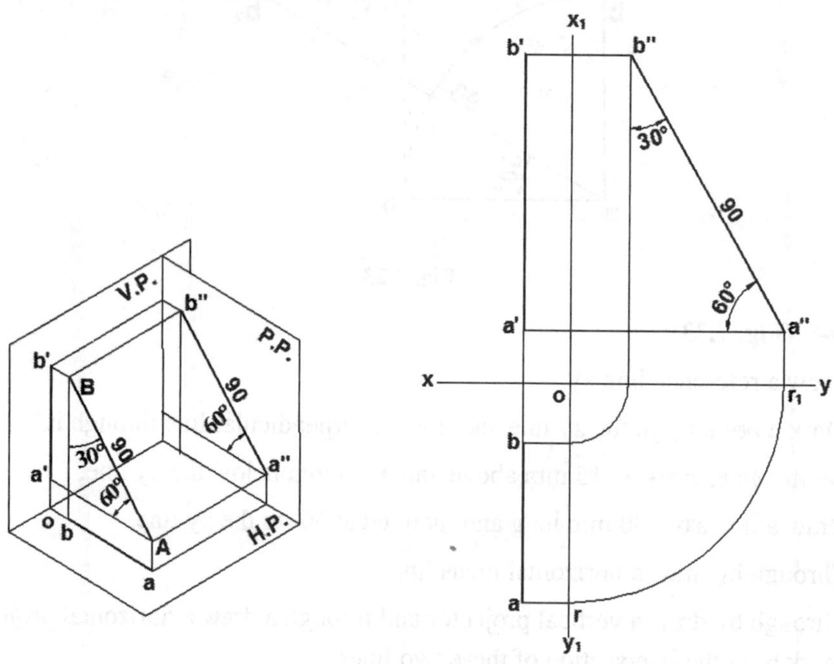

(a) (b)

Fig. 7.24

4. At a" draw a line a"b", 90 mm long and inclined at 60° to horizontal. The line a"b" represents side view of AB.
5. Through b", draw a horizontal and vertical projections to obtain b and b'.
6. Join a' to b' and a to b by thick lines. Thus, a'b' and ab represent front view and the top view of line AB respectively.

7.14 Determination of True Length and True Inclination of a Line when its Projections are given

In the previous section it has been discussed that when a line is inclined to both the H.P. and the V.P., neither its front view nor its top view shows true length and true inclinations with the reference planes. It has also been observed that when a line is parallel to a reference plane, its projection on that plane represents its true length and true inclination. Keeping these concepts in mind, true length and true inclinations of a line can be determined. Following methods are useful to determine true length and true inclination of a line.

Method –I (By making view parallel to the reference line)

In this method, given front view or top view is rotated about one of its ends to make it parallel to the reference line. Other view is projected from it.

Method –II (Trapezoid Method)

In this method, the line is rotated about its projection to bring it in the H.P. or V.P. where it gives true length and true inclination with the reference plane.

These methods are discussed with the help of following examples.

Problem 7.12 (Figs 7.25 & 7.26): *A straight line AB has its end A 15 mm above the H.P. and 25 mm in front of the V.P. and end B is 70 mm above the H.P. and 60 mm in front of the V.P. Its end projectors are 60 mm apart. Draw its projections and determine its true length and true inclination with reference planes.*

Construction Method –I (Refer Fig. 7.25):

(By making view parallel to the reference line)

1. Draw a reference line xy. On the reference line, mark two points r_1 and r_2, 60 mm apart.
2. Draw two vertical projectors through r_1 and r_2.
3. To draw the projections of end A, mark a' 15 mm above and a 25 mm below the reference line on the vertical projector through r_1.
4. Similarly, to draw projections of end B, mark b', 70 mm above and b, 60 mm below the xy line on the vertical projector through r_2.

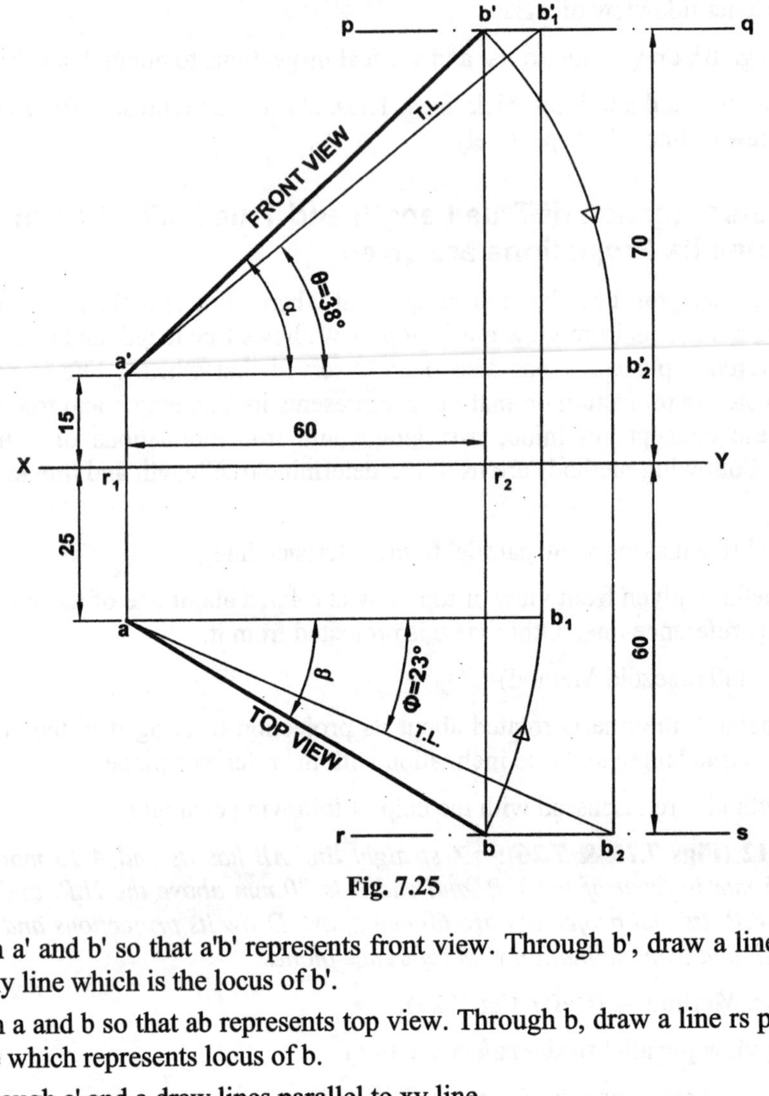

Fig. 7.25

5. Join a' and b' so that a'b' represents front view. Through b', draw a line pq parallel to xy line which is the locus of b'.
6. Join a and b so that ab represents top view. Through b, draw a line rs parallel to xy line which represents locus of b.
7. Through a' and a draw lines parallel to xy line.

To make front view parallel to xy line

8. Now with a' as centre and radius equal to a'b', draw an arc to intersect the horizontal line through a' at b_2'. Through b_2', draw a vertical projector to intersect locus rs at b_2.
9. Join a to b_2. The line ab_2 represents true length of the line AB. The angle ϕ is equal to the true inclination of line AB with the V.P. which is equal to 23° (by measurement).

To make top view parallel to xy line

10. With a as centre and radius equal to ab, draw an arc to intersect horizontal line through a at b_1. Through b_1, draw a vertical projector to intersect locus pq at b_1'.

11. Join a' to b_1'. The line $a'b_1'$ is the required true length and θ is true inclination with the H.P. which is equal to 38° (by measurement). The angle α and β are called as apparent angles of inclination with H.P. and V.P. respectively.

Construction Method –II (Fig. 7.26):
(Trapezoid Method)

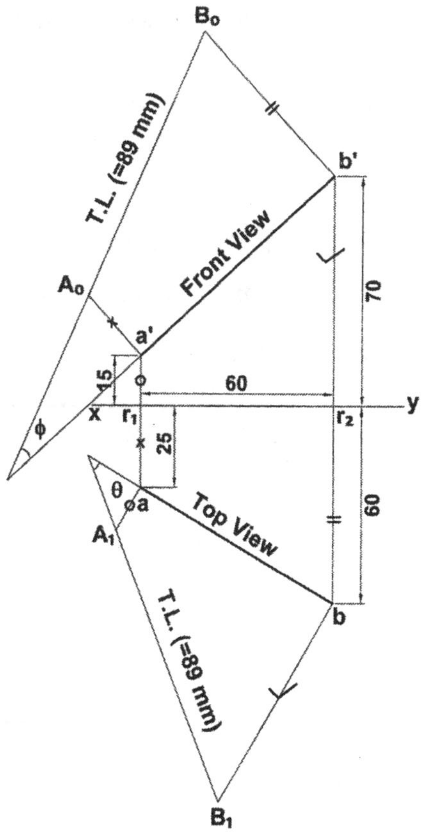

Fig. 7.26

1. Draw front view a'b' and top view ab as per steps 1-6 of Method-I.
2. Through a' and b' draw perpendicular $a'A_o$ and $b'B_o$ such that $a'A_o = r_1a$ and $b'B_o = r_2b$

   Join A_o to B_o such that A_oB_o represents true length of line AB. By measurement A_oB_o is found to be equal to 90 mm and φ to be equal to 25°. The angle φ is the angle of inclination of line AB with the V.P.

3. Through a and b, draw perpendiculars aA_1 and bB_1 such that $aA_1 = r_1a'$ and $bB_1 = r_2b'$.

   Join A_1 to B_1 such that A_1B_1 represents true length of the line AB. By measurement A_1B_1 is found to be equal to 90 mm and θ to be equal to 38°. The angle θ is the required angle of inclination of the line AB with the H.P.

7.15 Traces of Straight Lines

When a line parallel to both the planes is extended, it will never meet the reference planes (H.P. and V.P.). On the other hand, a line inclined to the H.P. will meet the H.P. if produced. Similarly, a line inclined to V.P. will meet the V.P. if produced. Thus, a point of intersection where a straight line or line produced (if necessary) meets the reference plane is called trace of the line.

If the line produced (extended if necessary) meets the H.P., the point of intersection is called as horizontal trace (H.T.) and if it meets the vertical plane, the point of intersection is called vertical trace (V.T.).

7.16 Traces of Lines in Different Positions

Traces of a line in different positions are discussed as follows:

7.16.1 Line Parallel to both H.P. and V.P.

When a line is parallel to both H.P. and V.P., it will never meet the reference planes if produced. So the line in this position will have no traces as shown in Fig. 7.27.

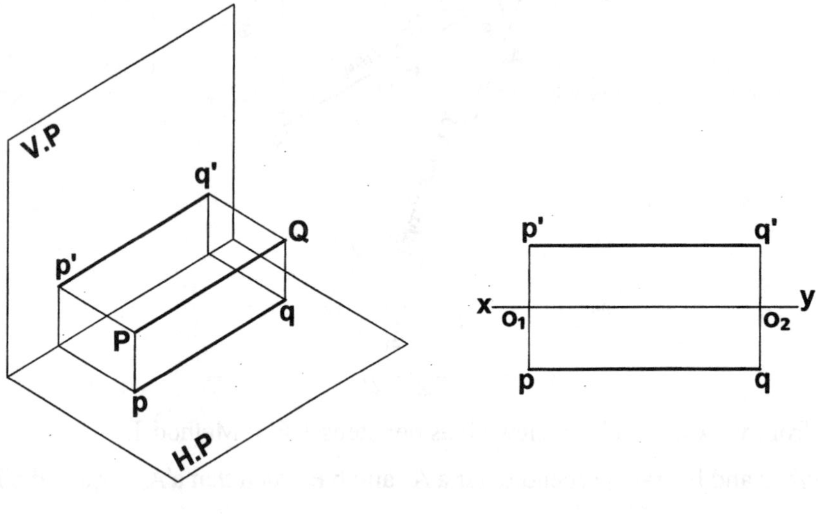

(a) (b)

Fig. 7.27

7.16.2 Line Inclined to H.P. Parallel to V.P.

Consider a line PQ inclined to H.P. and parallel to the V.P. as shown in Fig. 7.28. If the line pq is extended, it will meet the H.P. So the line has Horizontal Trace (H.T.) only.

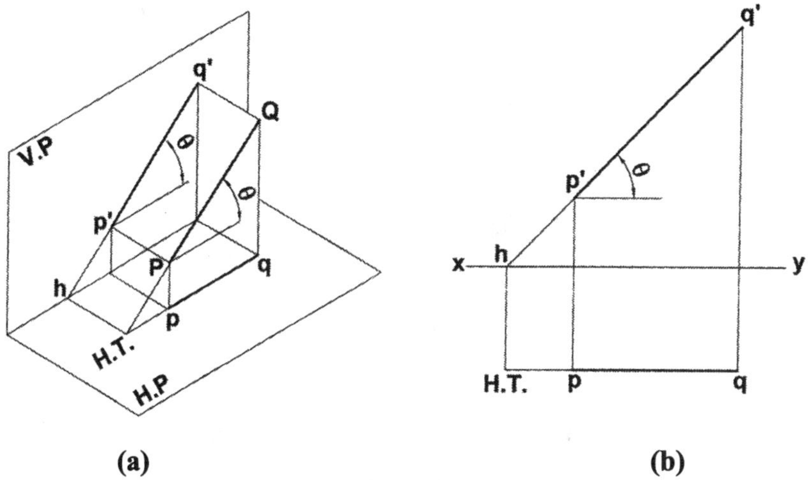

Fig. 7.28

7.16.3 Line Inclined to V.P. and Parallel to H.P.

When a line is inclined to the V.P. and parallel to the H.P., it will intersect at a point on the V.P., if produced. In this case the line will have only V.T. as shown in Fig. 7.29.

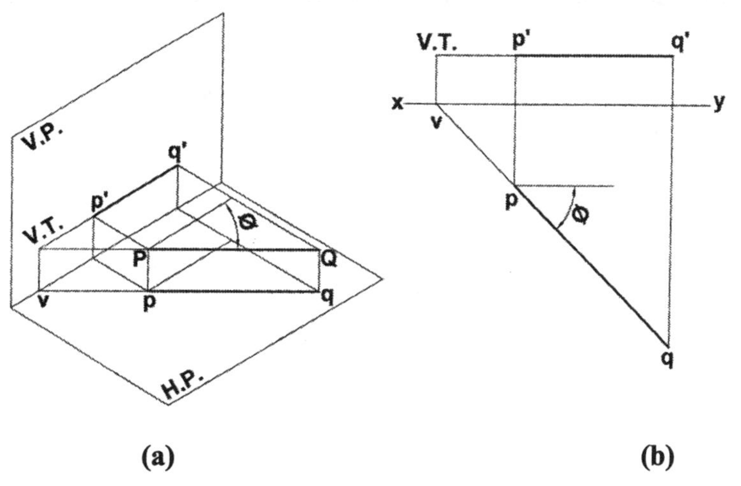

Fig. 7.29

7.16.4 Line Perpendicular to H.P. and Parallel to V.P.

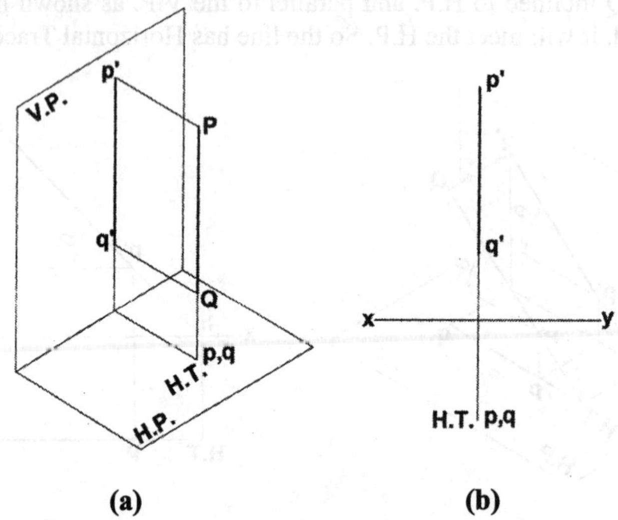

Fig. 7.30

Fig.7.30 shows position of a line PQ which is perpendicular to H.P. and parallel to V.P. If it is extended downward, it will meet the H.P. Thus, the line PQ will have only H.T.

7.16.5 Line Perpendicular to V.P. and Parallel to H.P.

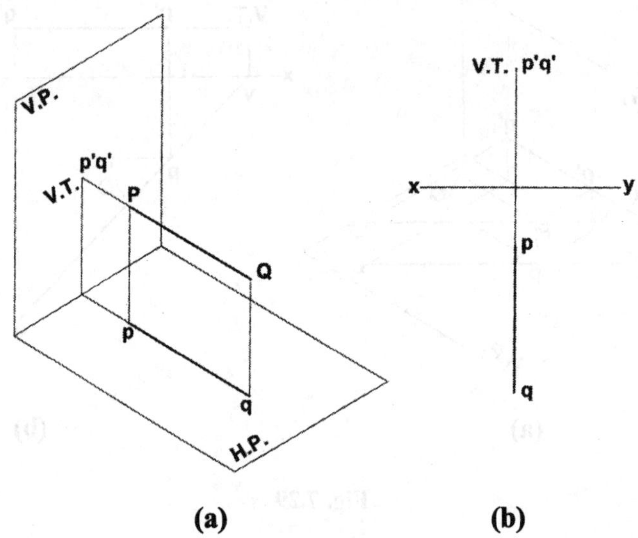

Fig. 7.31

Consider a line PQ parallel to H.P. and perpendicular to V.P. as shown in Fig.7.31. If it is extended, it will meet the V.P. So it has V.T. only.

7.16.6 Line Contained by H.P. and Inclined to V.P.

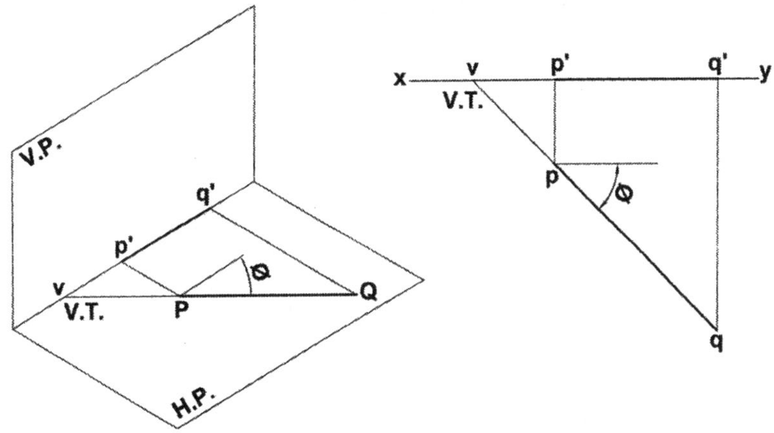

(a) (b)

Fig. 7.32

Fig.7.32 shows a line PQ contained by H.P. and inclined at ϕ with V.P. When the line is extended, it will meet the reference line. So the V.T. is obtained on the xy line. The line has no H.T.

7.16.7 Line Contained by V.P. and Inclined to H.P.

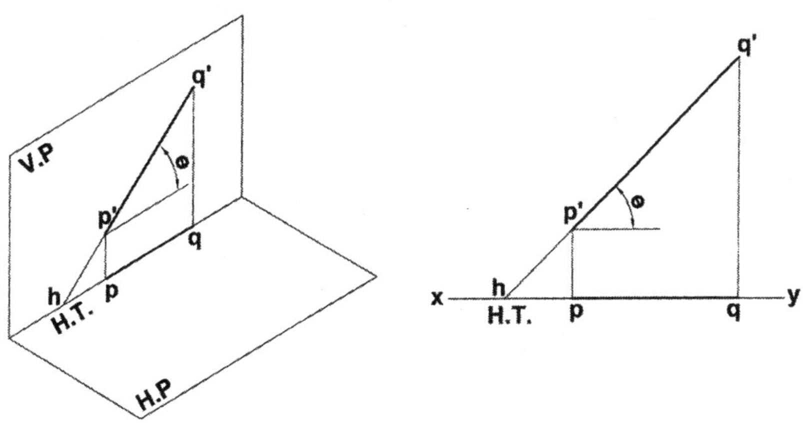

(a) (b)

Fig. 7.33

Fig.7.33 shows a line PQ contained by V.P. and inclined at θ with H.P. When the line is extended, it will meet the reference line. So the H.T. is obtained on the xy line. The line has no V.T.

7.16.8 Line Contained by both H.P. and V.P.

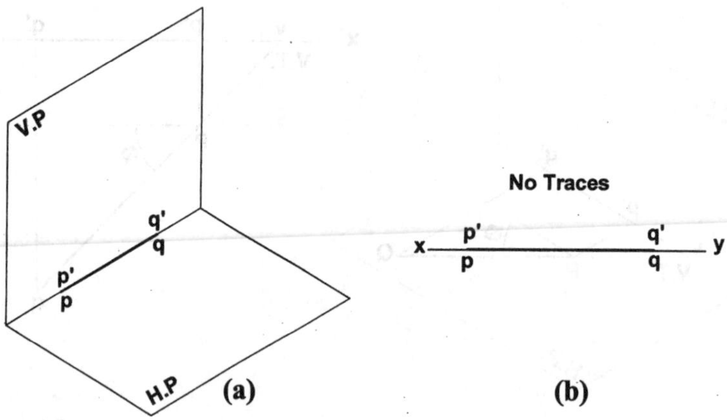

Fig. 7.34

Consider a line PQ contained by both H.P. and V.P. as shown in Fig. 7.34. When it is extended, it will never meet the H.P. or V.P. So, it has no traces.

7.16.9 Line Inclined to both the Planes

Case A: When $\theta + \phi \neq 90^{\circ}$

In this case, since the line is inclined to both the planes, if extended it will meet both the reference planes and therefore, it will have both H.T. and V.T. as shown in Fig. 7.35.

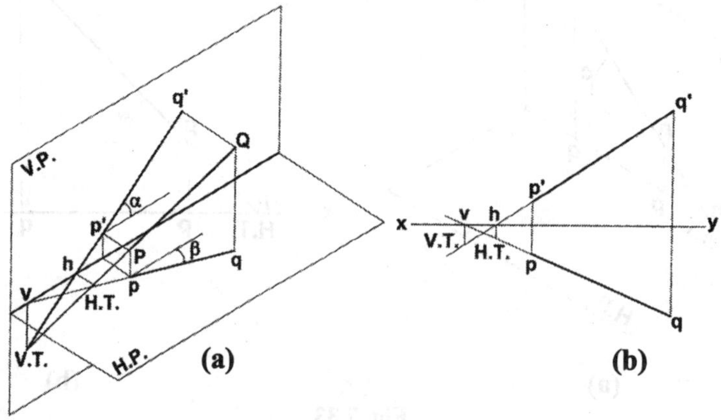

Fig. 7.35

Case B: When $\theta + \phi = 90^\circ$ (The line contained by a profile plane)

As shown in Fig. 7.36, a line PQ is inclined to both the planes with $\theta + \phi = 90^\circ$, the projections and traces of such line lie on a straight line.

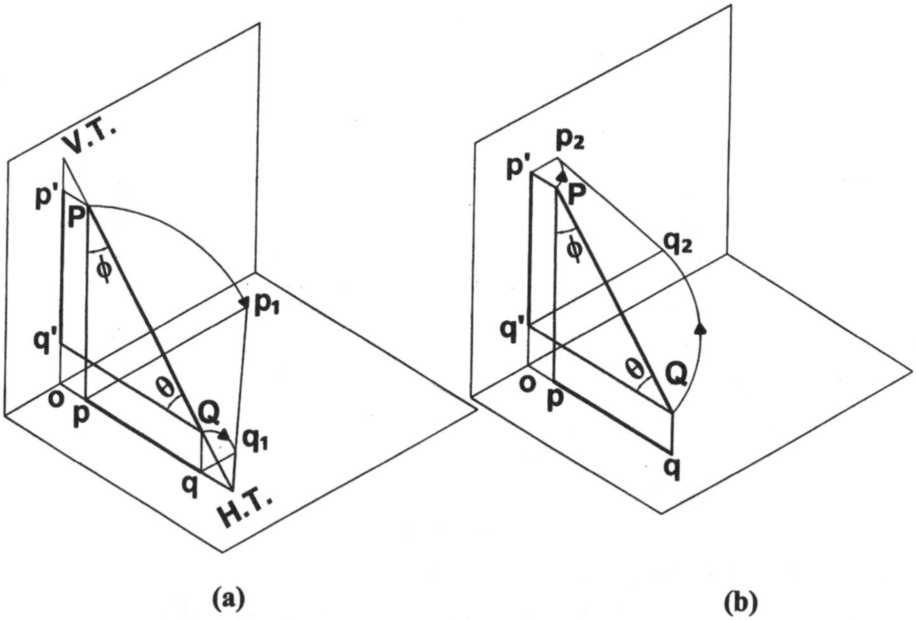

Fig. 7.36

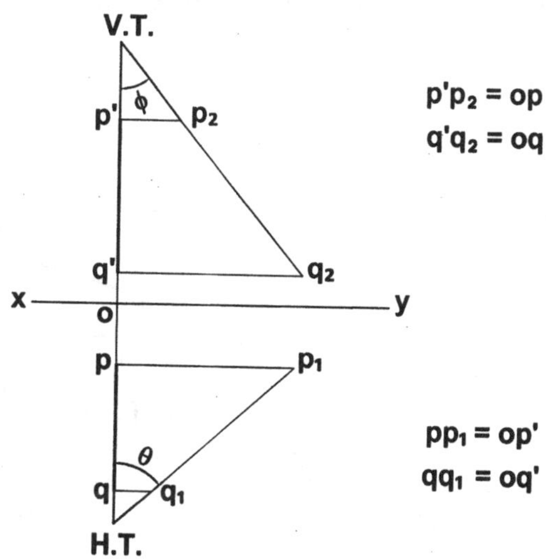

Fig. 7.36 (c)

7.17 Methods of Determining Traces of a Straight Line

Consider a line PQ inclined to both the reference planes. The traces of line PQ can be obtained by following methods.

Method -I

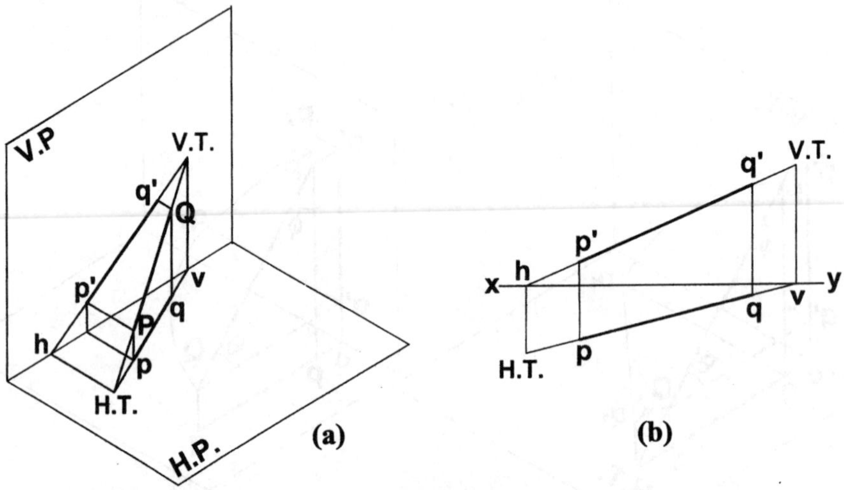

Fig. 7.37

1. Extend the front view p'q' to meet xy line at h.
2. Through h, draw a perpendicular to meet the extended pq at H.T.
3. Similarly extend the top view to meet the xy line at V.
4. Through V, draw a perpendicular to meet the extended front view at V.T.

Method-II

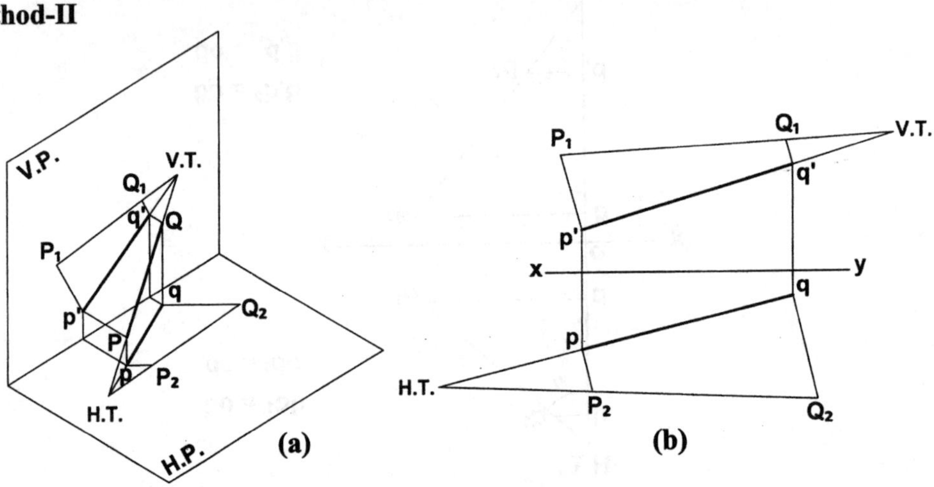

Fig. 7.38

1. In Fig. 7.38, p'q' and pq are front view and top view of line PQ respectively. Determine P_1Q_1 and P_2Q_2 as true length of the line using trapezoid method.
2. Obtain the point of intersection of p'q'-produced and P_1Q_1-produced which is the required V.T.
3. Obtain point of intersection of pq-produced and P_2Q_2-produced, which is the required H.T.

Additional Problems

Problem 7.13 (Fig. 7.39): *The top view of a 70 mm long line PQ measures 60 mm while the length of its front view is 45 mm. Its end P is in the H.P. and 15 mm in front of the V.P. Draw its projections and determine its inclinations with the H.P. and the V.P.*

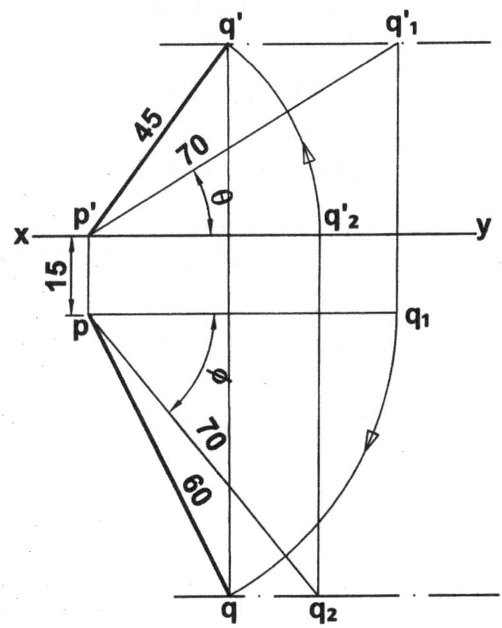

Fig. 7.39

Construction (Fig. 7.39):

1. Draw a xy line and mark a point p' on it to represent front view of end P. Mark its top view p, 15 mm below the xy line.
2. Through p, draw a line pq_1 equal to the length of top view (i.e., 60 mm) and parallel to the xy line. Draw a vertical projector through q_1.

3. With p' as centre and radius equal to 70 mm, draw an arc cutting the projector through q_1 at q_1'. Draw a horizontal (locus) line through q_1'.
4. Similarly, draw $p'q_2'$ coinciding with xy line equal to length of front view (i.e., 45 mm) and draw a vertical projector through q_2'.
5. With p as centre and radius equal to 70 mm, draw an arc cutting the projector through q_2' at q_2. Draw a horizontal (locus) line through q_2.
6. With p' as centre and radius equal to $p'q_2'$, draw an arc cutting locus line through q_1 at q'. Join p' to q'. Thus p'q' is the required front view.
7. With p as centre and radius equal to pq_1 draw an arc cutting the locus line through q_2 at q. Join p to q. Thus pq is required top view. Measure the angles θ and φ which represent inclination of line PQ with the H.P. and the V.P. respectively.

Problem 7.14 (Fig. 7.40): *A line PQ, 70 mm long has its end P, 20 mm above H.P. and 25 mm in front of V.P. The end Q is 50 mm above the H.P. and 60 mm in front of the V.P. Draw its projections and show its inclinations with H.P. and V.P.*

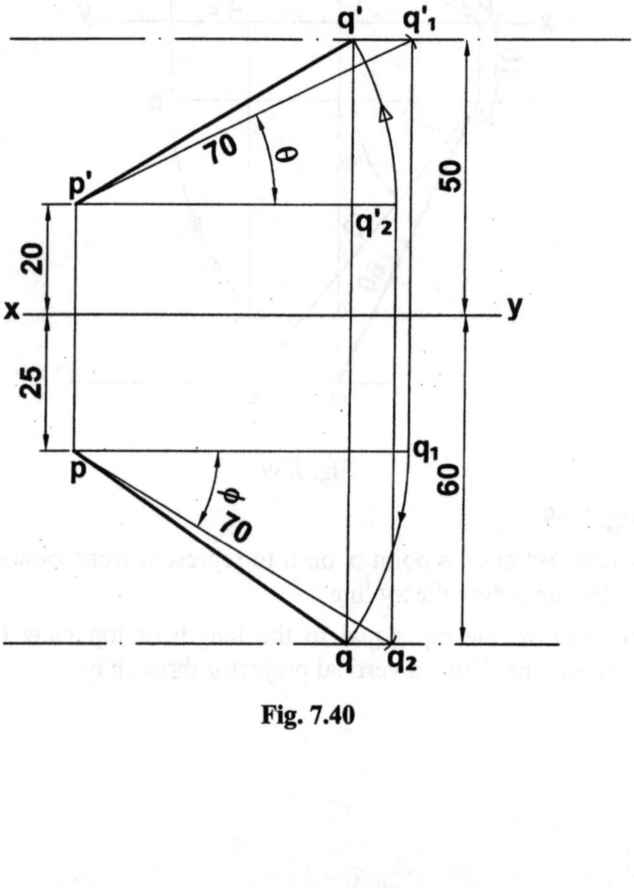

Fig. 7.40

Construction (Fig. 7.40):

1. Draw xy line and mark p' and p 20 mm above and 25 mm below the xy line respectively.
2. Draw a horizontal line 50 mm above xy line as locus of Q in front view.
3. Similarly draw another horizontal line 60 mm below the xy line to represent locus of end Q in the top view.
4. With p' as centre and radius equal to 70 mm, draw an arc cutting locus line at q_1'. Similarly, with p as centre and radius equal to 70 mm draw an arc cutting locus at q_2. Join p to q_2.. Join p' with q_1'. Project q_1' to q_1. pq_1 is the length of top view. With p as centre and radius equal to pq_1, draw an arc cutting the locus at q. Join p with q. Thus pq is required top view and ϕ is the angle of inclination of the line with the V.P.
5. Draw a vertical projector through q to intersect locus line at q'. Join p' with q'. Thus p'q' is the required front view and θ is the angle of inclination of the line PQ with the H.P.

Problem 7.15 (Fig. 7.41): *A line AB 80 mm long is inclined at 45° to the V.P. Its end A is in the H.P. and 15 mm in front of the V.P. The length of its top view is 60 mm. Draw the projections of line AB and determine its inclination with the H.P.*

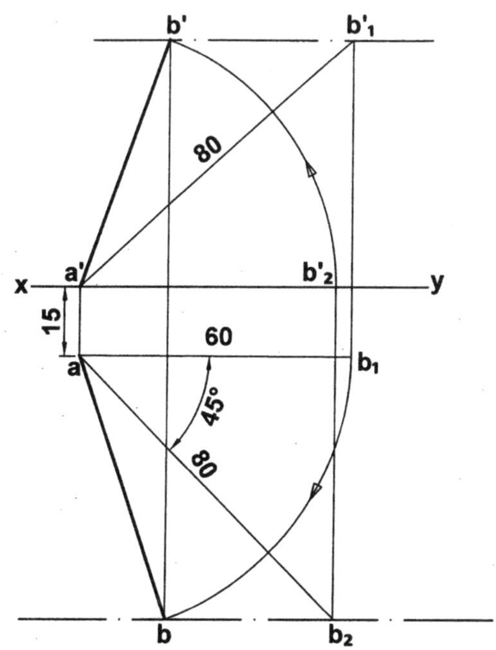

Fig. 7.41

Construction (Fig. 7.41):

1. Draw xy line. Mark a' on xy line and a, 15 mm below the xy line.
2. Through a, draw horizontal line ab_1 equal to length of top view (i.e., 60 mm).
3. Through a, draw line ab_2 80 mm long and inclined at 45° to ab_1.
4. Through b_2, draw a horizontal line representing locus of end B in the top view.
5. With a as centre and radius equal to ab_1, draw an arc cutting the locus line at b.
6. Join a with b. Thus ab represents top view of the line AB.
7. Through b_1, draw a vertical projector. Now with a' as centre and radius equal to 80 mm, draw an arc cutting the vertical projector at b_1'.
8. Through b_1', draw a horizontal line as locus of end B in the front view.
9. Project b upward to intersect locus line at b'. Join a' with b'. Thus, a'b' is the required front view and inclination of $a'b_1'$ with the xy line is the angle of inclination of the line AB with the H.P.

Problem 7.16 (Fig. 7.42): *A line PQ, 80 mm long is inclined at 30° to the H.P. Its end P is 12 mm above the H.P. and 20 mm in front of the V.P. Its front view measures 60 mm. Draw its projections and determine its inclination with the V.P.*

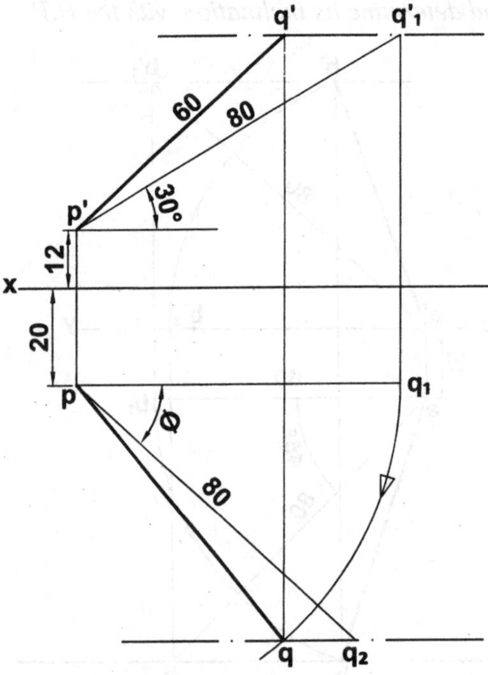

Fig. 7.42

Construction (Fig. 7.42):

1. Draw xy line and mark p' and p 12 mm above and 20 mm below the xy line respectively.
2. Through p' draw a line $p'q_1'$ 80 mm long and making an angle of 30° with the xy line. Through q_1', draw a horizontal locus line.
3. With p' as centre and radius equal to 60 mm, draw an arc cutting the locus line at q'. Thus p'q' represents the front view.
4. Project q_1' to q_1 such that pq_1 is parallel to xy line.
5. With p as centre and radius equal to pq_1, draw an arc cutting the vertical projector through q' at q. Join p with q. Thus pq is the required top view.
6. With p as centre and radius equal to 80 mm, draw an arc cutting locus line at q_2. Thus inclination of pq_2 with xy is the angle of inclination of line PQ with the V.P.

Problem 7.17 (Fig. 7.43): *The end A of a line AB is 20 mm behind the V.P. and is below the H.P. The end B is 15 mm in front of the V.P. and above the H.P. End projectors are 60 mm apart. The line is inclined at 45° to the H.P. and its H.T. is 15 mm behind the V.P. Draw its projections and find its true length. Also locate its V.T.*

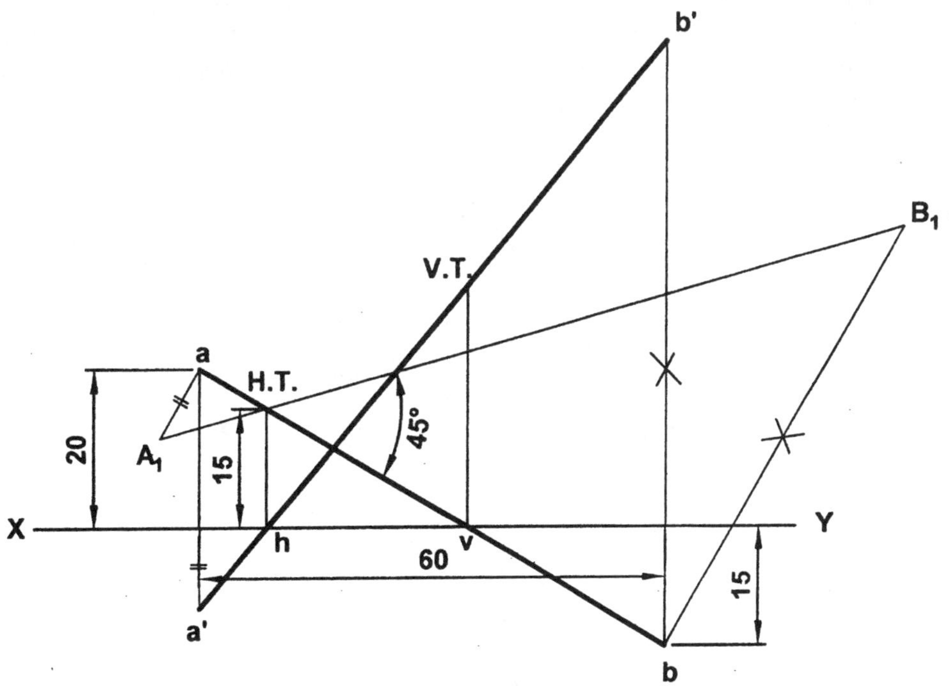

Fig. 7.43

Construction (Fig. 7.43):

1. Draw xy line. Draw end projectors 60 mm apart. Mark a, 20 mm above and b, 15 mm below the xy line on these end projectors. Join a to b so that ab represents top view of AB. Mark H.T. 15 mm above xy and on ab.
2. We have observed that the line representing the true length (obtained by trapezoid method) meets the top view or top view produced line at the H.T. making an angle equal to the inclination of the line with V.P. (i.e., ϕ). Hence at ends a and b, draw perpendiculars in opposite directions (because both ends are in different quadrants).
3. Through H.T., draw a line inclined at 45° with the ab intersecting the perpendiculars at A_1 and B_1. Thus line A_1B_1 represents true length of line AB. aA_1 and bB_1 are distances of the ends A and B respectively from the H.P.
4. Mark a' and b' on the vertical projectors such that their distances from the xy line are equal to aA_1 and bB_1 respectively.
5. Join a' with b'. Thus, a'b' is the required front view.
6. Mark v at the point of intersection of ab with xy line.
7. Draw a perpendicular through v to intersect a'b' at V.T.

Problem 7.18 (Fig. 7.44): *The top view of a 70 mm long line AB measures 50 mm. The end A is 12 mm above the H.P. and 50 mm in front of the V.P. The end B is 20 mm in front of the V.P. and above the H.P. Draw the projections of line AB and determine its angle of inclination with the H.P. and V.P.*

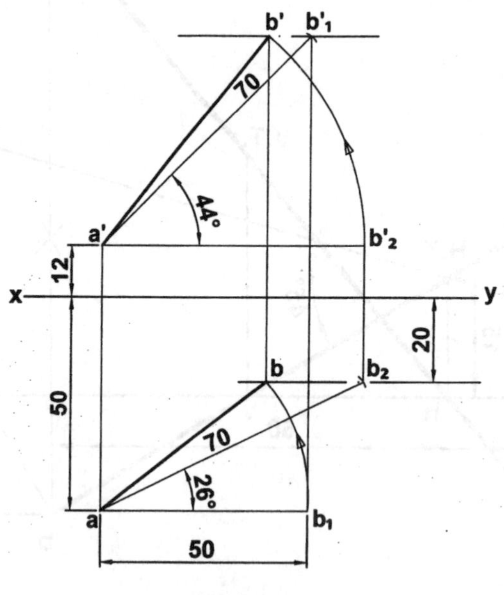

Fig. 7.44

Construction (Fig. 7.44):

1. Draw xy line and mark a point a' and a, 12 mm above and 50 mm below the xy line.
2. Draw a horizontal line 20 mm below the xy line as locus of end B in the top view.
3. With a as centre and radius equal to 70 mm, draw an arc to meet the horizontal line at b_2. Join a to b_2.
4. Draw line ab_1 as top view equal to 50 mm and parallel to xy line.
5. Through b_1, draw a vertical projector.
6. With a' as centre and radius equal to 70 mm, draw an arc to intersect vertical projector at b_1'. Join a' and b_1'.
7. Draw a horizontal line through b_1' as locus of B in the front view.
8. Project b_2 to b_2'. With a' as centre and radius equal to $a'b_2'$, draw an arc to intersect the locus line at b'. Join a' with b'. Thus a'b' represents the front view.
9. Through b', draw a vertical projector to meet the locus line at b. Thus ab is the required top view. Angles of inclination (θ and ϕ) can be measured directly.

Problem 7.19 (Fig. 7.45): *The end A of a line AB is 15 mm above the H.P. and 25 mm in front of the V.P. The end B is 10 mm below the H.P. and 40 mm behind the V.P. Draw the projections of the line AB if its end projectors are 50 mm apart. Determine its true length and locate its traces.*

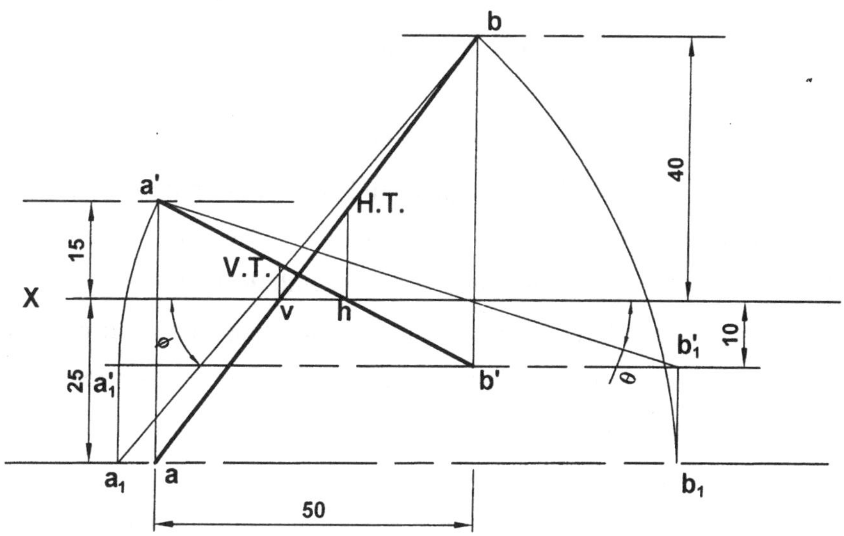

(a)

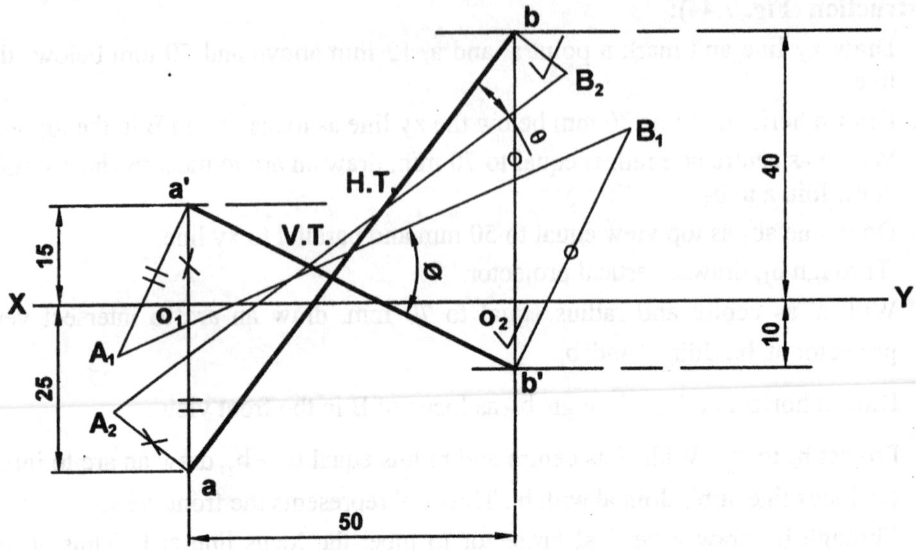

(b)

Fig. 7.45

Construction (Fig. 7.45):

This problem can be solved using one of the following two methods.

Method I – By making view parallel to xy line

(Refer Fig. 7.45 (a))

1. Draw front view a'b' and top view ab using the given information.
2. Keeping end a fixed rotate end b to a position b_1 thus making it parallel to xy line. Project b_1 to b_1' on the locus of b'. Join a' with b_1'. The line $a'b_1'$ represents true length of the line AB and θ is the true inclination with the H.P.
3. Similarly, keeping end b' fixed, rotate end a' to a_1', thus making it parallel to xy line. Project a_1' to a_1 on the locus of a. Join a_1 to b. The line a_1b represents true length and ϕ inclination of the line AB with the V.P.

To find the traces

4. Mark v at the point of intersection of the line ab with xy line then through v, draw a perpendicular line to intersect a'b' at V.T.
5. Mark h at the point of intersection of a'b' and xy line. Then through h, draw a perpendicular line to intersect top view ab at H.T.

Method II- By rotating the line about its projections till it lies in the H.P. or V.P. (Refer Fig. 7.45 (b))

1. Draw a'b' and ab as front view and top view respectively using given information.
2. At the ends a' and b' of the front view, draw perpendiculars on opposite sides of it (because a and b are on the opposite sides of xy line) such that $a'A_1 = o_1a$ and $b'B_1 = o_2b$. Join A_1 with B_1. The line A_1B_1 represents true length and ϕ is the true inclination with the V.P. The point of intersection of a'b' and A_1B_1 is the V.T. of the line AB.
3. Similarly at the ends a and b of the top view, draw perpendiculars on the opposite sides of ab such that $aA_2 = o_1a'$ and $bB_2 = o_2b'$. Join A_2 with B_2. The line A_2B_2 represents true length of the line AB and θ represents true inclination with the H.P. The point of intersection of ab and A_2B_2 is H.T. of the line AB.

Problem 7.20 (Fig.7.46): *The front view and the top view of a 70 mm long line AB measures 60 mm and 50 mm respectively. The end A is in the H.P. and end B is in the V.P. Draw its projections and locate its traces.*

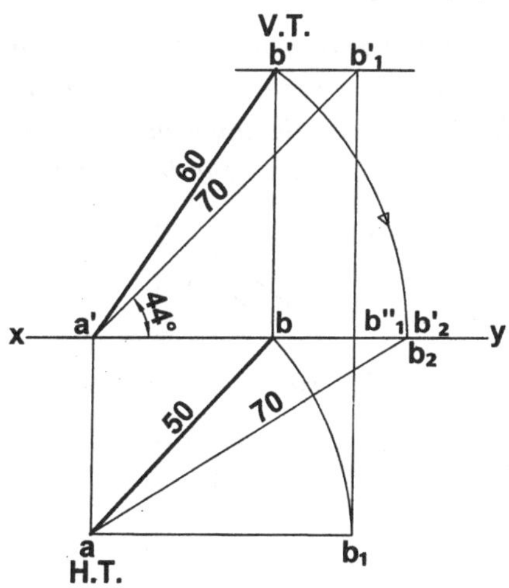

Fig. 7.46

Construction (Fig. 7.46):

1. Draw xy line and mark point a' on it (since one end is on the H.P.).
2. Through a', draw a horizontal line $a'b''_1$ equal to the length of top view (i.e., 50 mm)
3. Draw a vertical projector through b''_1.

4. With a' as centre and radius equal to 70 mm, draw an arc cutting the vertical projector through b_1'' at b_1'. Join a' with b_1'. Inclination of $a'b_1'$ with xy line represents inclination of line AB(θ) with the H.P.
5. Draw a horizontal line through b_1' as locus of b'. Now with a' as centre and radius equal to 60 mm (length of front view) draw an arc cutting locus line at b'. Join a' to b'. Thus a'b' is required front view.
6. Through b', draw a vertical projector to meet xy line at b (because end B is in the V.P.)
7. Draw a vertical projector through a'. With b as centre and radius equal to 50 mm (equal to the length of top view), cut an arc on the vertical projector through a' at a. Join a with b. Thus ab represents top view.
8. With a' as centre and radius equal to a'b', draw an arc to meet xy line at b_2'. Join a with b_2'. Inclination of ab_2' with xy represents the true inclination of the line AB (ϕ)with the V.P.
9. Since, front view a'b' meets the xy line at a' the point a also represents H.T. Similarly top view ab meets the xy line at b, the point b' also represents the V.T.

Problem 7.21 (Fig. 7.47): *The end P of a line PQ is at 40 mm in front of the V.P. and 30 mm above the H.P. The end Q is 10 mm in front of the V.P. and above the H.P. The line is inclined at 30° to the V.P. If the distance between the end projectors is 40 mm, draw its projections, determine its true length and true inclination with the H.P.*

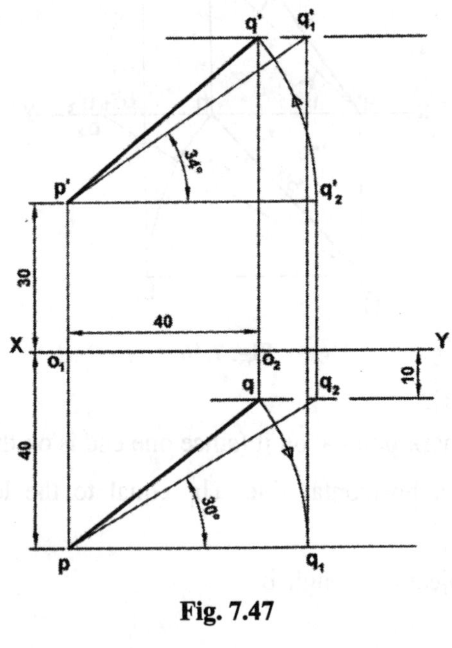

Fig. 7.47

Construction (Fig. 7.47):

1. Draw xy line and mark points o_1 and o_2 40 mm apart on it. Draw two vertical projectors through o_1 and o_2.
2. On the vertical projector through o_1, mark p' and p, 30 mm above and 40 mm below the xy line respectively.
3. On the vertical projector through o_2, mark a point q, 10 mm below the xy line. Join p with q thus pq represents top view.
4. Draw a horizontal line through q as locus of q.
5. Through p, draw a line making an angle of 30° with the horizontal (or xy line) intersecting the locus at q_2. Inclination of pq_2 with xy is the true inclination of the line PQ with the V.P.(i.e., ϕ).
6. Project q_2 to q_2'. With p' as centre and radius equal to $p'q_2'$, draw an arc cutting the vertical projector at q'. Join p' with q'. The line p'q' represents the front view. Draw a horizontal line through q' as locus of q'.
7. Through p, draw a horizontal line. Keeping end p fixed, turn end q to q_1.
8. Through q_1, draw a vertical projector to meet the locus of q' at q_1'.
9. Join p' to q_1'. The line $p'q_1'$ represents the true length of line PQ and its inclination with xy represents the true inclination (θ) of the line PQ with the H.P.

Problem 7.22 (Fig. 7.48): *A line PQ is inclined at 30° to the V.P. Its ends P and Q are 40 mm and 20 mm above the H.P. The length of its front view is 70 mm and its V.T. is 10 mm above the H.P. Draw the projections of the line PQ and determine its true length and true inclination with the H.P. and the V.P. Also locate its H.T.*

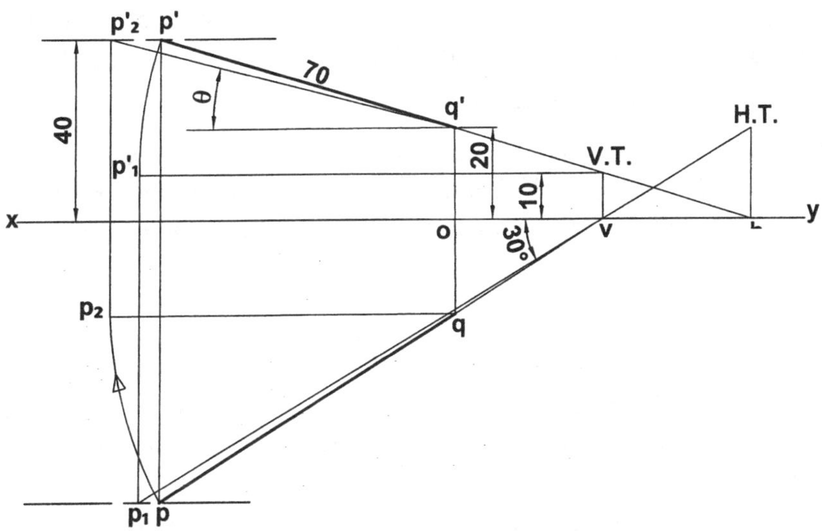

Fig. 7.48

Construction (Fig. 7.48):

1. Draw xy line and at any point o on the xy line, draw a vertical projector and mark q', 20 mm above xy line.
2. Draw a horizontal locus line, 40 mm above the xy line. With q' as centre and radius equal to 70 mm (equal to length of front view), draw an arc to cut the locus line at p'. Join p' to q' thus p'q' shows the front view.
3. Draw a line parallel to and 10 mm above the xy line. The V.T. will lie on this line. Produce p'q' back to intersect this line at V.T. and produce further to intersect xy at h.
4. Assuming V.T. p' to be front view of a line which is inclined at 30° to the V.P. and whose one end is at v, let us find its true length as follows:
5. With V.T. as centre and radius equal to V.T. p', draw an arc to cut a horizontal line through V.T. at p_1'. (This is required to make the line parallel to xy line). Through p_1', draw a vertical projector. Through v, draw a line making an angle of 30° with xy line cutting the vertical projector at p_1. Draw horizontal locus through p_1 and project p' to p on this locus line.
6. Since, top view (pq), v and H.T. will be collinear so draw a line joining p and v and produce further. Project q' to q on the line pv.

   Thus we have drawn p'q', as front view and pq as top view on the line PQ its true length and true inclination can be obtained in the usual manner as follows.
7. Keeping the end q fixed, turn p to p_2 to make the top view pq parallel to xy line. Project p_2 to p_2' on the locus of end P. Hence $q'p_2'$ is the true length and θ is true inclination of the line PQ with the H.P.

Problem 7.23 (Fig. 7.49): *A line AB inclined at 30° to the H.P., has its ends A and B, 32 mm and 66 mm in front of the V.P. respectively. The top view measures 62 mm and its H.T is 15 mm in front of the V.P. Draw its projections and determine its true length. Also locate its V.T.*

Construction (Fig. 7.49):

1. Draw the top view ab with the given information.
2. Draw a line parallel to and 15 mm below the xy line. Produce the top view ab to cut the xy line at v. Mark H.T. on this line where it cuts the horizontal line (i.e., 15 mm below xy).

   Assuming H.T. b to be top view of a line which is inclined at 30° to the H.P. and whose one end to be h, let us determine its true length as follows.

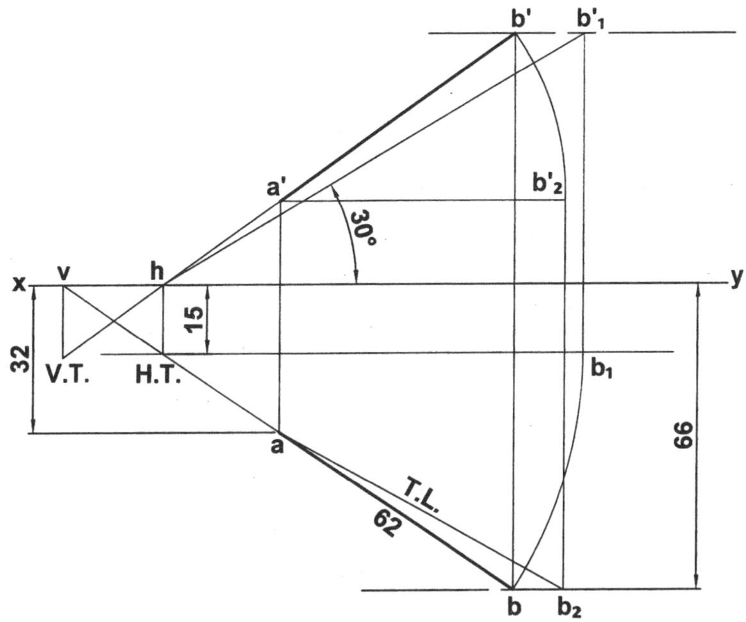

Fig. 7.49

3. With H.T. as centre and radius equal to H.T. b, draw an arc to cut horizontal line through H.T. at b_1. Project b_1 upward. Through h, draw a line making an angle of 30° with xy and cutting the projector through b_1 at b_1'. Draw a horizontal line through b_1' as locus of b' in front view.
4. But given top view is ab, so project b to b' on the locus line.
5. Join b' to h and extend further to meet the vertical projector through v at V.T.
6. Project a to a' on the line hb'. Thus a'b' is the required front view. In order to get true length and true inclination, make a'b' parallel to xy. Keeping end a' fixed, turn b' to b_2'. Project b_2' to b_2 on the locus of b.
7. Join a to b_2. Thus ab_2 represents true length of line AB and its inclination with xy is the true inclination of the line with V.P.

Problem 7.24 (Fig. 7.50): *A line AB inclined at 30° to the V.P. has its end A 12 mm above the H.P. The front view measures 65 mm and makes an angle of 40° with the reference line. The V.T. of the line is 20 mm below the H.P. Draw the projections of line AB, determine its true length and locate its H.T.*

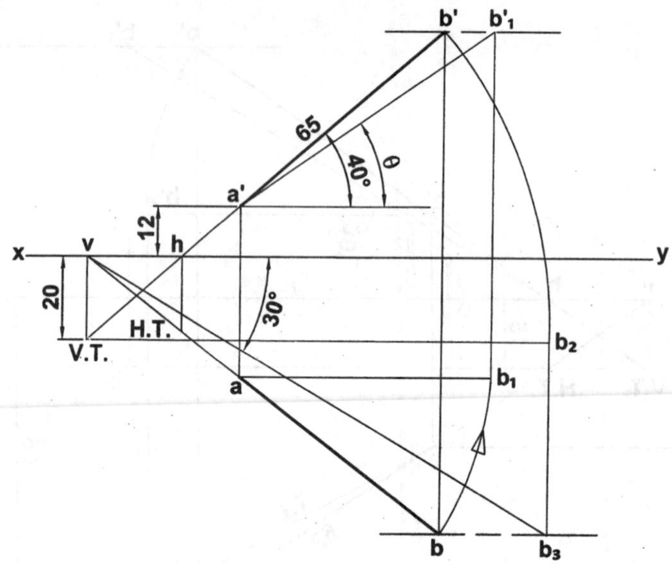

Fig. 7.50

Construction (Fig. 7.50):

1. Draw xy line and mark point a', 12 mm above it. Through a', draw a line a'b' (as front view) 65 mm long and making an angle of 40° with the reference line.
2. Draw a line 20 mm below and parallel to xy line. The V.T will lie on this line. Produce the front view a'b' to intersect the horizontal line at V.T. Project V.T. to v on the xy line.
3. Through v, draw a line inclined at 30° to the xy. Now with V.T. as centre and radius equal to V.T. b', draw an arc cutting the horizontal line through V.T. at b_2. Project b_2 to meet inclined line through v at b_3.

   Draw a horizontal line through b_3 which represents locus of b in the top view.
4. Project b' to b on the locus line. Join b with v (because top view, H.T. and v should be collinear). Project a' to a on the line vb. Thus ab represents top view of the line AB.
5. With a as centre and radius equal to ab, draw an arc to cut horizontal line through a at b_1. Project b_1 to b_1' on the locus of b'. Join a' with b_1'. Thus $a'b_1'$ represents true length of the line AB and its inclination (θ) is the true inclination with the H.P.

Problem 7.25 (Fig. 7.51): *A line AB, 65 mm long has its end A 30 mm above the H.P. and 40 mm in front of the V.P. Its V.T. is 10 mm above the H.P. and projectors through V.T. and end A are 40 mm apart. Draw the projections of line AB and determine its inclination with the H.P. and V.P. Also locate its H.T.*

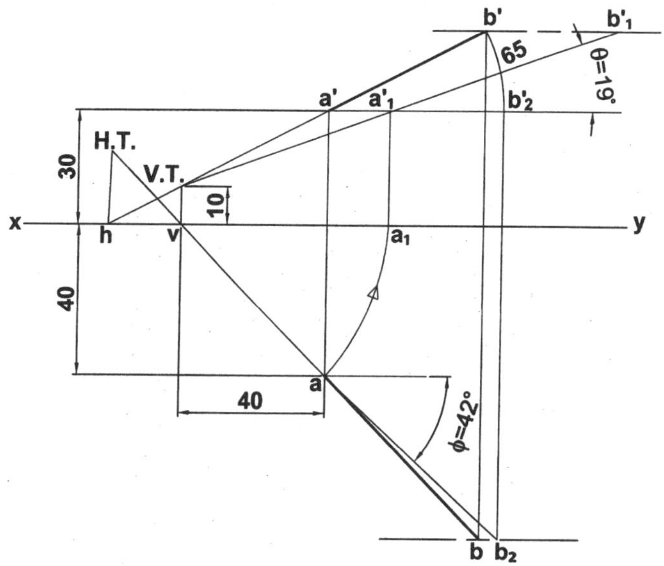

Fig. 7.51

Construction (Fig. 7.51):

1. Draw xy line and mark a' and a 30 mm above and 40 mm below the xy line respectively on a vertical projector.
2. Draw a vertical projector 40 mm away from the projector through a or a'. On it, mark V.T. 10 mm above xy and v on the xy.
3. Join V.T. with a' and v with a.
4. With v as centre and radius equal to va, draw an arc to cut horizontal line through v at a_1. Project a_1 to a'_1.

   Join V.T. to a'_1 and produce to b'_1 such that $a'_1b'_1 = 65$ mm. The inclination of $a'_1b'_1$ (i.e., θ) represents the true inclination of the line with the H.P.
5. Draw a horizontal line (locus) from b'_1 to meet V.T. a'-produced line at b'.

   Join a' with b'. Thus a'b' represents front view.
6. Join v with a and extend it to meet the vertical projector through b' at b. Join a with b. The line ab is the top view of the line AB.
7. Keeping end a' fixed, turn b' to b'_2 to make it parallel to xy line. Project b'_2 to b_2 on the locus. Join b_2 with a. Thus line ab_2 represents true length and its inclination (ϕ) is the true inclination of the line AB with the V.P.

Problem 7.26 (Fig. 7.52): *The front view of a line AB is inclined at 30° to the reference line. The end A is 12 mm above the H.P. and end B is 90 mm in front of the V.P. The H.T.*

of the line is 25 mm in front of the V.P. and V.T. is 20 mm below the H.P. Draw its projections and determine its inclination with the H.P. and V.P.

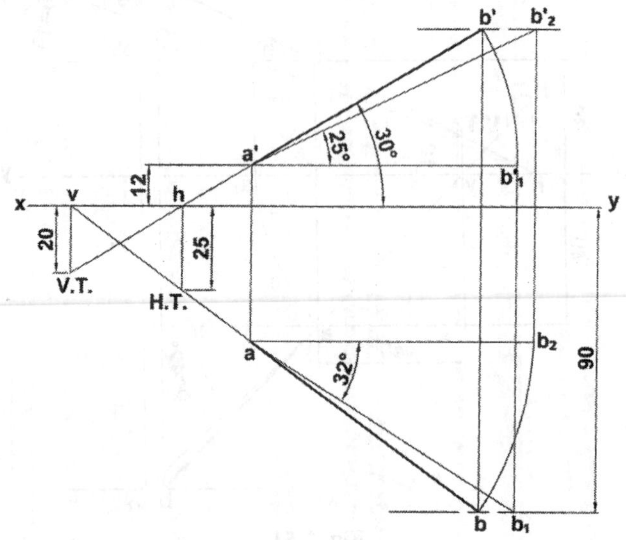

Fig. 7.52

Construction (Fig. 7.52):

1. Draw xy line. Mark a point a', 12 mm above the xy line.
2. Through a', draw a line inclined at 30° to xy line and produce it back to meet the xy line at h and V.T. 20 mm below xy line.
3. Through V.T., draw a vertical projector to meet xy line at v. Through h draw a vertical projector and mark H.T. on it 25 mm below xy.
4. Join v to H.T. and produce it to meet a point 90 mm below xy line at b. Draw a horizontal line through b as locus of b.
5. Draw a vertical projector through b to meet inclined line through a' at b'. Join a' with b'. The a'b' is the required front view.
6. Draw a horizontal line through b' to represent locus of b'.
7. Project a' to a on the line vb, Thus ab is the required top view.
8. Determine true length and true inclination by making the views parallel to xy line as described in previous problems.

Problem 7.27 (Fig. 7.53): *The front view of a line AB measures 50 mm and is inclined at 30° to the reference line. The end A is 20 mm above the H.P. and H.T. is 15 mm below the H.P. If the line is inclined at 23° to the H.P., draw its projections and determine its true length and true inclination with the V.P. Also locate its V.T.*

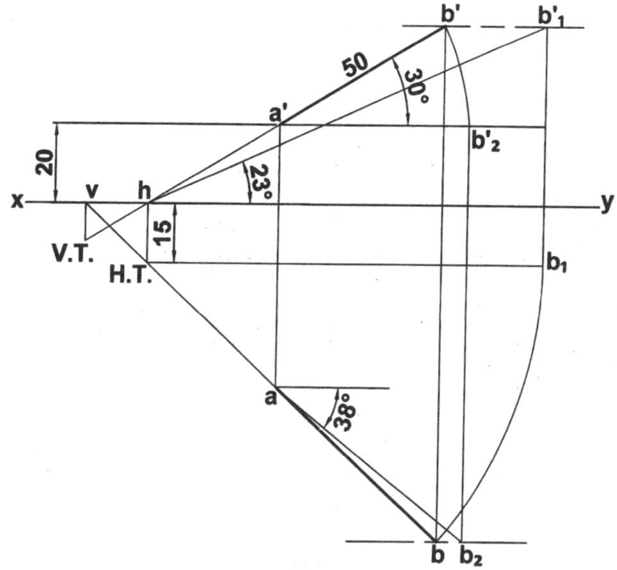

Fig. 7.53

Construction (Fig. 7.53):

1. Draw xy line. Draw a'b' 50 mm long and making an angle of 30° with xy line and produce it to meet xy line at h. Also draw a horizontal line through b' to represent locus of B in the front view.

   Assume hb' to be front view of a line inclined at 23° to the H.P. and one end of its top view is in H.T. Let us determine true length of it as follows.

2. Through h, draw a line inclined at 23° to xy line and cutting the locus of B in the front view at b'_1. Project b'_1 to meet the horizontal line through H.T. at b_1.

3. Draw an arc with H.T. as centre and radius equal to H.T. b_1 to cut the vertical projector through b' at b. Draw a horizontal line through b as its locus. Join HT to b. But given front view is a'b', so project a' to a on H.T. b. Thus ab is required top view.

4. Produce b H.T. to meet xy line at v. Also produce b'h to meet vertical projector through v at V.T. With a' as centre and radius equal to a'b', draw an arc to meet horizontal line through a' at b'_2.

5. Project b'_2 to b_2 on the locus of end B in the top view and join a with b_2.

   Thus ab_2 is true length of the line AB and its inclination with xy line represents true inclination of the line AB with the V.P.

Problem 7.28 (Fig. 7.54): *The ends of a line AB are on the same projector. The end A is 35 mm below the H.P. and 15 mm behind the V.P. the end B is 50 mm above the H.P. and*

40 mm in front of the V.P. Draw the projections of the line AB, find its true length and true inclinations with H.P and V.P. Also find its traces.

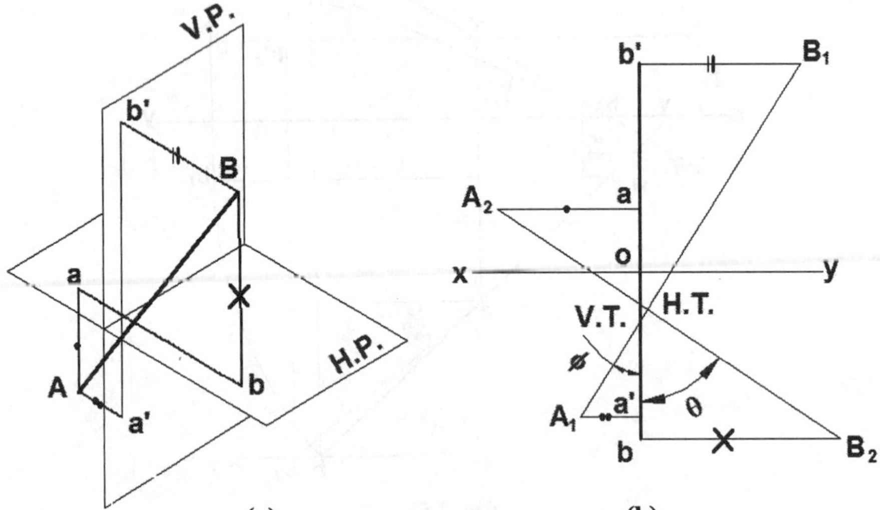

(a) (b)

Fig. 7.54

Construction (Fig. 7.54):

1. Draw xy line and draw the projections of end A and B of the line as per the given information. Here, a'b' and ab are front view and top view of the line respectively.
2. At ends a' and b', draw perpendiculars in opposite directions such that $a'A_1 = oa$ and $b'B_1 = ob$. Join A_1 to B_1. A_1B_1 is the true length of the line AB and ϕ is the inclination of AB with the V.P.
3. Similarly, at ends a and b, draw perpendiculars in opposite directions such that $aA_2 = oa'$ and $bB_2 = ob'$.

Join A_2 with B_2. The line A_2B_2 represents true length and θ is the true inclination of AB with the H.P. The point where A_1B_1 cuts a'b' is the V.T. of AB and the point where A_2B_2 cuts the ab is the H.T. of AB.

Problem 7.29 (Fig. 7.55): *A straight line AB 110 mm long is inclined at 30° to the H.P. and 45° to the V.P. Its mid point is in the V.P. and 25 mm above the H.P. If end A is in the third quadrant and end B is in the first quadrant, draw its projections.*

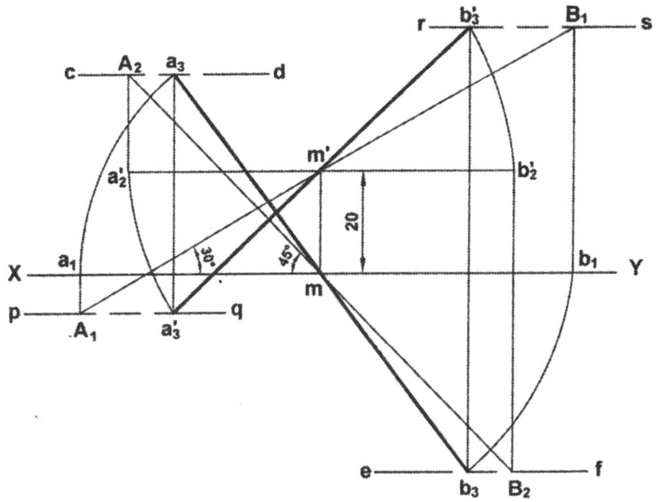

Fig. 7.55

Construction (Fig. 7.55):

1. Draw xy line and mark the projections of mid-point of the line as m', 20 mm above xy and m on the xy line.
2. Through m', draw a line inclined, at 30° (=θ) to the xy line. With m' as centre and radius equal to 55 mm (i.e., equal to ½ AB) draw arcs on both sides of m' on this line at A_1 and B_1. Draw horizontal lines pq through A_1 and rs through B_1 as locus of ends. Project A_1B_1 to a_1b_1 on the line xy line. Thus a_1b_1 represents length of top view.
3. Similarly through m, draw a line inclined at 45° (i.e., equal to ϕ) with xy line. With m as centre and radius equal to 55 mm (i.e., equal to ½ AB), cut arcs to get end points A_2 and B_2. Draw horizontal lines cd and ef through A_2 and B_2 respectively.

   Project A_2 and B_2 on a horizontal line through m' at a'_2 and b'_2 respectively. Thus $a'_2b'_2$ represents length of front view.
4. With m as centre and radius equal to ma_1 or mb_1 draw arcs cutting cd at a_3 and ef at b_3. Thus a_3b_3 is required top view.
5. With m' as centre and radius equal to $m'a'_2$ or $m'b'_2$ draw arcs cutting pq at a'_3 and rs at b'_3. Thus $a'_3b'_3$ is required front view.

Problem 7.30 (Fig. 7.56): *A lamp is hanging at the roof of a room having size 4 m × 4 m × 3.5 m centrally such that its vertical distance from the roof is 1 metre. Determine graphically its distance from one of the top corners.*

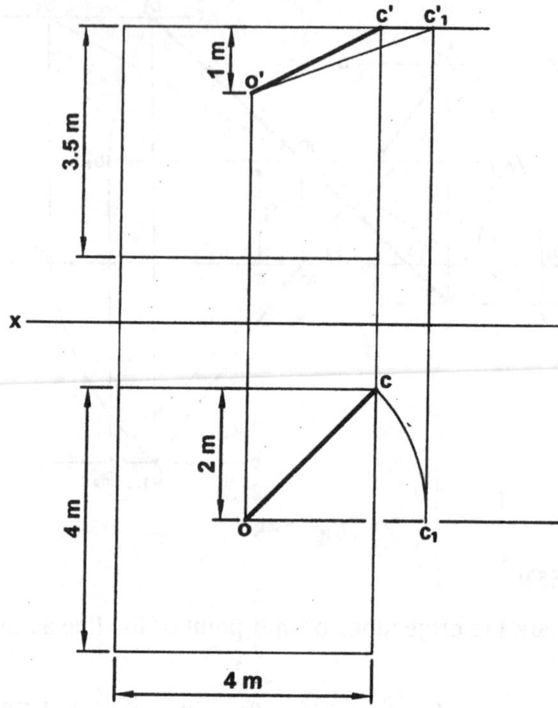

Fig. 7.56

Construction (Fig. 7.56):

1. Draw front view and top view of the room as per given dimensions using some suitable scale.
2. Mark o' 1 m below roof surface as front view of lamp.
3. Mark o in the centre of the square as top view of the lamp.
4. Join o' to c' and o to c. The line o'c' represents the front view of line joining lamp and the corner of room whereas oc represent its top view.
5. With o as centre and radius equal to oc draw an arc meeting horizontal line through o at c_1. Project c_1 to c'_1 on the horizontal line through c'. The line $o'c'_1$ represents actual distance between lamp and the corner of the room.

Problem 7.31 (Fig. 7.57): *The end P of a straight line PQ is 15 mm above the H.P. and 35 mm in front of the V.P. The end Q is 15 mm below the H.P. and 45 mm behind the V.P. The end projectors are 55 mm apart. Draw its projections and determine its true length and true inclination with the H.P. and the V.P. Also locate its traces.*

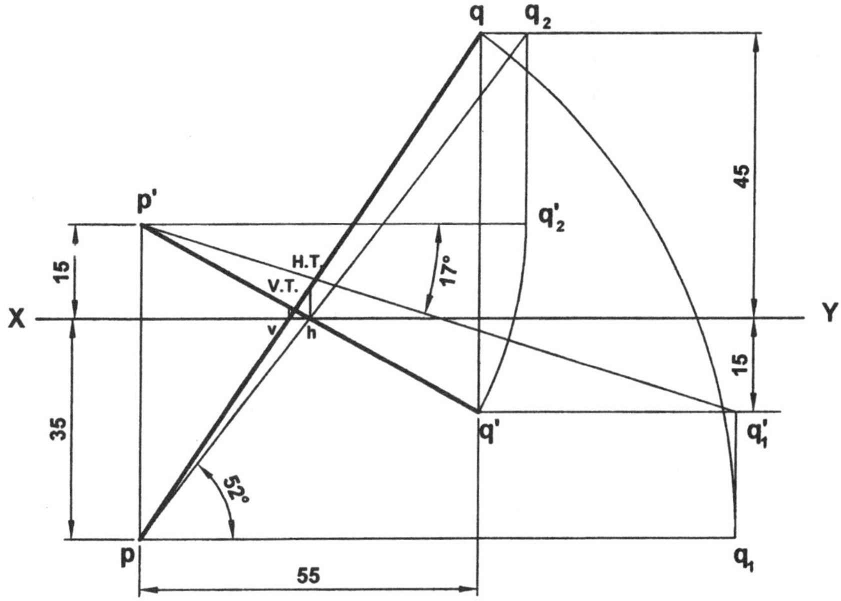

Fig. 7.57

Construction (Fig. 7.57):

1. Draw two vertical projectors 55 mm apart on the xy line.
2. Mark p' and p, 15 mm above and 35 mm below xy line respectively on a projectors.
3. Mark q' and q, 15 mm below and 45 mm above the xy line respectively on another vertical projector.
4. Join p' with q' and p with q, thus p'q' and pq represent front view and top view respectively.
5. With p' as centre and radius equal to p'q', draw an arc to meet horizontal line through p' at q'_2.
6. Project q'_2 to q_2 on the locus of q. Join p with q_2. Thus pq_2 represents true length of PQ and its inclination represents true inclination of the line PQ with the V.P.
7. With p as centre and radius equal to pq, draw an arc to meet horizontal line through p at q_1. Project q_1 to q'_1 on the locus of q'. Join p' with q'_1. The line $p'q'_1$ represents true length of line PQ and its inclination with xy is true inclination of the line AB with the H.P.
8. Mark v at the intersection of pq and xy line. Through v, draw a vertical projector to meet p'q' at V.T.
9. Mark h at the intersection of p'q' and xy line. Through h, draw a vertical projector to meet pq at H.T.

Problem 7.32 (Fig. 7.58): *Two apples on a tree are 1.5 m and 2 m above the ground and 0.5 m and 1.25 m away from a 0.25 m thick wall but on the opposite sides of it. The distance between the apples measured along the ground and parallel to the wall is 2.5 m. Find graphically the true distance between two apples.*

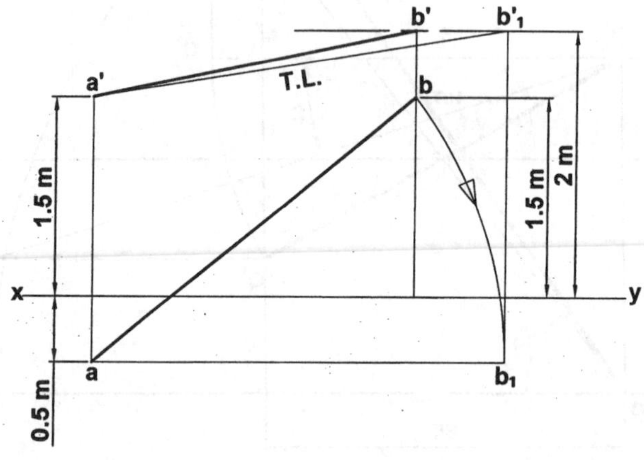

Fig. 7.58

Assume ground to be H.P. and front surface wall to be V.P. Since, wall is 0.25 m thick, its thickness is to be added in the given distance of apple to the opposite side of the wall.

If two apples are designated as A and B then,

Position of apple A: 1.5 m above the H.P. and 0.5 m in front of V.P.

Position of apple B: 2 m above the H.P. and 1.25 + 0.25 = 1.5 m behind the V.P.

Distance measured along the wall indicates that the end projectors are 2.5 m apart.

Construction (Fig. 7.58):

1. Draw xy line and mark a' and a 1.5 m above and 0.5 m below the xy line on a vertical projector.
2. Draw another vertical projector 2.5 m away from the first and on it mark b' and b, 2 m and 1.5 m above the xy line respectively.
3. Join a' with b' and a with b, such that a'b' and ab are front view and top view of the line AB. Draw a horizontal projector through b' as locus of b'.
4. With a as centre and radius equal to ab, draw an arc to meet b_1 on the horizontal line through a.
5. Project b_1 to b'_1 on the horizontal locus line. Join a' to b'_1. The line $a'b'_1$ represents true length of the AB or true distance between two apples.

Problem 7.33 (Fig. 7.59): *The distance between end projectors of a line AB is 60 mm and projectors through its traces are 100 mm apart. Its end A is 15 mm above the H.P. The front view and the top view are inclined at 60° and 30° with the xy line respectively. Draw its projections and determine its true length and true inclination with the H.P. and the V.P. Also locate its traces.*

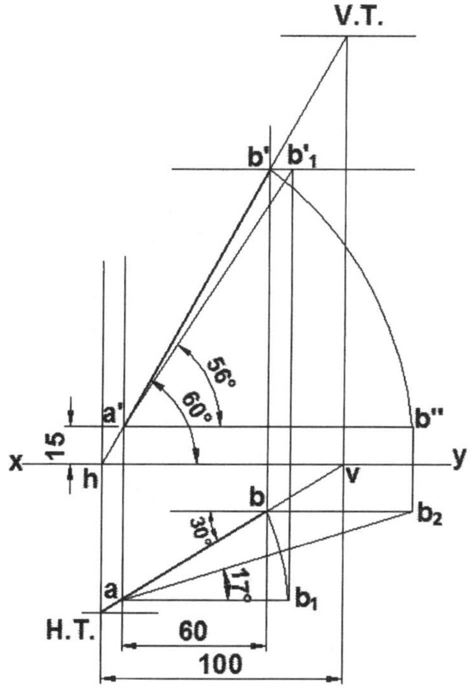

Fig. 7.59

Construction (Fig. 7.59):

1. Draw xy line and two vertical projectors 100 mm apart. Mark h and v at the intersection of xy and vertical projectors.
2. Through h, draw a line making an angle of 60° with xy line and through v, draw a line making an angle of 30° with the xy.
3. Mark a' 15 mm above the xy line and on the line through h. Draw a vertical projector through a' and another vertical projector 60 mm away from it.
4. Mark a, b, V.T. and H.T. on the vertical projectors as shown in the figure. Now a'b' and ab are front and top view respectively of the given line.

5. Keeping end a' fixed turn b' to b". Project b" to b_2 on the locus of b. Join a to b_2. The line ab_2 represents true length and its inclination with xy represents true inclination of the line with V.P.
6. Keeping end a fixed turn b to b_1. Project b_1 to b_1' on the locus of b'. Join a' with b_1'. The line $a'b_1'$ represents true length of the line AB and its inclination with xy line represents the true inclination of the line AB with the H.P.

Problem 7.34 (Fig. 7.60): *A room is 5 m × 4 m × 4 m high. Determine graphically the distance between top and bottom corners diagonally opposite.*

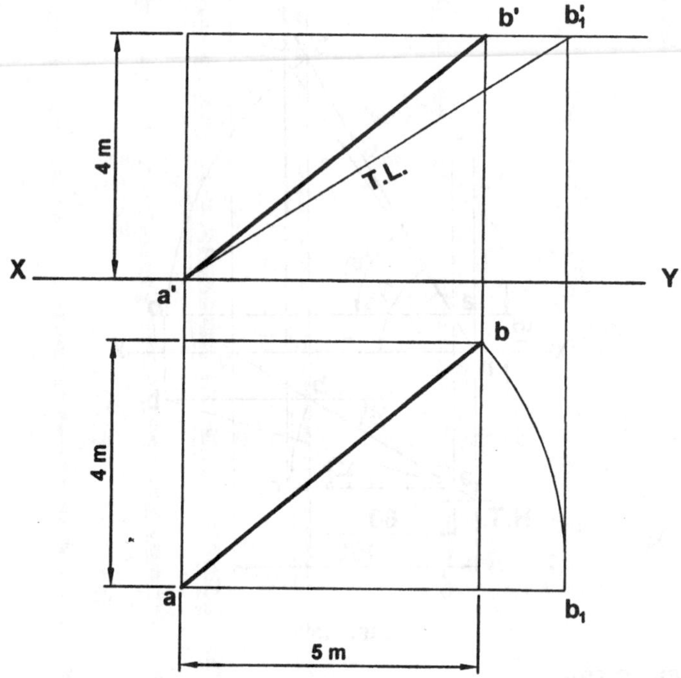

Fig. 7.60

Construction (Fig. 7.60):

1. Draw front view of the room as shown in the Fig. 7.60. Join top and bottom corners by a straight line. Mark a' and b' as front view of corner points. Draw a horizontal line through b' as its locus.
2. Draw top view of the room below xy line. Join corners a and b.

   The line joining diagonally opposite corners of the room is considered as line inclined to both H.P. and V.P. Let us find its true length as follows.

3. With a as centre and radius equal to ab, draw an arc to meet horizontal line through a at b_1.
4. Project b_1 to b'_1 on the locus of b'. The line $a'b'_1$ represents true length of AB.

Problem 7.35 (Fig. 7.61): *A room is 6m × 5m × 3.5 m high. An electric bulb B is above the centre of longer wall and 1.2 m below the ceiling. The bulb is 50 cm away from the longer wall. To operate the bulb, a switch 1 .3 m above the floor on the centre of adjacent wall is mounted. Find graphically the shortest distance between bulb B and switch S.*

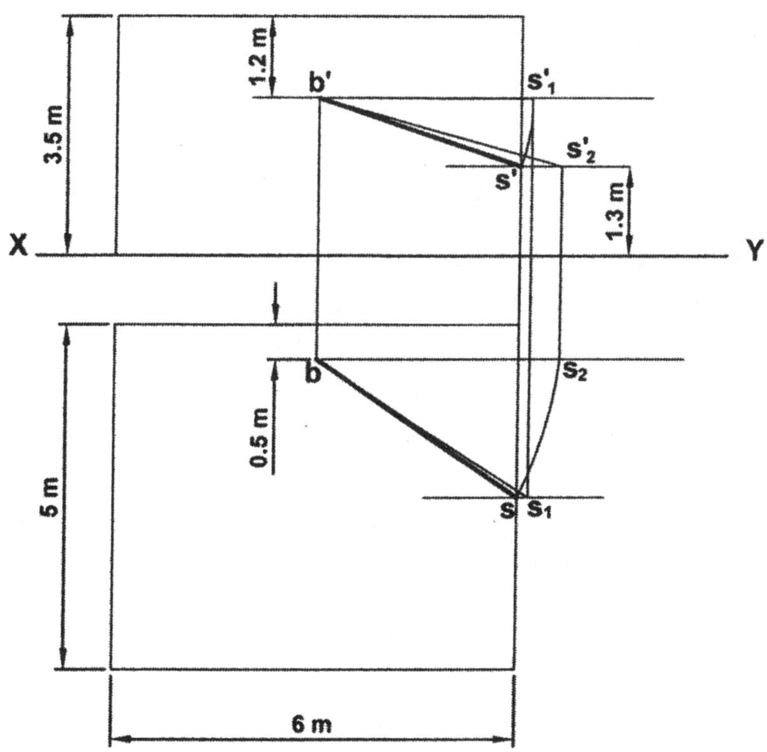

Fig. 7.61

Construction (Fig. 7.61):

1. Draw front view and top view of the room as shown in Fig. 7.61. (Select a scale of 1:100 or any suitable scale).
2. Mark b' as front view of the bulb 1.2 m below the top line (representing ceiling).
3. Mark s' as front view of switch on the right vertical line (representing wall) 1.3 m above the xy line.

4. Mark b as top view of bulb 0.5 m away from longer wall and on the vertical projector through b'.
5. Mark s as top view of switch on vertical projector through s' and in the centre of the 5 m long vertical line to the right.

   Thus b's' *and bs are front view and top view of line joining bulb and switch respectively.(i.e., BS). Let us find its true length as follows:*
6. Keeping end b' fixed turn s' to s'_1 on the horizontal projection through b'. Through s'_1, draw vertical projector to meet locus of s at s_1. Join b with s_1. Thus bs_1 represents true length of line BS. It can also be obtained by making top view parallel to xy and projecting s_2 upward on s'_2. Thus $b's'_2$ also represents true length of line BS.

Exercises - (a)

1. Draw the projection of a line AB, 70 mm long when it is:
   (a) Parallel to both the H.P. and the V.P. and 30 mm away from each plane.
   (b) Parallel to and 25 mm above the H.P. and 30 mm away from the V.P.
   (c) Parallel to and 30 mm in front of the V.P. and in the H.P.
   (d) Perpendicular to the V.P. and 30 mm above the H.P. Its end A is 20 mm away from the V.P.
2. Draw the projections of a line PQ, 65 mm long when its end P is 20 mm above the H.P. and 25 mm in front of the V.P. The line is inclined at 30° to the H.P. and parallel to the V.P.
3. A line PQ, 75 mm long is inclined at 45° to the H.P. and in the V.P. Its end P is 20 mm above the H.P. Draw its front and top view.
4. A line AB, 60 mm long is inclined at 45° to the V.P. and 30 mm above and parallel to the H.P. If end A is 20 mm in front of the V.P. Draw its projections.
5. A line PQ 70 mm long is inclined at 30° to the V.P. and is in the H.P. One of its ends is in the V.P. Draw its projections.
6. A line AB, 80 mm long is parallel to and 35 mm above the H.P. Its ends are 30 mm and 50 mm in front of the V.P. Draw its projections and determine its inclination with the V.P.
7. The top view of a 65 mm long line PQ measures 50 mm. The line is in the V.P. Its end P is 25 mm above the H.P. Draw its projections.

8. The front view of a line AB inclined at 30° to the V.P. measures 60 mm. The line is parallel to and 40 mm above the H.P. If end A is 25 mm in front of the V.P. Draw its projections.
9. A line AB is parallel to the V.P. and inclined at 30° to the H.P. Its end A is 25 mm away from both the planes. Its top view measures 60 mm. Draw its projections and determine its true length.
10. The front view of a 70 mm long line measures 50 mm. The line is parallel to the H.P. and its one end is in the V.P. and 20 mm below the H.P. Draw the projections of the line and determine its inclination with the V.P.

Exercises - (b)

1. Draw the projections of a line AB, 70 mm long when the line is
   (a) parallel to both the H.P. and the V.P. and 30 mm away from each.
   (b) parallel to and 25 mm above the H.P. and 30 mm away from the V.P.
   (c) parallel to and 30 mm in front of the V.P. and in the H.P.
   (d) perpendicular to H.P., 25 mm in front of the V.P. and end A is 10 mm above the H.P.
   (e) perpendicular to V.P., 30 mm above the H.P. and end A is 20 mm away from the V.P.
2. Draw the projections of a line PQ, 65 mm long when its end P is 20 mm above the H.P. and 25 mm in front of the V.P. The line is inclined at 30° to the H.P. and parallel to the V.P.
3. A line PQ, 75 mm long is inclined at 45° to the H.P. and in the V.P. Its end P is 20 mm above the H.P. Draw its front view and top view.
4. A line AB, 60 mm long is inclined at 45° to the V.P. and 30 mm above and parallel to the H.P. Its one end is 20 mm in front of the V.P. Draw its projections.
5. A line PQ 70 mm long, is inclined at 30° to the V.P. and is in the H.P. Its one end is in the V.P. Draw its projections.
6. A line AB, 80 mm long is parallel to and 35 mm above the H.P. Its two ends are 30 mm and 50 mm in front of the V.P. respectively. Draw its projections and determine its inclination with the V.P.
7. The top view of a 65 mm long line PQ measures 50 mm. The line is in the V.P. Its end P is 25 mm above the H.P. Draw its projections.

8. The front view of a line AB, inclined at 30° to the V.P. measures 60 mm. The line is parallel to and 40 mm above the H.P. If end A is 25 mm in front of the V.P. Draw its projections.
9. A line AB is parallel to the V.P. and inclined at 30° to the H.P. Its end A is 25 mm away from both the planes. Its top view measures 60 mm. Draw its projections and determine its true length.
10. The front view of a 70 mm long line measures 50 mm. The line is parallel to H.P. and one of its ends is in the V.P. and 20 mm below the H.P. Draw the projections of the line and determine its inclination with the V.P.

Exercises - (c)

1. A line AB, 70 mm long is inclined at 45° to the H.P. and 30° to the V.P. Its end A is in the H.P. and 35 mm in front of the V.P. Draw its projections and determine its traces.
2. A 90 mm long line PQ, has its end P both in the H.P. and the V.P. The line is inclined at 30° to the H.P. and 45° to the V.P. Draw the projections of line PQ. Also determine its traces.
3. A line AB, 90 mm long has its mid point m 45 mm above the H.P. and 30 mm in front of the V.P. The end A is 25 mm above the H.P. and 15 mm in front of the V.P. Draw its projections, locate its traces and find its inclination with H.P. and the V.P.
4. The top view of a 60 mm long line AB measures 52 mm. The length of its front view is 40 mm. Its end A is in the V.P. and 10 mm above the H.P. Draw its projections and determine its inclinations with the H.P. and the V.P.
5. The end A of a line AB is 60 mm above the H.P. and 35 mm behind the V.P. The end B is 20 mm above the H.P. and 70 mm in front of the V.P. Its end projectors are 65 mm apart. Draw the projections of line AB and determine its inclination with the H.P. and the V.P.
6. A line PQ, 90 mm long has its end P in the H.P. and end Q in the V.P. The line is inclined at 45° to the H.P. and 30° to the V.P. Draw its projections.
7. The front view of a 100 mm long line PQ measures 80 mm and its top view measures 90 mm. Its end Q and the mid point M are in the first quadrant. The mid point M is 20 mm from both the planes. Draw its projections and find its traces.

8. A line AB, 65 mm long, has its end A in the H.P. and 12 mm in front of the V.P. The end B is in the third quadrant. If the line is inclined at 30° to the H.P. and 60° to the V.P., draw its projections.
9. The front view of a line AB measures 60 mm and is inclined at 45° with the xy line. Its end A is in the H.P. and its V.T. is 15 mm below the H.P. The line is inclined at 30 to the V.P. Draw its projections and find its true length and true inclination with the H.P. Also determine its H.T.
10. A line AB 60 mm long has its end A 20 mm away from both the H.P. and V.P. The end B is 25 mm and 30 mm away from the H.P. and V.P. respectively. Draw the projections of the line AB and determine its traces.
11. The ends of a line AB are 25 mm and 15 mm above the H.P. The length of its front view is 45 mm and its V.T. is 10 mm above the H.P. Draw its projections and determine its true length and true inclination with the H.P. Also find its traces.
12. A line AB inclined at 45° to the V.P., has a 55 mm long front view. The end A is 15 mm from both the principal planes while end B is 40 mm above the H.P. Draw its projections and determine its true length and inclinations with the H.P. and the V.P. Also locate its traces.
13. The front view and top view of a line AB are 50 mm and 60 mm long respectively. The end A is 15 mm above the H.P. and 25 mm in front of the V.P. If the end projectors are 50 mm apart, draw its projections and determine its true length and true inclinations with the H.P. and the V.P. Also locate its traces.
14. A line AB is inclined at 30° to the H.P. The end A is 12 mm in front of the V.P. and its mid point M is 40 mm above the H.P. The front view measures 60 mm which is inclined at 45° to the xy line. Draw its projections and find its true length and inclination with the V.P. Also locate its H.T. and V.T.
15. A line PQ, 70 mm long has its end P in the H.P. and 12 mm in front of the V.P. The end Q is in the third quadrant. The line is inclined at 30° to the H.P. and 60° to the V.P. Draw its projections and locate its traces.
16. The front view of a line AB makes an angle of 30° with xy line. The H.T. of the line is 40 mm in front of the V.P. and its V.T. is 30 mm below the H.P. The end A is 10 mm above the H.P. and end B is 90 mm in front of the V.P. Draw its projections and determine its true length and inclinations with the reference planes.
17. A 100 mm long line AB, has its end A in the H.P. and 70 mm in front of the V.P. The other end B is 10 mm in front of the V.P. Draw its projections when the sum of its inclination with H.P. and V.P. is 90°.

18. The top view of a 70 mm long line AB measures 50 mm. The end A is 12 mm above the H.P. while its V.T. is 15 mm below the H.P. The line is inclined at 45° to the H.P. Draw its projections and locate its traces.

19. A line AB is inclined at 30° to the H.P. and 45° to the V.P. and its top view measures 70 mm. The end A is 75 mm in front of the V.P. and its V.T. is 12 mm above the H.P. Draw its projections and determine its inclination with the H.P. Locate its H.T. and V.T.

20. The distance between the end projectors of a line CD is 60 mm and the distance between its traces is 90 mm. The H.T. and V..T. of the line are 40 mm below and 60 mm above the reference line respectively. Draw its projections and determine its true length and inclinations with the H.P. and the V.P.

21. The front view of a line AB measures 65 mm and makes an angle of 30° with the xy line. The end A is in the H.P. and V.T. of the line is 12 mm below the H.P. If the line is inclined at 45° to the V.P., draw its projections. Also find its true length and inclination with the H.P.

22. A 100 mm long line is inclined at 45° to the H.P. and 30° to the V.P. Its mid point is 25 mm above the H.P. and 35 mm in front of the V.P. Draw its projections. Also locate its traces.

23. The end A of a line AB, is in front of the V.P. and 25 mm above the H.P. and end B is behind the V.P. and 30 mm below the H.P. A point on this line is in V.P. and 12 mm below the H.P. The distance between its end projectors is 70 mm. If the line is equally inclined with the H.P. and V.P., draw its projections and determine its inclinations with the H.P. and V.P.

24. A line AB, inclined at 30° to the V.P. has its ends 20 mm and 10 mm above the H.P. The length of its front view is 50 mm and its V.T. is 12 mm above the H.P. Draw its projections and determine its true length and inclination with the H.P. Also locate its H.T.

25. The front view of a line PQ makes an angle of 30° with the xy line. The H.T. of the line is 40 mm in front of the V.P. while its V.T. is 25 mm below the H.P. The end P is 15 mm above the H.P. and end Q is 100 mm in front of the V.P. Draw its projections and find its true length and inclinations with the reference planes.

26. A room is 5m × 4m × 4m high. Determine graphically the distance between the diagonally opposite top and bottom corners of the room.

27. Two apples on a tree are respectively 1.8 metres and 4 metres above the ground and 1.4 metres and 2.2 metres from a 0.25 metre thick wall but on the opposite

sides of it. The distance between the apples measured along the ground and parallel to the wall is 3 metres. Find the actual distance between the two apples.

28. A room is 10m × 6m × 4m high. An electric lamp is hanging in the centre of the room by a wire 1.5 metres long. The switch is situated at one corner of the room and 1.5 metre above the ground level. Draw the projections of a line joining switch and the lamp.
29. The end projectors of a line is 60 mm apart and projectors through the traces are 100 mm apart. Its one end is 10 mm above the H.P. Top view and front view make angles of 30° and 60° with the xy line respectively. Draw the projections of the line and find its true length and inclination with the H.P. and V.P. Also locate its traces.
30. A line AB, 90 mm long is inclined at 30° to the H.P. and 45° to the V.P. Its mid point M being 40 mm above the H.P. and 35 mm in front of the V.P. Draw its projections.
31. Two lines AB 42 mm long and AC 50 mm long make an angle of 120° between their front view and top view. The line AB is parallel to both the planes. Find the real angle between them.
32. An object O is placed 1.5 metres above the ground and in the centre of the room 4.2m × 3.5m × 3.5m high. Find the distance of O from one of the corners between the roof and two adjacent walls.
33. A wireless aerial tower 20 metres high is tied at the top by two guy ropes having angles of depression of 30° and 45°. Other ends of guy ropes are tied to two poles at the height of 5 metres and 2.5 metres. The two poles are 10 metres apart. Draw the projections of the arrangement and find the length of guy ropes.
34. Find the length of a solid diagonal of a cube of side 40 mm graphically.
35. A room is 6m × 5m × 4m high. An electric lamp is above the centre of the longer wall and 1 metre below the ceiling. The bulb is 0.25 metre away from the wall. The switch for the light is on an adjacent wall 1.4 metres above the floor and 1 metre away from the other longer wall. Find graphically the shortest distance between the switch and bulb.

CHAPTER 8

Projections of Planes

8.1 Introduction

Two dimensional objects having negligible thickness is known as plane figure or a plane. A plane has definite length or side or diameter. In this chapter, we shall study the method of obtaining projections of planes which are commonly used in engineering practices such as square, rectangle, pentagon, hexagon, circle, ellipse etc., as shown in Fig. 8.1.

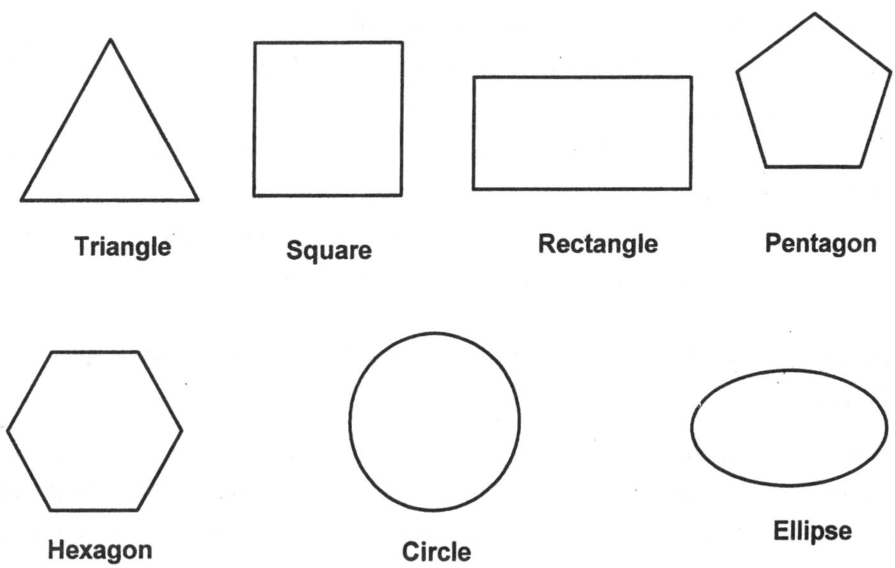

Fig. 8.1

8.2 Positions of Planes

Projections of any plane can be drawn if its position with respect to principal reference planes (i.e., H.P. and V.P.) is known. The Planes can be classified as follows:

1. **Perpendicular Planes:** Perpendicular planes have following positions.

   Plane perpendicular to both the planes.

   Plane parallel to one plane and perpendicular to other plane.

   Plane inclined to one plane and perpendicular to other plane.

2. **Oblique Planes:** Plane inclined to both the reference planes.

8.3 Traces of Planes

When a plane is extended (if necessary) it will intersect one of the reference planes or both the planes in lines depending upon its orientation.

Horizontal Trace: When a plane extended, if necessary, meets the horizontal plane in a line, then the line is called Horizontal Trace (H.T.).

Vertical Trace: When a plane extended, if necessary, meets the vertical plane in a line, then the line is called Vertical Trace (V.T.).

Following important points should be remembered:

1. When a plane is parallel to a reference planes (i.e., H.P. and V.P.), there will be no trace on that plane.
2. When a plane is perpendicular to both the reference planes, it will have both H.T. and V.T. lying on a straight line perpendicular to xy line.
3. When a plane is perpendicular to one of the reference planes (i.e., H.P. and V.P.) then its trace on other plane is perpendicular to xy line.

8.4 Projections of Planes Perpendicular to both the Reference Planes

When a plane is perpendicular to both the H.P. and the V.P. neither front view nor top view represents true shape of the plane. For such case first of all side view is obtained on a profile plane which represents its true shape. Then front view and top view are projected from side view. Since, plane is perpendicular to both the reference planes, if extended it will meet on both H.P. and V.P. Therefore, it will have both H.T and V.T.

The H.T and V.T lie on a line perpendicular to xy line. Fig. 8.2 shows projections of a square plane ABCD which is perpendicular to both H.P. and V.P.

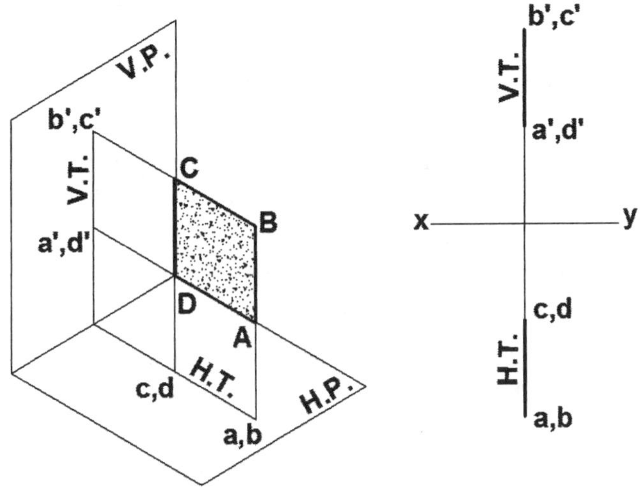

Fig. 8.2

Problem 8.1 (Fig. 8.3): *A triangular plane (in the form of equilateral triangle with 40 mm side) is perpendicular to both H.P. and the V.P. If it is considered to be in first quadrant, draw its projections and determine its traces.*

Construction (Fig. 8.3):

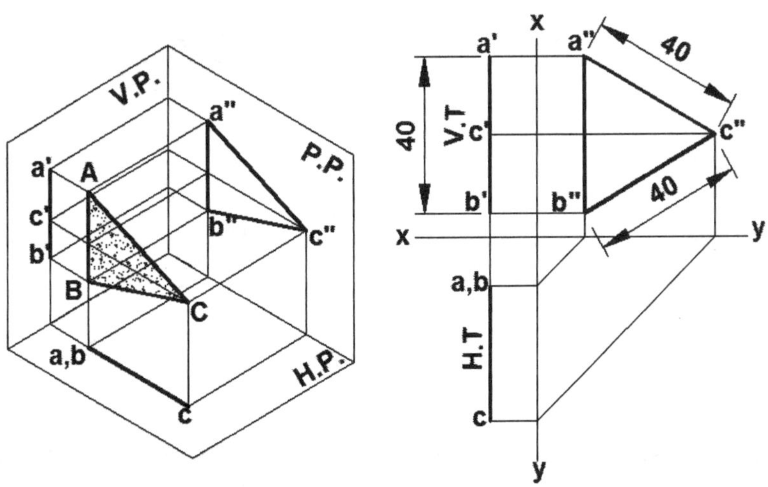

Fig. 8.3

1. Since, the plane is perpendicular to both the H.P. and V.P., its side view will show its true shape. So draw an equilateral triangle a"b"c" of side 40 mm as a side view.
2. Project the corners horizontally and mark points a', b' and c' on a vertical line. The line a'c'b' is the required front view and V.T. also.
3. Through a", b" and c" draw vertical projections to meet xy line and turn them at 45°.
4. At the intersection of vertical projections and corresponding projectors from side view, mark a, b and c. The line ac is required top view and H.T.

8.5 Projections of Planes Parallel to one Plane and Perpendicular to other Plane

(a) Plane Parallel to the V.P. and Perpendicular to H.P.: Fig.8.4 (a) shows a square plane ABCD parallel to the V.P. and perpendicular to the H.P. Since, it is parallel to the V.P. its projection on the V.P. will represents true shape. So first of all, the front view is drawn above the xy line. The top view is drawn by projecting front view vertically.

(b) Plane Parallel to the H.P. and Perpendicular to V.P.: When the plane is parallel to the H.P. and perpendicular to the V.P. its top view will show true shape as shown in Fig. 8.4 (b). In this case top view is drawn first with actual size. Then front view is obtained by projecting top view.

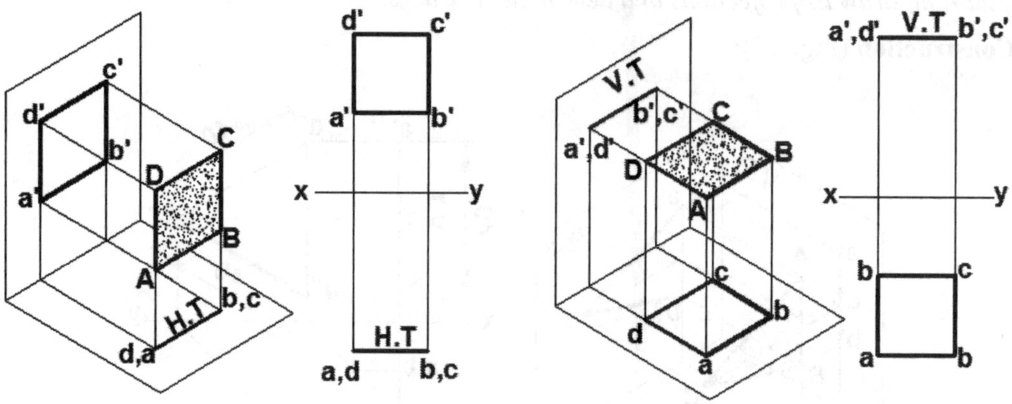

(a) (b)

Fig. 8.4

Problem 8.2 (Fig. 8.5): *A square plane ABCD of 50 mm side is kept on its corner on the H.P. such that its all sides are equally inclined to the H.P. and parallel to the V.P. If the plane is 25 mm away from the V.P. Draw its projections and show its traces.*

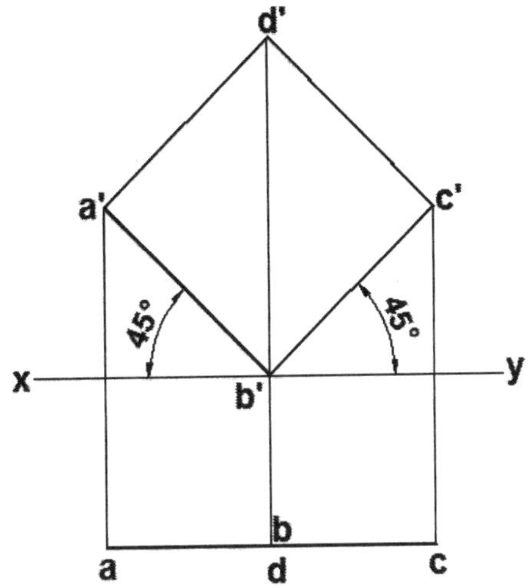

Fig. 8.5

Construction (Fig. 8.5):

1. Draw xy line.
2. Draw a square a'b'c'd' of side 50 mm with one corner on xy line and sides inclined a 45° to xy line. Through all corners a', b', c' and d', draw vertical projectors.
3. Draw a horizontal line parallel and 25 mm away from the xy line. On this line obtain points a, b, c and d at the intersection of vertical projectors. The line abc represents top view.

Problem 8.3 (Fig. 8.6): *A thin hexagonal plate of 25 mm side is resting on the H.P. with one of the sides inclined at 45° to the V.P. Draw its projections and show its trace.*

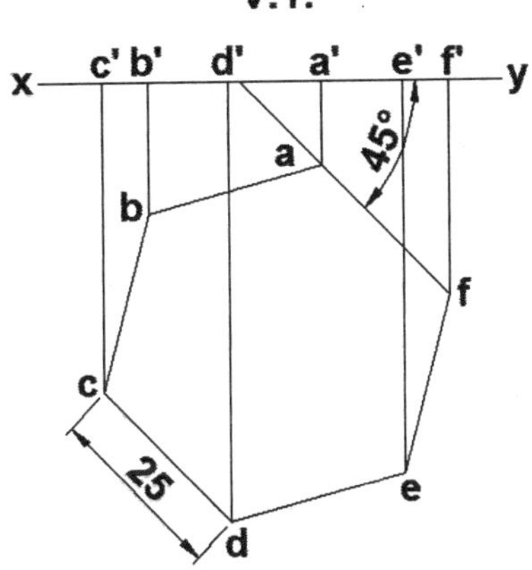

Fig. 8.6

Construction (Fig. 8.6):

1. Draw a xy line.
2. Draw a regular hexagon abcdef below the xy line such that side af is inclined at 45° to xy line. This represents top view.
3. Draw vertical projectors through all the corners of top view to intersect xy line.

   Horizontal line c'b'a'e'f' coinciding xy line represents the front view. This is also V.T.

Problem 8.4 (Fig. 8.7): *A square plane ABCD of side 40 mm has its plane parallel to H.P. and 15 mm away from it. Draw its projections when two of its sides are*

(a) *Parallel to the V.P.*

(b) *Inclined at 30° to the V.P.*

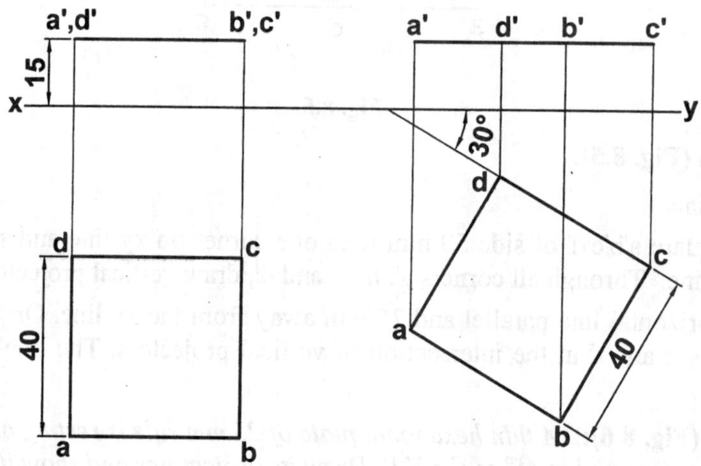

Fig. 8.7

Construction (Fig. 8.7):

Case 1: (When two sides are parallel to xy line)

1. Draw a xy line.
2. Draw a square abcd of side 40 mm such that two sides are parallel to xy line.
3. Draw a horizontal line parallel and 15 mm above the xy line.
4. Draw vertical projectors through all the corners in top view to intersect the horizontal line above xy. The horizontal line a'c' represents front view.

Case 2: (When two sides inclined at 30° to the xy)

1. Draw a xy line.

2. Below xy line, draw a square abcd such that a side is inclined at 30° to the xy line.
3. Obtain the front view by projecting the top view.

8.6 Projections of Plane Inclined to V.P. and Perpendicular to H.P.

(a) Plane Inclined to V.P. and Perpendicular to the H.P.

Figure 8.8 shows a square plane ABCD inclined at ϕ^o with the V.P. and perpendicular to the H.P. Its front view a'b'c'd' is smaller than its actual size because it is not parallel to V.P. But the top view is obtained as a line inclined at ϕ with xy line having its length equal to the true side. Its V.T. is perpendicular to the xy line and H.T. coincides with the top view. In such case, projections are drawn in two stages. In first stage the plane is assumed to be parallel to the V.P., and front view and top view are drawn. In second stage, top view is turned at required angle ϕ and new front view is obtained by joining point of intersection of horizontal and vertical projectors.

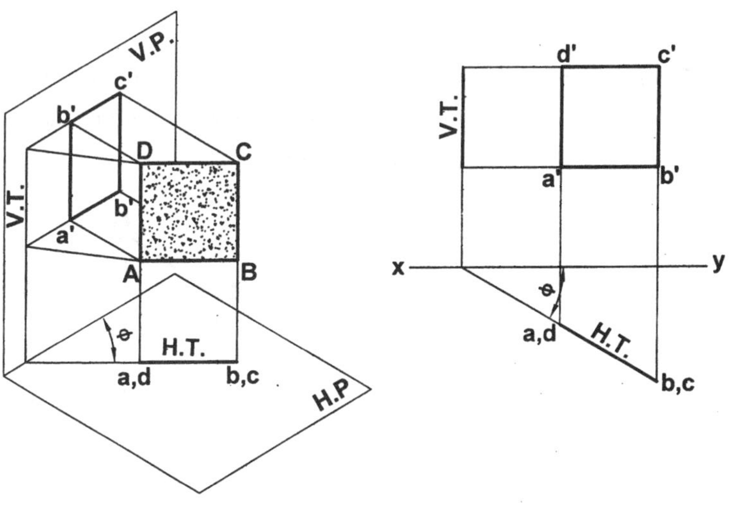

Fig. 8.8

(b) Plane Inclined to the H.P. and Perpendicular to the V.P.

Figure 8.9 shows a square plane ABCD inclined at θ^o to the H.P. and perpendicular to the V.P. The top view is shortened than its true size because the plane is not parallel to the H.P. But front view is a line inclined at θ^o with xy line having its length equal to its true side. The H.T. is perpendicular to xy line but V.T. coincides with front view.

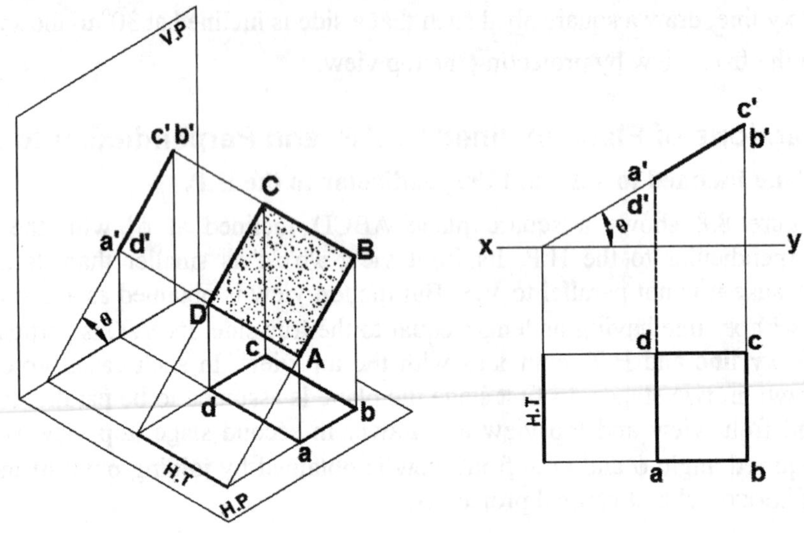

Fig. 8.9

Problem 8.5 (Fig. 8.10): *A hexagonal lamina of 25 mm side and negligible thickness is inclined at 30° to the V.P. and perpendicular to the H.P. One of its sides is parallel and 20 mm away from the V.P. Draw its projections.*

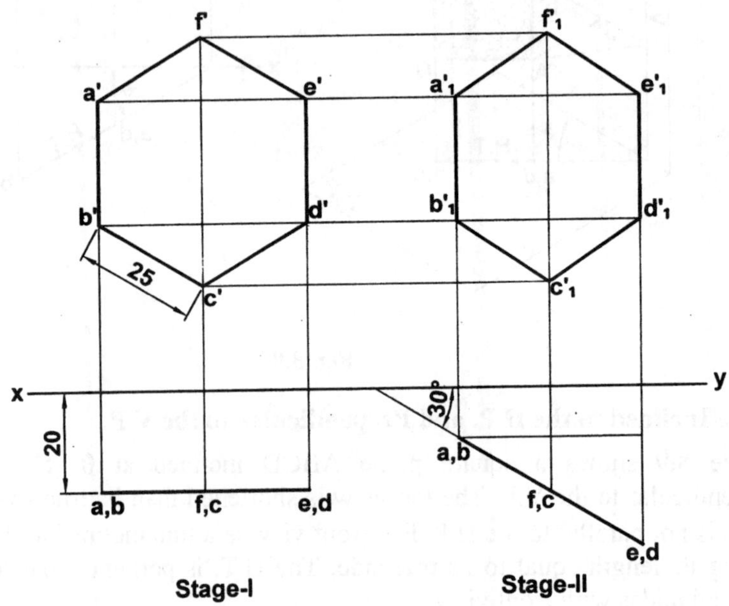

Fig. 8.10

Construction (Fig. 8.10):

Stage I

1. Assuming plane to be parallel to the H.P. draw a regular hexagon of 25 mm side above xy line.
2. Draw a horizontal line 20 mm below and parallel to xy line.

   Through a', b', c'...etc., draw vertical projections to meet horizontal line at a, b, c,.. etc.

Stage II

3. Reproduce the top view of first stage such that it makes an angle of 30° with xy.
4. Draw horizontal projectors through corners of front view of first stage and vertical projectors through corner points of new top view.
5. At the intersection of corresponding points mark a'_1, b'_1, c'_1 ... etc.

   Join $a'_1b'_1c'_1$ ... etc., which is the required front view.

Problem 8.6 (Fig. 8.11): *A circular plane having 50 mm diameter is inclined at 30° to the V.P. A point on its circumference is in the V.P. and centre is 30 mm above the H.P. Draw its projections.*

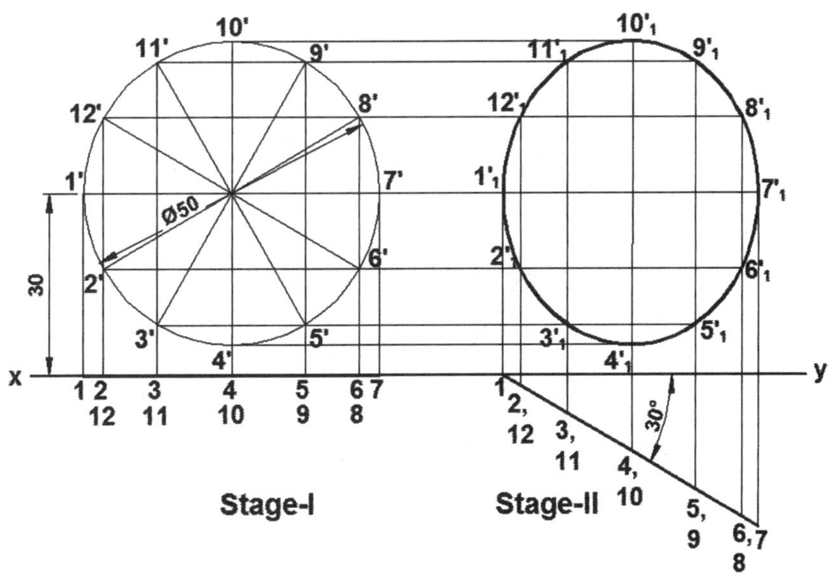

Fig. 8.11

Construction (Fig. 8.11):

Stage I

1. Assuming the plane initially in the V.P., draw a circle 50 mm diameter with its centre 30 mm above the xy line. Divide the circle into 12 equal parts and mark divisions as 1', 2', 3'... etc.
2. Through 1', 2' ... etc., draw vertical projectors to intersect xy line and obtain the top view which coincides xy line.

Stage II

3. Reproduce the top view keeping it inclined at 30° to the xy line.
4. Draw horizontal projectors through 1', 2', 3'.... etc., and vertical projector through top view of second stage.
5. Mark the points $1'_1$, $2'_1$, $3'_1$,... etc., at the intersection of horizontal projectors and vertical projectors of corresponding points.
6. Join $1'_1$, $2'_1$ .... etc., to obtain front view.

Problem 8.7 (Fig. 8.12): *A pentagonal plate of 25 mm side has an edge on the H.P. The surface of the plate is inclined at 45° to the H.P. and perpendicular to the V.P. Draw its projections.*

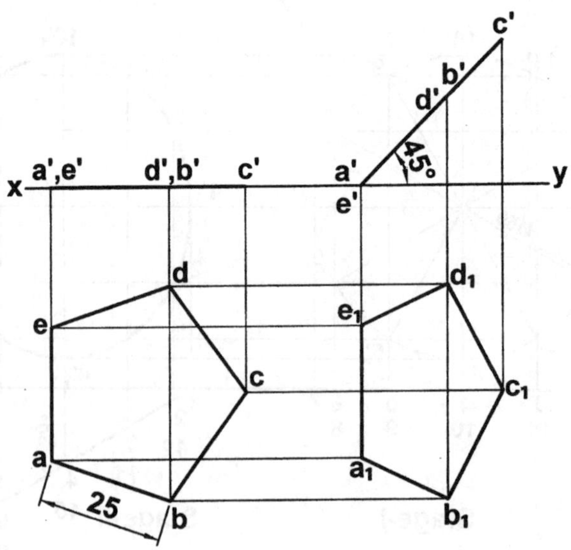

Fig. 8.12

Construction (Fig. 8.12):

Stage I

1. Draw a reference line xy. Below the xy line, draw a pentagon abcde.

2. Assuming it to be parallel to the H.P. Draw vertical projections through corners points in the top view, to intersect xy line. The front view is coinciding with xy line.

Stage II

3. Now rotate the top view by 45° and obtain the final top view as mentioned in previous case.

Problem 8.8 (Fig. 8.13): *A thin square plate with 50 mm side stands on one of its corners on H.P. and the opposite corner is raised so that one of the diagonals is twice of other. One diagonal is parallel to H.P. and V.P. Draw its projections and find the inclination of the plate with the H.P.*

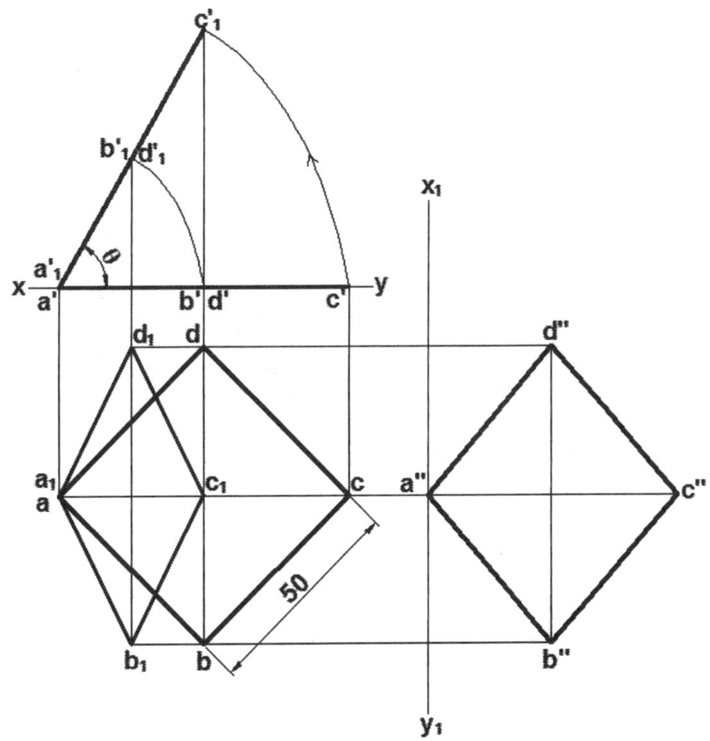

Fig. 8.13

Construction (Fig. 8.13):

Stage I

1. Assume initially, the square plane to be kept on the H.P. with its sides equally inclined with the V.P. and draw its top view abcd as square of 50 mm sides.
2. From the top view project front view on the xy line.

Stage II

3. Keeping end a fixed, rotate the front view (a'c') to $a'c_1'$ such that the length of diagonal ac_1 is half the length of ac.
4. Also rotate b'd' to cut the line $a'c_1'$ at b_1' or d_1'. Measure the angle θ which is the required angle.
5. Draw a new reference line x_1y_1 parallel to diagonal b_1d_1. Mark points a", b", c" and d" keeping their distance from x_1y_1 equal to the corresponding distances of a_1', b_1', c_1' and d_1' from xy line.

Problem 8.9 (Fig. 8.14): *A circular lamina of 60 mm diameter has its centre 15 mm above H.P. and 40 mm in front of the V.P. Its plane is inclined at 30° to the H.P. and perpendicular to the V.P. Draw its projections and traces.*

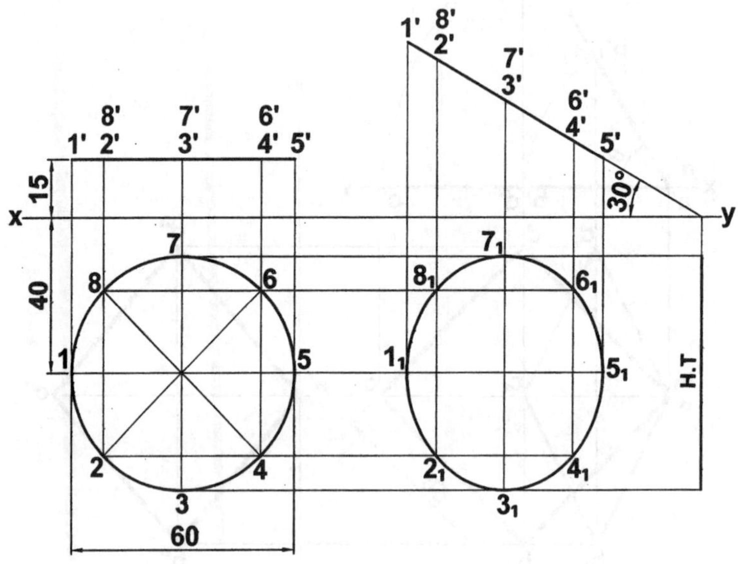

Fig. 8.14

Construction (Fig. 8.14):

Stage I

1. Draw a circle of 60 mm diameter as top view such that its centre is 40 mm below the xy line.
2. Project the front view from the top view which is a line 15 mm above and parallel to xy line.

Stage II

3. Reproduce the front view of stage-I making it inclined at 30° to xy line.
4. Draw vertical projectors through different points 1', 2'.... etc., on the front view and horizontal projectors through 1, 2, 3.... etc.
5. At the intersection of corresponding projectors, obtain points 1_1, 2_1, 3_1.... etc., and join them by a smooth curve which is the required top view.

Problem 8.10 (Fig. 8.15): *A rhombus having major and minor diagonals 50 mm and 38 mm respectively is resting on one of its corners on the H.P. such that its major diagonal is inclined at 45° with the H.P. and minor diagonal remains perpendicular to the V.P. Draw its projections.*

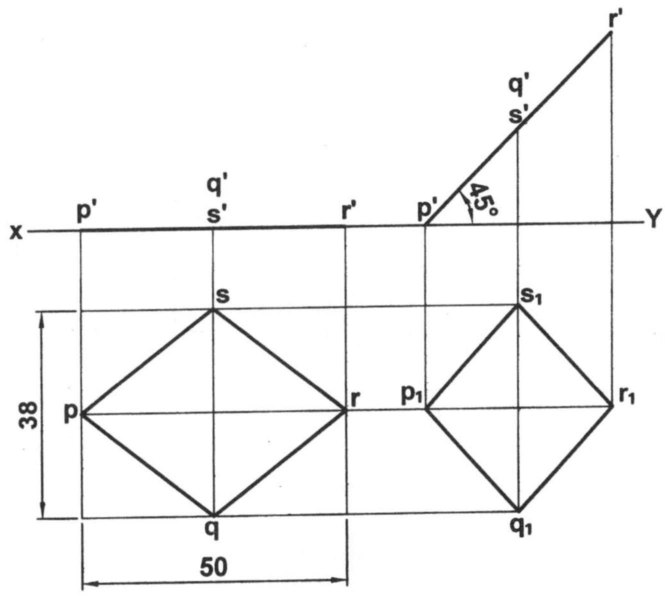

Fig. 8.15

Construction (Fig. 8.15):

Stage I

1. Since the plane has its corner on the H.P., it is first assumed to be on the H.P. So, draw a rhombus of given dimension below the xy line as top view.
2. Project its corner points on the xy line to obtain front view.

Stage II

3. Reproduce the front view such that p'r' makes an angle of 45° with xy line.
4. Draw vertical projectors through different corner points of front view of second stage and draw horizontal projectors through corner points of top view.

5. At the intersection of corresponding horizontal and vertical projectors, obtain points p_1, q_1, r_1 and s_1. Join these points to obtain top view.

Problem 8.11 (Fig. 8.16): *The top view of a plate of negligible thickness, whose surface is perpendicular to the V.P. and inclined at 45° to the H.P. is a circle of 50 mm diameter. Draw its projections and find its true shape.*

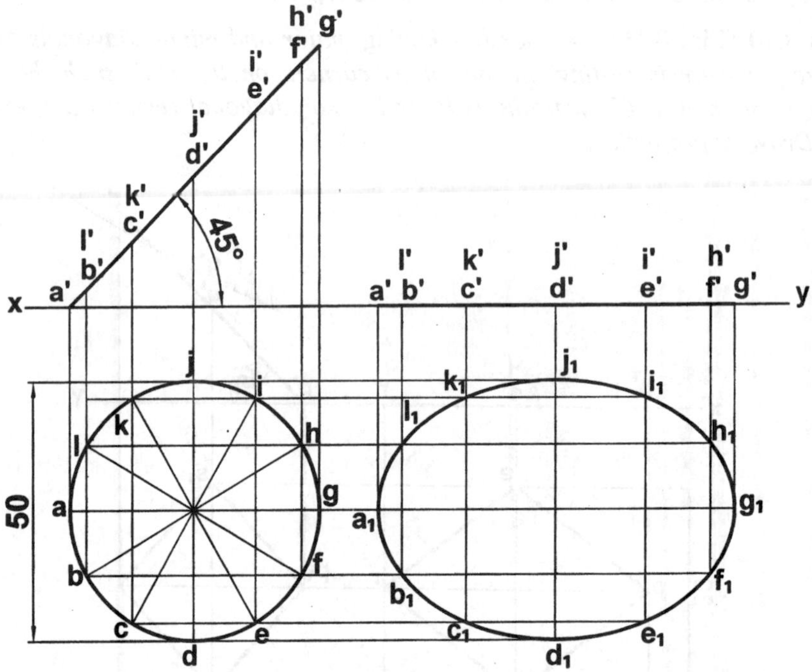

Fig. 8.16

Construction (Fig. 8.16):

Stage I

1. Draw a circle of 50 mm diameter below xy line and divide it into 12 equal parts. Project a to a' on the xy line.
2. Through a', draw a line inclined at 45° to xy line.
3. Project points a, b, c... etc., on the inclined line to obtain front view.

Stage II

4. Reproduce the front view, coinciding with xy line.
5. Obtain the new top view using the method as mentioned in the previous problems.

8.7 Projections of Oblique Planes

Plane whose surface is inclined to both the reference planes, is known as oblique plane. To draw projections of such planes placed in different positions, following procedure must be considered.

Case 1: Plane Inclined to the H.P. and An Edge or Diameter is Parallel to H.P. and Inclined V.P.

Projections of the given plane (in above mentioned position) are drawn in three stages.

Stage I (Initial Position)

Assume plane to be parallel to the H.P. and an edge (or diameter or diagonal) perpendicular to the V.P. Draw front view and top view in initial position.

Stage II (Intermediate Position)

Front view (a line) is tilled with the required angle (θ) and new top view is drawn by projecting top view of first stage and front view of second stage.

Stage III (Final Position)

Turn the top view with the required inclination with the V.P. and obtain the new front view by projecting front view of second stage and top view of third stage.

Note: In the above case if an edge is in the H.P., in the initial position, the plane is assumed to be lying in the H.P.

Case 2: Plane Inclined to the V.P. and An Edge (or Diameter or Diagonal) Parallel to the V.P. and Inclined to the H.P.

Projections of the plane given in above positions will also be drawn in three stages.

Stage I (Initial Position)

Assume the plane to be parallel to the V.P. and an edge (or diameter or diagonal) perpendicular to the H.P. and draw front view and top view.

Stage II (Intermediate Position)

Tilt the top view with the required angle and project new front view from it.

Stage III (Final Stage)

Tilt the front view of second stage with the required angle and draw new top view by projecting final front view.

Note: In this case if an edge is in the V.P., the plane is assumed to be lying in the V.P. with an edge perpendicular to the H.P. in initial position.

Problem 8.12 (Fig. 8.17): *A square lamina PQRS with 50 mm sides has a corner in the H.P. One of its diagonal is inclined at 30° to the H.P. and other diagonal is parallel to the H.P. and inclined at 45° with the V.P. Draw the projections of square PQRS.*

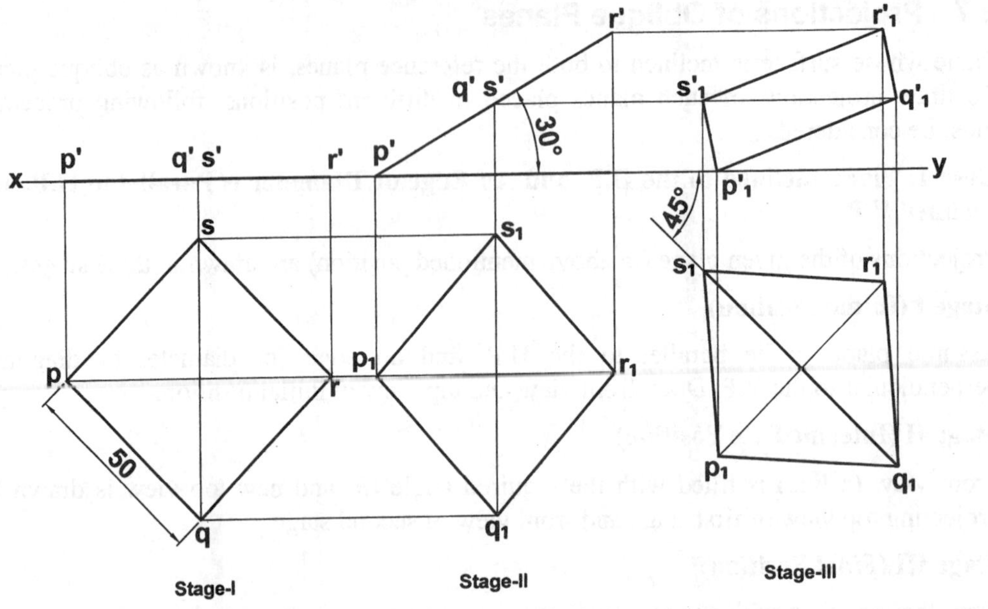

Fig. 8.17

Construction (Fig. 8.17):

Stage I

1. Assume the square lamina to be kept on the H.P. having its sides equally inclined to the V.P. Draw square pqrs below xy line as top view.
2. Project the corner p, q, r and s on the xy line. Thus line p'r' coinciding with xy line represents front view.

Stage II

3. Reproduce the front view of stage-I such that p'r' makes an angle of 30° with xy line.
4. Draw vertical projectors through p', q', r' and s' and horizontal projectors through p, q, r and s.
5. At the intersection of corresponding horizontal and vertical projectors, mark points p_1, q_1, r_1 and s_1. Join these points to obtain top view of second stage.

Stage III

6. Reproduce the top view of stage-II with a diagonal q_1s_1 inclined at 45° to xy line.
7. Draw vertical projectors through p_1, q_1, r_1 and s_1 and horizontal projectors through p', q', r' and s'.

8. At the intersection of vertical and horizontal projectors through corresponding points, mark $p_1'q_1'$ ...etc. and join them by straight lines to obtain final front view.

Problem 8.13 (Fig. 8.18): *A rectangular plane 50 mm long and 40 mm broad has its shorter edge on the H.P. The plane is inclined in such a way that its top view appears a square of 40 mm side. If an edge resting on the H.P. is inclined at 30° to the V.P. Draw its projections.*

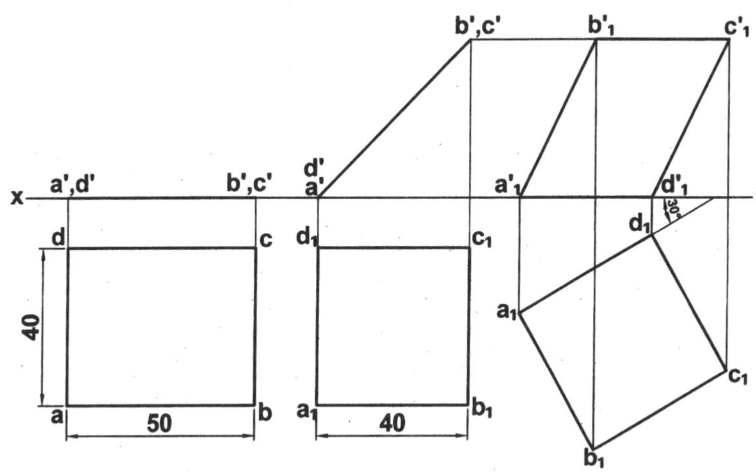

Fig. 8.18

Construction (Fig. 8.18):

Stage I

Assume the given plane to be on the H.P. with a side ad perpendicular to the V.P.

1. Draw rectangle abcd of given size below xy line with a side ad perpendicular to xy line.
2. Project corners a, b, c, d on the xy line so that line a'b' coinciding with xy line represents front view.

Stage II

3. Draw a square $a_1b_1c_1d_1$ below xy line. Project its corner points on the xy line.
4. Mark a'd' on the xy line. Now with a' as centre and radius equal to 50 mm, draw an arc to cut vertical projector through c_1 or b_1 at point b' or c'.

   Join a' with b' which is new front view.

Stage III

5. Reproduce top view of stage-II with a side a_1d_1 inclined at 30° to xy line.
6. Obtain final front view as mentioned in previous problems.

Problem 8.14 (Fig. 8.19): *Draw the projections of a regular pentagon of 25 mm sides having one of its sides on the H.P. and inclined at 45° to the V.P. The surface of the plane makes an angle of 30° with the H.P.*

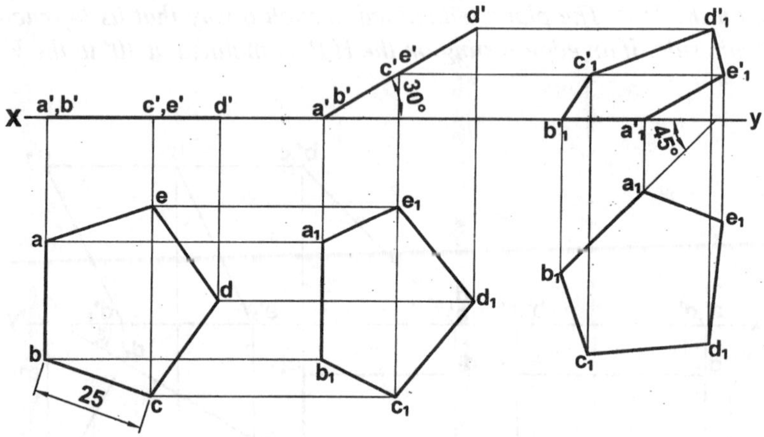

Fig. 8.19

Construction (Fig. 8.19):

Stage I

1. Assuming pentagonal plane to be placed on the H.P. with one of its sides perpendicular to the V.P., draw a regular pentagon of 25 mm sides as top view.
2. Project all corner of top view to xy line to obtain line a'd' as front view.

Stage II

3. Reproduce the front view of stage-I making an angle of 30° with xy line.
4. Draw vertical projectors through corner points of front view of stage-II and horizontal projectors from top view of stage-I and obtain points a_1, b_1, c_1 etc. Join them to obtain top view.

Stage III

5. Reproduce top view of second stage with its side a_1b_1 inclined at 45° with xy line.
6. Project the final front view $a_1'b_1'c_1'd_1'e_1'$ as shown.

Problem 8.15 (Fig. 8.20): *A hexagonal plane of 30 mm side is resting on one of its sides on the H.P. which makes an angle of 45° with the V.P. The surface of the plane also makes an angle of 45° with the H.P.*

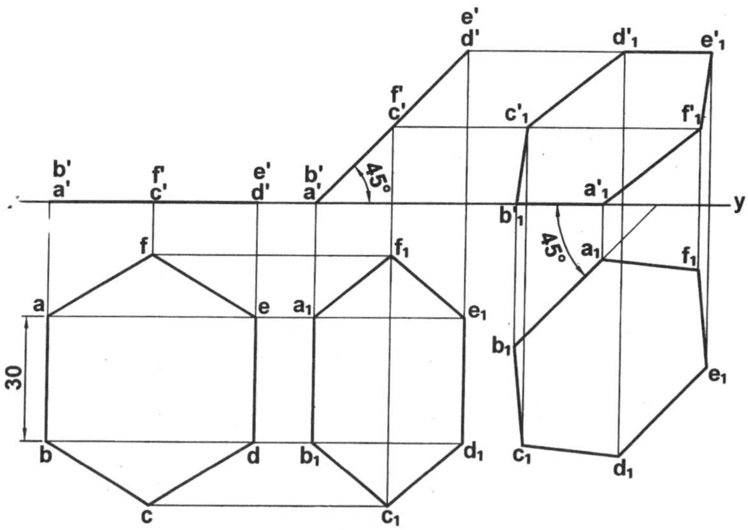

Fig. 8.20

Construction (Fig. 8.20):

Stage I

Assume the hexagon to be placed on the H.P. with a side ab perpendicular to the V.P. Draw a regular hexagon abcdef below xy line as top view and project the front view from it.

Stage II

Reproduce the front view a'e' inclined at 45° with xy line and project the second top view from it.

Stage III

Reproduce the top view of stage-II with a side a_1b_1 inclined at 45° to the xy line and project the final front view from it.

Problem 8.16 (Fig. 8.21): *A hexagonal plane of 25 mm side is resting on its corner on the H.P. with its surface making an angle of 45° to the H.P. If the diagonal containing that corner is inclined at 40°, draw its projections.*

Construction (Fig. 8.21):

Stage I

Assume the hexagon to be placed on the H.P. with a side parallel to xy line. Draw a regular hexagon abcdef of 25 mm side below the xy line and project the front view from it.

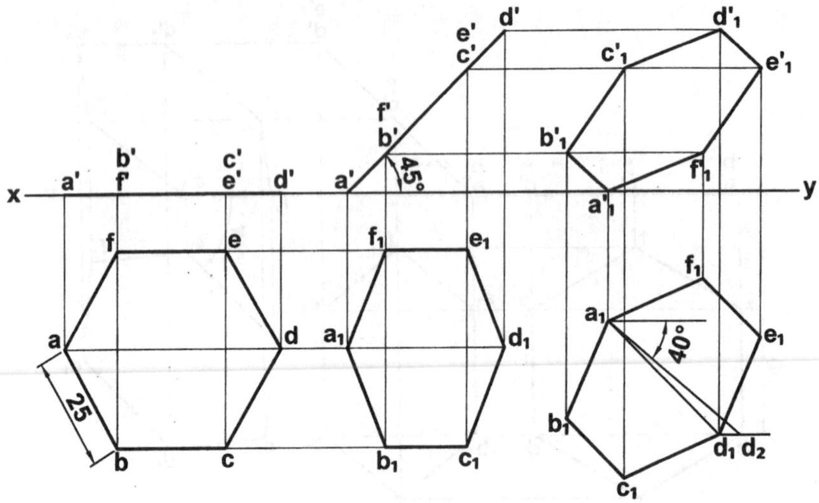

Fig. 8.21

Stage II

Draw a'd' inclined at 45° to xy line keeping a' on the xy line and project the second top view from it.

Stage III

Since, diagonal ad is inclined at 45° to H.P. and 40° to the V.P., it is required to find apparent angle as follows:

1. Through any point say a_1, draw a line a_1d_2 equal to true length of the diagonal ad of first stage and inclined at 40° to xy line.
2. Through d_2, draw a horizontal line as locus.
3. With a_1 as centre and radius equal to a_1d_1 (second stage) draw an arc to meet horizontal line through d_2 at d_1.
4. With a_1d_1 as diagonal reproduce top view of stage-II. Project the front view from it.

Problem 8.17 (Fig. 8.22): *A circular plate of 48 mm diameter appears as an ellipse in the top view having its major and minor axis 48 mm and 30 mm respectively. If the major axis of the ellipse is parallel to xy line, draw its projections.*

Construction (Fig. 8.22):

Stage I

Draw a circle of 48 mm diameter as top view and divide its circumference into twelve equal parts. Project its front view on xy line.

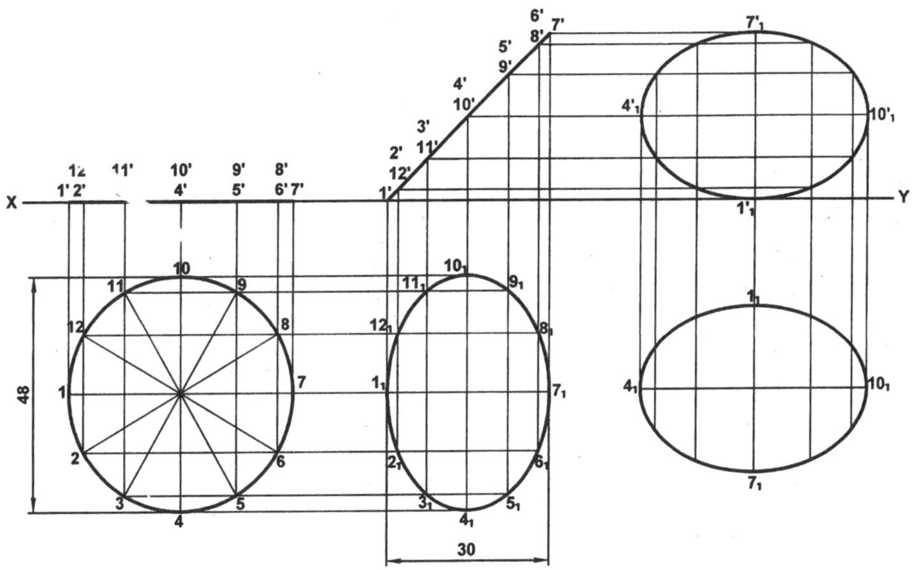

Fig. 8.22

Stage II

Draw minor axis of ellipse $1_1 7_1$ equal to 30 mm. Through 7_1, draw a vertical projector. Project 1_1 to 1' on xy line. With 1' as centre and radius equal to 48 mm draw an arc to meet vertical projector through 7_1 at 7'. Join 1' with 7' to obtain front view of second stage. Complete the top view.

Stage III

Reproduce top view of stage-II with its major axis $4_1 10_1$ parallel to xy line. Project the final front view from it as shown.

Problem 8.18 (Fig. 8.23): *A rhombus having its longer and smaller diagonal, 90 mm and 40 mm respectively. If the longer diagonal is inclined at 30° to the H.P. and smaller diagonal is parallel to both the reference plane, draw its projections.*

Construction (Fig. 8.23):

Stage I

Draw a rhombus abcd with the given dimensions as top view below xy line. Project the front view p'r' from the top view.

Stage II

Reproduce the front view of stage-I as p'r' inclined 30° to xy line and project the top view $p_1q_1r_1s_1$.

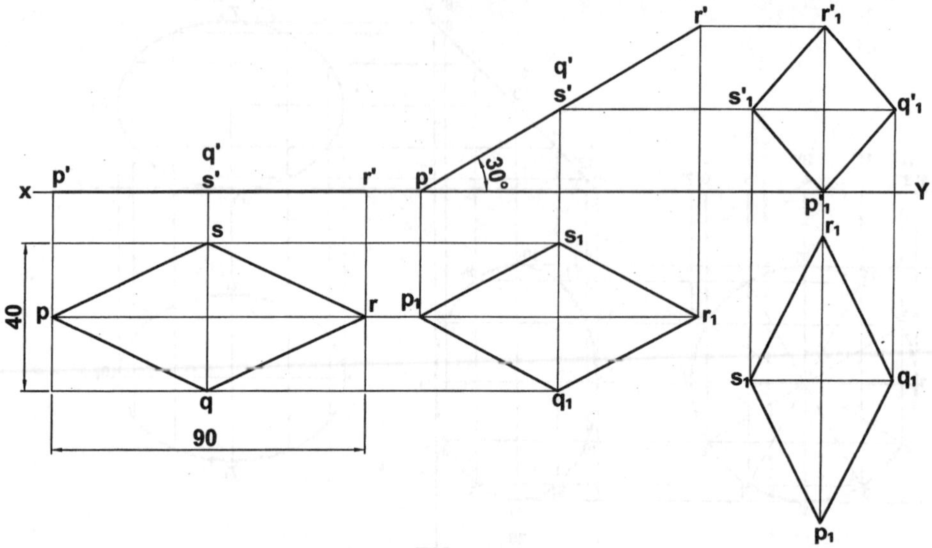

Fig. 8.23

Stage III

Reproduce the top view of stage-II keeping its diagonal s_1q_1 parallel to xy line and project final front view from it.

Problem 8.19 (Fig. 8.24): *A 30-60 set-square of negligible thickness has its longest edge 60 mm long, in the V.P. and inclined at 45° to the H.P. Its plane is inclined at 45° to the V.P. Draw its projections.*

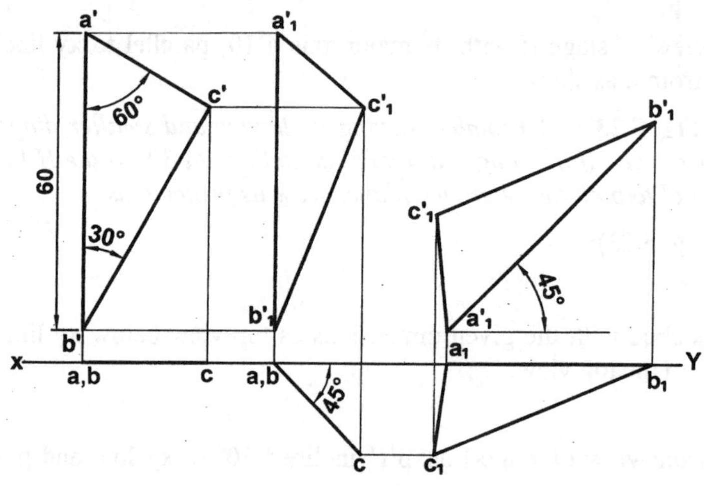

Fig. 8.24

Construction (Fig. 8.24):

Stage I

Initially assuming the set square to be in V.P. having its longest edge perpendicular to H.P., draw a triangle a'b'c' as front view. Project its top view ac on xy line.

Stage II

Reproduce top view of stage-I keeping its inclination 45° with xy line and project its front view a'_1 b'_1 c'_1.

Stage III

Reproduce the front view a'_1 b'_1 c'_1 of stage-II in such a way that its longest edge $a'_1b'_1$ makes an angle of 45° with xy line. Project the final top view a_1 b_1 c_1.

Problem 8.20 (Fig. 8.25): *Draw the projections of a circular plane of 48 mm diameter resting on a point P of its circumference on the H.P. with its plane inclined at 45° to the H.P. and*

(a) *top view of diameter PQ makes an angle of 30° with the V.P.*

(b) *the diameter PQ makes an angle of 30° with the V.P.*

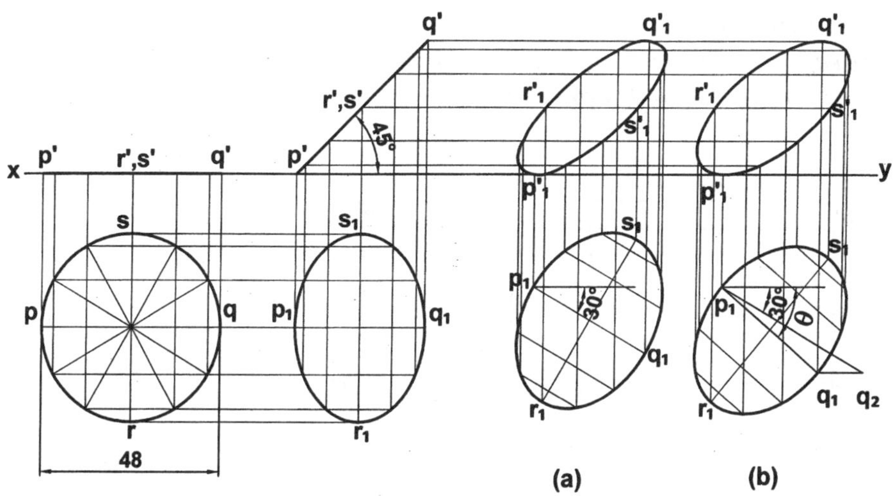

Fig. 8.25

Construction (Fig. 8.25):

Stage I

Initially, assume the circular plane to be on the H.P. So draw a circle of 48 mm diameter below the xy line as top view. Divide its circumference into twelve equal parts. Project its front view as a line p'q' coinciding with xy line.

Stage II

Reproduce the front view p'q' such that it makes an angle of 45° with the xy line. Project second top view.

Stage III(a)

Reproduce second top view such that its minor axis p_1q_1 makes an angle of 30° with xy line. Project its front view.

Stage III (b)

Since the diagonal is inclined to both the planes, its apparent angle is, determined as follows:

1. Draw a line p_1q_2 equal to the length of pq and making an angle of 30° with horizontal.
2. Draw a horizontal line through q_2.
3. Now with p_1 as centre and radius equal to p_1q_1 (stage-II), draw an arc to cut the horizontal line at q_1.
4. Join p_1 with q_1. Inclination of p_1q_1 with horizontal is the apparent angle θ.
5. Reproduce top view of stage-II having p_1q_1 as minor axis. Project final top view.

Problem 8.21 (Fig. 8.26): *A thin rectangular lamina ABCD of size 50 mm × 30 mm has its shorter edge AB in the V.P. and inclined at 30° to the H.P. If its front view appears as square of 30 mm side, draw its projections.*

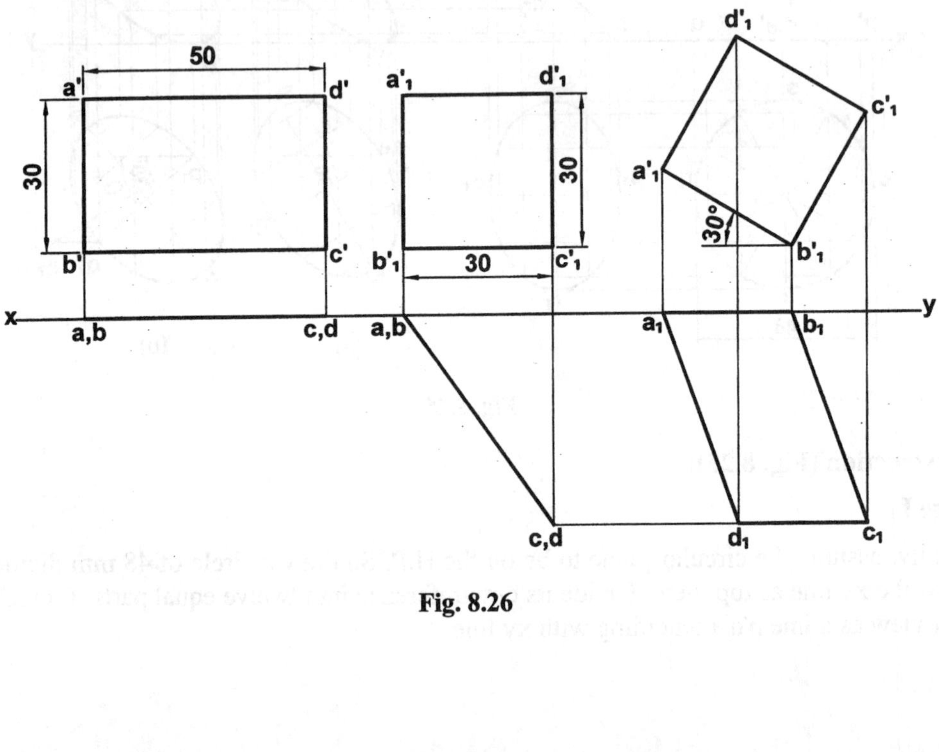

Fig. 8.26

Construction (Fig. 8.26):

Stage I

1. Initially, assume the plane to be in the V.P. So draw a rectangle a'b'c'd' of given dimensions above the xy line with its shorter edge a'b' perpendicular to xy line which represents front view.
2. Project its top view on xy line.

Stage II

3. Draw a square $a_1' \, b_1' \, c_1' \, d_1'$ of 30 mm side as the front view of second stage.
4. Project a_1' (or b_1') on the xy line at a (or b).
5. With a as centre and radius equal to ac (=50 mm) cut an arc on the vertical projector through c_1' at c. Thus, line ac is the second top view.

Stage III

6. Reproduce the front view of second stage with its side $a_1' \, b_1'$ making an angle of 30° with xy line.
7. Project final top view as shown.

Problem 8.22 (Fig. 8.27): *A thin circular plate of 60 mm diameter is kept inclined at 60° to the H.P. and 30° to the V.P. Draw its projections.*

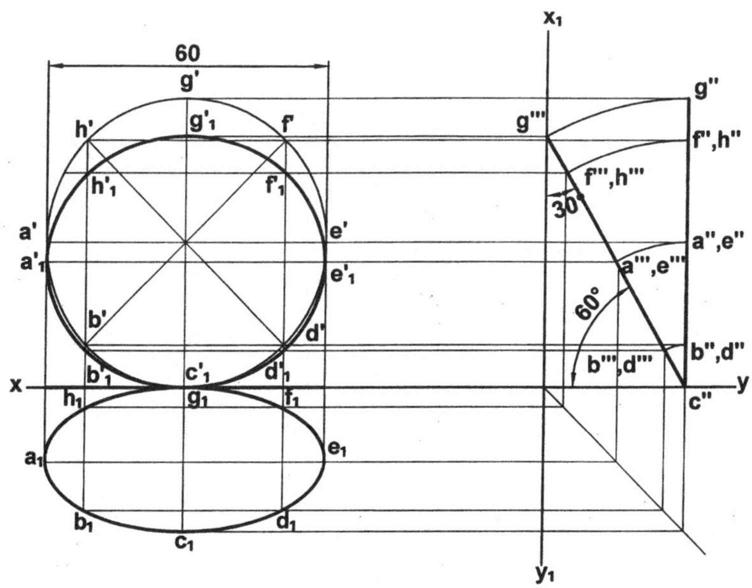

Fig. 8.27

Construction (Fig. 8.27):

1. Assuming circular plane parallel to the V.P., draw its front view as circle of 60 mm diameter.
2. Draw a reference line x_1y_1 perpendicular to xy line to represent an auxiliary plane perpendicular to both H.P. and V.P. Draw a line c"g" perpendicular to xy to represent side view of the given plane.
3. Divide the circle into eight equal parts and mark the points a', b', c'.....etc., on its circumference. Project these points on the side view as a"', b"', c"'... etc.
4. Tilt the side view such that it makes an angle of 60° with xy line and touches both xy and x_1y_1 lines.
5. Also transfer points a", b", c"... etc., on the new side view as a"', b"', c"'.... etc.
6. Through a"', b"', c"'... etc., draw horizontal projectors to meet the corresponding vertical projectors through corresponding on the initial front view. Join a'_1, b'_1, c'_1 ... etc., by a smooth curve which is the required front view.
7. Project the top view with the help of front view and side view.

Problem 8.23 (Fig. 8.28): *An equilateral triangle of 45 mm long sides has an edge on the H.P. and inclined at 60° to the V.P. If the plane is inclined at 45° with the H.P., draw its projections.*

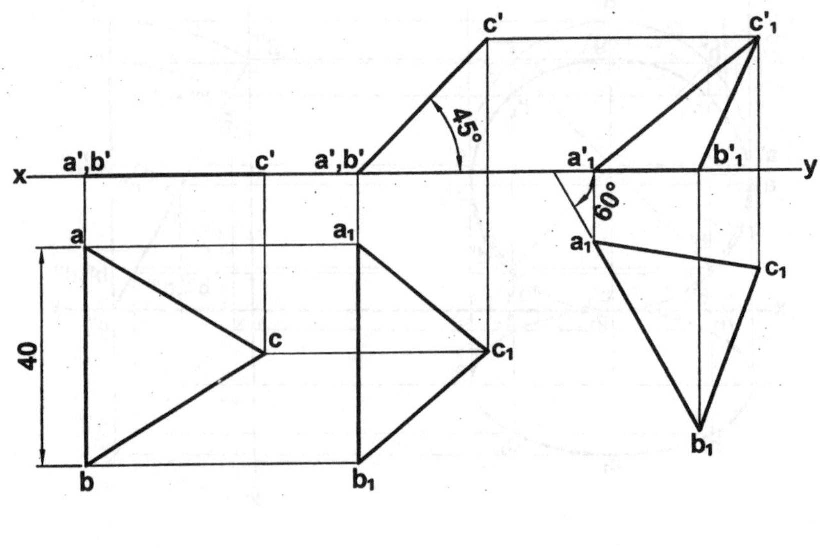

Fig. 8.28

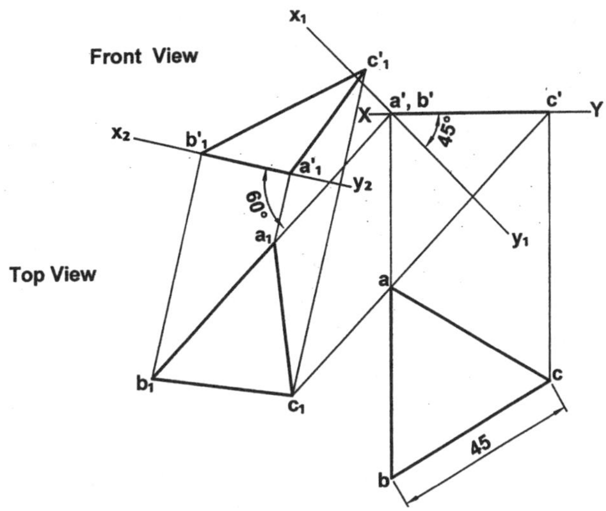

Fig. 8.29

Construction (Fig. 8.28 & 8.29):

Method I (General Method)

Stage I

Draw an equilateral triangle abc of 40 mm side with its side ab perpendicular to xy line. Project the front view on the xy line.

Stage II

Reproduce the front view a'c' inclined at 45° to the xy line. Project the top view from it.

Stage III

Reproduce the top view of stage-II with its side a_1b_1 inclined at 60° to xy line. Project final front view from it.

Method II (Auxiliary Plane Method)

In this method, instead of tilting the front or top view, projections are obtained on an auxiliary plane which is inclined at given angle with the reference plane.

1. Assuming plane to be placed on the H.P., draw an equilateral triangle abc of 45 mm side below xy line to represent top view. Project the front view on the xy line.
2. Draw a new reference line x_1y_1 passing through a' and inclined at 45° to it.
3. Project new top view such that distances of a_1, b_1, c_1 from x_1y_1 are equal to the distances of corresponding points of previous top view from the previous reference line. For example: $a'a_1 = a'a$, $b'b_1 = b'b$ etc.
4. Now draw a new reference line x_2y_2 inclined at 60° with the a_1b_1. Project the final front view a'_1, b'_1, c'_1 using method described in step 3.

Problem 8.24 (Fig. 8.30): *A regular hexagonal thin plate of 30 mm side has a square hole of 20 mm side centrally cut through it. It is resting on one of its sides on the H.P. with its surface inclined at 60° to the V.P. Its corner nearest to the V.P. is 20 mm away from the V.P. Draw its projections.*

Construction (Fig. 8.30):

Stage I

Assume the hexagonal plate to be placed on the H.P. on one of its side with its plane parallel to V.P. Draw a regular hexagon a'b'c'd'e'f' of 30 mm side with one side coinciding with xy line. Draw a square of 20 mm side inside the hexagon centrally.

Project its top view at a distance of 20 mm below xy line.

Stage II

Reproduce the top view of stage-I inclined at 60° with xy line. Draw horizontal projectors through initial front view and vertical projectors through top view of stage-II to obtain new front view $a'_1, b'_1, c'_1, d'_1, e'_1, f'_1$.

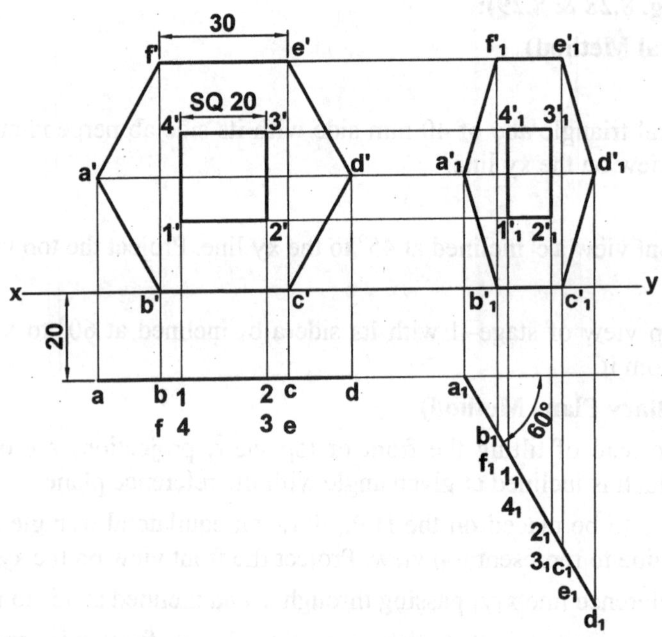

Fig. 8.30

Problem 8.25 (Fig. 8.31): *A regular hexagonal thin plate of 40 mm side has a circular hole of 40 mm diameter in its centre. It is resting on one of its sides on the H.P. with its surface inclined at 30° to the V.P. and perpendicular to the H.P. Draw its projections.*

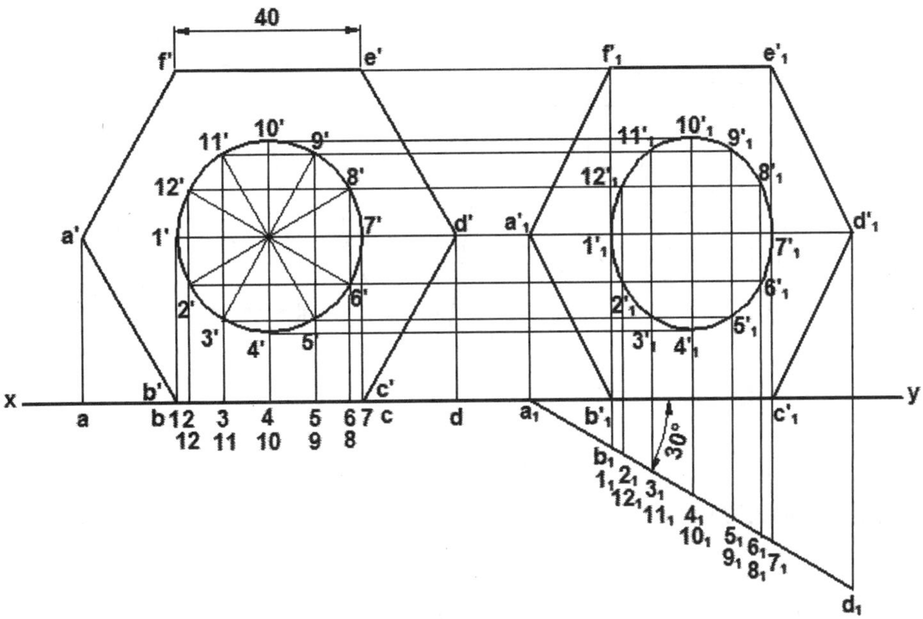

Fig. 8.31

Construction (Fig. 8.31):

Stage I

Assume the hexagonal plate to be in the V.P. with one of its sides on the H.P. Draw a regular hexagon a'b'c'd'e'f' having 40 mm side. Also draw a circle of 40 mm diameter inside the hexagon centrally.

Project its top view on the xy line.

Stage II

Reproduce the top view of initial stage inclined at 60° to the xy line. Project the final front view from new top view and initial front view.

Problem 8.26 (Fig. 8.32): *A rectangular plate ABCD measuring 50 mm × 30 mm has its diagonal AC inclined at 30° to the H.P., whereas the diagonal BD is inclined at 45° to the V.P. Draw its projections.*

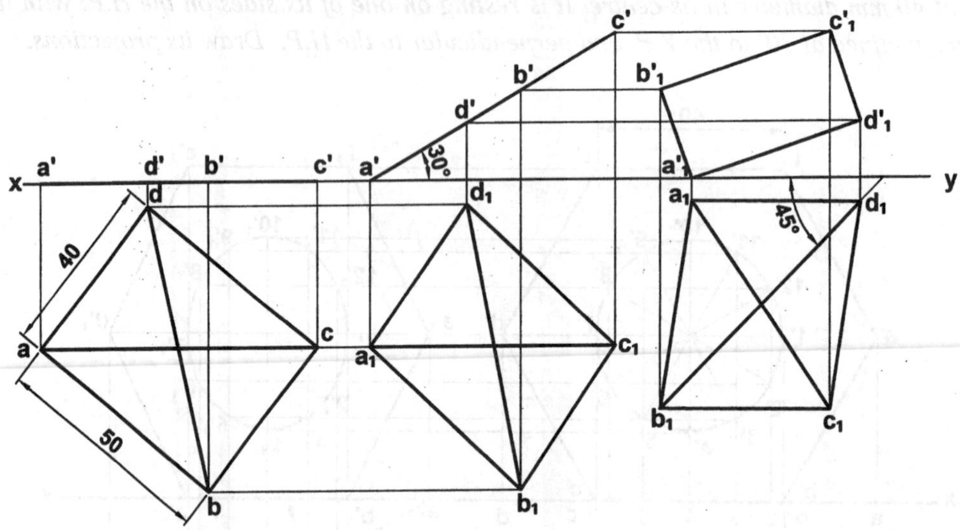

Fig. 8.32

Construction (Fig. 8.32):

Stage I

1. Assume the plate to be placed on the H.P. initially with its diagonal ac parallel to xy line. Draw top view as rectangle abcd with ac parallel to xy line. Project its front view on the xy line.

Stage II

3. Reproduce the front view a'c' making an angle of 30° with the xy line.
4. Project the second top view $a_1b_1c_1d_1$ from first top view and second front view.

Stage III

5. Reproduce the top view $a_1b_1c_1d_1$ such that diagonal b_1d_1 makes an angle of 45° with the xy line.
6. Project final front view a_1', b_1', c_1', d_1' from the second front view and third top view.

Problem 8.27 (Fig. 8.33): *The top view of a lamina whose surface is perpendicular to the V.P. and inclined at an angle of 45° to the H.P. appears as a regular hexagon of 25 mm side, having a side parallel to the xy line. Draw its projections and find its true shape.*

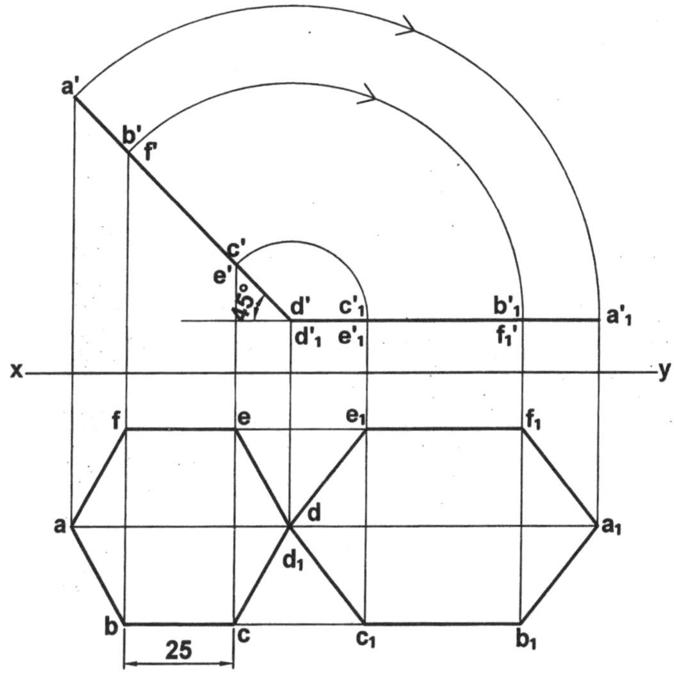

Fig. 8.33

Construction (Fig. 8.33):

Stage I

1. Draw a regular hexagonal abcdef of 25 mm side below xy line keeping one of its side parallel to xy line. This represents given top view.
2. Project its front view on a straight line inclined at 45° to xy line.

Stage II

3. To determine its true shape, the front view is required to be parallel to xy line. So, rotate the front view such that it is parallel to xy line.
4. Project final top view $a_1b_1c_1d_1e_1f_1$.

Problem 8.28 (Fig. 8.34): *A pentagonal plane of 30 mm side rests on its one of the corners on the H.P. If the side opposite to that corner is parallel to the V.P., draw its projections.*

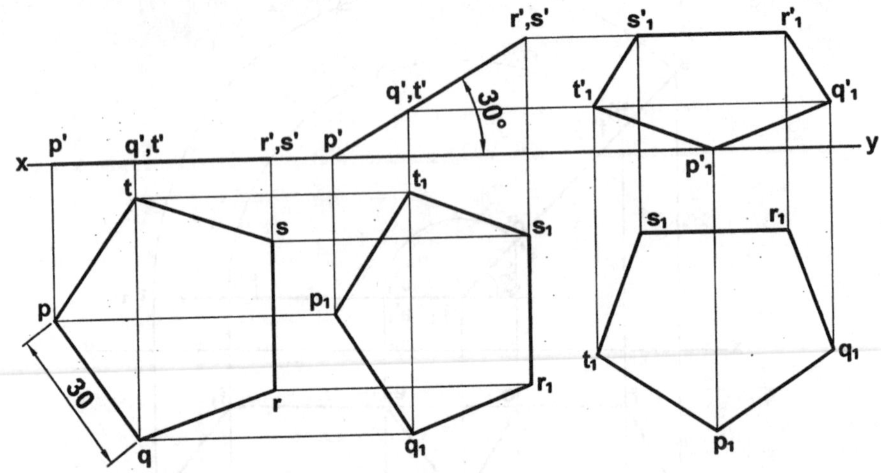

Fig. 8.34

Construction (Fig. 8.34):

Stage I

1. Draw a regular pentagon pqrst below xy line with one of its sides (say rs) perpendicular to xy line. This represents initial top view.
2. Project its front view p'q'r's't' on the xy line.

Stage II

3. Reproduce the front view making it inclined at 30° to the xy line.
4. Project second top view $p_1q_1r_1s_1t_1$ from second front view and initial top view.

Stage III

5. Reproduce the second top view with its side s_1r_1 parallel to xy line.
6. Project the final front view from the new top view and second front view.

Exercises

1. An equilateral triangle of 50 mm sides is resting on one of its sides on the H.P. The side on which it rests makes an angle of 60° to the V.P. and its surface is inclined at 45° to the H.P. Draw its projections.
2. A square ABCD having diagonals 50 mm long, is resting on a corner A on the H.P. The diagonal AC is inclined at 30° to the H.P. While diagonal BD makes an angle of 60° to the V.P. and parallel to the H.P. Draw its projections.

3. A thin pentagonal plate with 35 mm side has a side in the V.P. and inclined at 30° to the H.P. If its surface is inclined at 45° to the V.P., draw its projections.
4. A hexagonal plane of 25 mm side resting on its corner on the H.P. The top view of the diagonal through that corner is inclined at 45° to the V.P. and surface of the plane is inclined at 60° to the H.P. Draw its projections.
5. A hexagonal plane of 25 mm sides is resting on one of the sides on the H.P., which is inclined at 45° to the V.P. The surface of the plane is inclined at 30° to the H.P. Draw its projections.
6. Draw the projections of a regular pentagon of 40 mm side, having its surface inclined at 45° to the H.P. and a side parallel to the H.P. and inclined at an angle of 60° to the V.P.
7. A hexagonal plane of 25 mm side, has one of its corner in the V.P. and its surface is inclined at 45° to the V.P. Draw its projections when.
   (a) Front view of the diagonal through the corner in the V.P. is inclined 45° with the H.P.
   (b) The diagonal through corner in the V.P. makes an angle of 45° with the H.P.
8. A semicircular plate with 60 mm diameter has its straight edge in the H.P. and inclined at 45° to the V.P. If the surface of the plate makes an angle of 30° with the H.P., draw its projections.
9. Draw the projections of a rhombus having diagonals 100 mm and 50 mm long. The larger diagonal is inclined at 30° to the H.P. while smaller diagonal is parallel to both the reference planes.
10. A 30°-60° set square has its longest edge 120 mm which is in the H.P. and making an angle of 30° with the V.P. If the surface of set square is inclined at 45° to the H.P. Draw its projections.
11. A 30-60 set square has longest side equal to 90 mm. It is so kept that its longest side is in the V.P. and inclined at 45° to the H.P. If its surface is inclined at 30° to the V.P., draw its projections.
12. A thin pentagonal lamina with 30 mm side is resting on an edge on the H.P. and parallel to the V.P. The corner opposite to that side lies on the V.P. and 20 mm above the H.P. Draw the projections of the plane and find its inclination with the H.P.
13. The front view of an isosceles triangle having base 50 mm and altitude 70 mm is an equilateral triangle of 50 mm side whose one side is inclined at 45° to the xy line. Draw its projection.
14. A hexagonal plane of 30 mm side has a circular hole 30 mm diameter in its centre. When the surface is vertical and inclined at 30° to the V.P., draw its projections.

15. Draw the projections of a circle of 80 mm diameter when it is resting on a point on the circumference on the H.P. and its plane is inclined at 60° to the H.P. and 30° to the V.P.
16. A semicircular plate of 70 mm diameter has its straight edge in the V.P. and inclined at 30° to the H.P. If the surface of the plate is inclined at 30° with the V.P., draw its projections.
17. The top view of a square plate ABCD of 80 mm diagonals appears as a rhombus of 90 mm and 60 mm diagonals. If the plate is resting on one of its corner on the H.P. and one of its diagonal is parallel to both the H.P. and V.P., draw its projections.
18. A rectangular plate having shorter and longer sides, 40 mm and 70 mm respectively, is resting on its shorter side on the H.P. Its top view appears as another rectangle having shorter and longer side of 40 mm and 60 mm respectively. If the longer side makes an angle of 45° with the V.P., draw its projections.
19. A rectangular plate of 40 mm and 60 mm sides appears as a square with 40 mm side in the front view. The shorter side of the plane is in the V.P. and surface is inclined to the V.P. Draw its projections and determine its inclination with the V.P.
20. A plate having shape of isosceles triangle with 50 mm base and 70 mm altitude, is placed such that it appears as equilateral triangle of 50 mm sides in the front view and one of its sides makes an angle of 45° with the xy line. Draw its projections.
21. A circular disc of 60 mm diameter is resting on a point A on its circumference on the H.P. with its surface inclined at 45° to the H.P. and
    (a) The top view of the diameter through A is inclined at 30° to the V.P.
    (b) The diameter through A is inclined at 30° to the V.P.

CHAPTER 9

Projections of Solids

9.1 Introduction

An object having three dimensions is called as solid. Three dimensions include length, breadth and height or thickness. Solid is bounded by a number of planes or curved surfaces. To describe a solid graphically, at least two orthographic views are required such as front view and top view. Sometimes two views of a solid may not be sufficient to describe it completely, then additional view i.e., side view is also required. If solid is more complicated in design then a view projected on an auxiliary plane is helpful.

In this chapter, methods of obtaining projections when solid is placed in different positions with reference to horizontal and vertical reference planes (i.e., H.P and V.P), are described with the help of several problems.

9.2 Classification of Solids

Solids can be classified into two main categories i.e., polyhedron and solids of revolution.

9.2.1 Polyhedron

A polyhedron is defined as a solid bounded by plane figures, generally called faces. It can be further subdivided as follows:

1. Regular polyhedron
2. Prism
3. Pyramid

1. Regular Polyhedron

The solids having all the faces equal and regular are called as regular polyhedron. Some regular polyhedrons (as shown in Fig. 9.1) are described below.

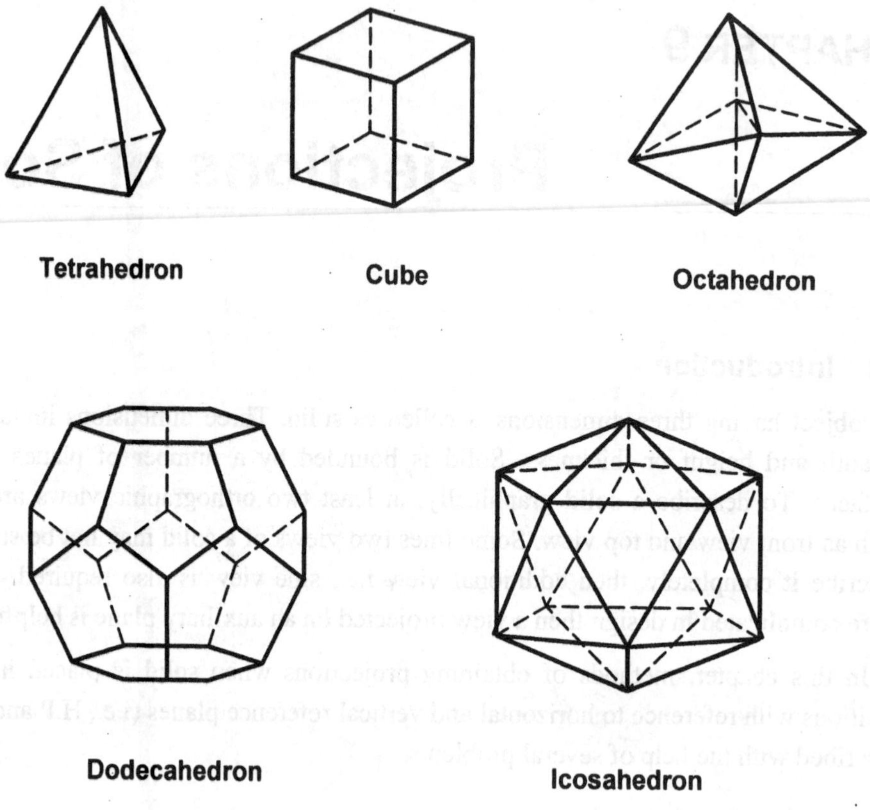

Fig. 9.1

(a) **Tetrahedron:** A solid bounded by four equilateral triangles is called tetrahedron.

(b) **Cube:** A solid bounded by six square faces is called as cube.

(c) **Octahedron:** A solid bounded by eight equilateral triangular faces is called as octahedron.

(d) **Dodecahedron:** A solid bounded by twelve regular pentagonal faces is known as dodecahedron.

(e) **Icosahedron:** A solid bounded by twenty triangular faces is known as Icosahedron.

2. **Prism:** A prism is a polyhedron having two equal and similar regular polygons parallel to each other. These regular polygons are called as its ends or bases. These ends are joined by number of rectangular faces (parallelogram faces in case of oblique solids). The imaginary line joining centres of the bases is called the axis. Prisms are named on the basis of shape of its bases. If base is triangular, the prism is called triangular prism. Similarly square prism, pentagonal prism, hexagonal prism etc., are named. Different types of prism are shown in Fig. 9.2.

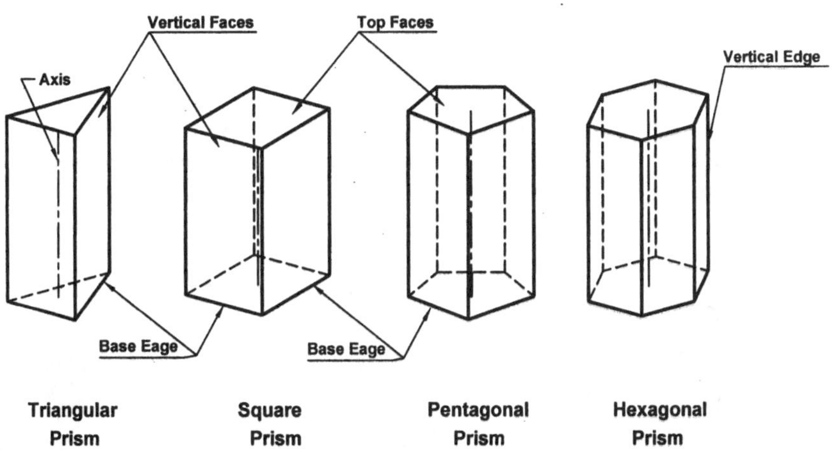

Fig. 9.2

3. **Pyramid:** A pyramid is also a polyhedron which has a regular polygon (such as triangle, square, pentagon etc.,) as a base and a number of triangular faces meeting at a point called the apex or vertex. The imaginary line joining its vertex and centre of the base is called as axis, which is perpendicular to the base for right and regular pyramid. All the faces of a pyramid are isosceles triangles. A Pyramid is named on the basis of shape of its base e.g., triangular pyramid, square pyramid, pentagonal pyramid etc., as shown in Fig. 9.3.

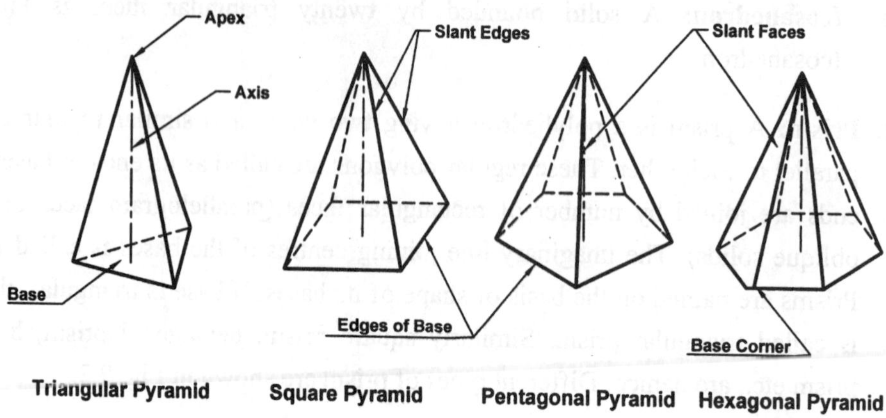

Fig. 9.3

9.2.2 Solid of Revolution

The solid formed by the revolution of a plane figure about its fixed edge, is known as solid of revolution. The cylinder, cone and sphere are some examples of solids of revolutions as shown in Fig. 9.4.

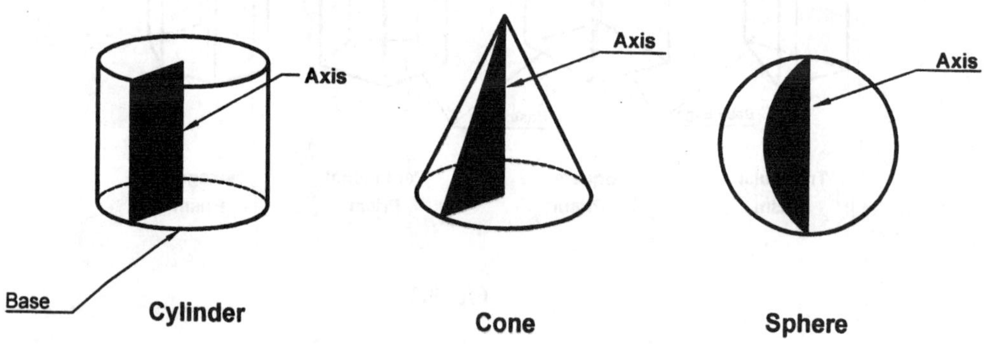

Fig. 9.4

(a) **Cylinder:** A cylinder is obtained when a rectangle is revolved about its fixed edge.

(b) **Cone:** A cone is obtained when a right triangle is revolved about one of its perpendicular sides.

(c) **Sphere:** When a semi circle is revolved about its diameter which is fixed, the solid generated is known as sphere.

9.3 Oblique Solid

An oblique solid has its axis inclined to its base. The oblique solids such as oblique cylinder, oblique cone and oblique prism are illustrated in Fig. 9.5.

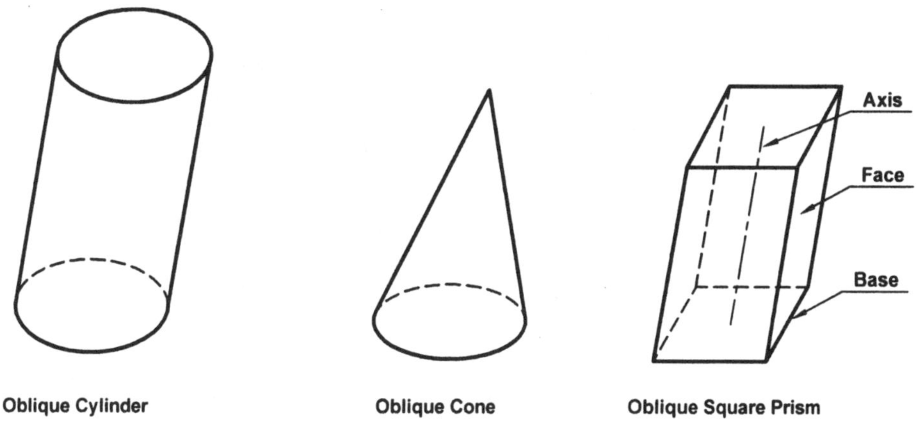

Fig. 9.5

9.4 Truncated Solid

When a solid is cut by a plane which is not parallel to its base and remaining portion after removing top portion, is known as truncated solid as shown in Fig. 9.6.

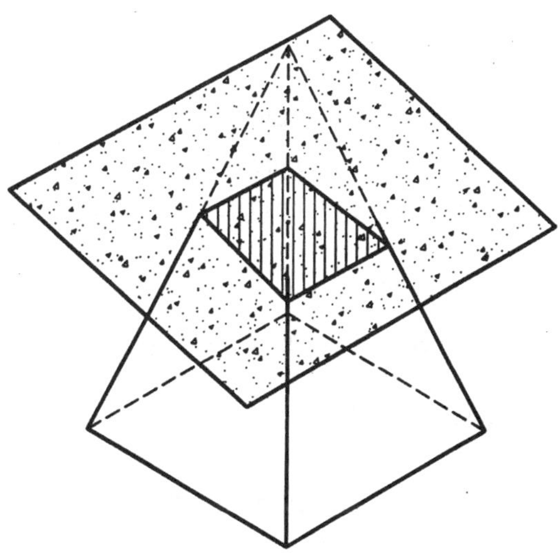

Fig. 9.6

9.5 Frustum

When a cone or a pyramid is cut by a plane which is parallel to its base, the remaining portion after removing top portion is known as frustum. The frustum of cone and pyramid are shown in Fig. 9.7.

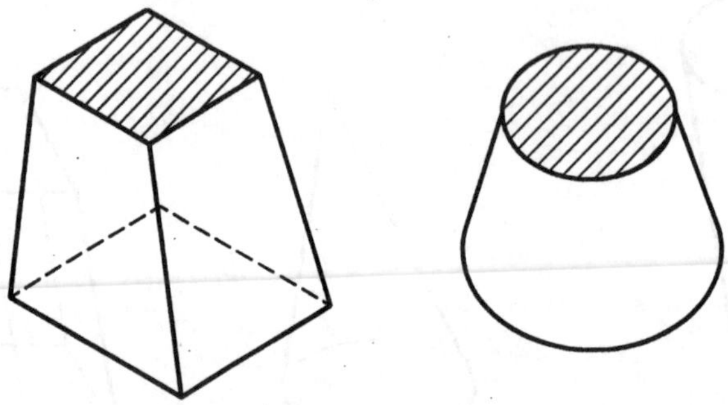

Frustum of square pyramid **Frustum of cone**

Fig. 9.7

9.6 Positions of Solid with Respect to Reference Planes (i.e., H.P and V.P)

A solid can have one of the following positions with respect to H.P and V.P.

1. Axis perpendicular to the H.P.
2. Axis perpendicular to the V.P.
3. Axis parallel to H.P and V.P.
4. Axis inclined to the H.P and parallel to the V.P.
5. Axis inclined to the V.P and parallel to the H.P.
6. Axis inclined to both the reference planes (i.e., H.P and V.P)

9.7 Axis Perpendicular to H.P

When right solid is placed on the H.P, its axis becomes perpendicular to H.P and parallel to the V.P. The position of solid in this case is also called as simple position. Since the base is parallel to H.P. its true shape and size will be seen in top view. Hence, in this case top view is drawn first and front view is obtained by projecting from top view.

Problem 9.1 (Fig. 9.8): *A square pyramid of base edge 40 mm and axis 55 mm long is resting on its base on the H.P. with one of its base side parallel to the V.P. Draw its projections.*

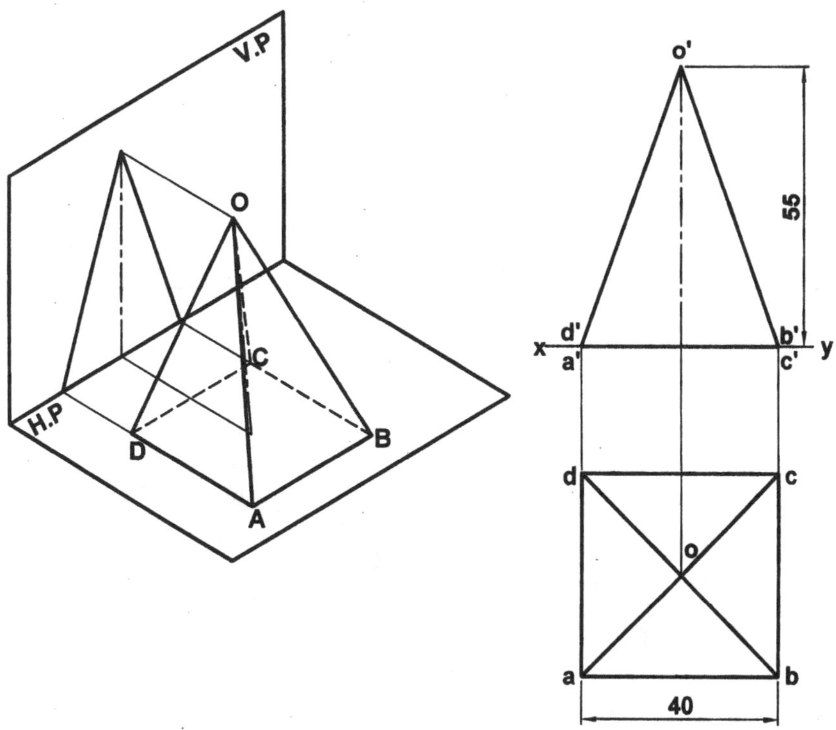

Fig. 9.8

Construction (Fig. 9.8):

Top View

1. Draw xy line. Below xy line, draw a square abcd of 40 mm side.
2. Join a-c and b-d and locate its centre o.

Front View

3. Through o, draw a vertical centre line to represent the axis of solid.
4. Project corners a, b, c and d on the xy line as a', b', c' and d' respectively.
5. Mark o' on the axis, 55 mm above the xy line.
6. Join o'-a' and o'-b'. The figure obtained above xy line is the required front view.

Problem 9.2 (Fig. 9.9): *A square pyramid of base edge 40 mm and axis 60 mm long is resting on its base on the H.P. Draw its projections when:*

(a) *All sides of its base are equally inclined to V.P.*

(b) *A side of its base is inclined 35° with the V.P.*

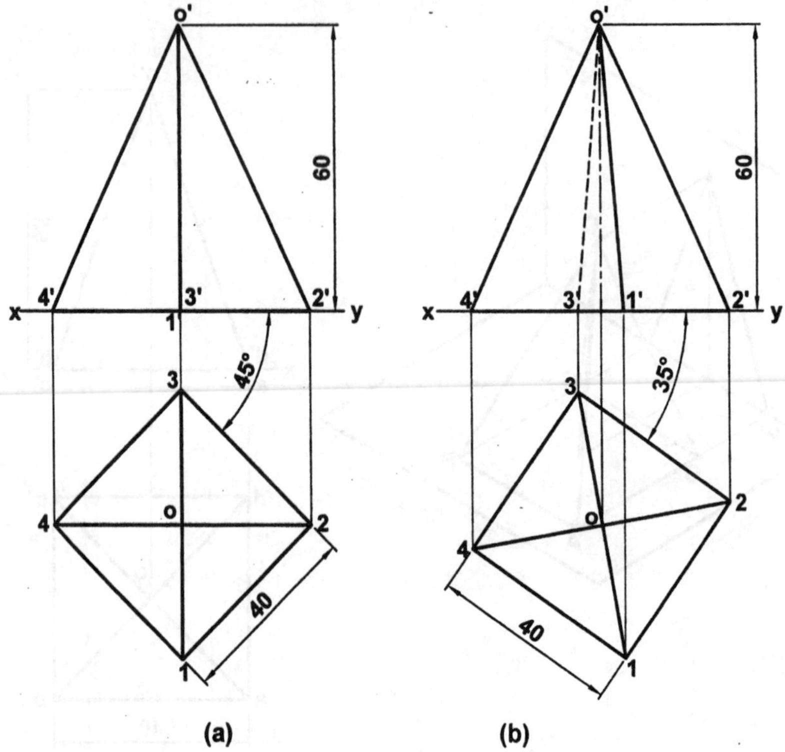

Fig. 9.9

Construction (Fig. 9.9):

Case(a) Fig. 9.9(a):

Top View

1. Draw xy line and below it, draw a line 40 mm long and inclined at 45° to the xy line.
2. Considering this line as one of the sides, complete the square 1234 (In this case all sides are equally inclined to V.P.)
3. Draw the diagonals 1-3 and 2-4. Locate its centre o. This is the required top view.

Front View

4. Draw vertical projectors through corner 1, 2, 3 and 4 intersecting xy line at 1', 2', 3' and 4'.
5. Through o, draw a vertical line representing the axis. Mark o', 60 mm above the xy line.
6. Join 1', 2', 3' and 4' with o'. This is the required top view.

Case (b) Fig. 9.9(b):

Top View

1. Draw a square 1234 of 40 mm side with a side inclined 35° to the xy line.
2. Draw its diagonal and locate its centre o.

Front View

3. Project corners 1, 2, 3 and 4 on the xy line and mark them 1', 2', 3' and 4' respectively.
4. Through o, draw a vertical axis. Mark apex o' 60 mm above xy line and on the axis.
5. Join 1', 2' and 4' with o' by Continuous line and 3' with o' by dotted line. Dotted line indicates hidden edge of the pyramid.

Problem 9.3 (Fig. 9.10): *A cube of 40 mm long edges is resting on the H.P. Draw its projections when*

(a) *Vertical faces are equally inclined to the V.P.*

(b) *One of the vertical faces is parallel to the V.P.*

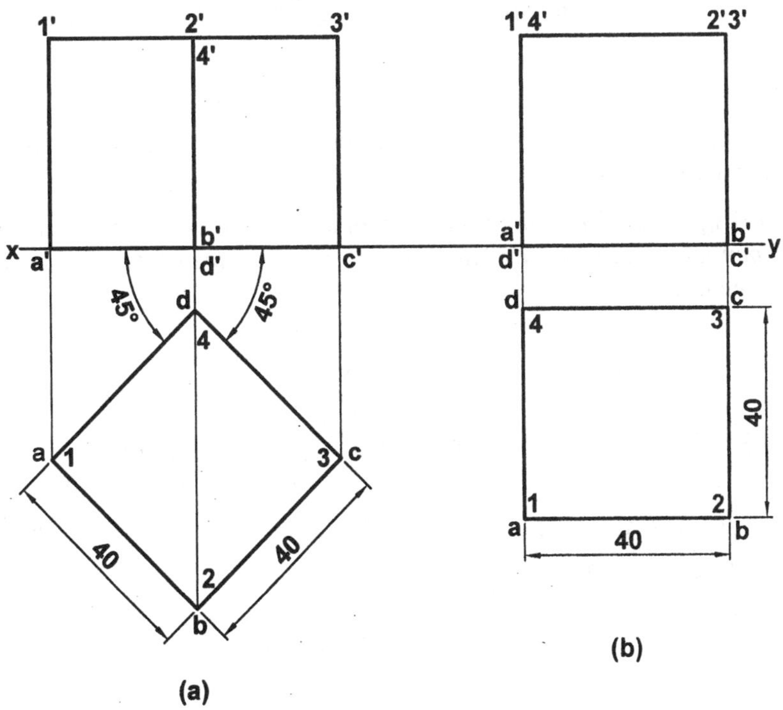

Fig. 9.10

Construction (Fig. 9.10):

Case (a) Fig.9.10 (a):

Top View

1. Draw a square abcd having 40 mm side with its sides equally inclined to xy line. (i.e., 45° with xy).
2. Since, top and bottom corners coincide each other in top view, also mark 1, 2, 3 and 4 at corners as shown. This is required top view.

Front View

3. Project bottom corners a, b, c and d on the xy line and mark, a', b', c' and d'.
4. Project top corners 1, 2, 3 and 4, 40 mm above the xy and mark them 1', 2', 3' and 4' respectively.
5. Join a'-1', b'-2', c'-3', d'-4', 1'-3' and a'-c' which is the required front view.

Case (b) Fig.9.10 (b):

Top View

Draw a square abcd of 40 mm side with a side say ab parallel to V.P. also mark 1, 2, 3 and 4 at the corner.

Front View

Project all the corners 1, 2, 3....etc., on the xy line and draw a square above xy line as explained in case (a). Mark the corners as 1', 2', 3'....etc.

Problem 9.4 (Fig. 9.11): *Draw the projection of a pentagonal pyramid base 35 mm edge and axis 55 mm long resting on its base on H.P. and an edge of its base parallel to V.P. Draw its side view also.*

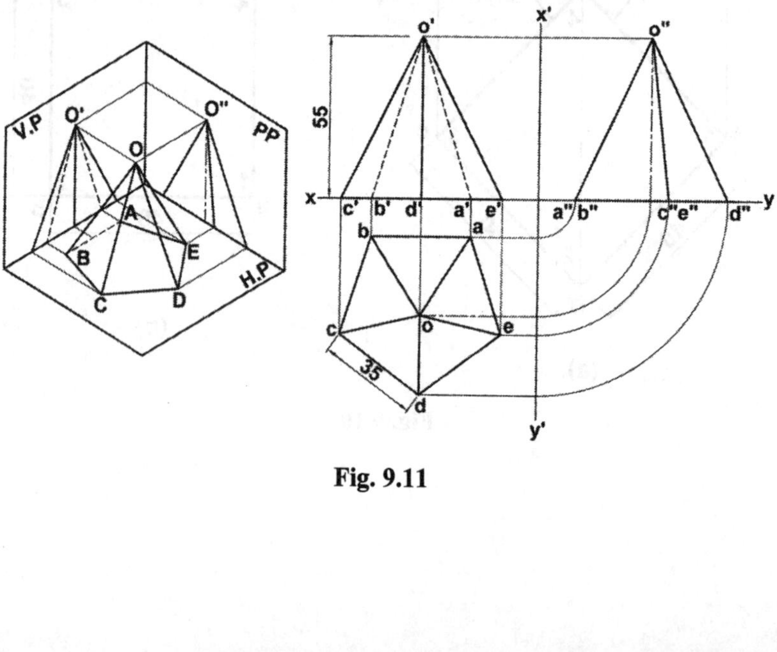

Fig. 9.11

Construction (Fig. 9.11):

Top View

1. Draw xy line. Draw a line ab 35 mm long parallel to xy line.
2. Draw a regular pentagon abcde with ab as one of the sides which is nearer to xy line. Locate its centre o and join it with corners a, b, c....etc.

Front View

3. Through o, draw a vertical line to represent the axis of solid. Mark the apex o' on the axis, 55 mm above the xy line.
4. Project all the corner a, b, c....etc., on the xy line and mark them, as a', b', c'....etc.
5. Join o'-c', o'-d' and o'-e' by straight lines to show visible edges and Join o'-b' and o'-a' by dashed lines to show hidden slant edges.

Side View

6. To draw the left side view, draw a new reference line x_1y_1, perpendicular to xy line and project the side view on it from the front view. It should always be remembered that distances of all the corner points from x_1y_1 should be equal to their corresponding distances in the top view from xy line. The side view can be drawn as explained below.
7. Draw horizontal lines through all the corner points and apex.
8. Draw horizontal lines through all the corners in the top view and up to x_1y_1 and turn these lines at 45° to intersect xy at a'', b'', c''...etc.
9. Also draw horizontal line through centre o in the top view and up to x_1y_1 and then project it to represent axis in the side view.
10. Draw horizontal projector through o' intersecting the axis in side view at o''.
11. Join a'', c'' and d'' with o'' by straight lines. This is the required side view.

Problem 9.5 (Fig. 9.12): *A hexagonal pyramid having base 35 mm side and axis 55 mm long is resting on its base on the H.P. Draw its projections when one of the base edges is inclined 45° to the V.P.*

Construction (Fig. 9.12):

Top View

1. Draw a line af 35 mm long and inclined at 45° to xy line.
2. On af construct a regular hexagon abcdef.
3. Locate its centre o and join it with all the corners.

Front View

4. Project all the corner points on the xy as a', b', c'...etc. Through o, draw a vertical line representing the axis. Mark apex o' on the axis, 55 mm above xy.
5. Join o'-a', o'-b', o'-c' and o'-d' by straight lines as visible edges.
6. Join o'-f' and o'-e' by dashed line to show hidden edges which are nearer to V.P.

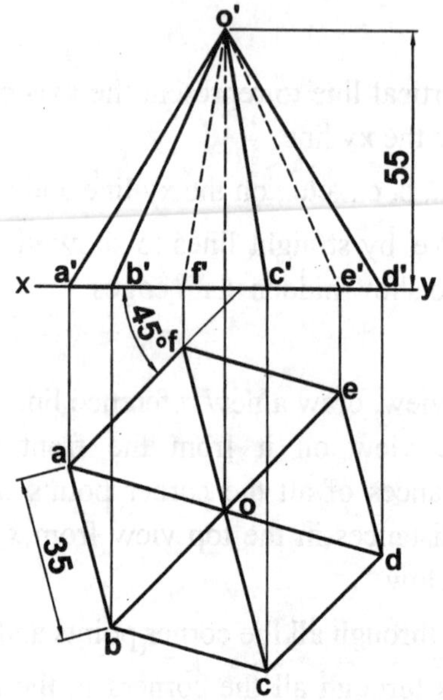

Fig. 9.12

Problem 9.6 (Fig. 9.13): *Draw the projections of the following solids when solids rest on their bases on H.P with their axis perpendicular H.P.*

(a) *A cylinder having a base diameter 40 mm and axis, 60 mm long.*

(b) *A cone having base diameter of 40 mm and axis, 60 mm long.*

Construction (Fig. 9.13):

The pictorial view of a cylinder and a cone are shown in Fig.9.13 (a) and (b).

Case (a) Fig. 9.13(c):

1. Draw a xy line and below it draw a circle of radius 40 mm.
2. Divide the circle into four equal parts. Name the division points as 1, 2, 3, 4, a, b, c and d.
3. Through o, draw a line perpendicular to xy line representing axis of the cylinder.
4. Draw vertical projectors through 1, 2, 3.... etc.

5. Mark a', b', c' and d' on the xy line where corresponding generators intersect.
6. Mark 1', 2', 3', and 4' on the corresponding generators 60 mm above the xy line.
7. Join a'-1', c'-3', 1'-3' and a'-c' to form a rectangle, which is required front view.

Case (b) Fig. 9.13(d):

1. Draw a circle of 40 mm diameter in the top view. Locate its centre at o.

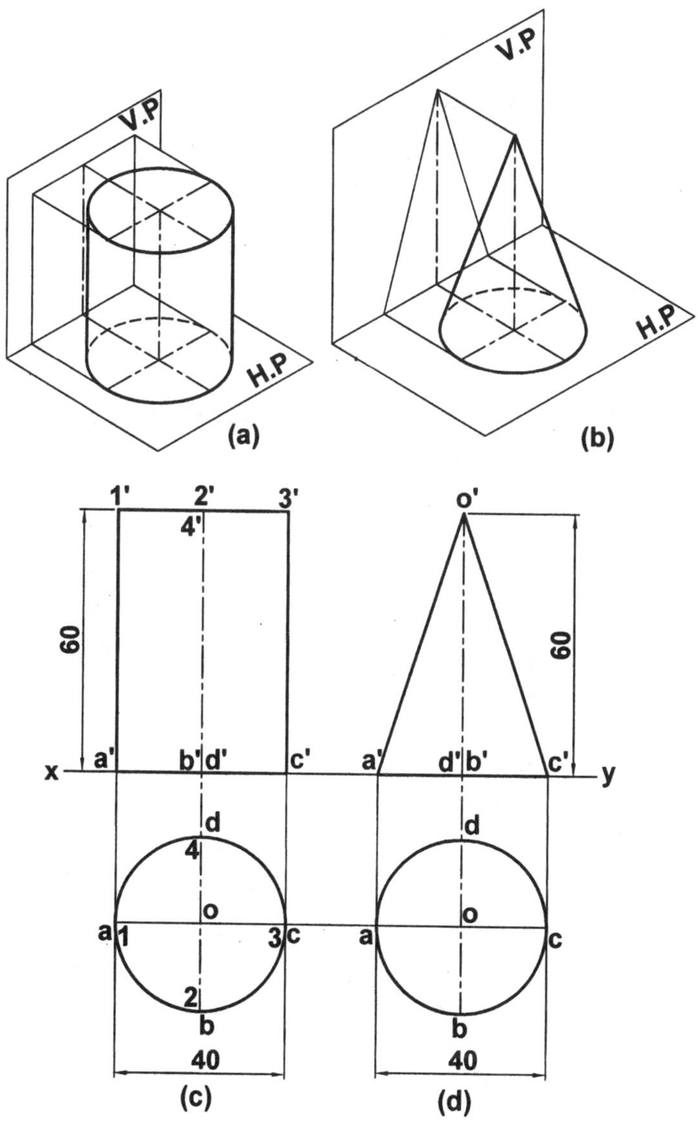

Fig. 9.13

2. Divide the circle into four equal parts and mark the divisions as a, b, c and d.
3. Project these points in the front view as a', b', c' and d'.
4. Through o, draw a vertical line representing axis of the cone.
5. Mark apex o', 60 mm above xy line.
6. Join o'-a' and o'-c' to form a triangle, which is the required top view.

9.8 Axis Perpendicular to V.P.

When the axis of a right solid is kept perpendicular to the V.P., its base becomes parallel to V.P. Therefore, front view represents true size and shape. In such cases, front view is drawn first. Top view is projected from the front view.

Problem 9.7 (Fig. 9.14): *A pentagonal prism, side of base 30 mm and axis 60 mm long is resting on one of its rectangular faces on the H.P with its axis perpendicular to the V.P. Draw its projections.*

Construction (Fig. 9.14):

In this case, prism is resting on its rectangular face with its axis perpendicular to V.P. therefore its pentagonal faces will be parallel to V.P. and its projection on the V.P (i.e., front view) will be a regular pentagon with its true shape as shown in Fig.9.14 (a). So begin with front view as follows:

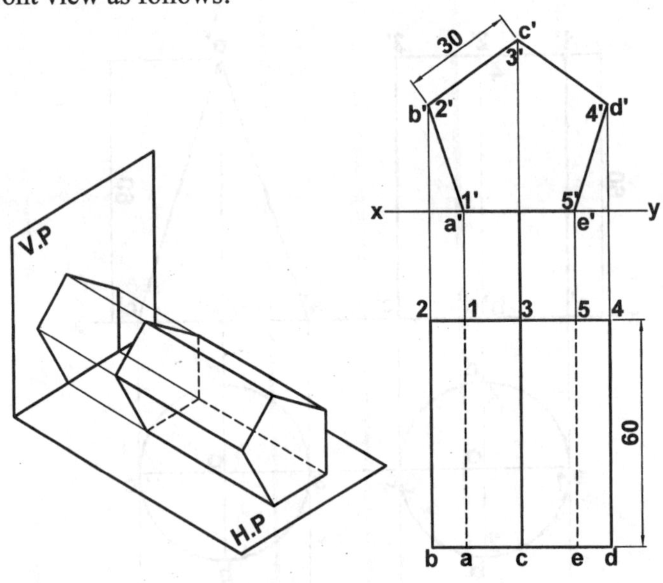

(a) (b)

Fig. 9.14

Front view

1. Draw a regular pentagon a'b'c'd'e' of 30 mm side above xy line.

Top View

2. Project all the corner of front view down and mark the corner points as 1, 2, 3...etc., for the face nearer to V.P. and a, b, c....etc., for the other face each 60 mm away from 1, 2, 3....etc., respectively.
3. Complete rectangle and show visible edges 2-b, 3-c and 4-d by a dark full line and hidden edges 1-a and 5-e by dashed lines.

Problem 9.8 (Fig. 9.15): *A pentagonal pyramid, side of base 30 mm and axis 70 mm long has its base parallel to V.P. and one of the base edges in H.P. Draw front view, top view and side view.*

Construction (Fig. 9.15):

Front View

1. Draw a regular pentagon of 30 mm side. Locate its centre o'. Join all the corners a', b', c'...etc., with o'.

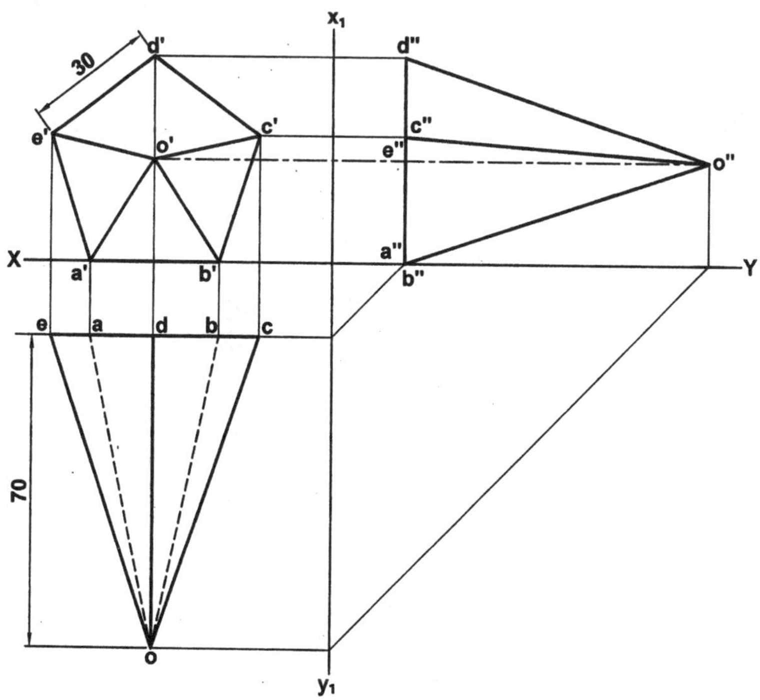

Fig. 9.15

Top View

1. Project o', a', b', c'....etc., down below xy line.
2. Mark a, b, c....etc., at the intersection of vertical projectors with a line parallel to xy.
3. Mark apex o on the line through o'.
4. Join a, b, c....etc., with apex o as shown in Fig.9.15.

Side View

5. Draw a new reference line x_1y_1, perpendicular to xy line.
6. From each point in the front view, draw horizontal projection line.
7. From each point in the top view, draw horizontal projection line to meet x_1y_1 and then turn them at 45° to meet xy line. Beyond xy line draw vertical projectors.
8. At the intersection of corresponding vertical and horizontal projectors mark a", b", c".....etc. Join o" with a", b", c"....etc., by straight lines.

9.9 Axis Parallel to the both the H.P and V.P.

When a solid (excluding cube and cuboid) rests on its face on the H.P. having its axis parallel to both the H.P and V.P., neither top view nor front view provides actual size and shape. Therefore, in such problems side view is required to be drawn first. The side view is drawn on an auxiliary plane (perpendicular to both H.P and V.P). Front view and top view are drawn by projecting from side view.

Problem 9.9 (Fig. 9.16): *A triangular prism, side of base 30 mm and axis 55 mm long, is resting on its rectangular face on the H.P and with its axis parallel to both the H.P and the V.P. Draw its projections.*

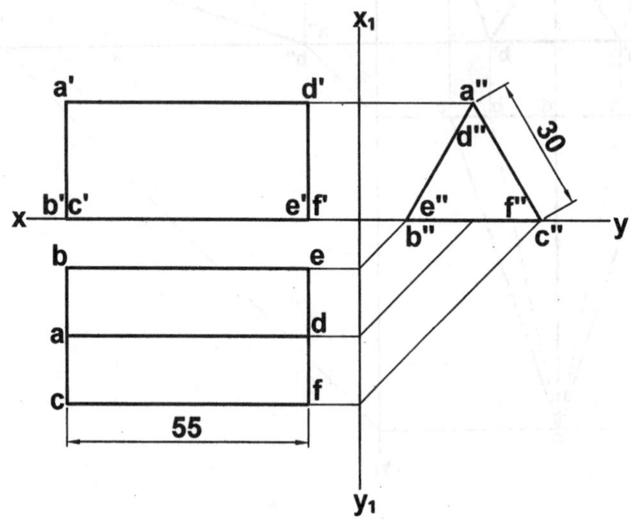

Fig. 9.16

Construction (Fig. 9.16):

Side View

1. Draw an equilateral triangle of 30 mm side keeping one of the side coinciding with xy line.
2. Mark . ¬ corner points as a", b", c"....etc.

Top View

3. Project a", b", c"....etc., on xy line. Through these points, draw projectors inclined at 45° to xy intersecting the x_1y_1 and then make them parallel to xy.
4. Draw two vertical and parallel lines 55 mm apart crossing above generators. Mark a, b, c, d, e and f on the corresponding generators.
5. Join b-e, a-d and c-f by dark lines as they represent visible edges and complete the top view as shown.

Front View

6. Project side view horizontally and top view vertically upward and obtain the front view as shown.

Problem 9.10 (Fig. 9.17): *A pentagonal prism, edge of base 30 mm and axis 50 mm long is resting on one of its rectangular faces on the H.P. with axis parallel to both H.P and V.P. Draw its projections.*

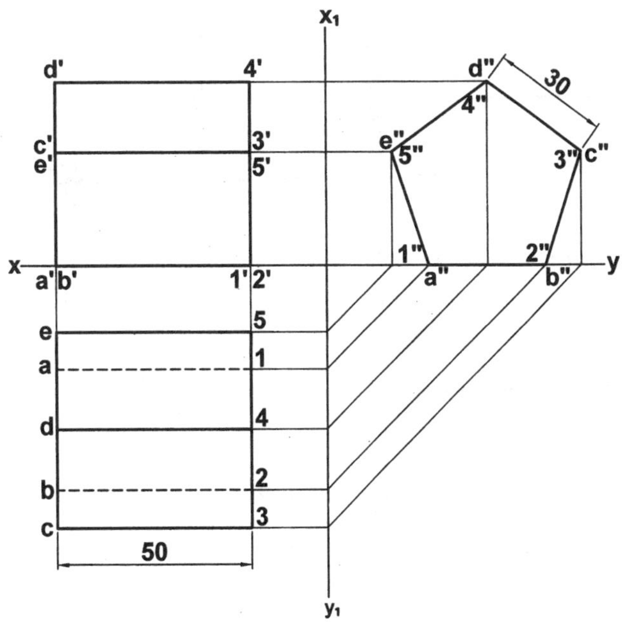

Fig. 9.17

Construction (Fig. 9.17):

Side View

1. Draw a regular pentagon a"b"c"d"e" of 30 mm side with one of the sides a"b" coinciding with xy line.

Top View

2. Project down the side view and obtain top view as explained in problem 9.9.

Front View

3. Project front view from side view and top view as explained in problem 9.9.

Problem 9.11 (Fig. 9.18): *A hexagonal prism, edge of base 30 mm and axis 65 mm long is resting on one of the longer edges such that its axis is parallel to both the H.P and the V.P. Draw its projections.*

Construction (Fig. 9.18):

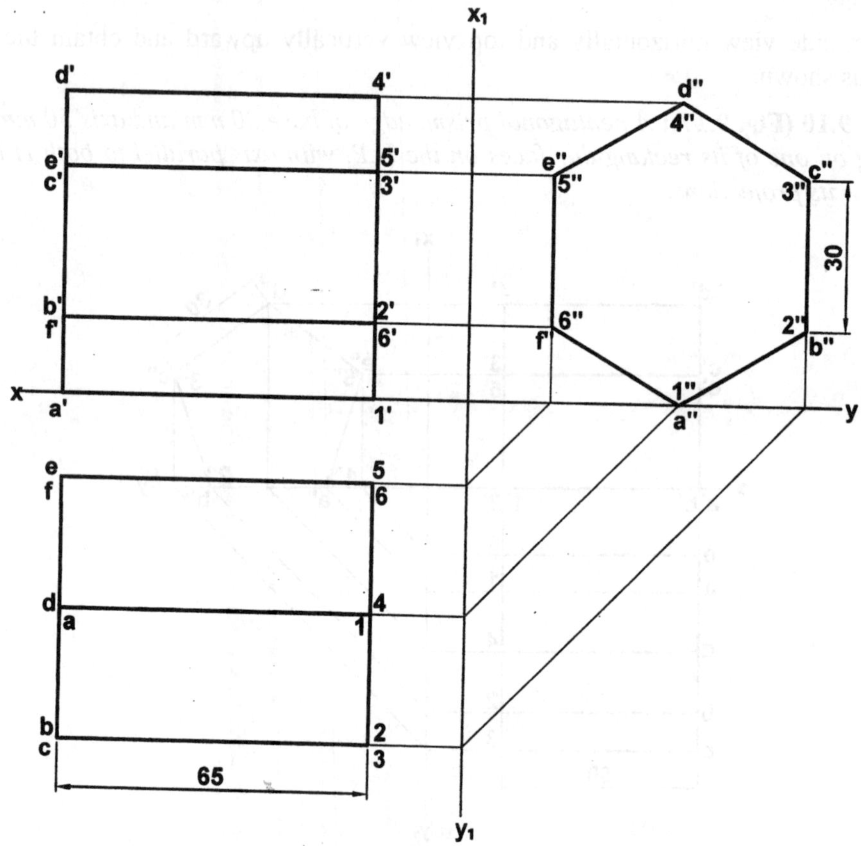

Fig. 9.18

Side View

1. Draw a regular hexagon a"b"c"d"e"f" of 30 mm side with one of its corner on the xy line and two sides perpendicular to xy line.

Top View

2. Project the top view below xy line as explained in problem 9.9

Front View

3. Project front view from side view and top view as explained in problem 9.9

9.10 Axis Inclined to H.P. and Parallel to V.P.

When the axis of the solid is inclined to the H.P. and parallel to the V.P. then the projections are drawn in two stages.

Stage-I

In this initial stage, solid is first assumed to be in simple position (i.e., placed on its base on the H.P.) since, top view gives its true size and shape, it is drawn first and then front view is drawn by projecting from top view.

Stage-II

After drawing projections of solid in simple position final projections can be obtained by using one of the following two methods.

Method I (by altering the position of the view):

The front view as drawn is stage-I is reproduced with its axis inclined to the H.P. (or xy line) and the new top view is drawn with the help of top view of stage-I and tilted front view.

Method II (by altering reference line or auxiliary plane):

In this method, a new reference line is drawn as per given conditions. This reference line represents an auxiliary plane on which final view is projected.

The procedures of obtaining projections by these methods are explained with the help of following examples.

Problem 9.12 (Fig. 9.19): *A pentagonal prism having base 30 mm side and axis 60 mm long is resting on an edge of its base on the H.P. such that its axis is inclined at 60° to the H.P and parallel to the V.P.*

Method I

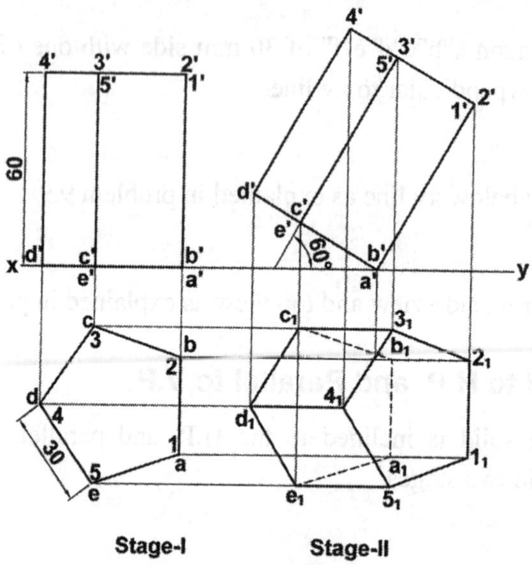

Fig. 9.19(a)

Method II

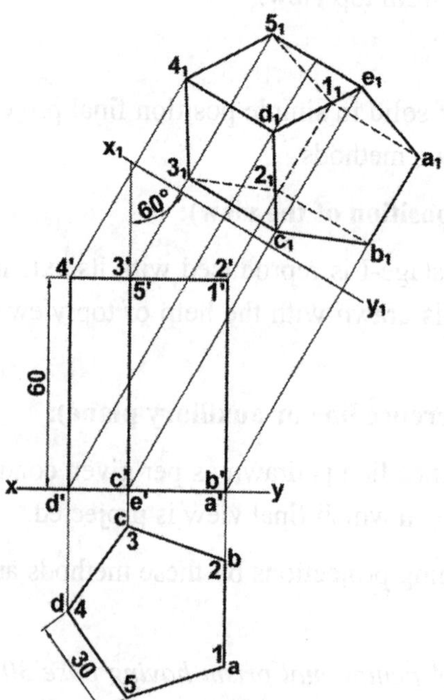

Fig. 9.19(b)

Construction (Fig. 9.19):

Stage I

Assuming the prism to be kept on its face on the H.P., with a side ab perpendicular to xy line. Draw front view and top view as follows:

1. Draw a regular pentagon abcde of 30 mm side, below xy line with a side (say ab) perpendicular to xy line.
2. Project up all the corners in top view and obtain front view as shown.

Stage II

Method I (By altering position of view (Fig. 9.19(a)):

1. Reproduce the front view of stage-I with its axis inclined at 60° to the xy line and edge a'b' on the H.P.
2. Draw vertical projectors through a', b', c'....etc., and horizontal projectors through top view of stage-I.
3. Mark a_1, b_1, c_1.... etc., at the intersection of horizontal and vertical projectors.
4. Join these points and complete the top view as shown.

Method II (By altering reference line (Fig. 9.19(b)):

1. Draw a new reference line x_1y_1 inclined at 60° to the axis to represent auxiliary inclined plane.
2. Draw projectors from all the points in the front view perpendicular to x_1y_1 and on them, mark points 1_1, 2_1, 3_1....etc., keeping their distances from x_1y_1 equal to their corresponding distances in the top view from xy line.

Problem 9.13 (Fig. 9.20): *A hexagonal pyramid base 30 mm side and axis 65 mm long, has its triangular face on the H.P. with its axis parallel to the V.P. Draw its projections.*

Construction (Fig. 9.20):

Stage I

1. Assuming the pyramid resting on its base on the H.P. with one of its sides perpendicular to V.P., draw a regular hexagon abcdef of 30 mm side to represent the top view.
2. Project front view above xy line.

Stage II

Method I (By altering position of a view):

1. Reproduce the front view of stage-I with the slant edge coinciding with xy line.
2. Project all the corners a_1', b_1', c_1' ...etc., vertically.
3. Project all the corners of top view of stage-I horizontally.

4. Obtain points a_1, b_1, c_1....etc., where the corresponding horizontal and vertical projectors intersect.
5. Join o_1-a_1 and o_1-b_1 by dashed lines as they represent invisible edges.
6. Join o_1-c_1, o_1-d_1, o_1-e_1 and o_1-f_1 by continuous lines as they represent visible edges.
7. Join o_1a_1 and o_1b_1 by dashed lines to represent hidden edges.

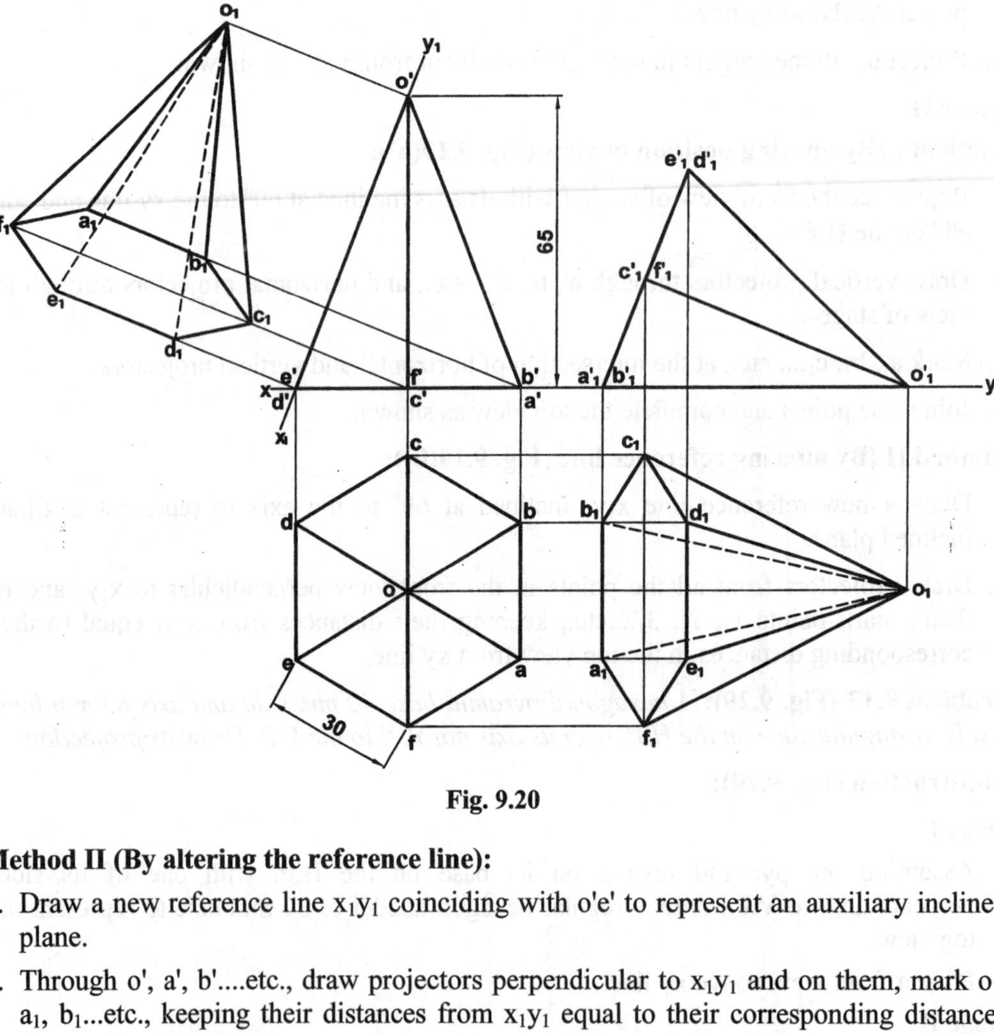

Fig. 9.20

Method II (By altering the reference line):

1. Draw a new reference line x_1y_1 coinciding with o'e' to represent an auxiliary inclined plane.
2. Through o', a', b'....etc., draw projectors perpendicular to x_1y_1 and on them, mark o_1, a_1, b_1...etc., keeping their distances from x_1y_1 equal to their corresponding distances from xy in the top view.
3. Join corner points a_1, b_1, c_1....etc., with o_1 and complete the top view as shown in the figure.

Problem 9.14 (Fig. 9.21): *A square pyramid of base 35 mm side and axis 65 mm long held in such a way that one of its edges connecting one of the corners of the base and the apex is perpendicular to the ground and parallel to the V.P. Draw its projections.*

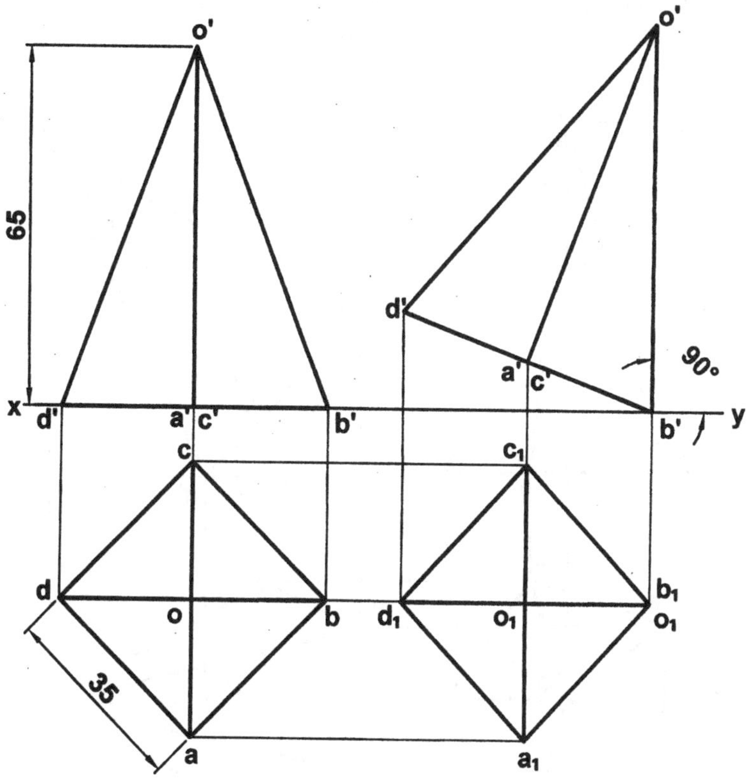

Fig. 9.21

Construction (Fig. 9.21):

Stage I

1. Draw a square abcd of 35 mm side. Locate its centre o and join it with all the corner. This represents top view in simple position.
2. Project front view from the top view.

Stage II

1. Reproduce the front view of stage-I keeping slant edge o'b' perpendicular to the xy line.
2. Draw vertical projectors through all the corners of front view of stage-II and horizontal projectors through all the corners of top view of stage-I.

3. Obtain a_1, b_1, c_1....etc., at the intersection of corresponding horizontal and vertical projectors and complete the top view as shown in figure.

Problem 9.15 (Fig. 9.22): *A right regular pentagonal pyramid side of base 25 mm and height 60 mm, rests on a corner of its base on H.P such that its axis is inclined at 45° to the H.P. and is parallel to the V.P. Draw its projections.*

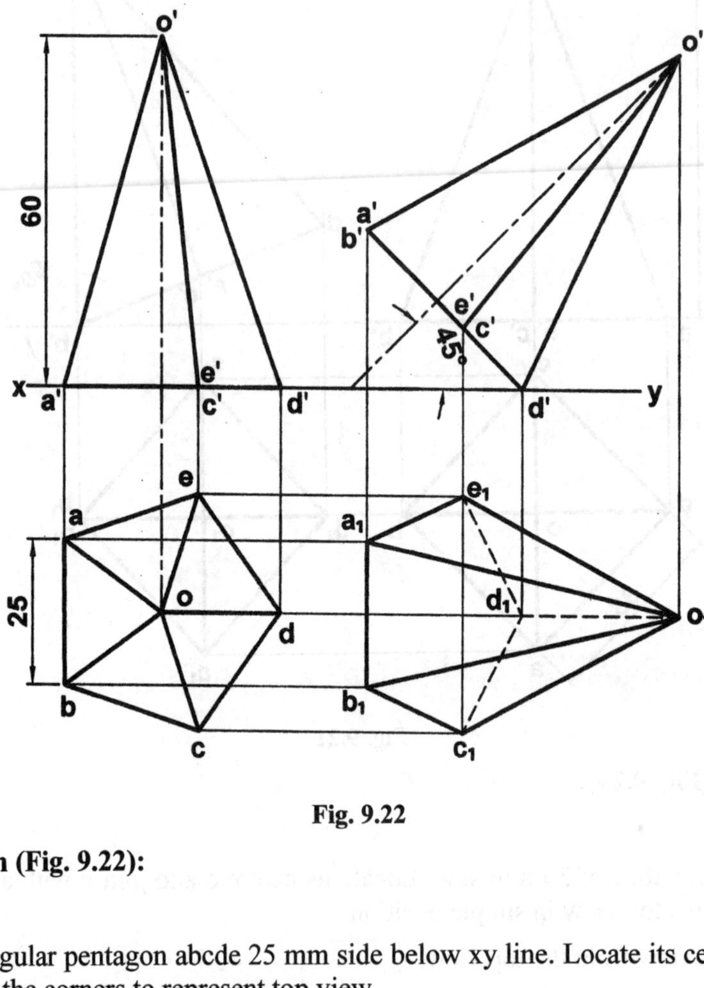

Fig. 9.22

Construction (Fig. 9.22):

Stage I

1. Draw a regular pentagon abcde 25 mm side below xy line. Locate its centre o and join it with all the corners to represent top view.
2. Project the front view from the top view.

Stage II

1. Reproduce the front view with its axis making an angle of 45° with xy line.
2. Project the final top view as explained in problem 9.14.

Problem 9.16 (Fig. 9.23): *A hexagonal prism of side of base 25 mm and axis 65 mm long is freely suspended from a corner of the top face. Draw the projections of the prism in suspended position.*

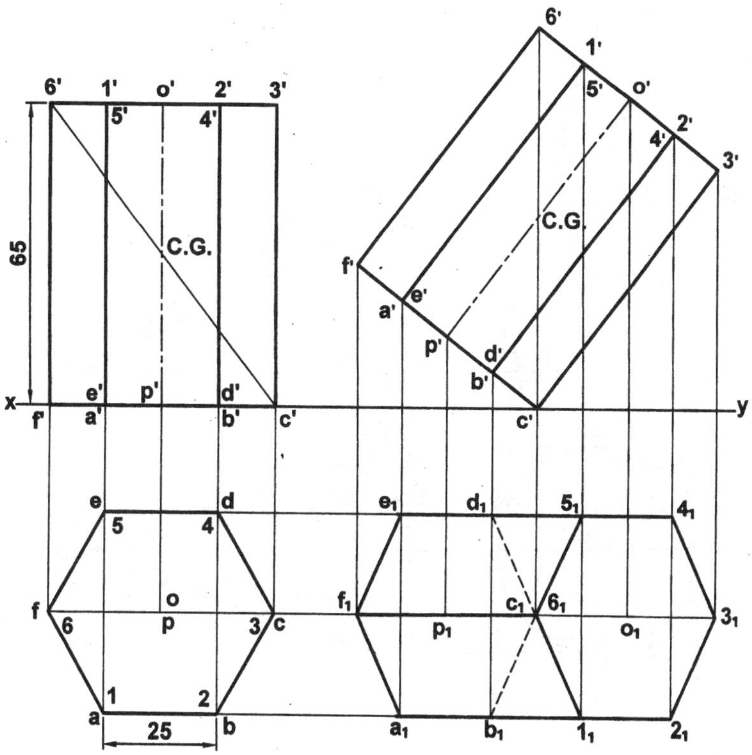

Fig. 9.23

Construction (Fig. 9.23):

Stage I

1. Draw a regular hexagon abcdef 25 mm side below xy line as top view.
2. Project front view from the top view as shown.

Stage II

1. Since, the prism is freely suspended from a corner the line joining opposite corners 6'c' and passing through its centre of gravity will be vertical.

   Therefore, reproduce the front view of first stage keeping diagonal 6'c' vertical.
2. Project the top view below xy line. Draw base edge b_1c_1 and c_1d_1 with dashed lines as they are invisible.

Problem 9.17 (Fig. 9.24): *A right circular cone having base diameter of 40 mm and height 70 mm lies on H.P. on one of its elements with its axis parallel to V.P. Draw its projections.*

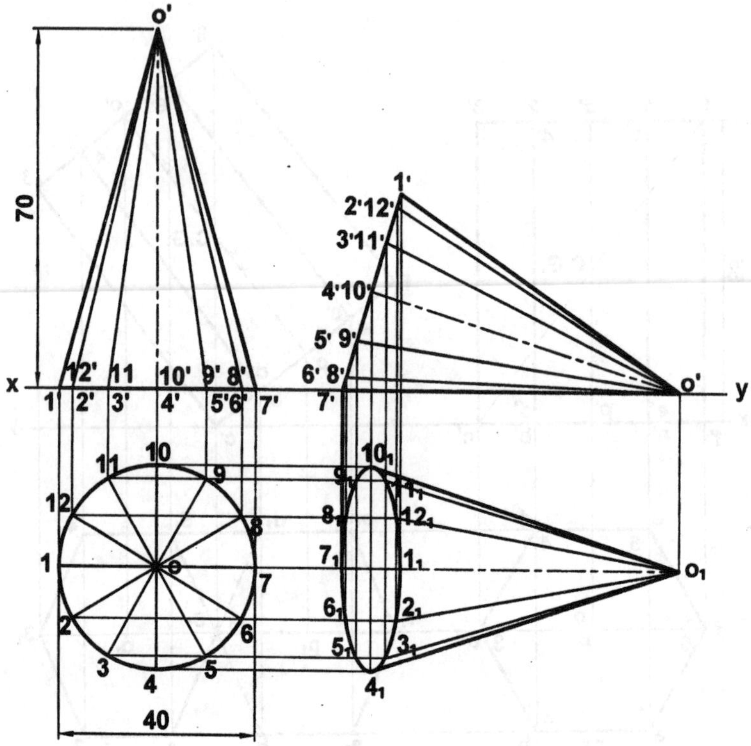

Fig. 9.24

Construction (Fig. 9.24):

Stage I

1. Draw a circle of 40 mm diameter and divide it into twelve equal parts. Name the division points as 1, 2, 3...etc. Join centre o with 1, 2, 3....etc., by radial lines. This represents top view in simple position. Draw vertical projectors through 1, 2, 3... etc., to meet xy line at 1', 2', 3' ... etc., respectively.
2. Through o, draw a vertical line representing axis of the solid. Mark apex o' on it 70 mm above the xy line.
3. Join 1', 2', 3'...etc. with o' by straight lines representing its generators.

Stage II

1. Reproduce the front view of stage-I with generator O'7' coinciding with xy line.
2. Project the top view from it.

Problem 9.18 (Fig. 9.25): *A right circular cone having base diameter of 40 mm and axis 70 mm long, rests on its base rim on H.P. with its axis parallel to V.P. and one of the elements perpendicular to H.P. Draw its projections.*

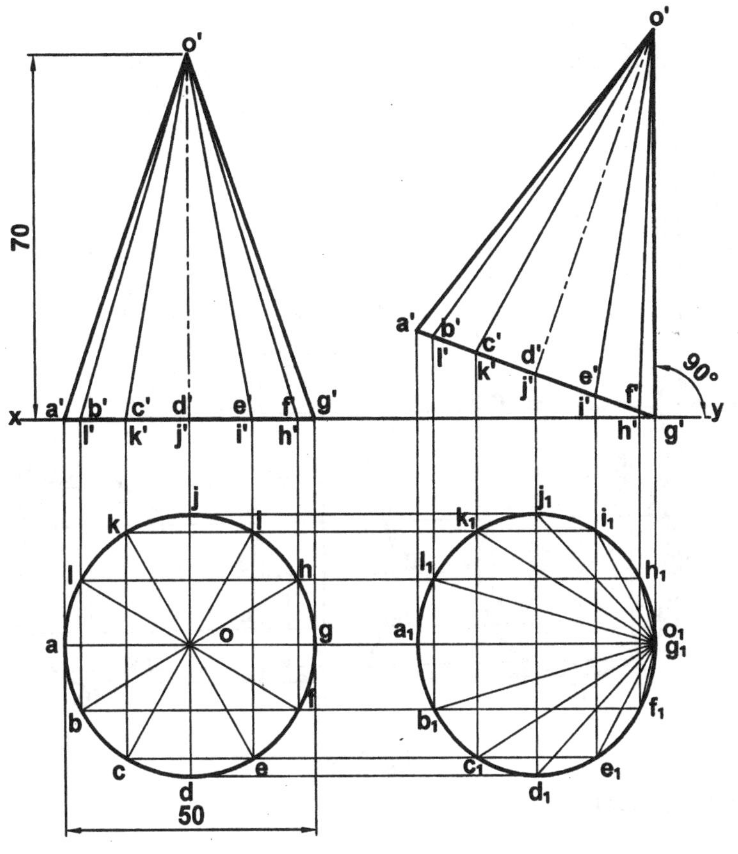

Fig. 9.25

Construction (Fig. 9.25):

Stage I

1. Draw the circle of 50 mm diameter below xy line as top view in simple position.
2. Divide the circle in twelve equal parts and draw radial lines o-a, o-b, o-c.....etc.

Stage II

1. Reproduce the front view of stage-I with its element o'g' perpendicular to xy line.
2. Project the new top view from it.

Problem 9.19 (Fig. 9.26): *A cylinder having 40 mm base diameter and 65 mm long axis rests on the H.P. with axis inclined at 45° to H.P. and parallel to V.P. Draw its projections. Follow the change of position method.*

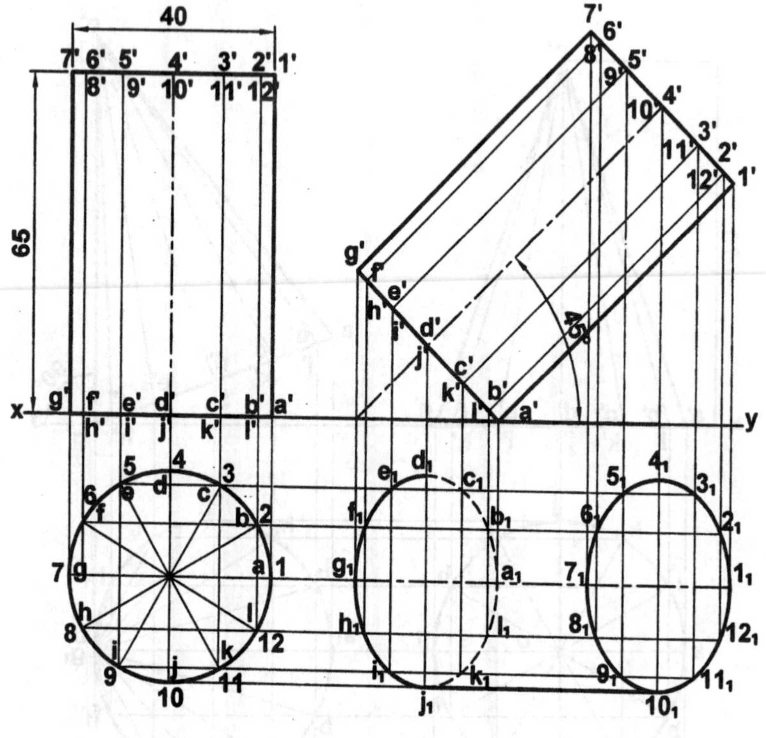

Fig. 9.26

Construction (Fig. 9.26):

Stage I

1. Draw a circle of 40 mm diameter below the xy line and divide it into twelve equal parts as shown in figure.
2. Project the front view and draw its generators.

Stage II

1. Reproduce the front view of stage-I with its axis inclined at 45° to xy line.
2. Project the new top view as explained in previous problems.
3. In the new top view, draw inner part of the ellipse representing base of the cylinder, by dashed curve as it is the invisible portion.

Problem 9.20 (Fig. 9.27): *A pentagonal pyramid having base 30 mm side and axis 65 mm long, has an edge of base parallel to the H.P and apex lies on the H.P. and axis is parallel to the V.P. and inclined at 45° to the H.P. Draw its projections.*

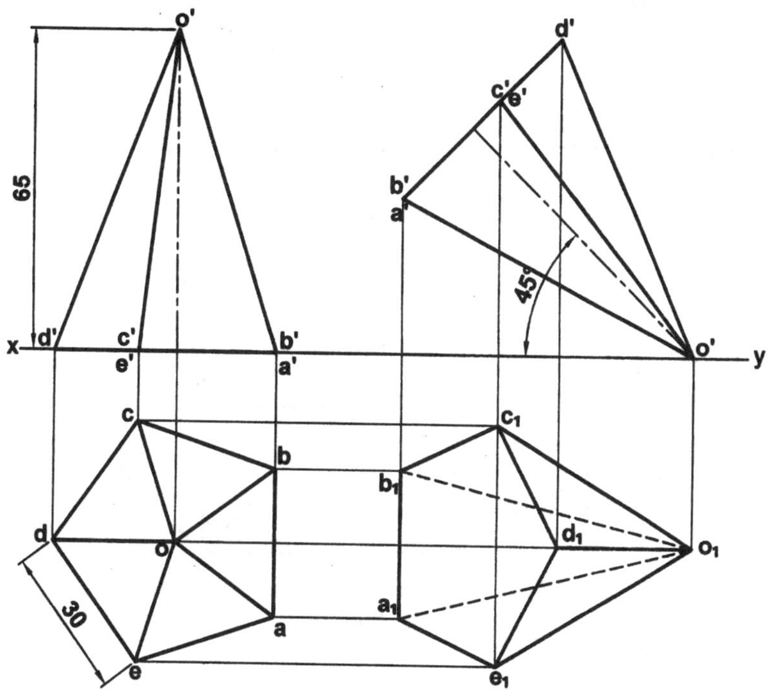

Fig. 9.27

Construction (Fig. 9.27):

Stage I

1. Draw a regular hexagon abcde of 30 mm side with a side ab perpendicular to xy line. Join a, b, c, d, e with o. This represents the top view in initial position.
2. Project front view above xy line.

Stage II

1. Reproduce the front view of stage-I with apex o' on the xy line and axis making an angle of 45° with xy line.
2. Project the final top view. Draw edges o_1a_1 and o_1b_1 by dashed lines as they represent invisible edges.

Problem 9.21 (Fig. 9.28): *A tetrahedron of 60 mm edge is resting on an edge on H.P. such that face containing that edges are equally inclined to the H.P. and edge lying on the H.P. is perpendicular to the V.P. Draw its projections.*

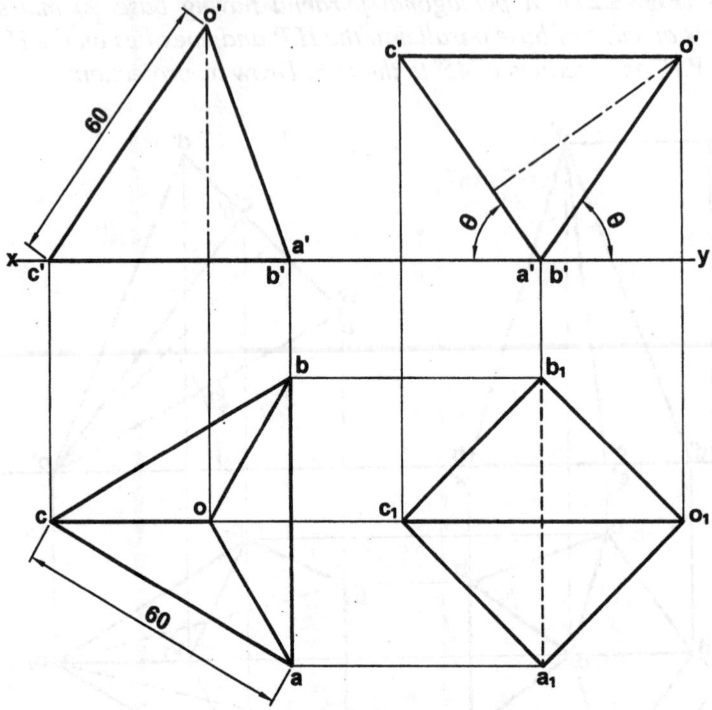

Fig. 9.28

Construction (Fig. 9.28):

Stage I

1. Draw an equilateral triangle abc of 60 mm side with a base side ab perpendicular to the xy line.
2. Locate o as centroid of triangle abc and join all the corners of the triangle with o.
3. Project its front view above xy line.

Stage II

1. Reproduce the front view of stage-I such that faces a'b'c' and a'b'o' make equal angles with xy line.
2. Obtain o_1, a_1, b_1 and c_1 at the intersection of horizontal projectors through top view of stage-I and vertical projectors through front view of stage-II.
3. Join o_1 to a_1, b_1 and c_1. Complete the top view as shown.

Problem 9.22 (Fig. 9.29): *A cube having 40 mm sides is resting on one of its corners on the H.P. with a solid diagonal vertical. Draw the projections of the cube.*

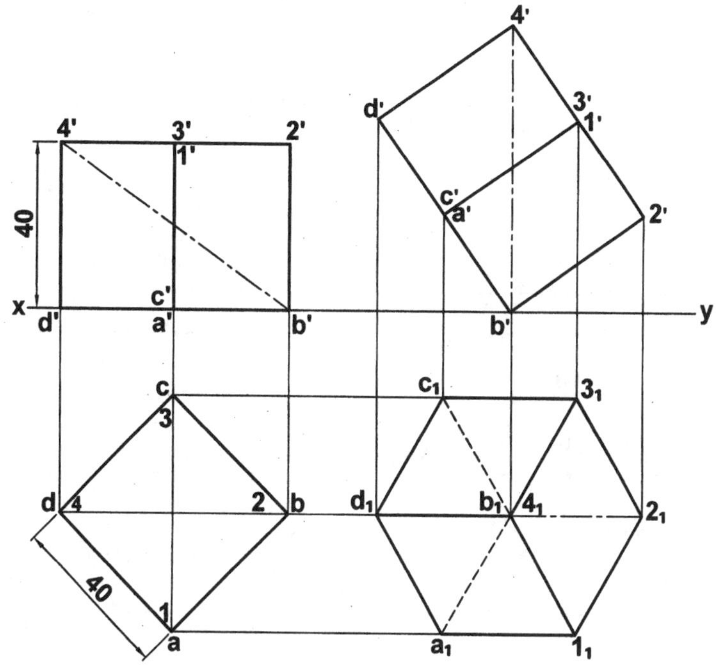

Fig. 9.29

Construction (Fig. 9.29):

Stage I

1. Draw a square abcd of 40 mm side below xy line as top view of the cube.
2. Project its front view and draw its solid diagonal 4'b'.

Stage II

1. Reproduce the front view of stage-I keeping solid diagonal 4'b' vertical.
2. Project the final top view below xy line.
3. Join a_1-b_1 and b_1-c_1 by dashed lines to represent invisible edges.

Problem 9.23 (Fig. 9.30): *A hexagonal prism base 40 mm side and 40 mm long axis has a centrally drilled hole of 40 mm diameter. Draw its projections when it is resting on one of its corners on the H.P. with its axis inclined at 50° to H.P. and two of its faces parallel to V.P.*

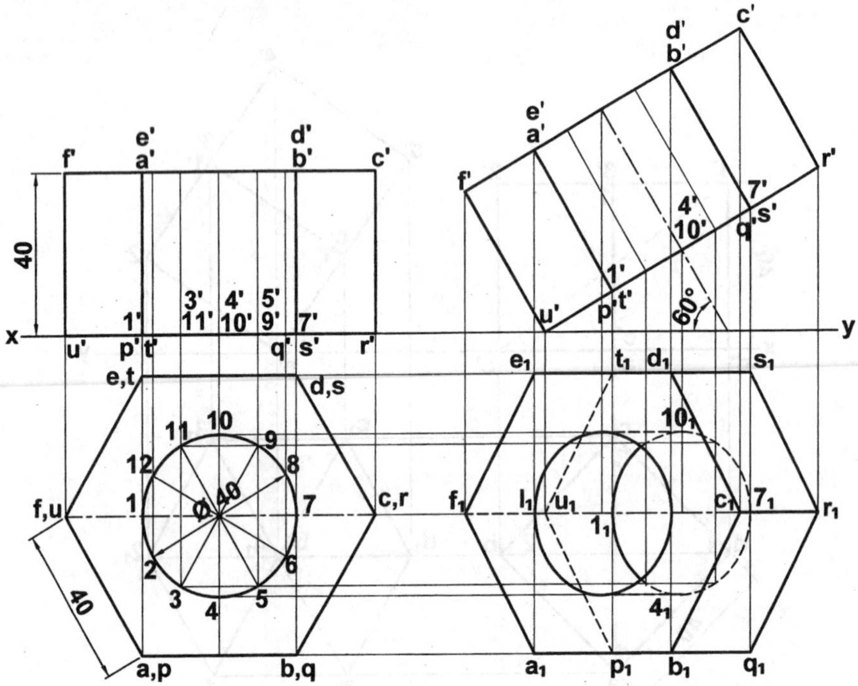

Fig. 9.30

Construction (Fig. 9.30):

Stage I

1. Draw a regular hexagon abcdef of 40 mm side below xy line.
2. Draw a circle of 40 mm diameter in the centre of the hexagon. Divide the circle into twelve equal parts. Name the divisions as 1, 2, 3....etc.
3. Project the front view and name the corner points as well as generators of cylindrical hole as shown in figure.

Stage II

1. Reproduce the front view with its axis inclined at 60° to xy line.
2. Project the final top view below xy line. Since, a part of the ellipse for lower end of the hole will be visible, draw it with continuous curve and its remaining part by dashed curve.

9.11 Axis Inclined to V.P. and Parallel to H.P.

Problem 9.24 (Fig. 9.31): *A pentagonal prism, base 25 mm side and axis 60 mm long is resting on one of its rectangular faces on the H.P. with the axis inclined at 45° to the V.P. Draw its projections.*

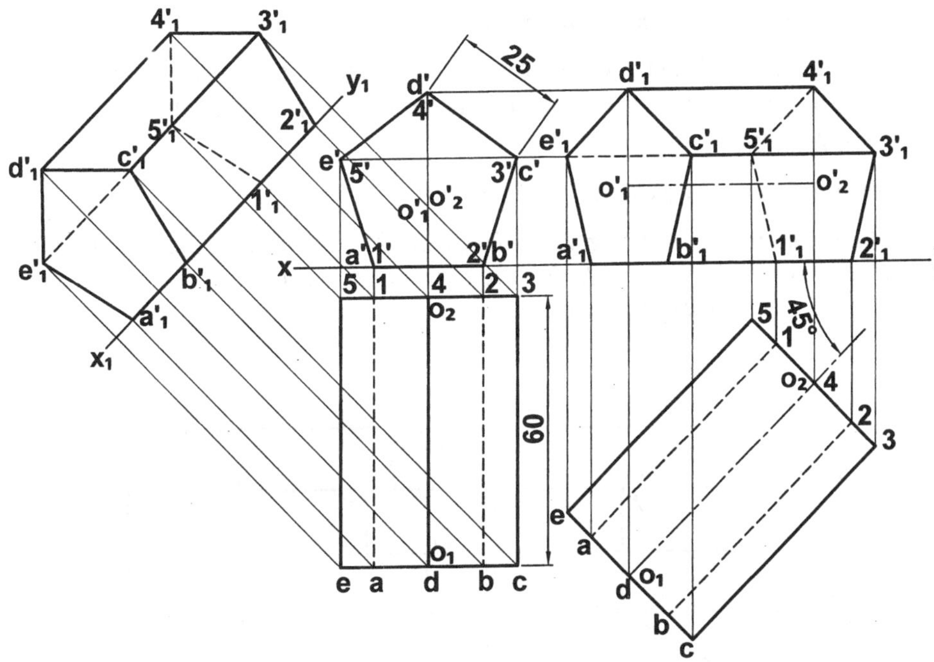

Fig. 9.31

Construction (Fig. 9.31):

Stage I

1. Draw a regular pentagon of 25 mm side with one of its side a'b' on the xy line.
2. Project its top view as shown in figure.

Stage II

1. Reproduce the top view with its axis inclined 45° to the xy line.
2. Project all the corners in the top view upward and all the corners in the front view of stage-I horizontally.
3. Mark a'_1, b'_1, c'_1 ....etc., where the corresponding horizontal and vertical generators intersect.
4. Join a'_1, b'_1, c'_1 ....etc., to obtain a pentagon.

5. Draw pentagon $1'_1$, $2'_1$, $3'_1$, $4'_1$, $5'_1$ in which draw sides $4'_1 5'_1$ and $1'_1 5'_1$ by dashed lines.
6. Draw lines $a'_1 1'_1$, $b'_1 2'_1$ ....etc., to represent longer edges.

Method II (By alternating position of reference line):

1. Draw a new reference line x_1y_1 making an angle of 45° with the axis.
2. Through a, b, c, 1, 2, 3....etc., draw projectors perpendicular to x_1y_1 and on them mark points a'_1, b'_1, c'_1 ....etc. such that their distance from x_1y_1 are equal to distances of a', b', c'....etc. from the xy line.
3. Complete the front view as shown in figure.

Problem 9.25 (Fig. 9.32): *A hexagonal prism, base 25 mm side and axis 60 mm long has an edge of its base in the V.P. such that the axis makes an angle of 35° to the V.P. and parallel to the H.P. Draw its projections.*

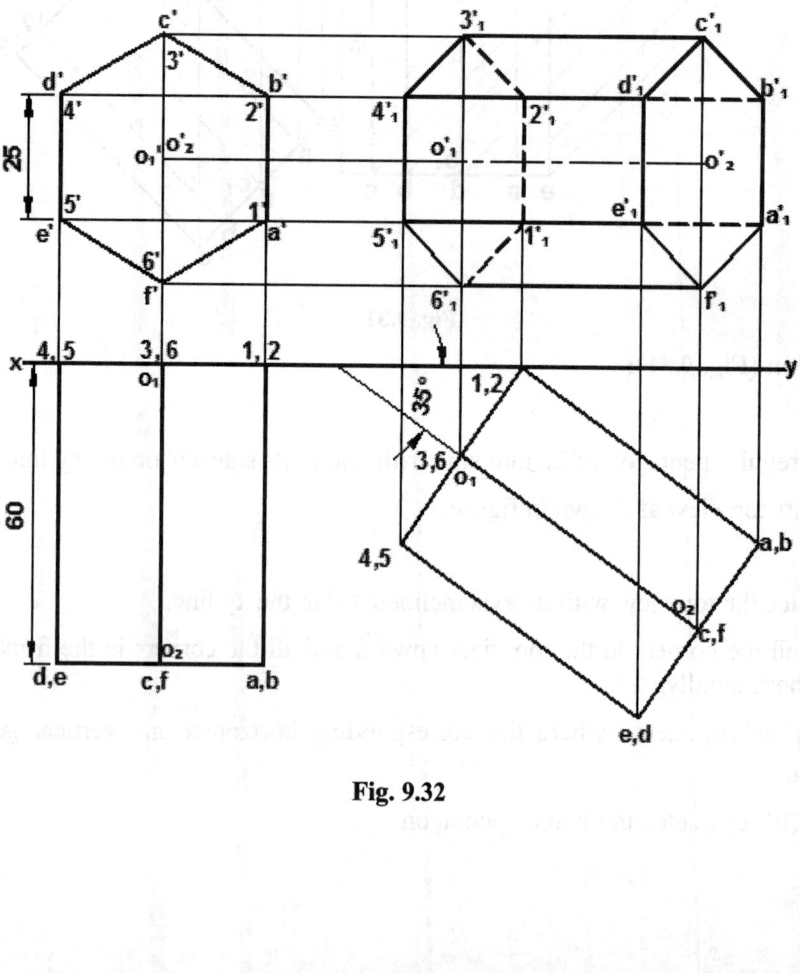

Fig. 9.32

Construction (Fig. 9.32):

Stage I

Assume base in the V.P. initially.

1. Draw a regular hexagon a'b'c'd'e'f' of 25 mm side as front view.
2. Project the top view and mark the corner points as shown in figure.

Stage II

1. Reproduce the top view with axis making an angle of 35° with xy line.
2. Project the final front view above xy line as explained in previous problems, show hidden edges by dashed line.

Problem 9.26 (Fig. 9.33): *Draw the projections of pentagonal pyramid base 25 mm side and axis 55 mm long having a triangular face in the V.P. and axis parallel to the H.P.*

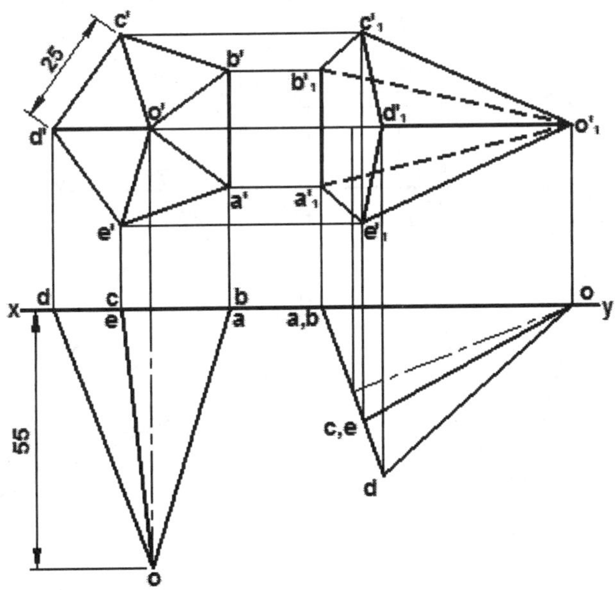

Fig. 9.33

Construction (Fig. 9.33):

Stage I (Begin with front view assuming base in the V.P.)

1. Draw a regular pentagon a'b'c'd'e' of 25 mm side above xy line with a side of base a'b' perpendicular to xy line.
2. Project top view keeping apex 55 mm below xy line.

Stage II

3. Reproduce the top view with slant edge oa coinciding with xy line.

2. Project the front view and draw $o'_1 b'_1$ and $o'_1 a'_1$ with dashed lines to represent invisible edges.

Problem 9.27 (Fig. 9.34): *Draw the projections of a cone having base diameter 40 mm and axis 65 mm long has a point on base circle in the V.P. Draw its projections when its axis inclined at 40° to the V.P. and parallel to the H.P.*

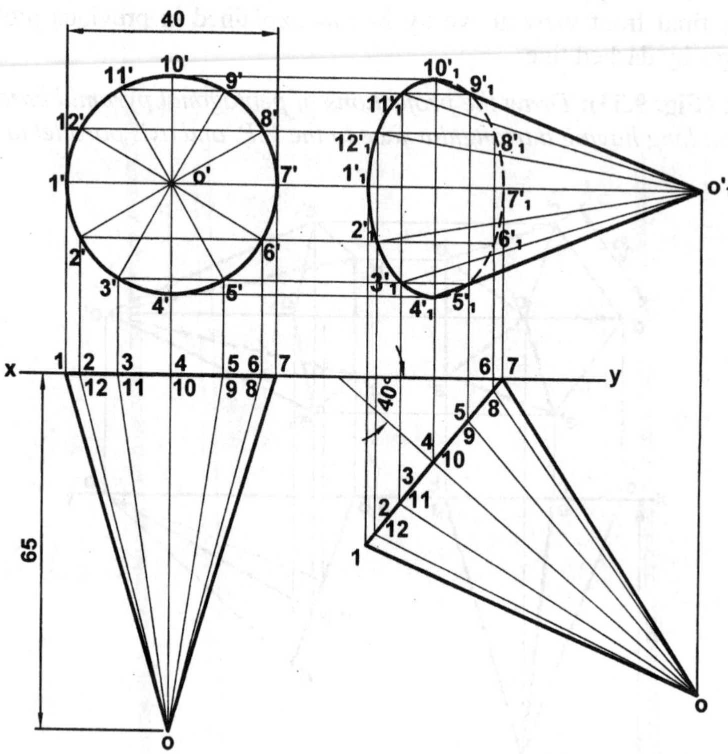

Fig. 9.34

Construction (Fig. 9.34):

Stage I (Assume base of the cone in the V.P. initially)

1. Draw a circle of 40 mm diameter above xy line as front view and divide this circle into twelve equal parts. Name the divisions as 1', 2', 3'....etc.
2. Project top view as shown in figure.

Stage II

1. Reproduce the top view with axis inclined at 40° to xy line and touching the base circle with xy line.
2. Project the final front view as shown in the figure.

Problem 9.28 (Fig. 9.35): *A cylinder having base diameter 50 mm and axis 80 mm long is lying on the ground with its axis inclined at 40° to the V.P. and parallel to the ground. Draw its projections.*

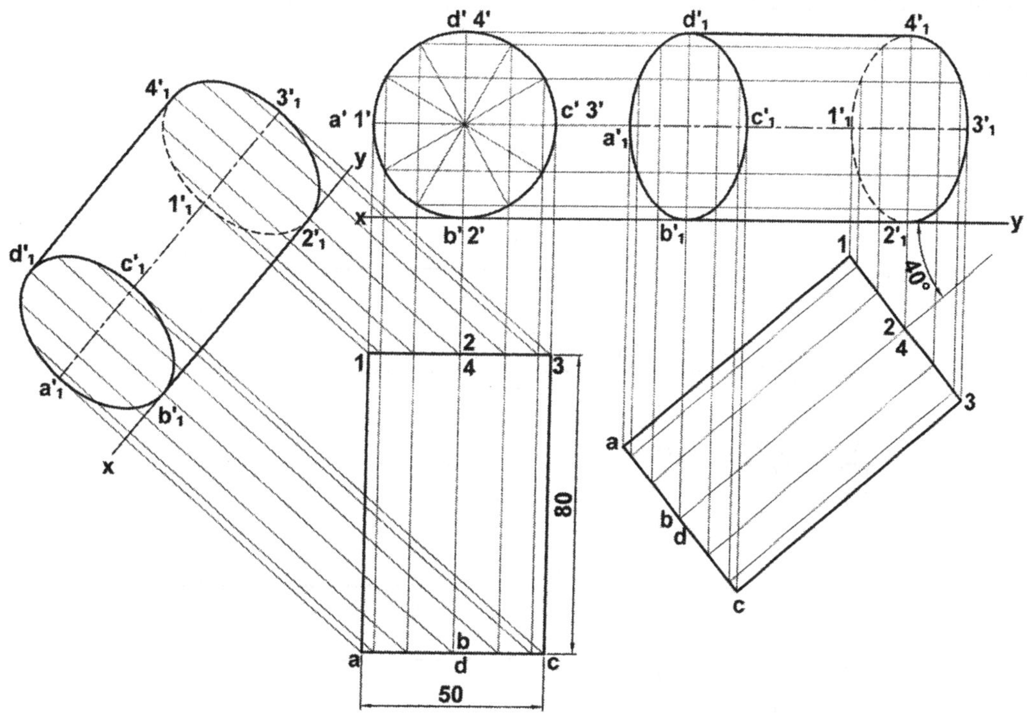

Fig. 9.35

Construction (Fig. 9.35):

Stage I

1. Draw a circle of 50 mm diameter touching xy line as front view.
2. Project its top view and draw its generators.

Stage II (By altering the position of top view)

1. Reproduce the top view with axis inclined at 40° to xy line.
2. Project front view as shown in figure. It may be noted that a portion of ellipse for base nearer to V.P., will be shown by dotted line.

Method II (By altering the position of the reference line)

1. Draw a new reference line x_1y_1 inclined at 40° to the xy line.
2. Through various points on the top view, draw projectors perpendicular to, x_1y_1 and on them mark points a'_1, b'_1, c'_1 ...etc., such that their distance from x_1y_1 are equal to the distances of corresponding points from xy in the initial front view.
3. Join the above points as shown in figure to complete final front view.

Problem 9.29 (Fig. 9.36): *Draw the projections of a pentagonal prism base 25 mm side and axis 60 mm long having one of the rectangular faces in the V.P. and axis parallel to H.P. and V.P.*

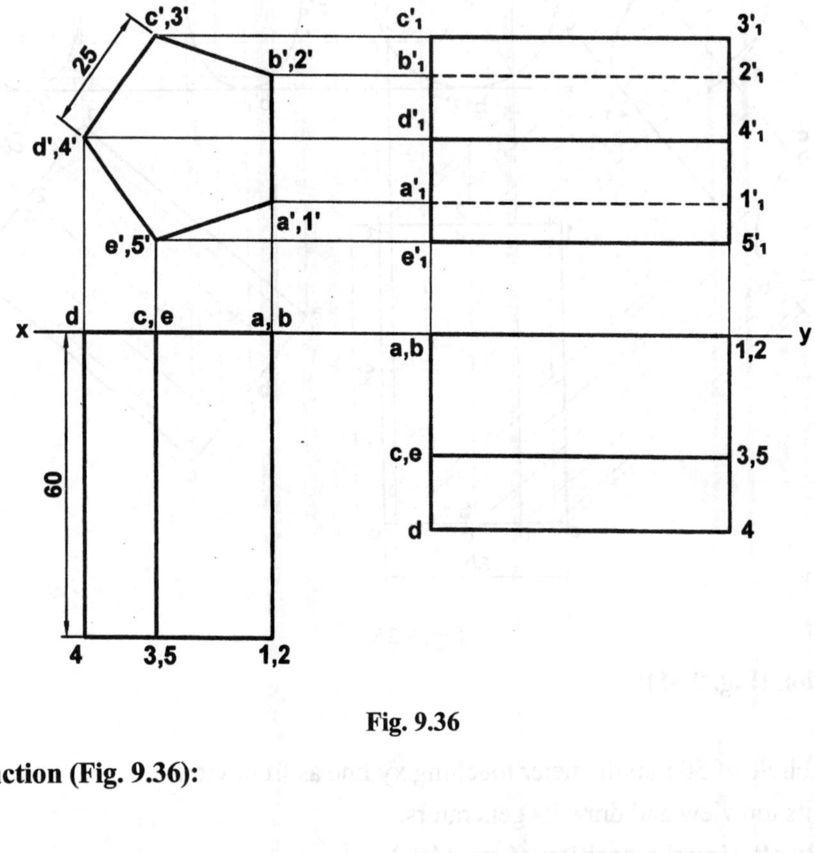

Fig. 9.36

Construction (Fig. 9.36):

Stage I

1. Draw a regular pentagon a'b'c'd'e' of 25 mm side with a base a'b' perpendicular to xy line as front view.
2. Project the top view of the prism as shown.

Stage II

1. Reproduce the top view keeping the rectangular face ab21 in the V.P. i.e., a1 coinciding with xy line.
2. Draw vertical projectors through all points in the top view of second stage and horizontal projectors through all points in the front view of stage-I.
3. Obtain points a'_1, b'_1, c'_1 ....etc., at the intersection of horizontal and vertical projectors.
4. Join these points in correct sequence to obtain final front view.

Problem 9.30 (Fig. 9.37): *A right circular cone having diameter of base 40 mm and height 56 mm rests on the H.P. such that its axis is inclined at 45° to the V.P. and is parallel to the H.P. Draw its projections when the apex is away from the V.P.*

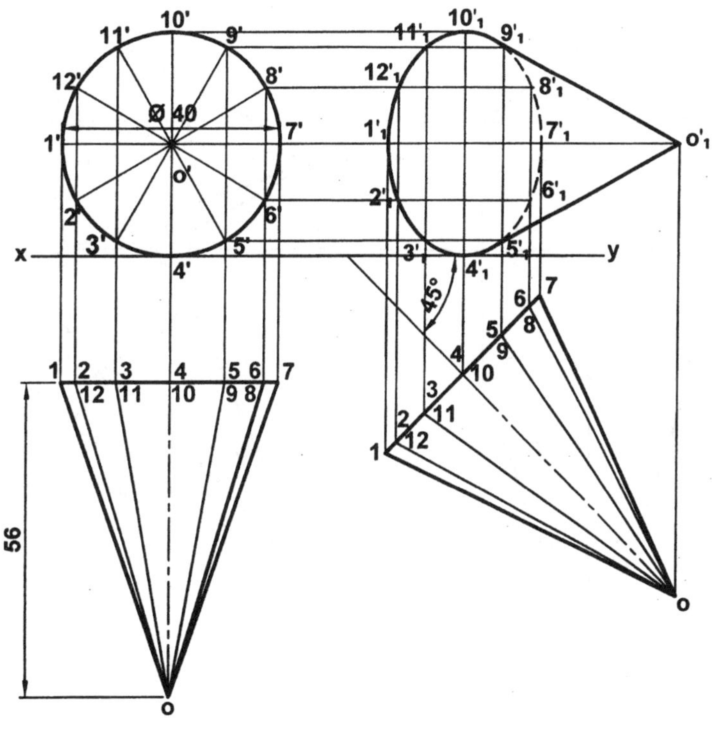

Fig. 9.37

Construction (Fig. 9.37):

Stage I (Begin with front view)

1. Draw a circle of 40 mm diameter such that it touches the xy line. Divide it into twelve equal parts and draw radial lines o'1', o'2'... etc.
2. Project its top view keeping axis parallel to H.P. Draw its generators.

Stage II

1. Reproduce the top view of stage-I so that its axis is inclined at 45° to xy line.
2. Project all the points upwards from the top view of stage-II and horizontally from the front view of stage-I.
3. Mark $1'_1, 2'_1, 3'_1$ ....etc., where the corresponding horizontal and vertical projectors intersect.
4. Join $1'_1, 2'_1$ ...etc., with o'_1.

Problem 9.31 (Fig. 9.38): *A pentagonal pyramid base 25 mm side and axis 55 mm long has one of its base edges on the H.P. if the axis is inclined at 45° to the V.P. and parallel to H.P., draw its projections.*

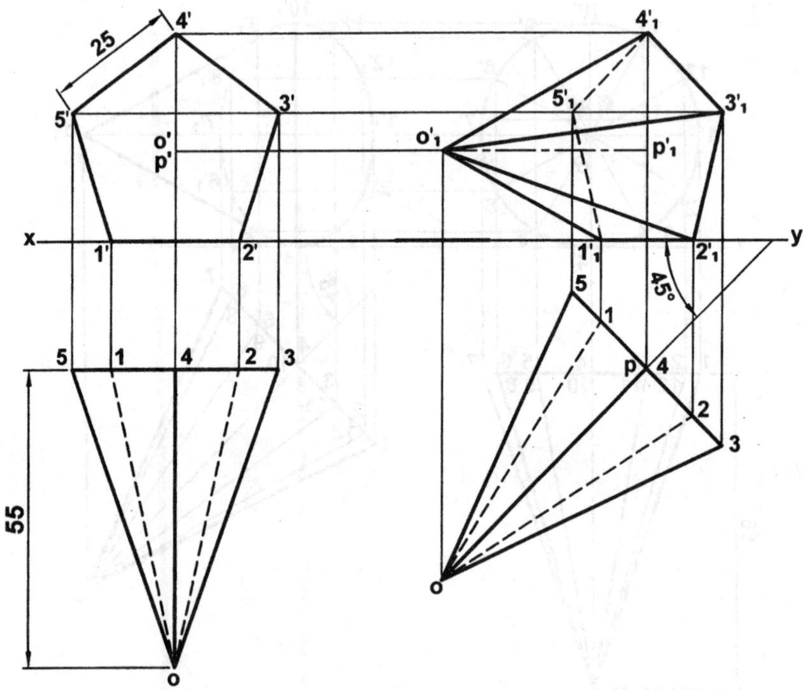

Fig. 9.38

Construction (Fig. 9.38):

Stage I

Since, axis is inclined with V.P. so initially assume it to be perpendicular to V.P.

1. Draw a regular pentagon of 25 mm side with a side on the xy line.
2. Project the top view from front view.

Stage II

1. Reproduce the top view of stage-I such that its axis makes an angle of 45° with xy line.
2. Draw the final front view as explained in problem 9.30.
3. Show invisible edges with dashed lines.

9.12 Axis Inclined to both the H.P. and the V.P.

When the axis of a solid is inclined to both the H.P. and V.P., projections are drawn in three stages. In first stage, the solid is assumed to be placed in simple position i.e., axis perpendicular to the H.P. In second stage solid is tilled to one plane at the required angle and front and top view are drawn. In third stage the final view is drawn by making required inclination with other plan.

Problem 9.32 (Fig. 9.39): *A square prism having side of the base 30 mm and axis 60 mm long is resting on an edge of its base which makes an angle of 30° to the V.P. and the axis is inclined at 45° to the H.P. Draw its projections.*

Construction (Fig. 9.39):

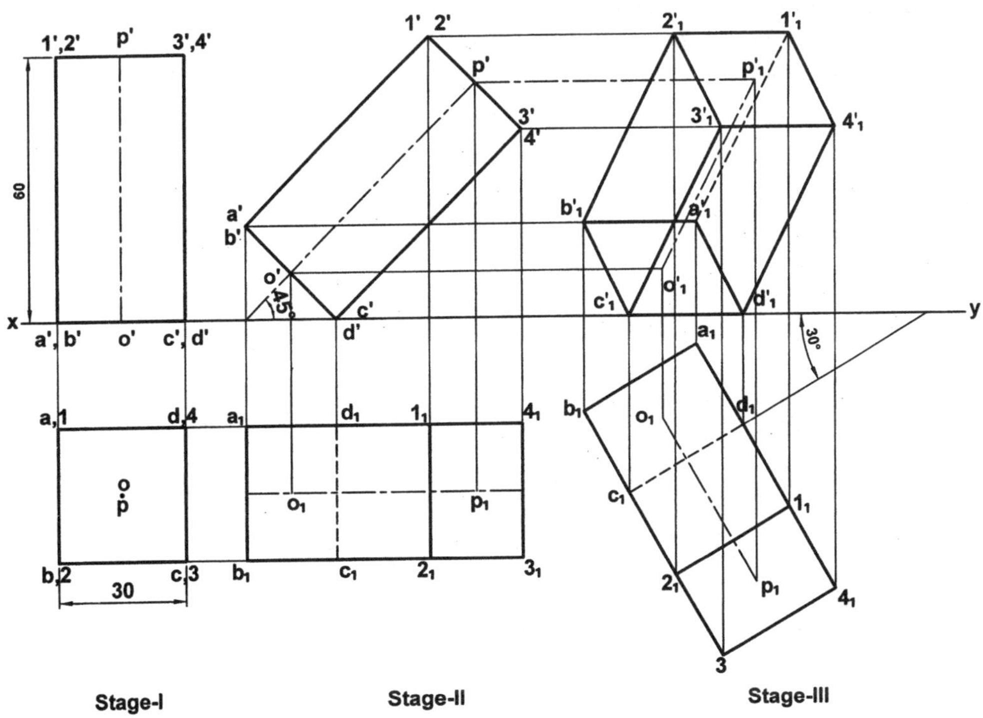

Fig. 9.39

Stage I

Begin with top view assuming the solid is resting on its base with one of its side perpendicular to xy line.

1. Draw a square abcd of 30 mm side below xy line to represent the top view in simple position.
2. Project its front view.

Stage II

Tilt the first-front view.

1. Reproduce the front view of stage-I so that its axis is inclined at 45° to xy line.
2. Project the second top view.

Stage III

Tilt the second top view.

1. Reproduce the second top view such that side c_1d_1 makes an angle of 30° with xy line.
2. Project the final front view from third top view and second front view.

Problem 9.33 (Fig. 9.40): *A pentagonal prism having edge of the base 25 mm and axis 65 mm long is resting on an edge of its base. Its axis is inclined at 45° to H.P. and the edge of its base on which the solid rests is inclined at 35° to the V.P. Draw its projections.*

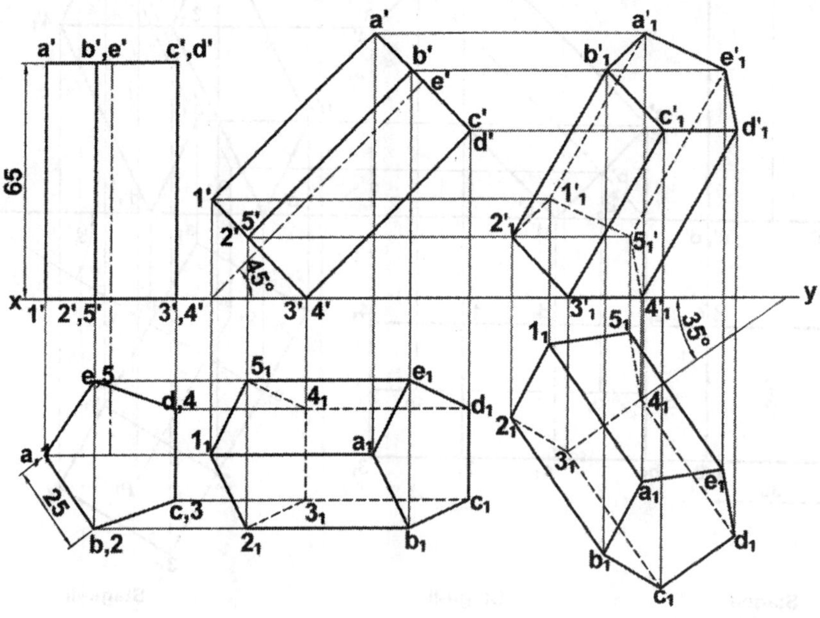

Fig. 9.40

Construction (Fig. 9.40):

Stage-I

Begin with top view.

1. Draw a regular pentagon abcde of 25 mm side to represent top view.
2. Project front view from the top view.

Stage II

1. Reproduce the front view of stage-I such that axis makes an angle of 45° with xy line.
2. Project the top view.

Stage-III

1. Tilt the second top view such that edge $3_1 4_1$ makes an angle of 35° with xy line.
2. Project the final front view as explained in problem 9.30.

Problem 9.34 (Fig. 9.41): *A square pyramid having edge of base 25 mm and axis 60 mm long rests on one of its slant faces on the H.P. and the edge of the base containing the slant edge is inclined at 45° to the V.P. Draw its projections when vertex is drawn towards the V.P.*

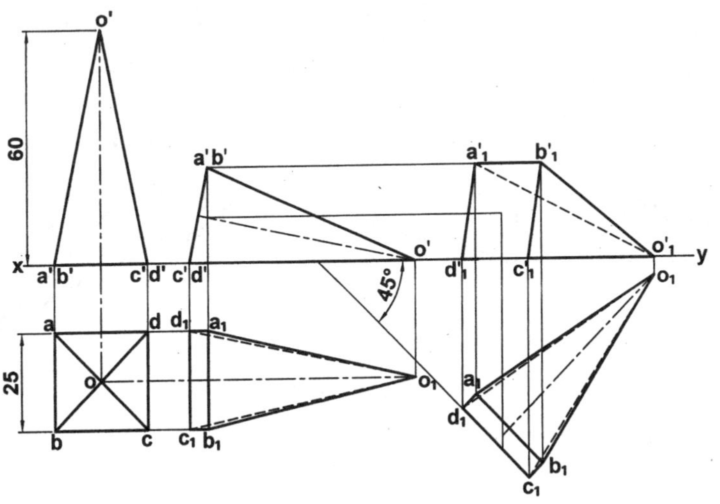

Fig. 9.41

Construction (Fig. 9.41):

Stage I

Begin with top view.

1. Draw a square abcd of 25 mm side. Locate its centre o and join it with a, b, c and d.
2. Project front view from the top view.

Stage II

1. Reproduce the front view of stage-I such that its triangular face o'c'd' coincides with xy line.
2. Project second top view as shown.

Stage III

1. Reproduce the second top view such that side of base c_1d_1 is inclined at 45° to xy line and apex is nearer to xy line.
2. Project the final front view as explained in problem 9.30.

Problem 9.35 (Fig. 9.42): *A square pyramid, edge of base 30 mm and axis 55 mm long is resting on one of its slant faces on the H.P. and vertical plane containing the axis is inclined 45° to the V.P. Draw its projections when vertex is away from V.P.*

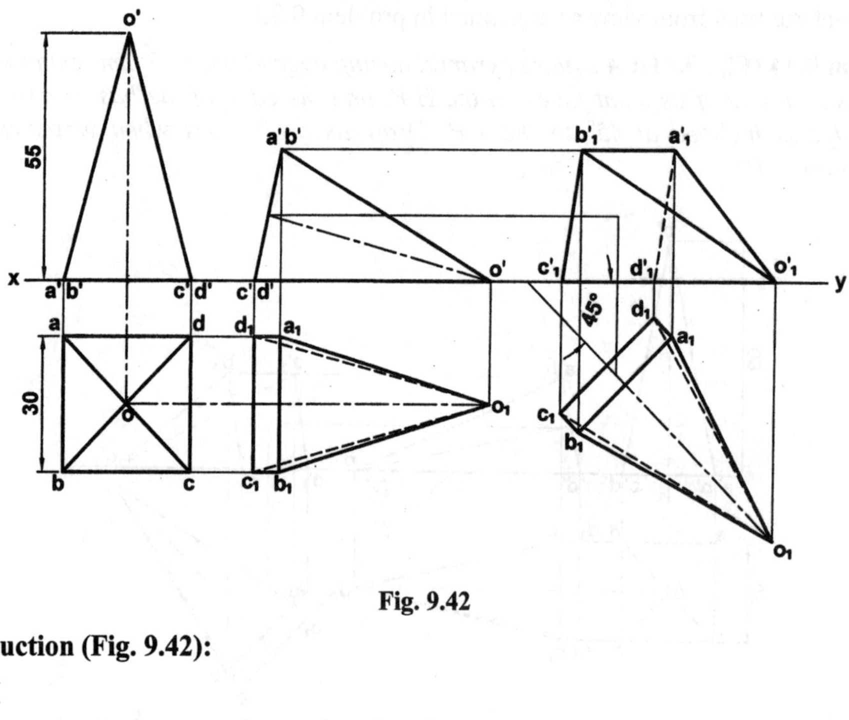

Fig. 9.42

Construction (Fig. 9.42):

Satge I

Draw a square abcd as top view and project front view.

Stage II

Tilt the front view of stage-I and project second top view as explained in problem.9.34.

Stage III

Tilt second top view so that the line representing plane containing the axis is inclined at 45° to xy line, but apex is away from the xy line. Project its final front view as shown in the figure.

Problem 9.36 (Fig. 9.43): *A pentagonal pyramid edge of base 25 mm and axis 60 mm long rests on an edge of its base on the H.P. such that the plane containing the base and resting edge makes an angle of 30° to the H.P. and vertical plane containing the axis makes an angle of 30° to V.P.*

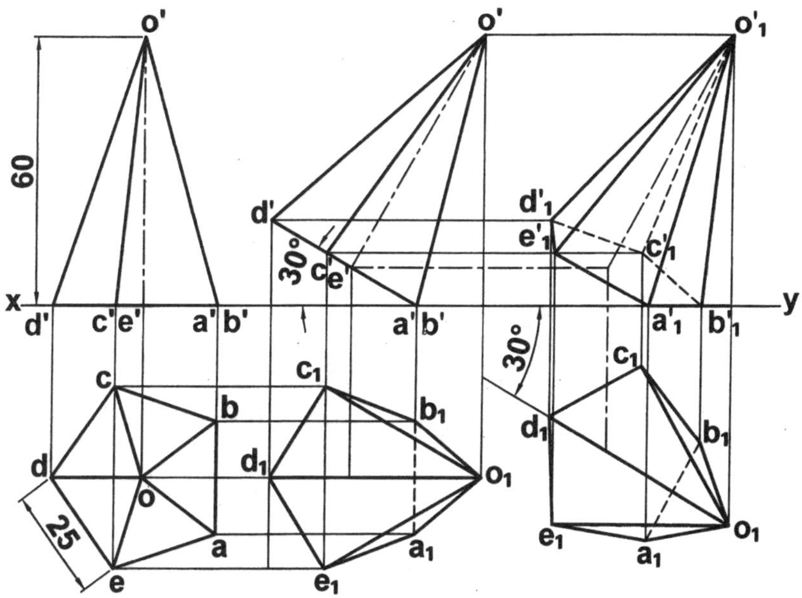

Fig. 9.43

Construction (Fig. 9.43):

Stage I

1. Draw a regular pentagon abcde of 25 mm side below xy line to represent top view in simple position.
2. Project front view as shown.

Stage II

1. Reproduce the front view of stage-I such that its base makes an angle of 30° with xy line.
2. Project second top view as explained in problem.9.30.

Stage III

1. Reproduce the top view of stage-II such that o_1d_1 is inclined at 30° to xy line.
2. Project its final top view. Draw edges $c_1' d_1'$, $b_1' c_1'$ and $o_1' c_1'$ by dotted lines as these lines represent invisible edges.

Problem 9.37 (Fig. 9.44): *A hexagonal pyramid having edge of base 25 mm and axis 60 mm long has an edge of its base on the H.P. and inclined at 45° to the V.P. The axis is inclined at 30° to the H.P. Draw its projections.*

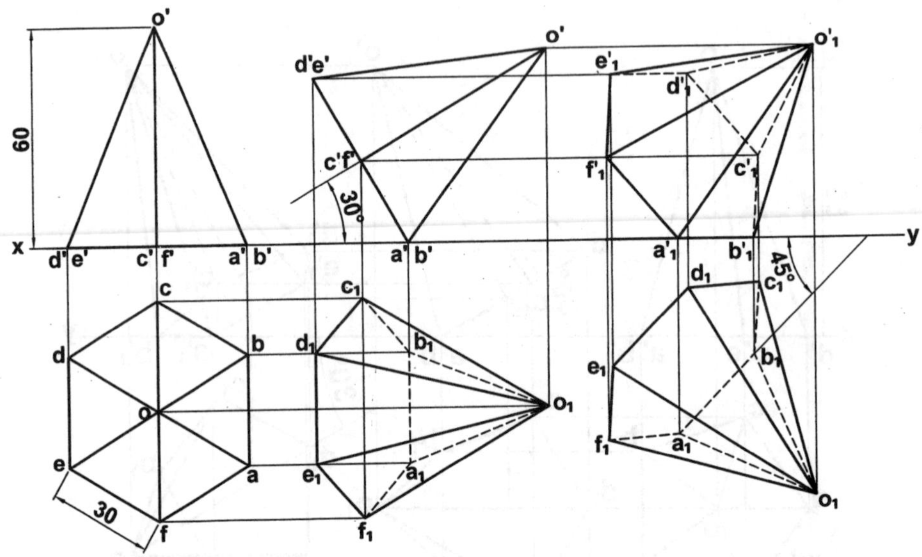

Fig. 9.44

Construction (Fig. 9.44):

Stage I

1. Draw a regular hexagon abcdef of 25 mm side as top view below xy line keeping one of the base edges perpendicular to xy line.
2. Project front view above xy line.

Stage II

1. Reproduce the front view of stage-I with its axis making an angle of 30° with xy line.
2. Project second top view as explained in problem 9.30.

Stage III

1. Reproduce the top view of stage-II such that its a side of base say a_1b_1 is inclined at 45° to xy line.
2. Project its final front view.

Problem 9.38 (Fig. 9.45): *A hexagonal pyramid having edge of base 30 mm and axis 65 mm long is resting on an edge of its base on the H.P, the triangular face containing the resting edge is perpendicular to the H.P. and parallel to the V.P. Draw its projections.*

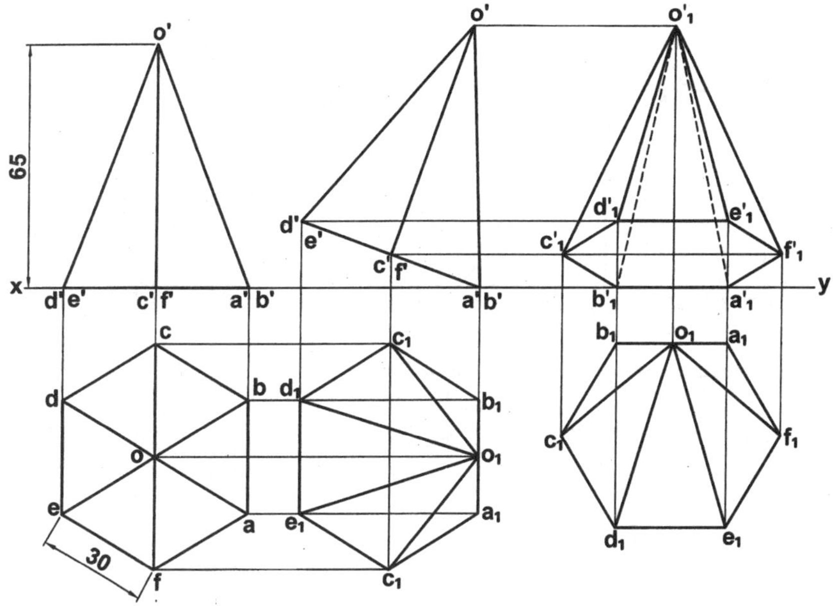

Fig. 9.45

Construction (Fig. 9.45):

Stage I

1. Draw a regular hexagon abcdef of 30 mm side below xy line such that a side of its base say ab is perpendicular to xy line. Join all the corners with centre o by straight lines.
2. Project its front view above xy line.

Stage II

1. Reproduce the front view of stage-I with line o'a' perpendicular to xy line.
2. Project its top view as explained in problem 9.30.

Stage III

1. Reproduce the top view of stage-II such that its side of base, a_1b_1 is parallel to xy line.
2. Project its final front view. Show invisible edges by dotted lines.

Problem 9.39 (Fig. 9.46): *A hexagonal prism having edge of base 20 mm and axis 50 mm long, is resting on a corner of its base. The axis makes on angle of 45° with the H.P. and the top view of axis is inclined at 60° to the V.P. draw its projections.*

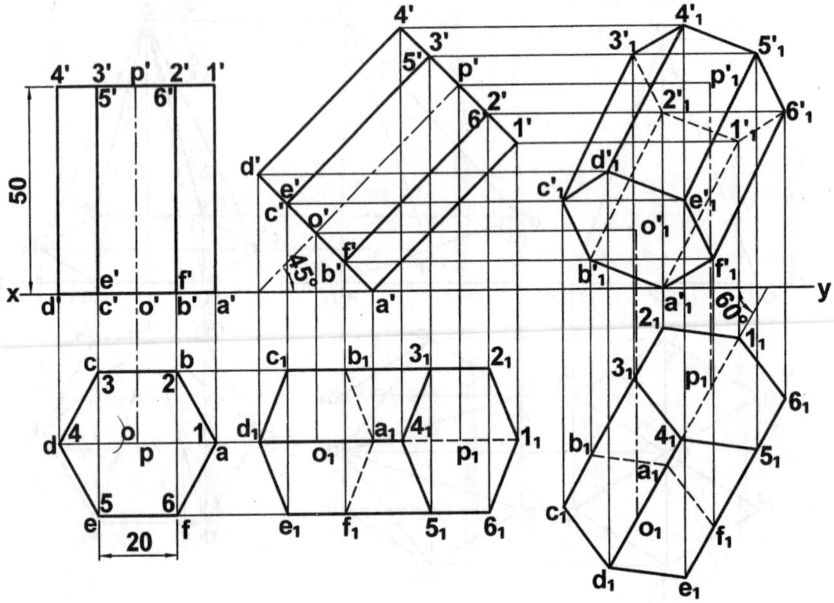

Fig. 9.46

Construction (Fig. 9.46):

Stage I

1. Draw a regular hexagon abcdef of 20 mm side below xy line with its two sides parallel to xy line. This represents the top view of given prism in simple position.
2. Project its front view above xy line.

Stage II

1. Reproduce the front view of stage-I with its axis inclined at 45° to xy line.
2. Project second top view as explained in problem 9.30.

Stage III

1. Reproduce the top view of stage-II such that its axis is inclined at 60° to xy line.
2. Project final front view. Show invisible edges by dotted lines.

Problem 9.40 (Fig. 9.47): *A hexagonal pyramid having edge of base 25 mm and axis 50 mm long is resting on one of its base corner on the H.P. The axis is inclined 30° to the H.P. and vertical plane containing the axis and that corner is inclined at 45° to the H.P. Draw its projections.*

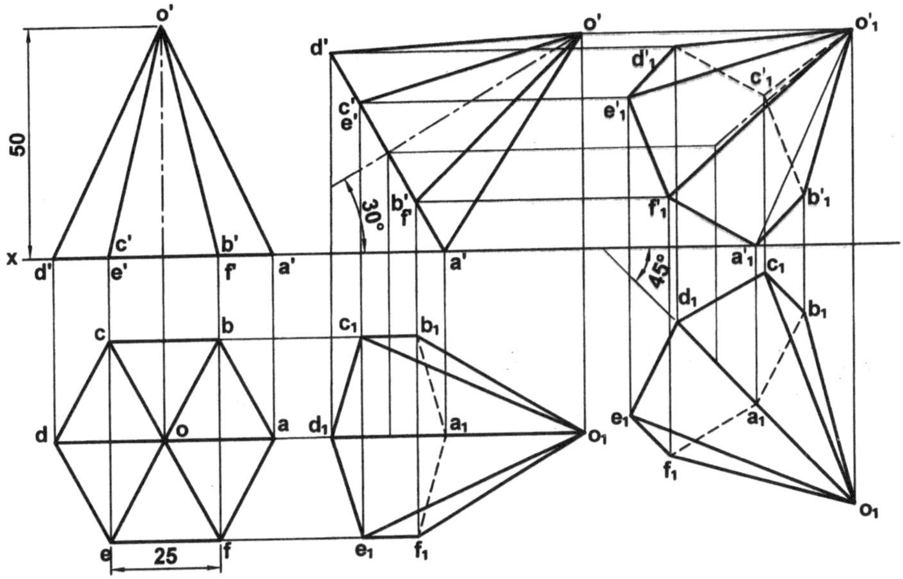

Fig. 9.47

Construction (Fig. 9.47):

Stage I

1. Draw a regular hexagon abcdef of 25 mm side below xy line with two of its base sides parallel to xy line. Join all the corners with centre o by dark lines.
2. Project its front view above xy line.

Stage II

1. Reproduce the front view of stage-I with its axis inclined at 30° to xy line.
2. Project second top view as explained in problem 9.30.

Stage III

1. Reproduce the top view of stage-II such that o_1d_1 makes an angle of 45° with xy line.
2. Project final front view above xy line as shown. Show invisible edges by dotted lines.

Problem 9.41 (Fig. 9.48): *A hexagonal pyramid having side base 25 mm and axis 60 mm long is resting on a slant edge on the H.P. such that plane containing that edge and axis is perpendicular to the H.P. and inclined at 45° to the V.P. Draw its projections.*

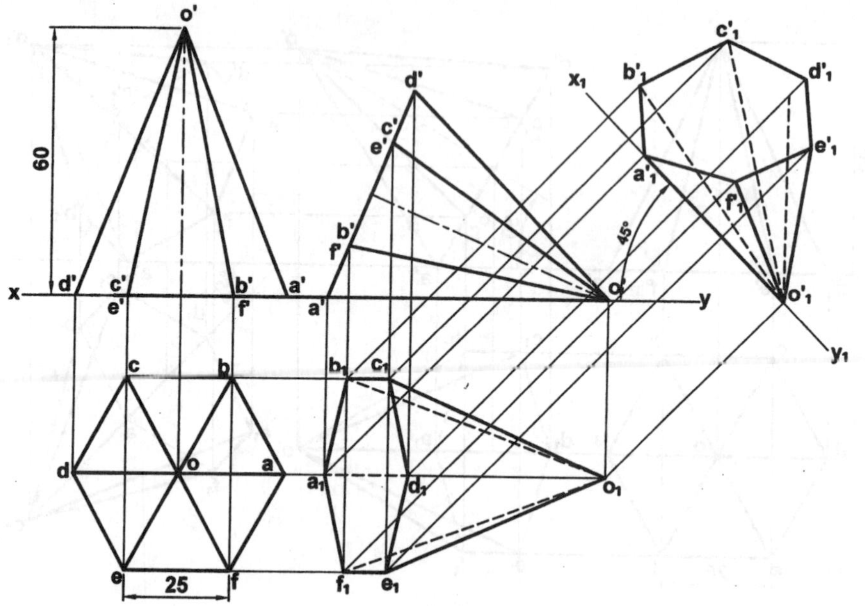

Fig. 9.48

Construction (Fig. 9.48):

Stage I

1. Draw a regular hexagon abcdef of 25 mm side below xy line with two of its base edges parallel to xy line. Join all the corners with centre o. This represents top view of the pyramid in simple position.
2. Project its front view above xy line.

Stage II

1. Reproduce the front view of stage-I with a slant edge o'a' coinciding with xy line.
2. Project top view as explained in problem 9.30.

Stage III (Using auxiliary plane method)

1. Draw a new reference line x_1y_1 inclined at 45° to the xy line.
2. Through all the corners in the second top view, draw projectors perpendicular to x_1y_1.
3. Mark a_1', b_1', c_1' ....etc., on these projectors such that their distances from x_1y_1 are equal to the distances of a', b', c'....etc., respectively from xy line.
4. Join a_1', b_1', c_1' ....etc., in correct sequence to obtain final front view.

Problem 9.42 (Fig. 9.49): *A cube having edges 30 mm long is resting on one of its corner on the H.P. Draw its projections when one of its solid diagonal is perpendicular to the V.P.*

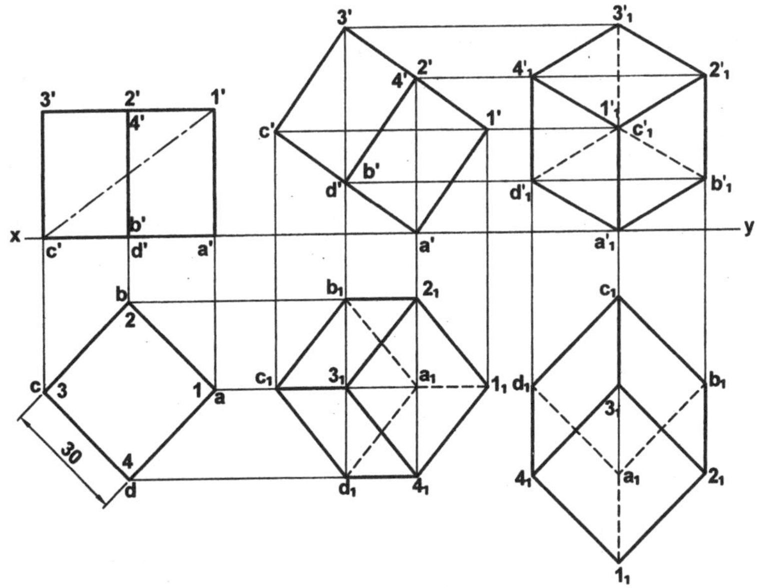

Fig. 9.49

Construction (Fig. 9.49):

Stage I

1. Draw a square abcd of 30 mm side with its sides equally inclined to xy line. This represents top view of cube in simple position.
2. Project the front view and draw its solid diagonal 1'-c'.

Stage II

1. Reproduce the front view of stage-I such that solid diagonal 1'-c' is parallel to xy line.
2. Project top view below xy line.

Stage III

1. Reproduce the top view of stage-II such that solid diagonal c_1-1_1 is perpendicular to xy line.
2. Project final front view and show invisible edges by dotted lines.

Problem 9.43 (Fig. 9.50): *A hexagonal prism having edge of base 25 mm and axis 55 mm long has an edge on its base in the V.P. and inclined 45° to the H.P. The corner of the base which is farthest away from the V.P. is 65 mm from the V.P. Draw its projections.*

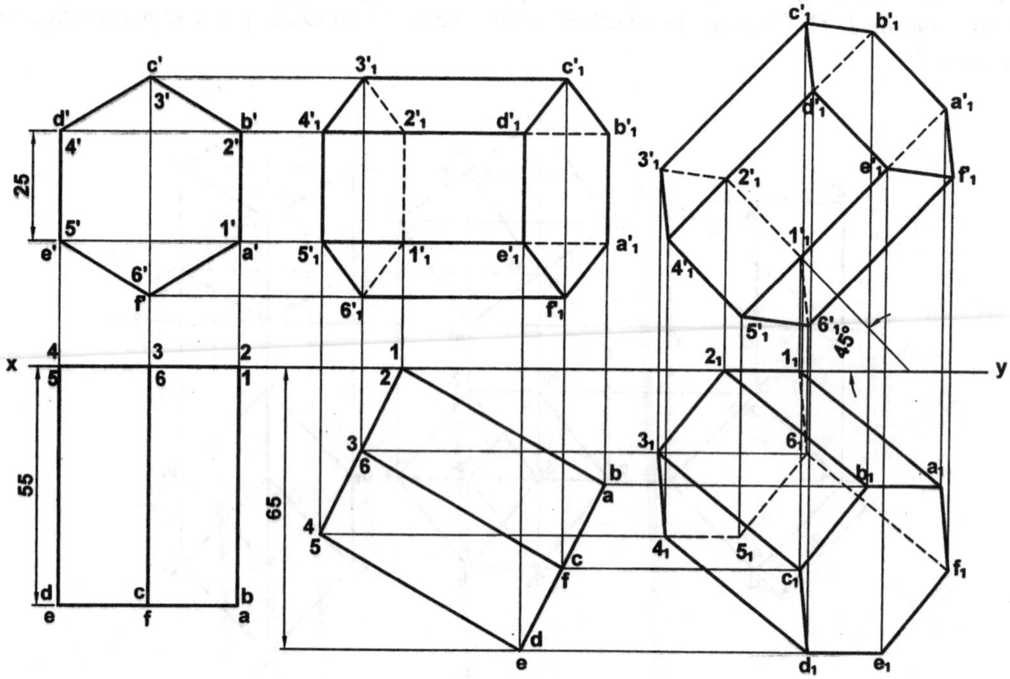

Fig. 9.50

Construction (Fig. 9.50):

Stage I

(Assume initially a base to be in contact with V.P.)

1. Draw a regular hexagon a'b'c'd'e'f' of 25 mm side above xy line such that two of its sides are perpendicular to xy line.
2. Project its top view such that its base touches with xy line.

Stage II

1. Reproduce the top view of stage-I such that its corner 1, 2 is in touch with xy line and corner d, e is 65 mm away from the xy line.
2. Project its new front view above xy line.

Stage-III

1. Reproduce the front view of stage-II with a side of base $1_1' \, 2_1'$ inclined at 45° to xy line.
2. Project the final top view. Show invisible edges by dotted line.

Problem 9.44 (Fig. 9.51): *A pentagonal prism having edge of base 30 mm and axis 65 mm long has an edge of its base in V.P. and inclined at 40° to the H.P. The axis makes an angle of 30° to the V.P. Draw its projections.*

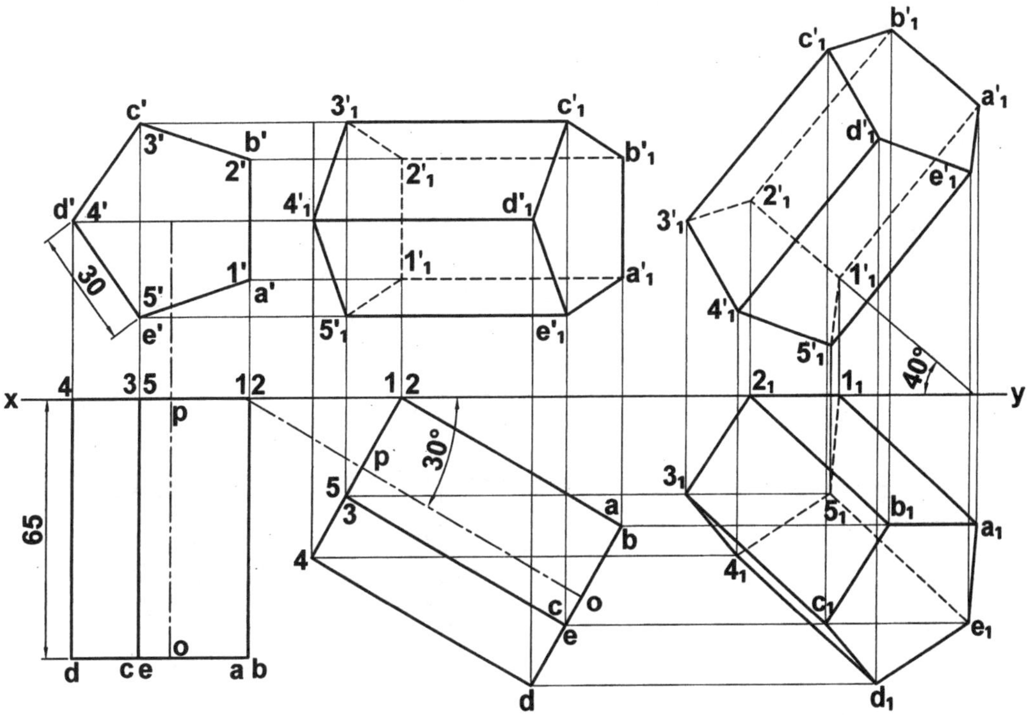

Fig. 9.51

Construction (Fig. 9.51):

Stage I

(Assume initially a base to be in the V.P.)

1. Draw a regular pentagon a'b'c'd'e' of 30 mm side above xy line to represent front view.
2. Project its top view.

Stage II

1. Reproduce the top view of stage-I with its axis inclined at 30° to xy line.
2. Project its second front view.

Stage III

1. Reproduce the front view of stage-II such that side $1_1' \, 2_1'$ makes an angle of 40° with xy line.
2. Project its final top view as shown.

Problem 9.45 (Fig. 9.52): *A pentagonal pyramid, base 25 mm side and axis 60 mm long has one of its triangular faces in the V.P. The edge of the base contained by that face is inclined at 45° with the H.P. Draw its projections.*

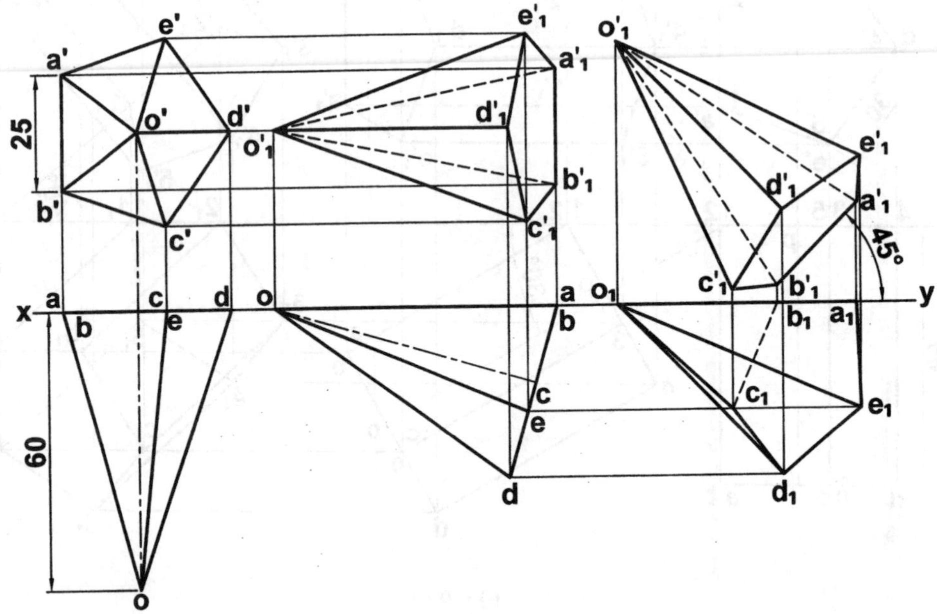

Fig. 9.52

Construction (Fig. 9.52):

Stage-I

(Assume initially the base to be in V.P.)

1. Draw a regular pentagon a'b'c'd'e' of 25 mm side. Join all the corners with centre o. This represents the front view in simple position.
2. Project its top view keeping the base line in touch with xy line.

Stage II

1. Reproduce the top view of stage-I with its slant face coinciding with xy line.
2. Project the front view as explained in problem.9.30.

Stage III

1. Reproduce the front view of stage-II such that its base side $a'_1 b'_1$ is inclined at 45° to xy line.
2. Project its final top view as shown in the figure.

Problem 9.46 (Fig. 9.53): *A square pyramid having an edge of the base 30 mm and the axis 65 mm long, has a corner of its base in the V.P. The slant edge containing that corner is inclined at 30° to the V.P. and the plane containing the slant edge and the axis is inclined at 45° to the H.P. Draw its projections.*

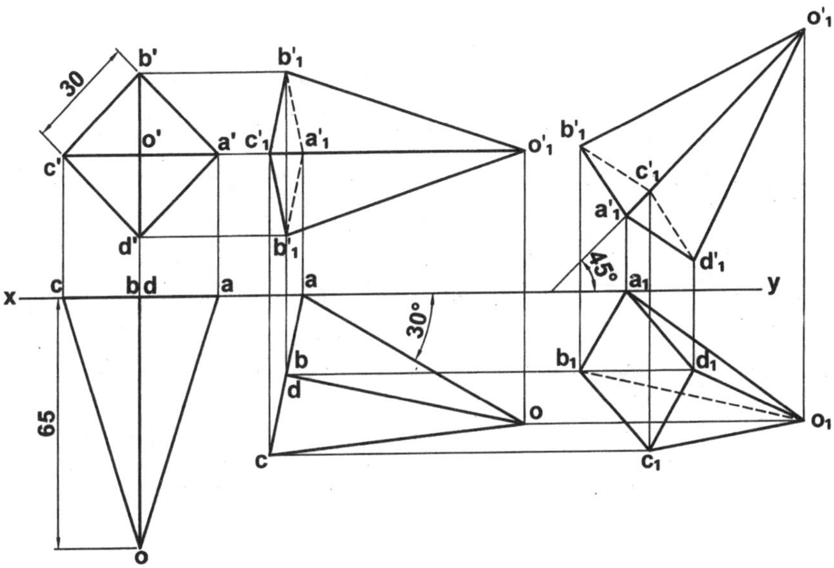

Fig. 9.53

Construction (Fig. 9.53):

Stage I

(Assume initially the base to be in the V.P.)

1. Draw a square a'b'c'd' of 30 mm side above xy line. Join all corner points with centre o'. This represents front view.
2. Project top view keeping base line coinciding with xy line.

Stage II

1. Reproduce top view of stage-I such that corner a lies on the xy line and slant edge ao makes an angle of 30° with xy line.
2. Project the front view as shown in figure.

Stage III

1. Reproduce the front view of stage-II with $o'_1 a'_1$ inclined at 45° to xy line.
2. Project its final top view. Show its hidden edges by dotted lines.

Problem 9.47 (Fig. 9.54): *A hexagonal prism having edge of base 25 mm and axis 60 mm long is resting on an edge of its base on the H.P. such that its axis is inclined at 30° the H.P and 45° to the V.P. Draw its projections.*

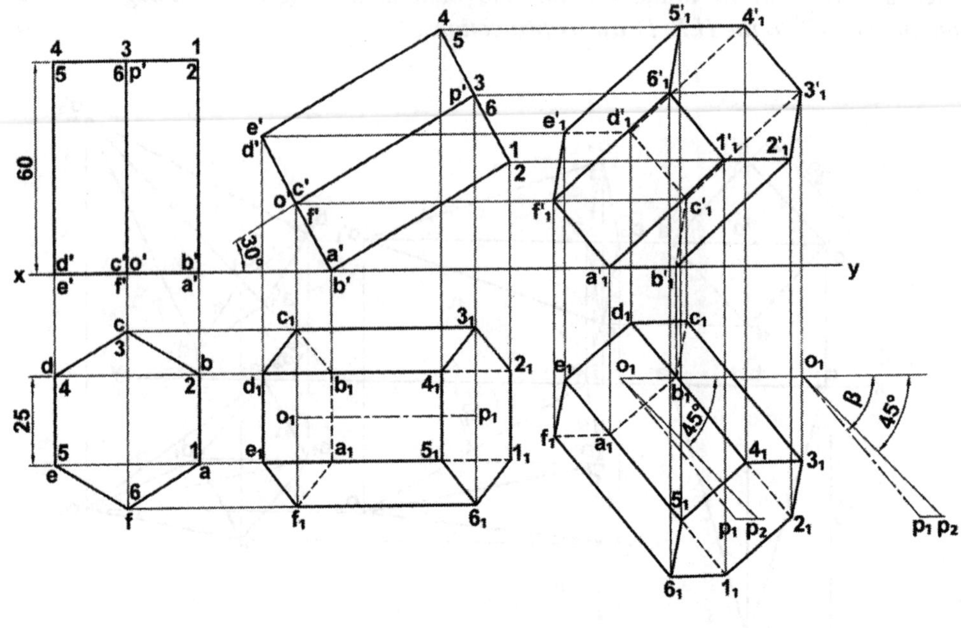

Fig. 9.54

Construction (Fig. 9.54):

Stage I

(Assume initially the base to be in the H.P.)

1. Draw a regular hexagon abcdef of 25 mm side below xy line to represent top view in simple position.
2. Project front view above xy line.

Stage II

1. Reproduce the front view of stage-I with its axis inclined at 30° to xy line and side a'-b' on xy line.
2. Project its top view as shown.

Stage III

Note: Since, axis is inclined to both H.P. and V.P., it is required to determine its apparent angle β which is greater than actual inclination (i.e., 45°).

1. Mark a point o_1 below xy line. Draw a line o_1p_2 equal to true length of the axis (i.e., 60 mm) and inclined at 45° to the xy line. With o_1 as centre and radius equal to o_1p_1 (length of top view of axis), draw an arc cutting a horizontal line through p_2 at p_1. The angle of inclination of o_1p_1 with horizontal is the required apparent angle.
2. Reproduce the top view of stage-II with o_1p_1 as axis.
3. Project the final front view. Show the invisible edges by dotted lines.

Problem 9.48 (Fig. 9.55): *A square pyramid having edge of base 40 mm and axis 60 mm long is lying on its triangular faces on the H.P. with its axis inclined at 30° to the V.P. Draw its projections.*

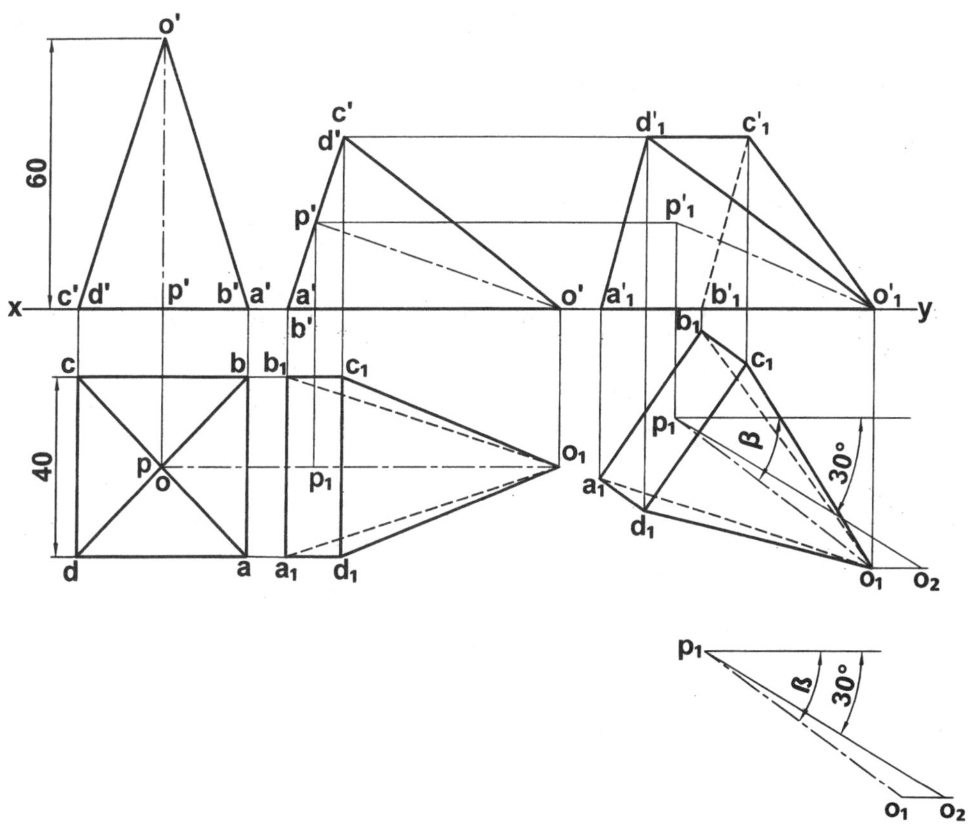

Fig. 9.55

Construction (Fig. 9.55):

Stage I

1. Draw a square abcd of 40 mm side below xy line. Join all the corner with centre o. This represents top view of given square pyramid in simple position.
2. Project its front view above xy line.

Stage II

1. Reproduce the front view of stage-II such that line representing triangular face o'a'b' coincides with xy line.
2. Project its top view.

Stage III

1. Since, axis is inclined to both H.P. and V.P., determine apparent angle β as explained in problem 9.47 and reproduce second top view considering p_1o_1 as axis.
2. Reproduce final front view above xy line.

Problem 9.49 (Fig. 9.56): *A pentagonal prism edge of base 30 mm and height 60 mm is resting on one of its corner on the H.P. The longer edge containing that corner is inclined at 45° to the H.P. and axis is inclined at 30° to the V.P.*

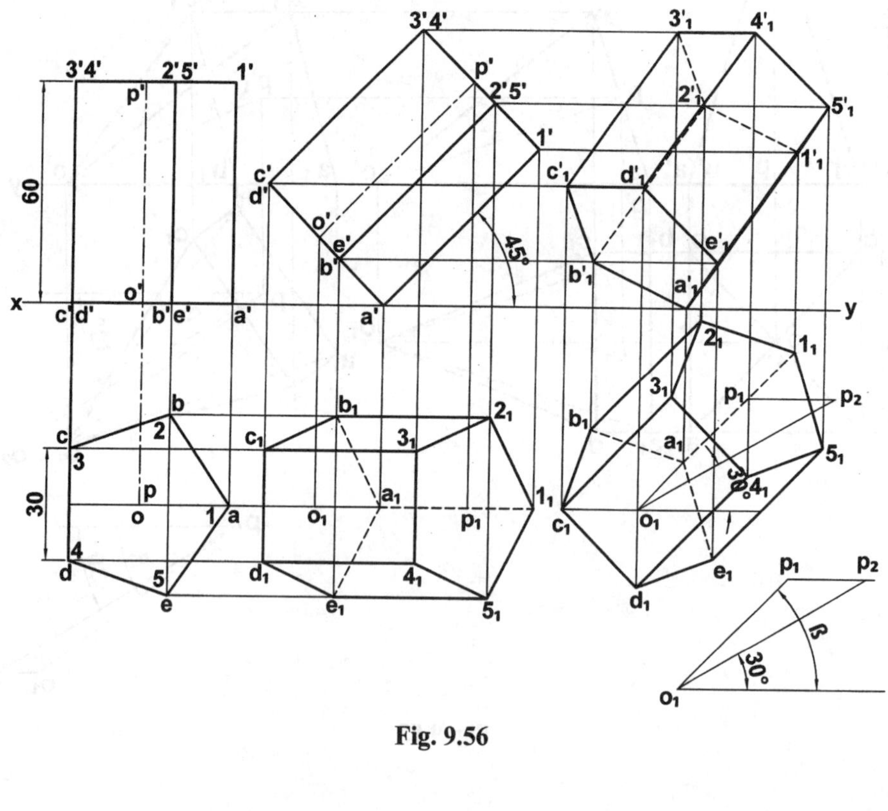

Fig. 9.56

Construction (Fig. 9.56):

1. **Stage I:** Draw a regular pentagon abcde of 30 mm side below xy line to represent top view. Project its front view.
2. **Stage II:** Reproduce the front view of stage-I such that the prism rests on its base corner a' and slant edge a'1' is inclined at 45° to xy line. Project the second top view.
3. **Stage III:** Determine the apparent angle β as explained in problem 9.47 with o_1p_1 as axis (inclined at β° with horizontal) reproduce the top view of stage-II. Project its final front view as shown in the figure.

Problem 9.50 (Fig.9.57): *A cone base diameter 40 mm and axis 60 mm long is resting on a point on the base circle on the H.P. Draw the projections.*

(a) *The axis makes an angle of 30° with the H.P. and 45° with the V.P.*

(b) *The axis makes an angle of 30° with the H.P. and its top view is inclined at 45° to the V.P.*

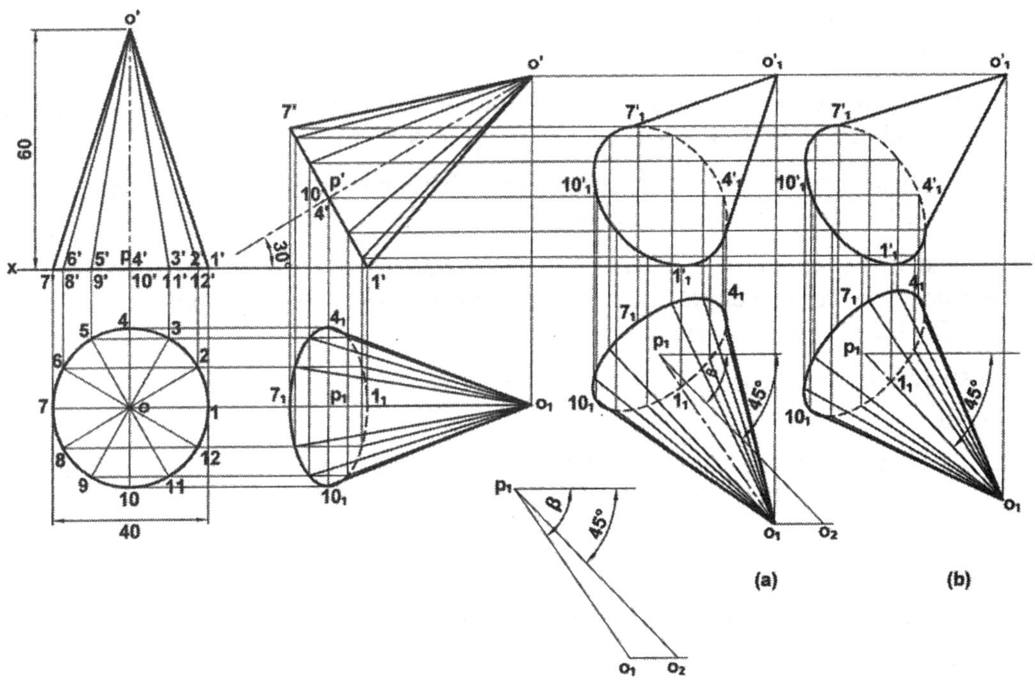

Fig. 9.57

Construction (Fig. 9.57):

Stage I

Draw a circle of 40 mm diameter below xy line. Divide it into twelve equal parts. Draw radial lines o1, o1, o3....etc. This represents top view of cone. Project front view and draw its generators.

Stage-II

Reproduce front view of stage-I such that its axis is inclined at 30° to xy line. Project its top view.

Stage III

(a) **When the axis makes an angle of 30° with the H.P. and 45° with the V.P. (Fig. 9.57 (a))**

Draw a line p_1o_2 equal to the true length of axis (i.e., 60 mm) and inclined at 45° with the xy line. Now with p_1 as centre and radius equal to o_1p_1 (apparent length of axis) draw an arc cutting the horizontal line through o_2 at o_1. Join p_1-o_1.

Reproduce, the top view of stage-II with o_1p_1 as axis. Project its final top view as shown in Fig. 9.57(a).

(b) **When the axis makes an angle of 30° with the H.P. and its top view is inclined at 45° to the V.P. of stage II (Fig. 9.57(b)).**

Reproduce the top view of stage-II such that top view of axis o_1p_1 is inclined at 45° to the horizontal. Project its final front view.

Problem 9.51 (Fig. 9.58): *A cylinder having base diameter 40 mm and axis 65 mm long, has a point on the base circle in the V.P. Its axis is inclined at 30° to the V.P. and 45° to the H.P. Draw its projections.*

Construction (Fig. 9.58):

Stage I

Draw a circle of 40 mm diameter and divide it into twelve equal parts. Draw radial lines o'1', o'2', o'3'...etc. This represents the front view. Project its top view and draw its generators.

Stage II

Reproduce top view of stage-I such that its axis is inclined at 30° to xy line. Project its front view.

Stage III

Since, the axis is inclined to both H.P. and V.P. determine apparent angle α as explained in problem 9.47. Now considering $p'_1\ o'_1$ as axis, reproduce the front view of stage II. Project the final top view.

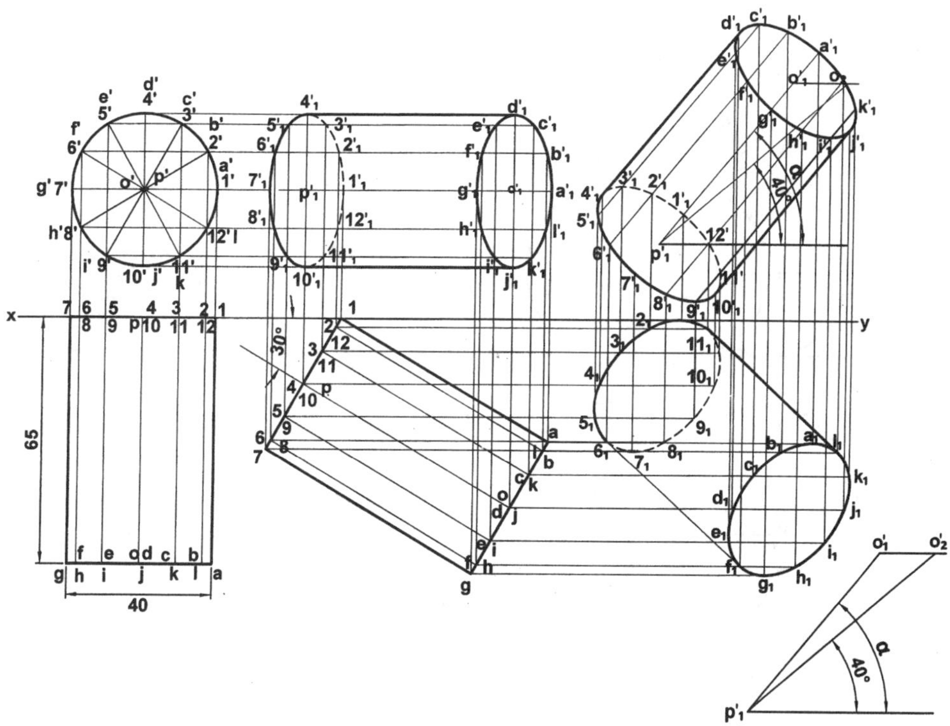

Fig. 9.58

9.13 Projections of Spheres

Problem 9.52 (Fig. 9.59): *Three spherical balls of equal diameters of 42 mm are resting on the H.P. so that each touches the other two and the line joining centres of any two balls is parallel to the V.P. A fourth spherical ball of diameter 52 mm is placed on top of the three balls so as to form a pile. Draw front view, top view and side view of the arrangement. Determine distance of the centre of fourth ball above the H.P.*

Construction (Fig. 9.59):

1. In the top view, lines joining centres of three spheres will form an equilateral triangle. So in the top view, draw an equilateral triangle $o_1o_2o_3$ of 42 mm side keeping a side say o_1-o_2 parallel to xy line.
2. With o_1, o_2 and o_3 as centres and radius equal to 21 mm, draw three circles.
3. Mark o_4 as centre of equilateral triangle and with o_4 as centre and radius equal to 26 mm, draw another circle. This gives the required top view.
4. Through o_1, o_2 and o_3, draw vertical projectors and on these projectors obtain o'_1, o'_2 and o'_3 respectively 21 mm above xy line.

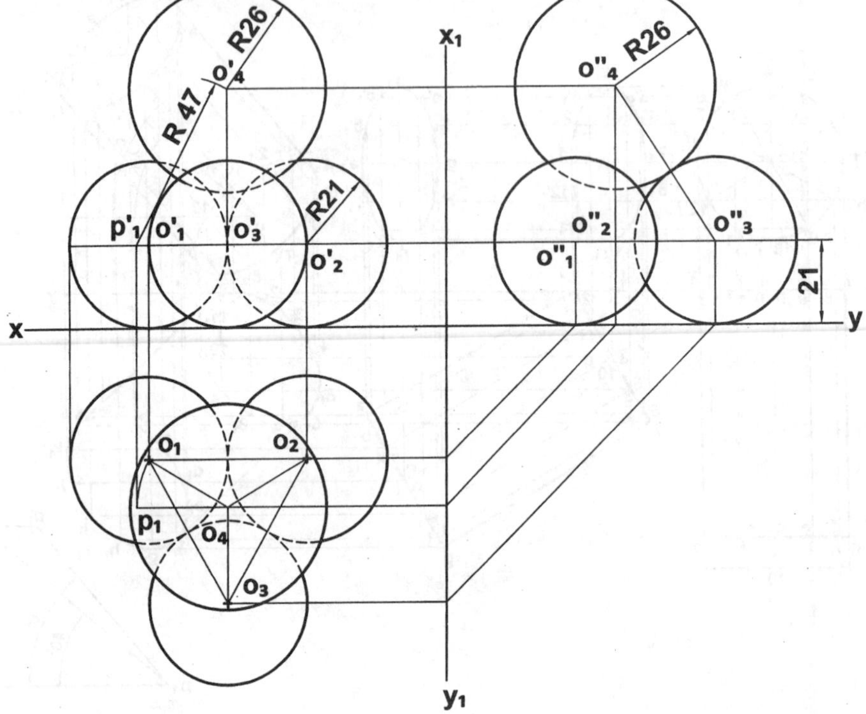

Fig. 9.59

5. With o'_1, o'_2 and o'_3 as centres and radius equal to 21 mm, draw three circles.
6. The centre of top sphere (o'_4) will lie on the projectors through o_4. In order to locate the centre of top sphere, make one of the lines say o_1o_4 parallel to xy line as follows.
   (i) With o_4 as centre and radius equal to o_4o_1, draw an arc intersecting horizontal line through o_4 at p_1.
   (ii) Project p_1 upward and obtain p'_1.
   (iii) With p'_1 as centre and radius equal to 47 mm (21 + 26 = 47 mm), draw an arc cutting vertical line through o_3 at o'_4 . o'_4 is the required centre of top sphere.
7. With o'_4 as centre and radius equal to 26 mm draw a circle.
8. Project the side view as shown in the figure.

Problem 9.53 (Fig. 9.60): *A square prism, having side of the base 25 mm and axis 65 mm long, is resting on its base on the H.P. such that its two faces remain parallel to* *the V.P. Find the radius of four equal spheres resting on the H.P. and each touching a face of the prism and other two spheres. Draw the projections of the arrangement.*

Construction (Fig. 9.60):

1. Draw top view and front view of the given square prism of 25 mm base side and 55 mm long axis.
2. In the top view, draw diagonals of square abcd and produce them on both sides.
3. Draw the bisectors of angle ecb and hbc intersecting each other at o_1. Draw the perpendicular o_1m. The length of o_1m is the required radius of circle.

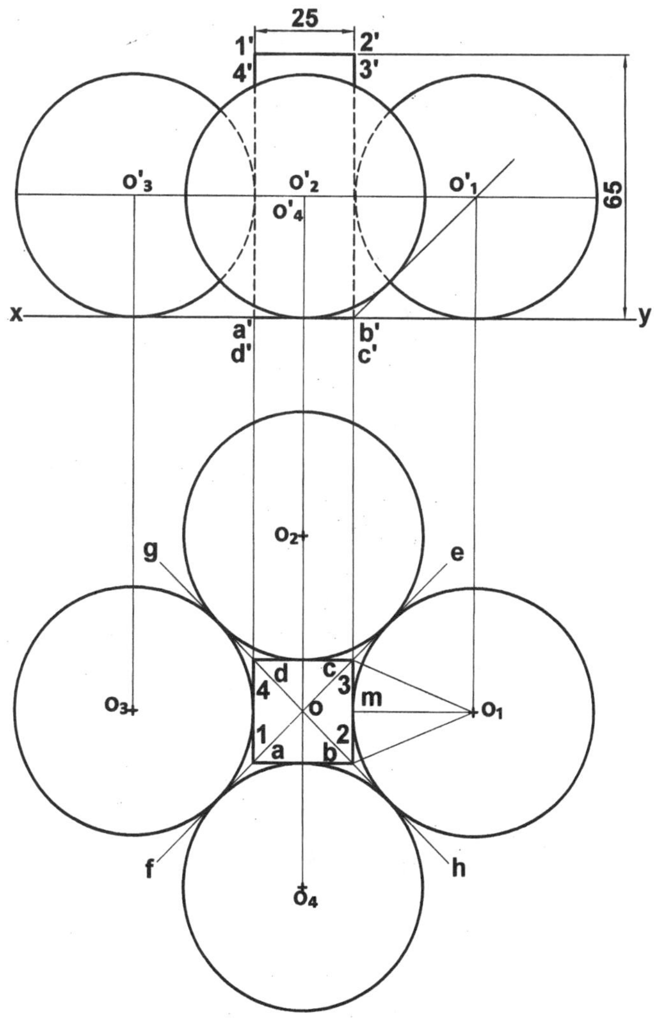

Fig. 9.60

4. Similarly obtain other centres o_2, o_3 and o_4.
5. With o_1, o_2, o_3 and o_4 as centres and radius equal to o_1m draw four circles.
6. Draw the bisector of angle 2'b'y intersecting the projectors through o_1 at o_1'.
7. Draw a horizontal line through o_1' vertical lines through o_1, o_2, o_3 and o_4. Mark centres o_2', o_3' and o_4' at the intersection of horizontal lines through o_1' and vertical lines through o_2, o_3 and o_4 respectively.
8. With o_2', o_3' and o_4' as centre and radius equal to o_1m draw three circles.
9. Show inner portion of circles by dotted lines.

Problem 9.54 (Fig. 9.61): *A square pyramid of base side 40 mm and axis 65 mm resting on the H.P. Four spherical balls of equal diameters are place on the H.P. so that each ball touches the other two balls and a triangular face of the pyramid. Determine the diameter of the spheres and draw the projections of the arrangement when two sides of the base are parallel to the V.P.*

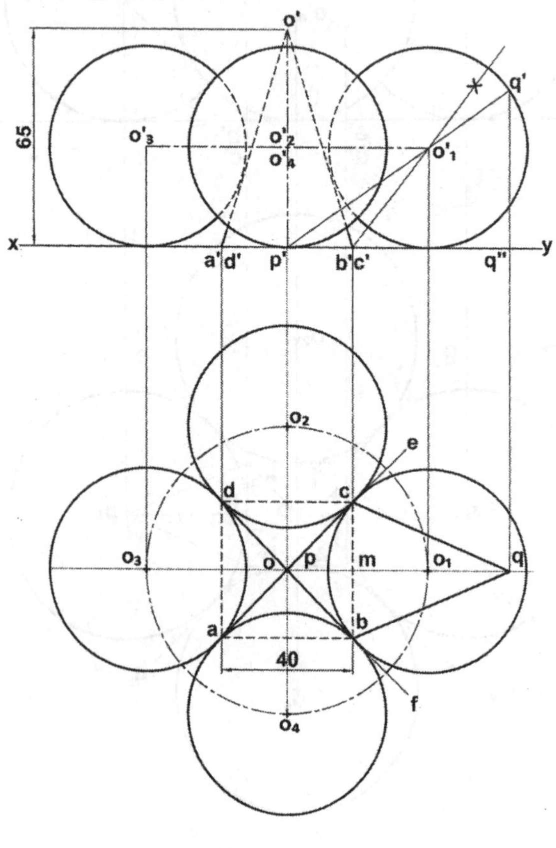

Fig. 9.61

Construction (Fig. 9.61):

1. Draw front view and top view of given square pyramid.
2. In the top view, mark the corners as a, b, c and d. Also mark p and o as base centre and apex respectively. Draw its diagonals and produce them as pe and pf.
3. Mark m at the intersection of bisector of angle epf and bc.
4. Bisect angle ecm and angle fbm and mark q at the intersection of these angle bisectors.
5. Project point q vertically to locate point q' such that q'q" = mq.
6. Join q' with p'.
7. Draw bisector of angle o'b'y intersecting the line p'q' at o_1'.
8. Through o_1', draw a vertical projector to intersect the centre line at o_1. Thus o_1 and o_1' represent projections of centre of a sphere. With o_1 and o_1' as centres and radius equal to o_1b or o_1c, draw circles.
9. With o_1 as centre and radius equal to oo_1 draw a circle to locate other centres o_2, o_3 and o_4 and o_2', o_3' and o_4' as explained in problem 9.52.
10. With o_2, o_3 and o_4 as centres and radius equal to o_1b, draw circles in the top view.
11. With o_2', o_3' and o_4' as centres and radius equal to o_1b draw circles in the front view.
12. Show invisible portion of spheres and pyramid by dotted curve/lines.

Problem 9.55 (Fig. 9.62): *Six equal spherical balls are resting on the H.P. in such a way that each touches the other two balls and a vertical face of a hexagonal prism having base side 25 mm and axis 60 mm long. Find the diameter of the spheres and draw the projections of the arrangement.*

Construction (Fig. 9.62):

1. Draw a regular hexagon abcdef of 25 mm side. Join all the corners with its centre. Project the front view of prism above xy line. Extend oc to m and ob to n.
2. Draw bisectors of angles mcb and cbn intersecting each other at o_1.
3. From o_1 draw a perpendicular o_1m_1 on bc. Thus o_1m_1 is the required radius of circle.
4. Draw the bisector of angle c'2'y which intersects the projector through o_1 at o_1'.
5. With centres o_1 and o_1' and radius equal to o_1m_1 draw two circles represents front view and top view of sphere 1.
6. Similarly obtain centres o_2, o_3, o_4, o_5 and o_6 in the top view and o_2', o_3', o_4', o_5' and o_6' in the front view. Draw projections of other spheres as shown in Fig. 9.62.

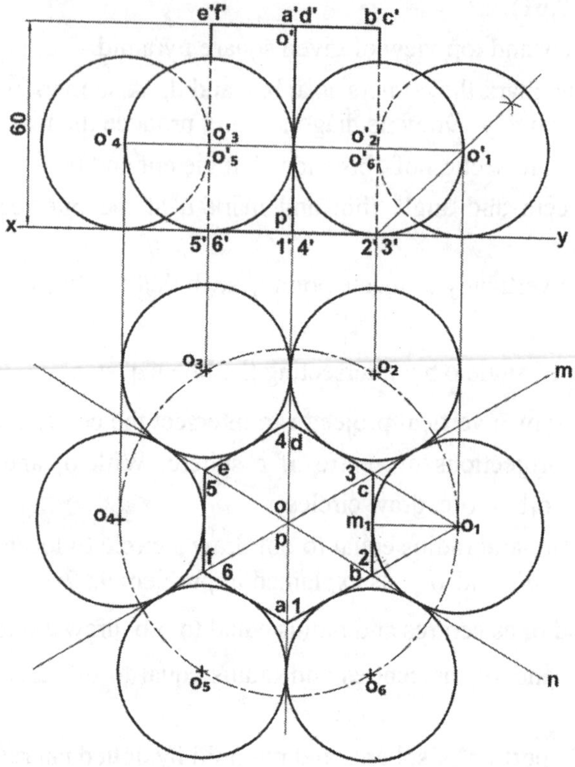

Fig. 9.62

7. Draw projections of other five circles as shown.

Exercises - (a)

1. Draw the projections of a triangular prism having side of its base 40 mm and axis 60 mm long resting on its base on the H.P. such that a side of its base is parallel to the V.P.
2. A cube with 40 mm edge is resting on the H.P. when
   (a) Vertical faces are equally inclined with the V.P.
   (b) A vertical face is inclined at 30° to the V.P.
3. A square prism having side of the base 40 mm and axis 65 mm long has its axis parallel to and 50 mm in front of the V.P. An edge of its base is perpendicular to the V.P., draw its front view and the top view.

4. Draw the projections of a hexagonal pyramid having side of the base 40 mm and axis 65 mm long, resting on its base on the H.P. such that a side is parallel to xy line.
5. A pentagonal pyramid with side of base 30 mm and axis 65 mm long has its base on the H.P. and axis is parallel to and 40 mm in front of the V.P. Draw its projections when a side of the base is
   (a) Parallel to the V.P.
   (b) Perpendicular to the V.P.
   (c) Inclined at 45° to the V.P.
6. A triangular prism having edge of base 50 mm and axis 70 mm long is resting on one of its rectangular faces on the H.P. and axis perpendicular to the V.P. Draw its projections.
7. Draw the projections of a square prism having side of base and axis 65 mm long has its base in the V.P. such that
   (a) A rectangular face is parallel to the H.P.
   (b) All the rectangular faces are equally inclined with the H.P.
8. A cone having base diameter 50 mm and axis 70 mm long is resting on its base on the H.P. such that its axis is vertical and 40 mm in front of the V.P. Draw its projections.
9. A cylinder with base diameter 40 mm and axis 60 mm long has its axis perpendicular to the V.P. and 40 mm above the H.P. Its one end is 20 mm in front of the V.P. Draw its projections.
10. Draw the projections of a square pyramid with side of base 30 mm and axis 65 mm long having all edges of the base equally inclined to the H.P. and axis parallel to and 50 mm away from the reference planes.

Exercises - (b)

1. A triangular prism having edge of the base 30 mm and axis 65 mm long, is resting on an edge of its base on the H.P. such that its axis is parallel to the V.P. and inclined at 45° to the H.P. Draw its projections.
2. A pentagonal pyramid having edge of base 30 mm and axis 70 mm long, is resting on the corner of its base on the H.P. Its axis is parallel to the V.P. and inclined at 45° to the H.P. Draw its projections.
3. A hexagonal prism having edge of base 30 mm and axis 70 mm long is resting on a corner of its base on the H.P. with longer edge through that corner is inclined at 45° to the H.P. Draw its projections.
4. A square pyramid with side of base 30 mm and axis 70 mm long has one of its triangular faces in the V.P. and axis parallel to and 35 mm above the H.P. Draw its projections.

5. Draw the projections of a cone having base diameter 40 mm and axis 70 mm long lying on a generator on the H.P. such that its axis makes an angle of 45° with the V.P.
6. Draw the projections of a cube of edge 40 mm which is resting on a corner on the H.P. and one of its solid diagonals is parallel to both the H.P. and the V.P.
7. A square pyramid having side of base 40 mm and axis 70 mm long is resting on a corner on the H.P. and slant edge containing that corner is vertical. Draw its three views.
8. A cone having base diameter 40 mm and axis 65 mm long, is resting on a point on its base circle on the H.P. The highest point of the base circle is 25 mm above the H.P. and 35 mm in front of the V.P. Draw its projections.
9. Draw the projections of a cone having base diameter 50 mm and axis 75 mm long when its axis is parallel to the H.P. and inclined at 45° to the V.P.
10. A cone, base diameter 40 mm and axis 70 mm long, is lying on one of its generator on the H.P with its axis making an angle of 30° with the V.P. Draw its projections.
11. A cylinder, base diameter 40 mm and axis 75 mm long, is lying on one of its generators on the H.P. when the axis is inclined at an angle of 30° with the V.P., draw its projections.
12. A hexagonal prism having side of base 30 mm and axis 70 mm long is resting on a triangular face on the H.P. and the longer edge is inclined at 60° with the V.P. Draw its projections.
13. A hexagonal pyramid with 30 mm side of base and 75 mm long axis has an edge of its base in the V.P. and apex is 55 mm away from both the reference planes. If axis is parallel to the H.P., draw its projections.
14. A hexagonal prism side of base 30 mm and axis 65 mm long has one of its rectangular faces in the V.P. If the axis is inclined at 60° to the H.P., Draw its projections.
15. A pentagonal pyramid, side of base 30 mm and axis 60 mm long, is resting on corner of its base on the H.P. in such a way that highest corner of the base is 25 mm above the H.P. Draw its projections.

Exercises - (c)

1. A hexagonal prism having side of base 30 mm and axis 70 mm long is resting on its edge of base on the H.P. which is inclined at 45° to the V.P. Its axis makes an angle of 60° with the H.P., Draw its projections.
2. A cone with base diameter 40 mm and axis 70 mm long, has one of its generators in the V.P. and inclined at 30° to the H.P. If the apex is in the H.P. Draw its projections.

3. A triangular prism, side of base 40 mm and axis 75 mm long, is resting on one of the corners of base on the H.P. with a longer edge containing that corner inclined at 45° to the H.P. The vertical plane containing that edge and axis is inclined at 30° to the V.P. Draw its projections.
4. Draw the projections of a square pyramid having side of the base 40 mm and axis 70 mm long, has one of its triangular faces on the H.P. and slant edge containing that face is parallel to the V.P.
5. A square pyramid, having side of base 30 mm and axis 70 mm long is kept on the H.P. on one of its slant edges so that the vertical plane passing through that edge and the axis makes an angle of 30° with the V.P. Draw its three views.
6. A pentagonal pyramid having side of its base 30 mm and axis 70 mm long is resting on one of the corners of its base on the H.P. with its axis inclined at 45° to the H.P. The vertical plane containing that corner and axis is inclined at 30° to the V.P. Draw its projections.
7. A pentagonal pyramid, side of base 30 mm and axis 75 mm long, has an edge of base in the V.P. and inclined at 30° to the H.P. while the triangular face containing that edge makes an angle of 45° with the V.P. Draw its three views.
8. A pentagonal pyramid side of base 30 mm and axis 75 mm long has a triangular face in the V.P. and slant edge containing that face is parallel to the H.P. Draw its projections.
9. A pentagonal prism, side of base 25 mm and axis 70 mm long is resting on edge of base on the H.P. The highest point of the base on which it rests and edge of the base on which it rests is parallel to the V.P. Draw its projections.
10. A hexagonal prism, edge of base 25 mm and axis 70 mm long, has an edge of base parallel to the H.P. and inclined at 45° to the V.P. If the axis inclined at 60° to the H.P. Draw its projections.
11. A hexagonal prism, side of base 25 mm and axis 70 mm long is resting on one of the edges of its base on the H.P. such that the edge on which solid rests, makes an angle of 45° with the V.P. and the base is inclined at 60° with the H.P. Draw its projections.
12. A hexagonal prism having side of base 25 mm and axis 75 mm long rests on one of the corners of the base on the H.P. If the solid diagonal is perpendicular to the V.P., draw its projections.
13. A hexagonal pyramid having side of base 30 mm and axis 70 mm long is resting on an edge of base on the H.P. which makes an angle of 30° with the V.P. If the slant face containing this edge makes an angle of 45° with the H.P., Draw its projections.
14. A hexagonal pyramid, side of base 25 mm and axis 70 m long is resting on one of its slant edges on the H.P. A vertical plane passing through this edge and axis is inclined at 60° to the V.P. Draw its projections.

15. A hexagonal pyramid having edge of base 25 mm and axis 70 mm long has a triangular face in the V.P. and the edge of base containing that face makes an angle of 30° to the H.P. Draw its projections.
16. A cone having diameter of base 40 mm and axis 70 mm long, is resting on one of its generators on the H.P. Axis is inclined at 30° to the V.P. If apex is nearer to the V.P., Draw its projections.
17. A cone having diameter of base 30 mm and axis 70 mm long, is resting on one of its generator on the H.P. Draw its projections when:
    (a) Its generator on which it rests is inclined at 30° to the V.P.
    (b) Its axis is inclined at 30° to the V.P.
18. A cylinder, base diameter 40 mm and axis 70 mm long, is resting on its circular rim in such a way that its axis makes an angle of 45° with the H.P. and top view of its axis is inclined at an angle of 60° with the V.P. Draw its projections.
19. A right circular cylinder with base diameter 30 mm and axis 70 mm long is resting on a point of its base circle on the H.P. such that the base makes an angle of 30° with the H.P. and its axis is inclined at 45° to the V.P. Draw its projections.
20. Three equal spheres of 50 mm diameter are resting on the H.P. such that each touches the other two spheres and the line joining the centres of two of them is parallel to the V.P. A fourth sphere of 60 mm diameter is placed on the top of three spheres so as to make a pile. Draw three views of the arrangement.
21. A square prism side of base 20 mm and height 50 mm is resting on its base on the H.P. with two of its vertical faces parallel to the V.P. Four spheres of equal diameters are resting on the H.P. in such a way that each sphere touches a face of the prism and two other spheres. Determine the diameter of fours spheres and draw its projections.
22. Six spheres of same diameters are resting on the H.P. in such a way that each sphere touches the other two spheres and a triangular face of a hexagonal pyramid having base with 30 mm side and axis 65 mm long. One of the base edge of the pyramid is parallel to the V.P. Find the diameter of the sphere and draw the projections of the arrangement.
23. Four equal spheres of 30 mm diameter are resting on the H.P. in such a way that each touches the other two spheres. The line joining the centres of two touching spheres is inclined at 30° to the V.P. A fifth sphere of 40 mm diameter is placed centrally on the top of four spheres so as to form a pile. Draw the projections of the arrangement.
24. Three spheres of 30 mm, 50 mm and 70 mm diameters are resting on the ground so that each touches the other two. If the top view of the line joining centres of any two spheres, makes an angle of 30° to the V.P., Draw the projections of the arrangement.

CHAPTER 10

Sections of Solids and Intersection of Cylinders

10.1 Introduction

Internal details of an objects are shown by drawing dotted lines if the object is simple but when the internal features are complicated then they cannot be interpreted by dotted lines because large number of dotted lines on the drawing make it more complicated. In such cases it is imagine that the object is cut by cutting plane. The part between cutting plane and the observer is assumed to be removed so as to show the internal detail of any object. After removing the cut portion, a view is obtain which is called as section or sectional view. The imaginary cutting plane is also known as section plane. The sectional view is shown by drawing section lines (hatching line) which are evenly spaced parallel lines (3 to 5 mm apart) inclined at 45° to the reference line.

10.2 General Terminology

1. **Section Plane or Cutting Plane:** These are imaginary planes which cut the object from the given position to show the internal details of it. These are also known as cutting planes. They may be perpendicular to one of the reference planes and parallel, perpendicular or inclined to other plane. Section planes are described by their traces e.g., horizontal trace (H.T.) or vertical trace (V.T.).
2. **Section or Cut Section:** The surface obtained by cutting an object by the cutting plane or section plane is called as section or cut section of the object.
3. **Sectional View:** The projection of a section on a plane of projection (H.P. or V.P.) is called sectional view. The sectional views may be of following types.

(i) **Sectional Front View:** The projection of a cut section obtained on vertical plane (V.P.) is called as sectional front view.

(ii) **Sectional Top View:** The projection of a cut section obtained on the horizontal plane (H.P.) is called as sectional top view.

(iii) **Sectional Side View:** The projection of a cut section obtained on a profile plane is called as sectional side view.

(iv) **Apparent Section:** The projection of a cut section obtained on the principal plane to which the section plane is inclined is called as apparent section.

(v) **True Section or True Shape of the Section:** The projection of a cut section obtained on a plane parallel to section plane is called true section. It shows actual size and shape of the section. When the section plane is inclined, the true shape is obtained by projecting the section on an auxiliary plane parallel to the section plane. When the section plane is perpendicular to both H.P. and V.P. then sectional side view is required to show the true shape of the section. The sectional side view is obtained on a profile plane (P.P.).

10.3 Types of Section Planes

The section planes may be classified as follows:

(i) **Section Plane Perpendicular to V.P. or Parallel to H.P. (Horizontal Section Plane):** The section plane perpendicular to vertical plane (V.P.) and parallel to the horizontal plane (H.P.) is known as horizontal section plane as shown in Fig. 10.1. It has Vertical Trace (V.T.) parallel to the xy line and has no horizontal Trace (H.T.).

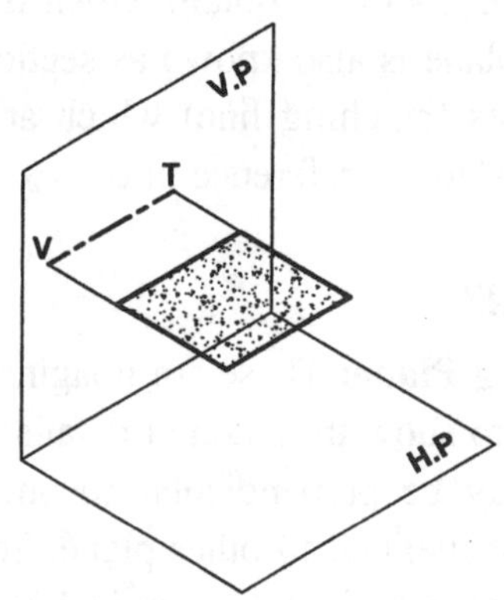

Fig. 10.1

(ii) Section Plane Perpendicular to the H.P. and Parallel to the V.P. : The section plane perpendicular to H.P. and parallel to the V.P. has H.T. parallel to xy line and has no V.T. as shown in Fig. 10.2.

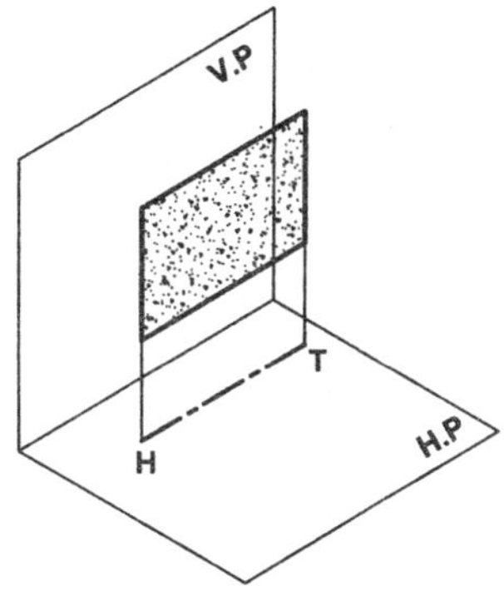

Fig. 10.2

(iii) Section Plane Perpendicular to H.P. and Inclined to V.P. (Auxiliary Vertical plane): The section plane which is perpendicular to H.P. and inclined to V.P. is also known as auxiliary vertical plane (A.V.P.). It has V.T. perpendicular or xy line and H.T. inclined at ϕ° with xy line as shown in Fig. 10.3.

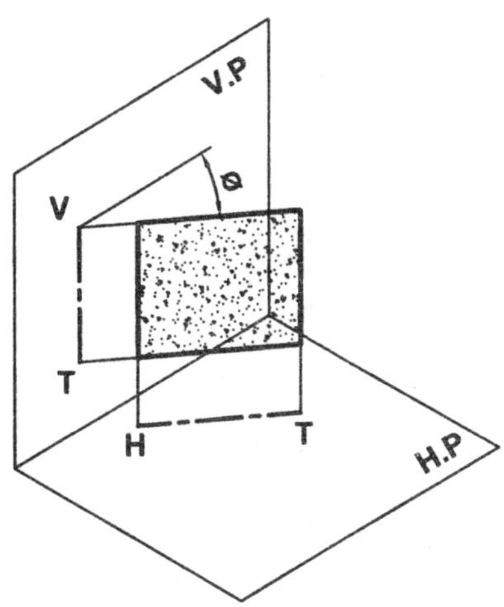

Fig. 10.3

(iv) Section Plane Perpendicular to V.P. and Inclined to H.P.: Section plane which is perpendicular to the V.P. and inclined to H.P. is also known as auxiliary inclined plane (A.I.P.). It has a V.T. perpendicular to the xy line and a H.T. inclined at θ° with xy line as shown in Fig. 10.4.

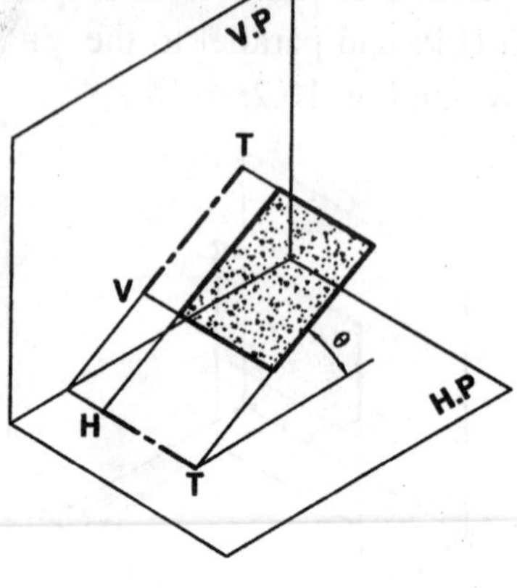

Fig. 10.4

(v) **Section Plane Perpendicular to both H.P. and V.P. (Profile Plane):** The plane perpendicular to both H.P. and V.P. is known as profile plane. It has both H.T. and V.T. perpendicular to xy line as shown in Fig. 10.5.

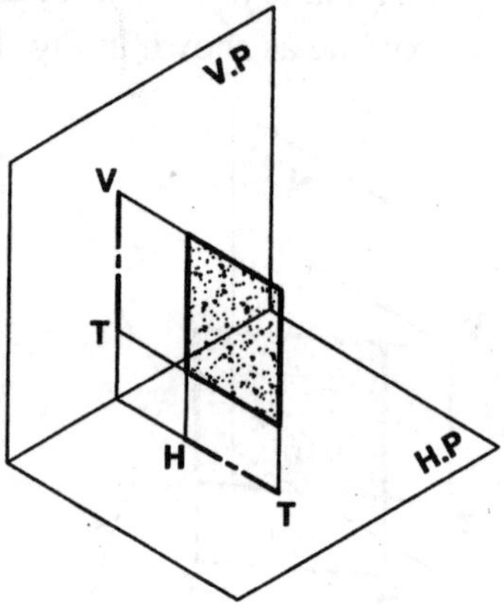

Fig. 10.5

In this chapter, methods of obtaining sections of following solids are discussed.

1. Sections of prism.
2. Sections of pyramid.
3. Sections of cylinders.
4. Sections of cones.
5. Sections of spheres.

10.4 Sections of Prisms

A prism can be sectioned by a section plane which can have following position.

(a) Section plane parallel to the V.P.

(b) Section plane parallel to the H.P.

(c) Section plane perpendicular to H.P. and inclined to V.P. (A.V.P.)

(d) Section plane perpendicular to V.P. and inclined to the H.P. (A.I.P.)

10.4.1 Section Plane Parallel to the V.P.

Problem 10.1 (Fig. 10.6-10.10): *A cube having 40 mm long edges is resting on one of its faces on the H.P. in such a way that a vertical face makes an angle of 30° with the V.P. It is cut by a section plane parallel to the V.P. and 10 mm away from the axis and farther away from the V.P. Draw its sectional front view and top view.*

Construction (Fig. 10.6-10.10):

The step-by-step procedure for solving this problem is explained as follows:

1. Draw a square of 40 mm long side below xy lines such that a side say cd makes an angle of 30° with xy line. (Refer Fig. 10.6)

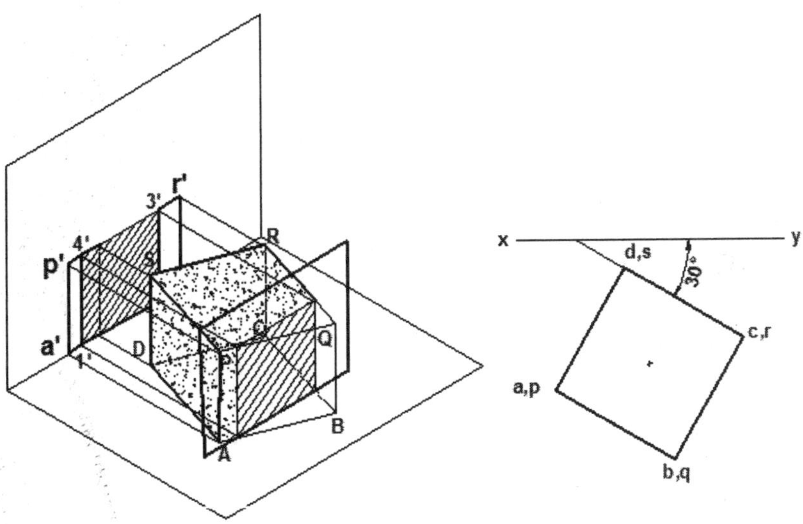

Fig. 10.6

2. Through corners a, b, c and d, draw vertical projectors to meet xy line at a', b', c' and d' draw vertical lines a'p', b'q', c'r' and d's' respectively such that a'p'=b'q'=c'r'=d's' = 40 mm. (Refer Fig. 10.7)

3. In the top view, draw a line HT parallel to xy line and 10 mm away from the centre of the square. Since, edge ab coincides with pq, there will be two points of intersection at which edges are cut by section plane. So mark the point 1, 4 at which HT intersects with ab or pq. Similarly mark 2, 3 at which HT intersects with bc or qr. (Refer Fig. 10.8)

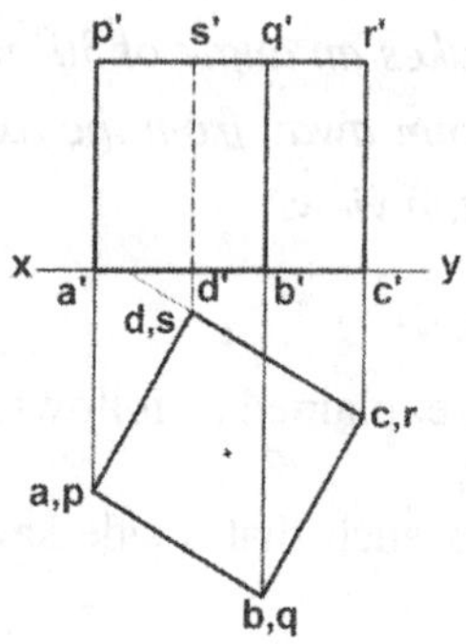

Fig. 10.7

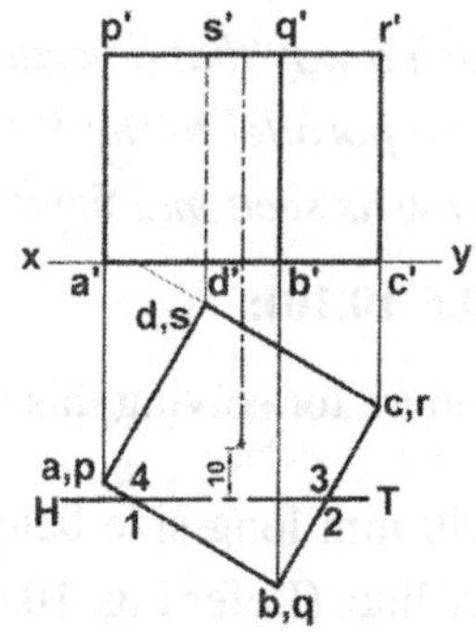

Fig. 10.8

4. Through points 1 and 2, draw vertical projectors to meet xy at 1' and 2'. Through 1' and 2', draw vertical lines 1'4' and 2'3' equal to 40 mm. (Refer Fig. 10.9)

5. Hatch the area 1'2'3'4' and complete the view as shown in Fig. 10.10.

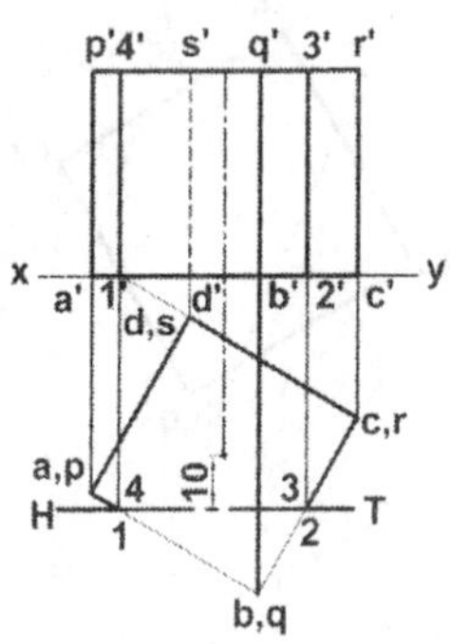

Fig. 10.9

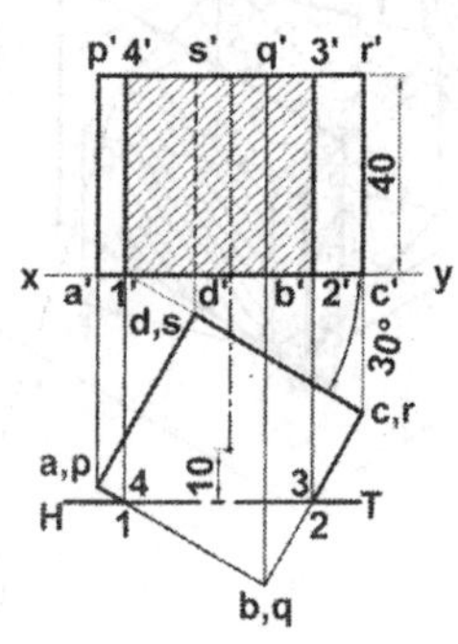

Fig. 10.10

Problem 10.2 (Fig. 10.11): *A square prism having edges of base 30 mm and axis 50 mm long is resting on its base on the H.P. in such a way that side of the bases are equally inclined to the V.P. It is cut by a section plane parallel to and nearer to the V.P. and 9 mm away from the axis of the prism. Draw sectional front view and top view.*

Construction (Fig. 10.11):

As section plane is parallel to the V.P., it is perpendicular to H.P. So in the top view, a line parallel to xy will represent its H.T.

1. Draw top view and front view of given square prism in the required position.
2. In the top view, draw a line H.T. parallel to xy line and 9 mm away from the axis. The line H.T. represents section plane.
3. Mark a point p at which edge ad is cut by section plane. Similarly mark q, r and s on edges cd, 14 and 34 respectively at which H.T. intersects.
4. Through p, q, r and s, draw vertical projectors to intersect corresponding edges in the front view at p', q', r' and s'.
5. Join p', q', r' and s' in proper sequence by dark lines and hatch the rectangle leaving lines for cut portion relatively fainter.

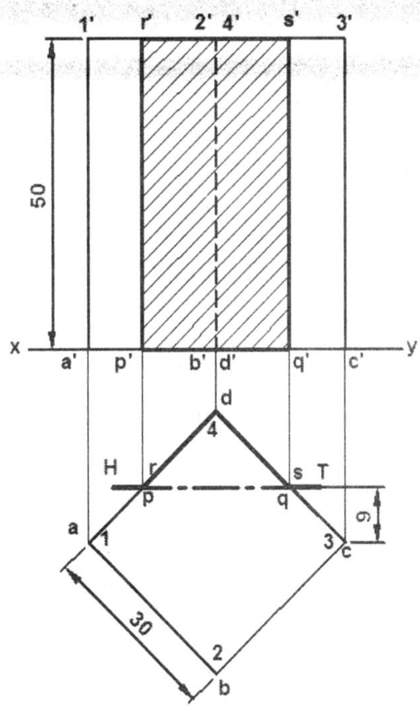

Fig. 10.11

Problem 10.3 (Fig. 10.12): *A triangular prism side of base 35 mm and axis 85 mm long is resting on one of its triangular faces on the H.P. with its axis inclined at 45° to the V.P. It is cut by a section plane parallel to the V.P. and bisecting the axis. Draw sectional front view and top view.*

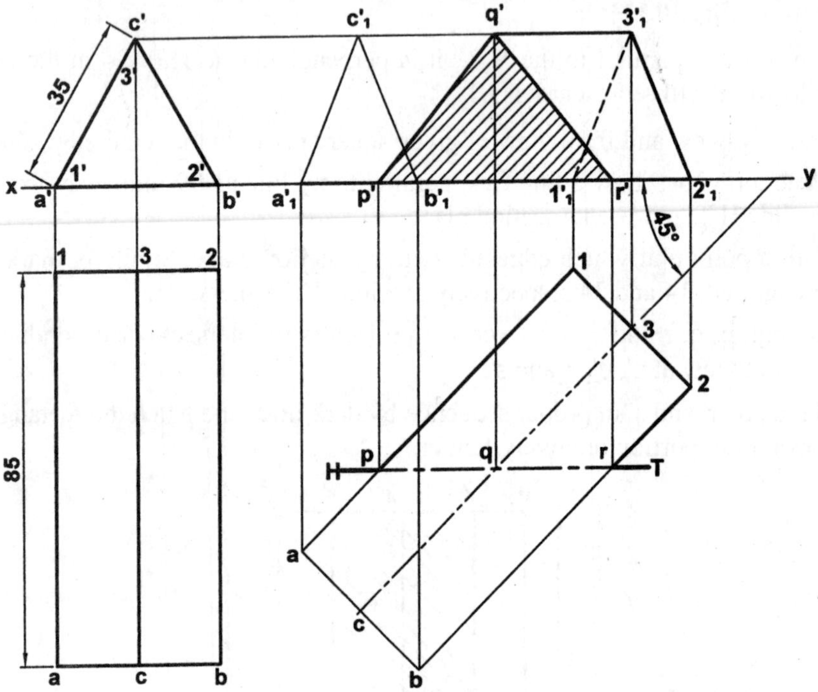

Fig. 10.12

Construction (Fig. 10.12):

1. Draw the projections of triangular prism in the required position.
2. In the top view, draw a line HT parallel to xy line and bisecting the axis. This represents top view of section plane.
3. Mark points p, q and r at which edges are cut by section plane viz. Edge 1-a at p, edge 3-c at q and edge 2-b at r.
4. Project these points to intersect the corresponding edges at p', q' and r' in the front view.
5. Join p', q' and r' and hatch the triangle p'q'r'.
6. Complete the sectional front view leaving the lines for cut portion fainter.

Problem 10.4 (Fig. 10.13): *A pentagonal prism of side of base 30 mm and axis 65 mm long is resting on one of its rectangular faces on the H.P. with its axis inclined at 30° to* *the V.P. It is cut by a section plane parallel to V.P. and bisecting the axis. Draw sectional front view and top view.*

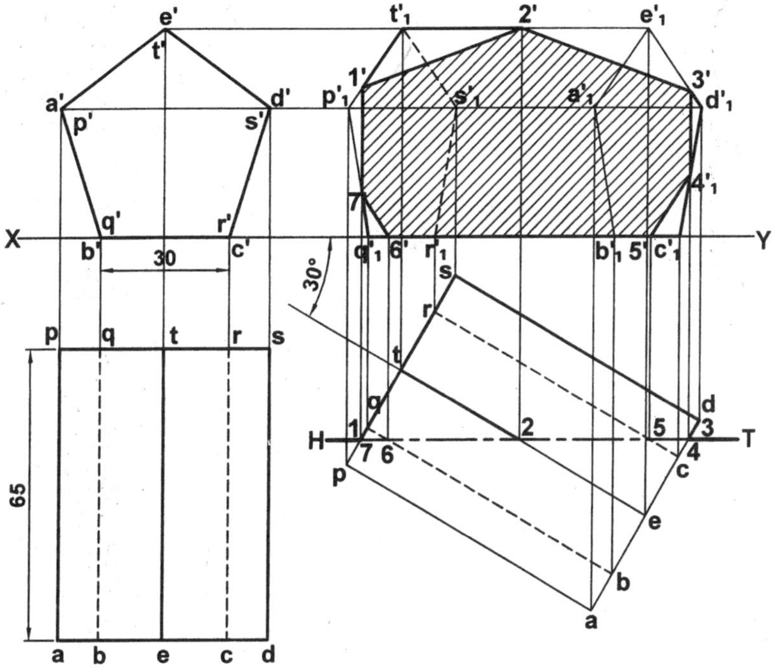

Fig. 10.13

Construction (Fig. 10.13):

1. Draw the front view and top view of pentagonal prism in the required position.
2. In the top view, draw a line HT parallel to xy line and dividing the axis in two equal parts.
3. Name the points at which section plane cuts edges viz. 1 at pr, 2 at te, 3 at ed, 4 at cd, 5 at rc, 6 at bq and 7 at pq.
4. project points 1, 2, 3....etc., on the corresponding edges at 1', 2', 3'....etc.
5. Join these points in a proper sequence.
6. Hatch the closed figure 1'2'3'4'5'6'7'.

Problem 10.5 (Fig. 10.14): *A cube having 40 mm long edge is resting on one of its faces on the H.P. with edges of the base, equally inclined to the V.P. It is cut by a section plane parallel to the H.P. and passing through a point on the axis 10 mm below the top face. Draw sectional top view and front view.*

Construction (Fig. 10.14):

1. Draw a square abcd of 40 mm long sides below xy line such that its sides are equally inclined with V.P. (i.e., at 45° to xy line). The corners a, b, c and d will also coincide with p, q, r and s respectively.
2. Through a, b, c and d, draw vertical projectors to meet xy at a', b', c' and d' respectively.
3. Through a', b', c' and d', draw vertical lines a'p', b'q', c'r' and d's' respectively equal to length of 40 mm.
4. Draw a line VT parallel to xy and 10 mm below the top face to represent section plane.
5. Since, the section plane cuts all the vertical faces, of the cube the complete area of square abcd will show the section. So hatch the square abcd which represents sectional top view. This also represents true shape of the section.

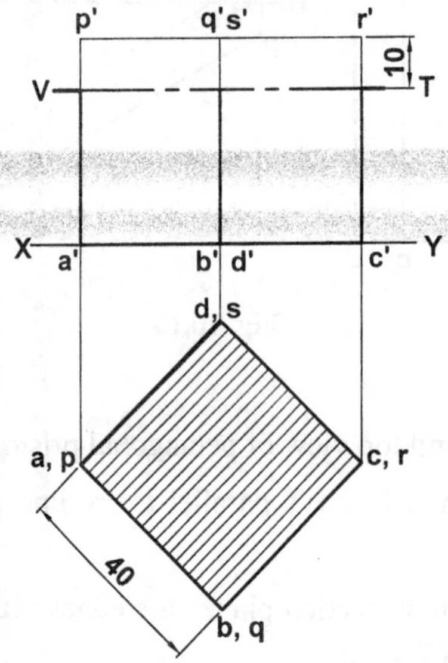

Fig. 10.14

Problem 10.6 (Fig. 10.15): *A triangular prism edge of the base 35 mm and axis 55 mm long lying on one of its triangular faces on the H.P. with its axis inclined as 30° to the V.P. It is cut by a horizontal section plane 10 mm above the H.P. Draw its front view and sectional top view.*

Construction (Fig. 10.15):

1. Draw the projections of triangular prism in the required position.
2. In the front view, draw a line VT parallel to and 10 mm above the xy line.

3. Mark the points p', q', r' and s' at which edges are cut by VT.
4. Project these points on the corresponding edges in the top view at p, q, r and s respectively.
5. Join p, q, r and s in the correct sequence.
6. Hatch the area pqrs.
7. Thus top view obtained is the required sectional top view.

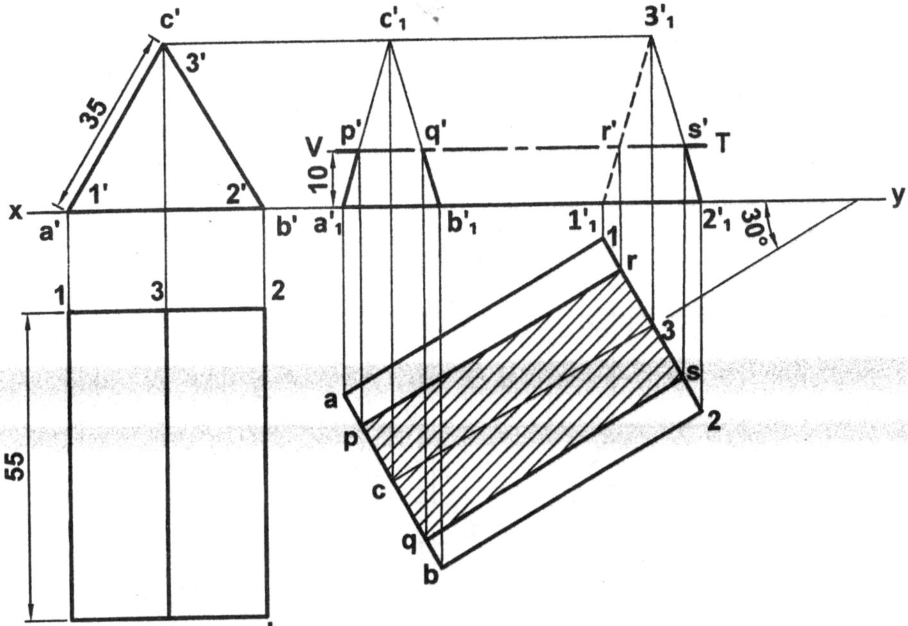

Fig. 10.15

Problem 10.7 (Fig. 10.16): *A hexagonal prism having edge of base 25 mm and axis 60 mm long is resting on one of its corner on the ground with the base making an angle of 60° with the ground. The axis is parallel to the V.P. It is cut by a section plane parallel to H.P. and perpendicular to V.P. in such a way that it is 15 mm from the base as measured along the axis. Draw the sectional top view and front view.*

Construction (Fig. 10.16):

1. Draw the projections of hexagonal prism in the required position.
2. In the front view, draw a line VT parallel to xy line and passing through a point on the axis 15 mm from the base. This represents front view of the sectional plane.
3. Mark the points p', q', r', s' and t' at which edges b'c', 2'b', 1'a', 6'f' and e'f' respectively are cut by section plane.

4. Project these points on the corresponding edges in the top view.
5. Join p, q, r, s and t is the proper order and hatch it. This is the required sectional top view.

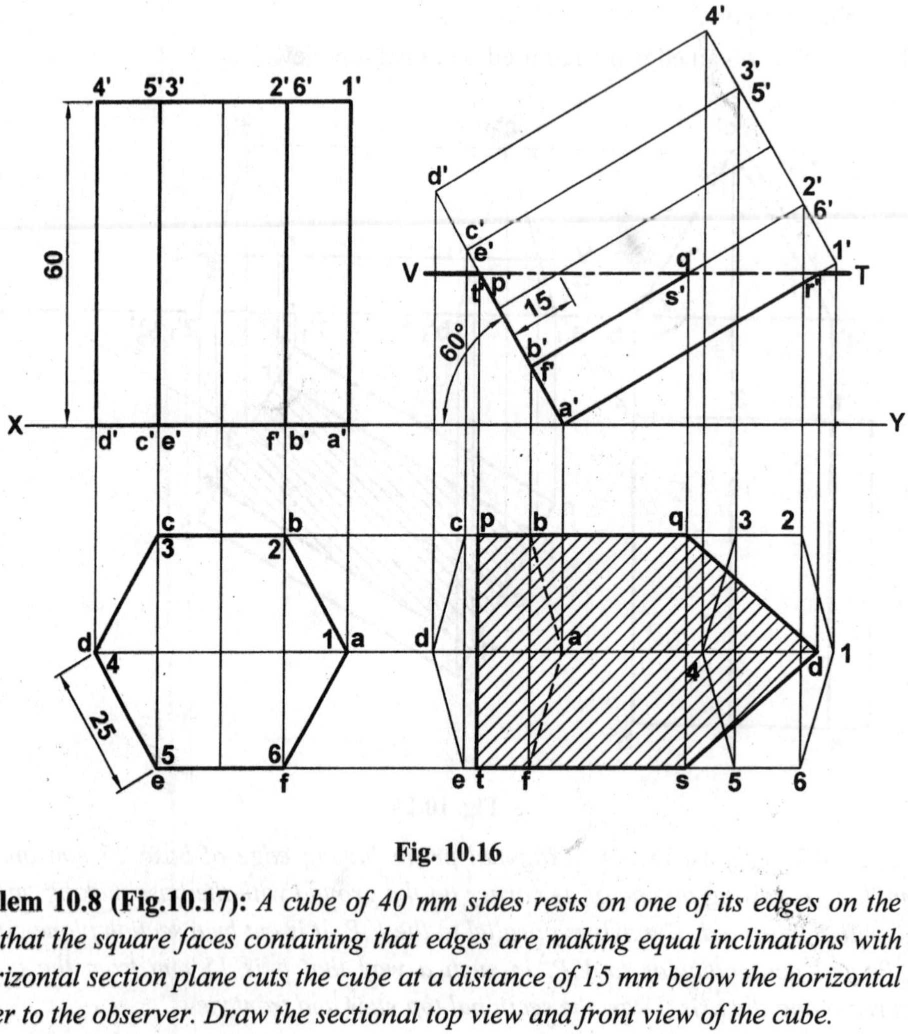

Fig. 10.16

Problem 10.8 (Fig.10.17): *A cube of 40 mm sides rests on one of its edges on the H.P. such that the square faces containing that edges are making equal inclinations with H.P. A horizontal section plane cuts the cube at a distance of 15 mm below the horizontal edge nearer to the observer. Draw the sectional top view and front view of the cube.*

Construction (Fig. 10.17):

1. Draw the projections of a cube of 40 mm side in the required position.
2. In the front view, draw a line VT parallel to xy line and 15 mm away from the corner farthest away from the xy line.
3. Mark points 1', 2', 3' and 4' at which edges p's', r's', c'd' and d'a' respectively are cut by section plane.

4. Project 1', 2', 3' and 4' on the corresponding lines in the top view at 1, 2, 3 and 4 respectively.
5. Join these points in a proper order and hatch rectangle 1234. This is required sectional top view.

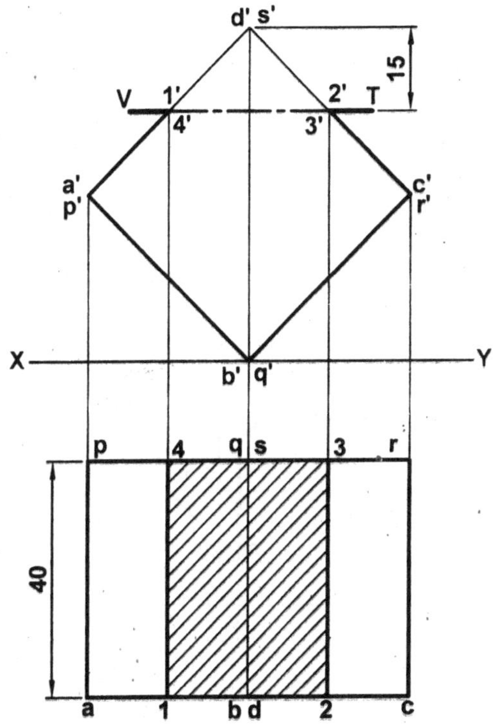

Fig. 10.17

Problem 10.9 (Fig. 10.18): *A pentagonal pyramid having base edge 30 mm and axis 70 mm long is resting on one of its triangular faces on the H.P. such that its axis is parallel to V.P. It is cut by a horizontal section plane which intersects its axis at a distance 15 mm from the base. Draw its front view and sectional top view.*

Construction (Fig. 10.18):

1. Draw the projections of a pentagonal pyramid in the required position.
2. In the front view, draw a line VT parallel to xy line and passing through a point on the axis 15 mm from the base.
3. Mark points 1', 2', 3', 4' and 5' at which edges b'c', a'e', o'e', o'd' and o'c' respectively are cut by section plane.

4. Project these points on the corresponding lines in the top view at 1, 2, 3, 4 and 5 respectively.
5. Join them in a proper order and hatch 12345 which is the required sectional top view.

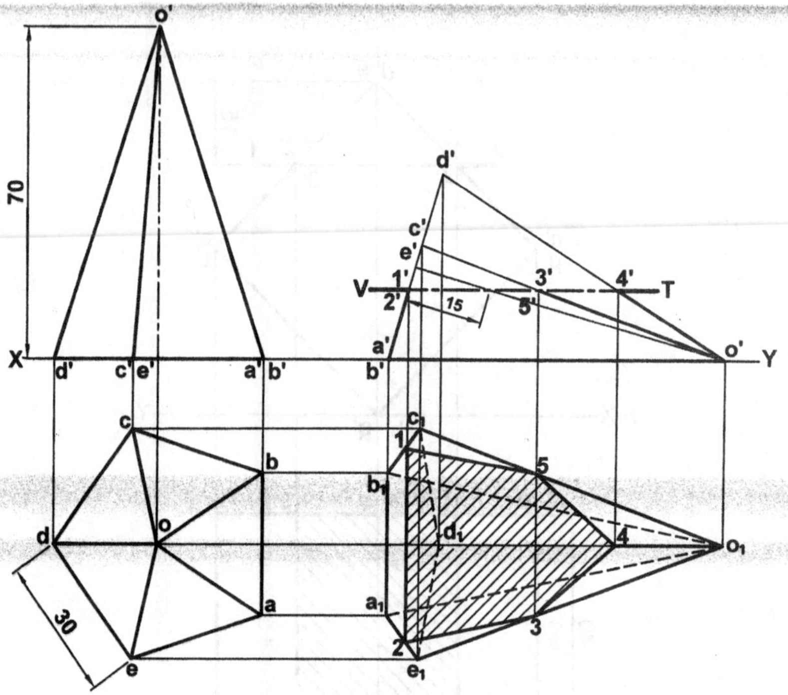

Fig. 10.18

10.4.2 Section Plane Perpendicular to the H.P. and Inclined to the V.P. (A.V.P)

Problem 10.10 (Fig. 10.19): *A cube having 40 mm long edges is resting on one of its faces on the H.P. in such a way that a vertical face is inclined at 30° to the V.P. It is cut a section plane inclined at 60° to the V.P. and perpendicular to the H.P. and is 10 mm way from the axis. Draw sectional front view and true shape of the section.*

Construction (Fig. 10.19):

1. Draw the projections of a cube of 40 mm side in the required position.
2. Draw a line HT in the top view such that it makes an angle of 60° with xy line and passes through a point 10 mm away from the axis.
3. Mark points 1, 2, 3 and 4 at which the edges ad, ab, pq and ps respectively are cut by section plane.

4. Project these points on the corresponding edges in the front view and join them in a proper order.
5. Hatch the rectangle 1', 2', 3' and 4'. Complete the sectional front view as shown in the figure.

   Since, section plane is not parallel to V.P., the above figure does not represent true shape of the section. So, to draw true shape of the section, it is projected on an auxiliary vertical plane parallel to the section plane as follows.
6. Draw a new reference line x_1y_1 parallel to HT. To project the section on x_1y_1, draw projectors through 1, 2, 3 and 4 perpendicular to it and on them, mark $1_1'$, $2_1'$, $3_1'$ and $4_1'$ such that their distances from x_1y_1 are equal to their corresponding distances from xy line in the front view viz. $2_1'3_1' = 2'3'$, $1_1'4_1' = 1'4'$ and $3_1'$ and $4_1'$ will on x_1y_1.
7. Join $1_1'2_1'3_1'4_1'$ by dark line and hatch the figure as shown. This is the required true shape of the section.

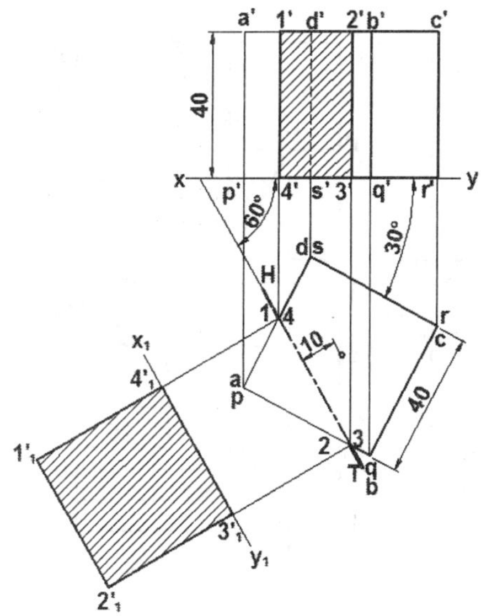

Fig. 10.19

Problem 10.11 (Fig. 10.20): *A hexagonal prism having sides of base 30 mm and axis 65 mm long is resting on its rectangular faces on the H.P. with its axis parallel to the V.P. It is cut by a vertical section plane, the HT of which makes an angle of 45° with xy line and cuts the axis at a point 20 mm from one of its ends. Draw sectional front view and true shape of the section.*

Construction (Fig. 10.20):

1. Draw the projections of a hexagonal prism in the required position.
2. In the top view, draw a line HT inclined at 45° to xy and passing through a point on the axis 20 mm from the base.
3. Mark points at which edges are cut by section plane.
4. Project these points on the corresponding lines in the front view. To locate points 4' and 5', first project points 4 and 5 on the base edge (Stage-I). Mark 4' and 5' in the front view (Second stage-II) such that their distance from corners are equal to their corresponding distances in the front view of stage-I.
5. Join 1'-2'-3'-4'-5'-6'-7' by straight lines and hatch the figure.
6. To draw true shape of the section, draw a new reference line x_1y_1 parallel to HT and project the true shape on this reference line.

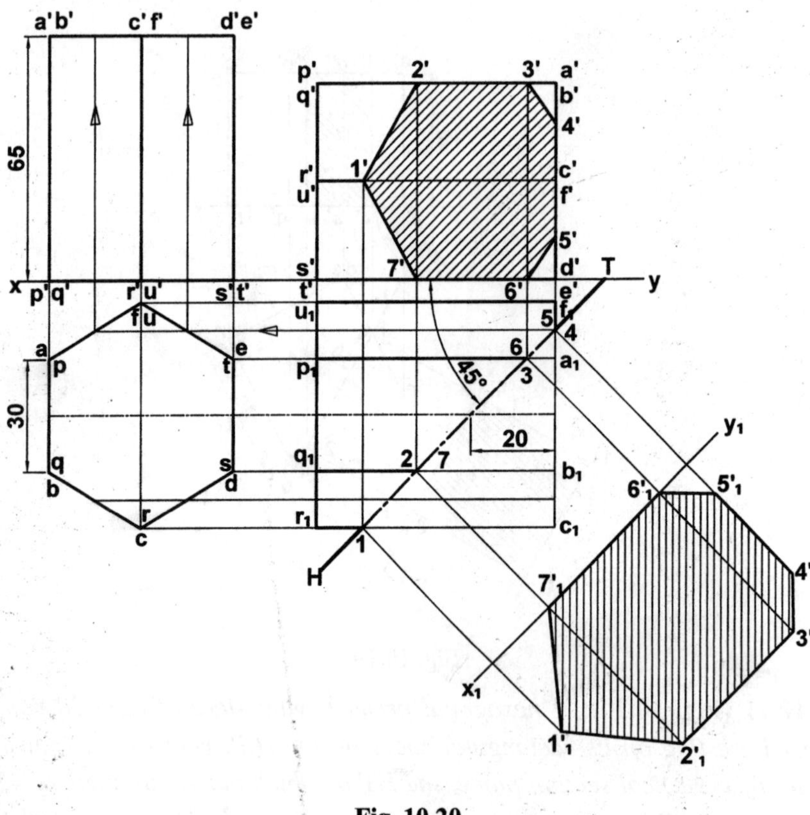

Fig. 10.20

Problem 10.12 (Fig. 10.21): *A hexagonal prism, edge of base 30 mm and axis 65 mm long is resting on one of its rectangular faces on the H.P. with its axis parallel to the V.P. It is cut by a plane, the H.T. of which makes an angle of 45° with the xy line and bisects the axis. Draw the sectional front view and top view of the section.*

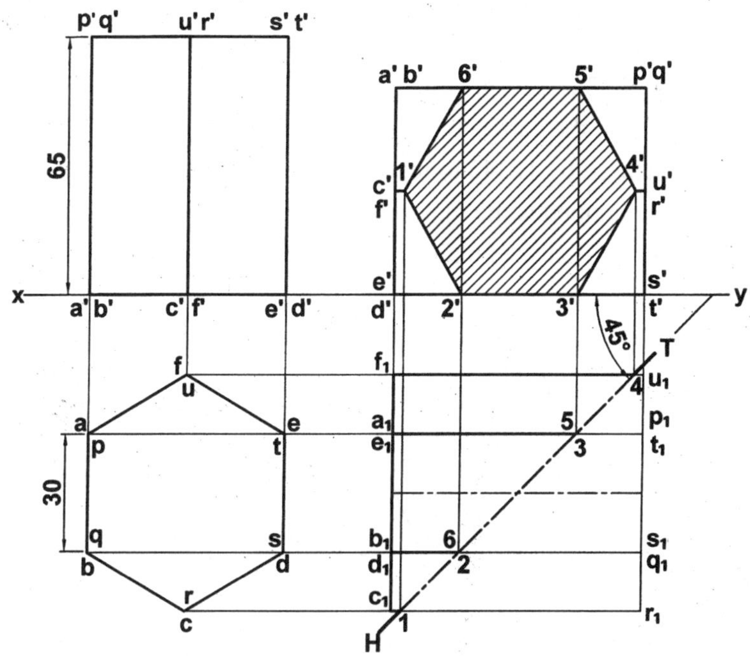

Fig. 10.21

Construction (Fig. 10.21):

1. Draw the projections of given hexagonal prism in the required position.
2. Draw a line HT inclined at 45° to xy line such that it bisects the axis.
3. In the top view mark the points 1, 2, 3....etc., at which edges are cut by section plane.
4. Project these points on the corresponding lines in the front view.
6. Join 1'-2'-3'-4'-5'-6' by straight lines and hatch the figure. This is the required sectional front view.

10.4.3 Section Plane Perpendicular to V.P. and Inclined to H.P.(A.I.P)

Problem 10.13 (Fig. 10.22): *A cube having 40 mm long edges is resting on one of its faces on the H.P. in such a way that a vertical face is inclined at 30° to the V.P. It is cut* *by a section plane inclined at 45° to the H.P. and passing through the axis at a point 10 mm from its top end. Draw front view sectional top view and true shape of the section.*

Construction (Fig. 10.22):

1. Draw projections of given cube in the required position.
2. Draw a line VT inclined at 45° to xy line and passing through the axis at a point 10 mm from the top end.
3. Mark the points 1', 2', 3'....etc., at which edges are cut by the VT.
4. Project these points on the corresponding edges in the top view.
5. Hatch the area enclosed by 12345 to represent section.
6. The top view thus obtained is the required sectional top view.
7. Draw true shape of section as explained previous problems.

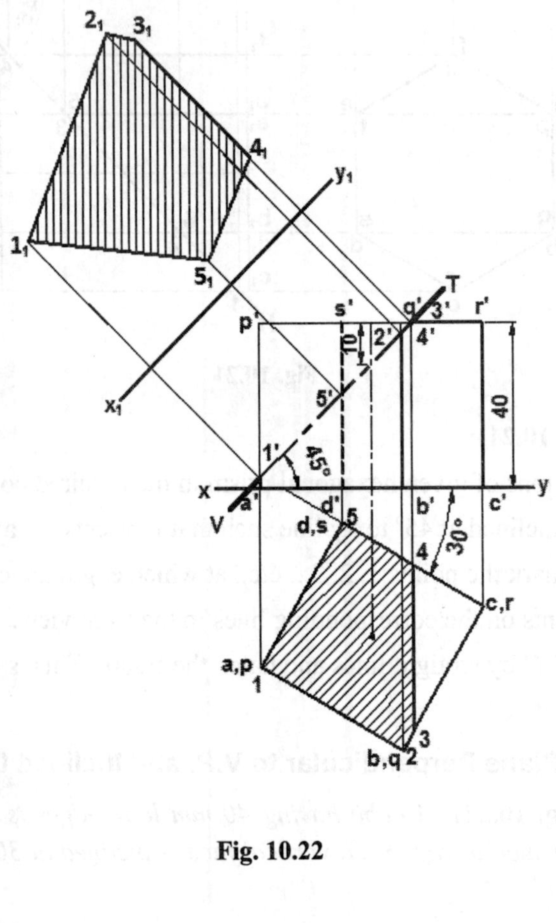

Fig. 10.22

Problem 10.14 (Fig. 10.23): *A pentagonal prism having side of the base 30 mm and axis 65 mm long is resting on an edge of its base on the H.P. with its axis parallel to the V.P. and inclined 60° to the H.P. It is cut by a section plane perpendicular to the V.P. and inclined at 60° to the H.P. (A.I.P.) and passing through highest corner of the prism. Draw front view, sectional top view and true shape of the section.*

Construction (Fig. 10.23):

1. Draw the projections of given pentagonal prism in the required position.
2. In the front view draw a line VT inclined at 60° to the xy line and passing through the highest corner s'.
3. Mark the points 1', 2', 3'....etc., at which edges are cut by VT. Project these points on the corresponding edges in the top view at 1, 2, 3....etc., respectively.
4. Join these points in a proper order and hatch the figure.
5. To draw the true shape of the section, draw a new reference line x_1y_1 parallel to VT and project the view on this line as shown in the figure.

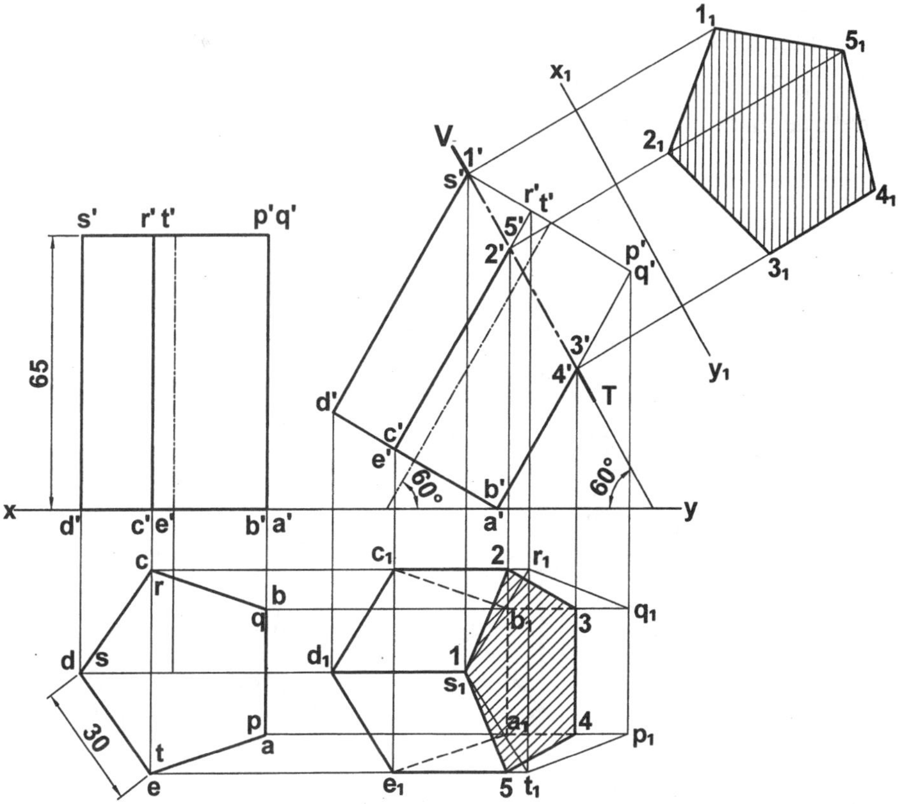

Fig. 10.23

Problem 10.15 (Fig. 10.24): *A cube having 40 mm long edge is resting on its base on the H.P. with its vertical faces equally inclined to V.P. It is cut by section plane, perpendicular to the V.P. so that the true shape of section is a regular hexagonal. Determine the inclination of the cutting plane with the H.P. and draw the sectional top view and true shape of the section.*

Construction (Fig. 10.24):

1. Draw the projections of given cube in the required position.
2. Since, true shape of the section is a regular hexagon, the section plane should cut all the edges of the base and top faces at their mid points.

   So, in the top view, mark mid points of edges as 1, 3, 4 and 6. Join 1-2-3-4-5-6 by straight lines and hatch the figure. This represents sectional top view.
3. Project points 1, 2, 3....etc., in the front view to obtain the V.T. of section plane.
4. Draw a new reference line x_1y_1 parallel to VT. Project the true shape on this line by keeping distances of points 1_1, 2_1, 3_1....etc., from x_1y_1 line equal to their corresponding from xy in the top view.
5. Join 1_1, 2_1, 3_1...etc., in a proper order and hatch the figure as shown. This is the required true shape of section.

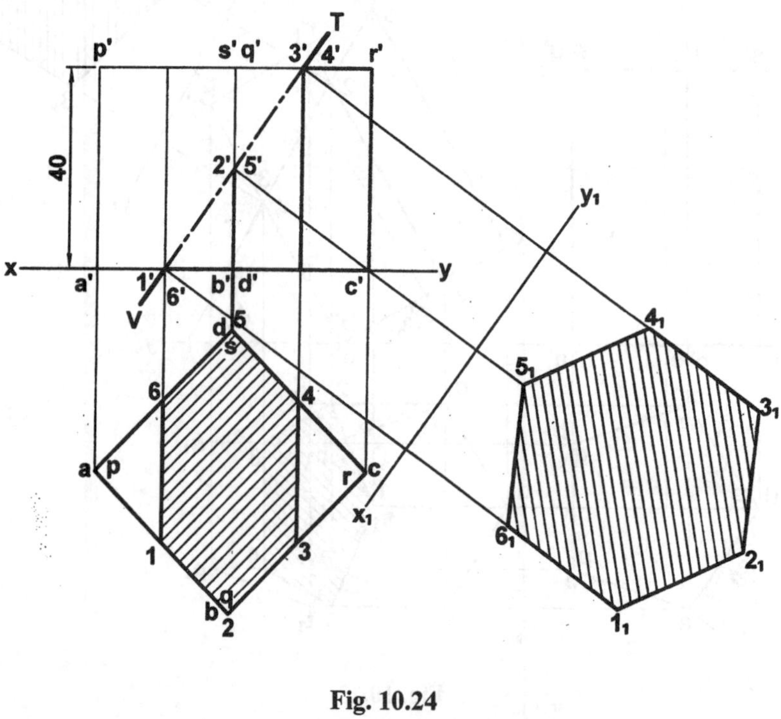

Fig. 10.24

Problem 10.16 (Fig. 10.25): *A square prism having edge of base 30 mm and axis 60 mm long is resting on its base on the H.P. with edges of base equally inclined to the V.P. It is cut by a section plane whose V.T. makes an angle of 30° with xy line and bisects the axis. Draw front view sectional top view and true shape of the section.*

Construction (Fig. 10.25):

1. Draw the projections of given square prism in the required position.
2. In the front view draw a line V.T. inclined at 30° to xy and bisecting the axis.
3. Mark the points 1, 2, 3 and 4 at which the edges are cut by section plane.
4. Project them on the corresponding edges in the top view. Since all vertical edges are cut the entire area of square will be hatched.
5. To draw true shape of the section, draw a new reference line x_1y_1 parallel to V.T. and project the true shape of the section on it as explained in previous problem.

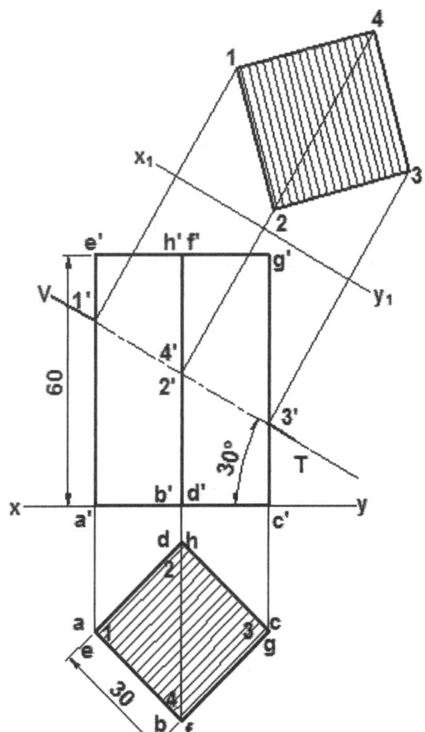

Fig. 10.25

Problem 10.17 (Fig. 10.26): *A square prism having side of base 30 mm and axis 70 mm long is resting on its base on the H.P. with edges of the base equally inclined to the V.P. It is cut by an auxiliary inclined plane (AIP) passing through the midpoint of the axis in such a way that the true shape of the section is rhombus having longer diagonal of*

60 mm. Draw sectional top view and true shape of the section and determine the inclination of AIP with H.P.

Construction (Fig. 10.26):

1. Draw top view and front view of given square prism.
2. In the front view, mark the midpoint of the axis.
3. With midpoint as centre and radius equal to 30 mm (half the longer diagonal of the rhombus) draw arcs cutting the opposite vertical edges of the prism. Join these points by a line which will pass through the midpoint of the axis.
   This line is the required VT of the section plane.
4. Mark 1', 2', 3' and 4' on the V.T. as shown. Project them to obtain true shape of the section as explained in problem 10.10.
5. Measure the inclination of VT with xy which shows the angle of inclination of AIP with the H.P.

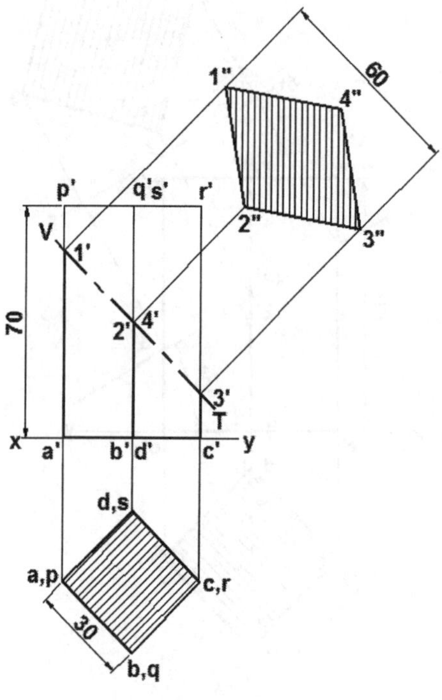

Fig. 10.26

Problem 10.18 (Fig. 10.27): *A square prism having side of base 30 mm and axis 70 mm long has its base on the H.P. with its faces equally inclined to V.P. It is cut by a section plane perpendicular to the V.P, inclined at 60° to the H.P. and passing through a point* *on the axis 20 mm below its top face. Draw its front view, sectional top view and another top view on an auxiliary incline plane (AIP) parallel to the cutting plane.*

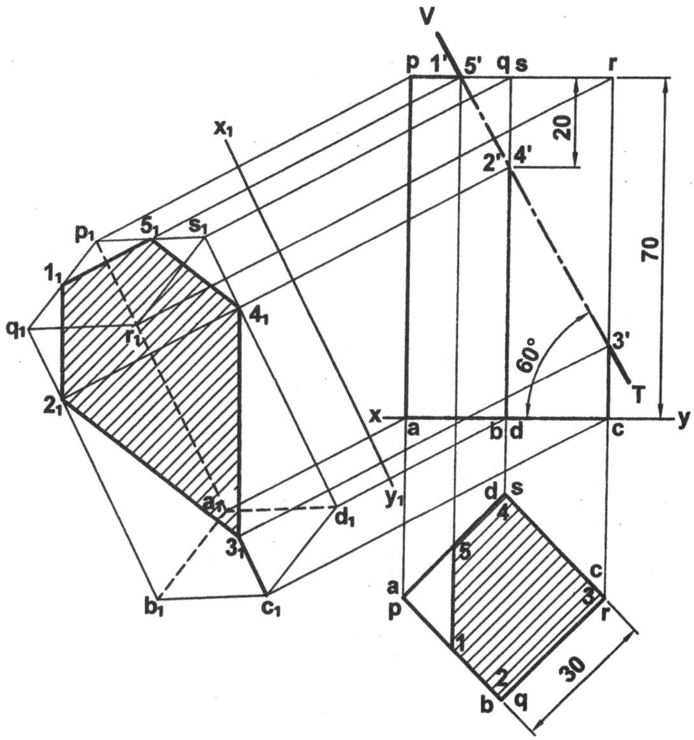

Fig. 10.27

Construction (Fig. 10.27):

1. Draw the projections of given square prism in the required position.
2. In the front view, draw a line VT inclined at 60° to the xy line and cuts the axis at a point 20 mm from the top face.
3. Mark points 1', 2', 3'....etc., at which the edges are cut by section plane.
4. Project them on the corresponding edges in the top view.
5. Hatch figure obtained by joining points 1-2-3-4-5.
6. To draw the true shape of the section, draw a new reference line x_1y_1 parallel to VT.
7. Project the true shape of the section on x_1y_1 as explained earlier.

Problem 10.19 (Fig. 10.28): *A pentagonal prism having edge of base 25 mm and axis 65 mm long is resting on an edge of base on the H.P. with its axis parallel to V.P. and inclined at 60° to the H.P. A section plane having its H.T. perpendicular to xy and the*

V.T. inclined at 60° to the xy line. The V.T. passes through the highest corner of the prism. Draw the front view sectional top view and true shape of the section.

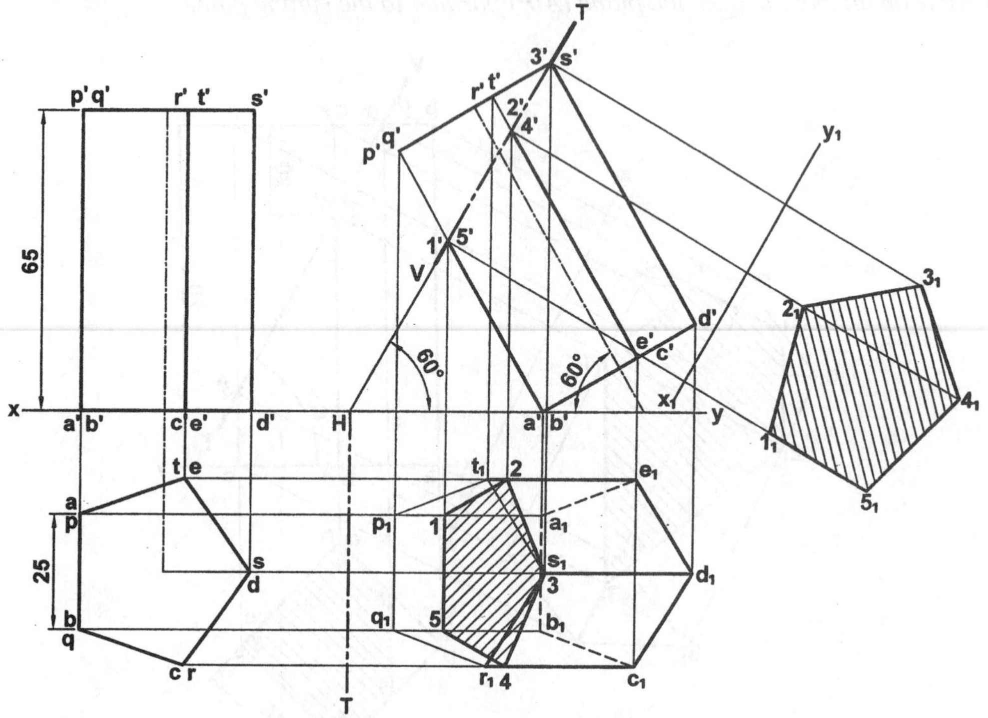

Fig. 10.28

Construction (Fig. 10.28):

1. Draw the projections of the given pentagonal prism in the required position.
2. In the front view draw a line VT inclined at 60° to xy line and passing through the highest corner of the prism and line HT perpendicular to xy line.
3. Name the points at which edges are cut by V.T.
4. Project them on the corresponding edges in the top view. Join 1-2-3-4-5 by dark lines and hatch the figure.
5. To draw, true shape of the section, draw a new reference line parallel to VT. Project the true shape of the section on it.

Problem 10.20 (Fig. 10.29): *A hexagonal prism having side of base 20 mm and axis 65 mm long has a square hole of side 15 mm through it centrally. It is cut by a section plane perpendicular to V.P. and inclined at 45° to the H.P. in such a way that it passes through a point on the axis, 40 mm from its bottom end. Draw its front view, sectional top view and true shape of the section.*

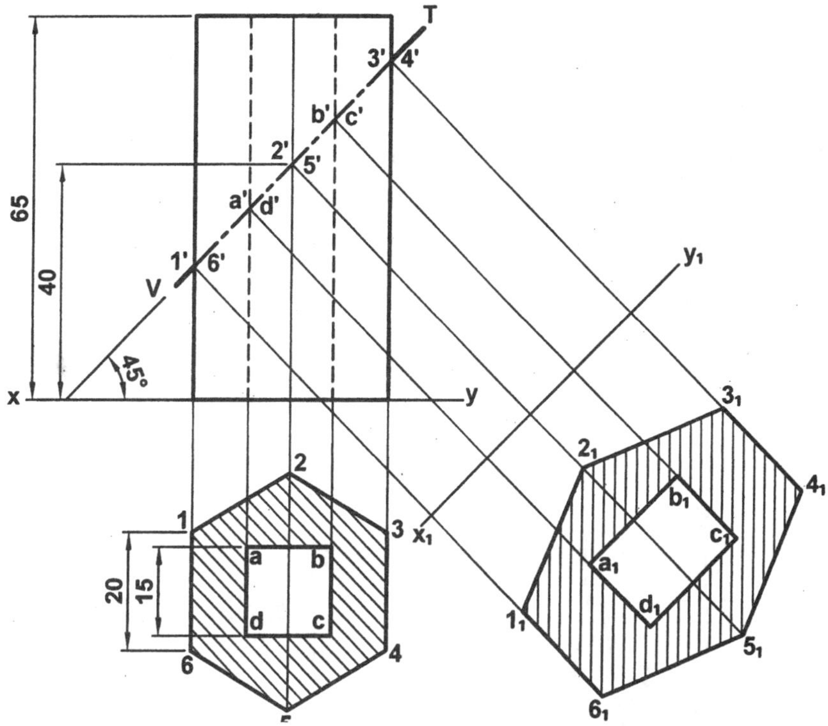

Fig. 10.29

Construction (Fig. 10.29):

1. Draw the projections of the given hexagonal prism in the required position.
2. Draw a line VT inclined at 45° to xy line which cuts the axis at a point 40 mm above the base.
3. Name the points 1', 2', 3'....etc., in a proper sequence at which edges are cut by the section plane.
4. Since, all the edges are cut by the section plane, hatch the hexagon 123456 leaving the square abcd blank which represents hollow portion in the top view.
5. To draw the true shape of the section draw a new reference line x_1y_1 parallel to VT. Project the true shape of the section on it. Hatch the figure as shown.

10.5 Sections of Pyramids

Sections of pyramids are explained with the help of several problems for the following positions of section plane with respect to the reference planes.

(a) Section plane parallel to the H.P.

(b) Section plane parallel to the V.P.

(c) Section plane perpendicular to the V.P. and inclined to H.P. (A.I.P.)

(d) Section plane perpendicular to the H.P. and inclined to the V.P. (A.V.P.)

(e) Section plane perpendicular to both H.P. and V.P. (P.P.)

10.5.1 Section Plane Parallel to the H.P.

Problem 10.21 (Fig. 10.30): *A square pyramid edge of base 30 mm and axis 65 mm long is resting on its base on the H.P. with edges of base equally inclined to the V.P. It is cut by a section plane parallel to the H.P. and passing through a point on the axis 20 mm from the base. Draw the front view and sectional top view.*

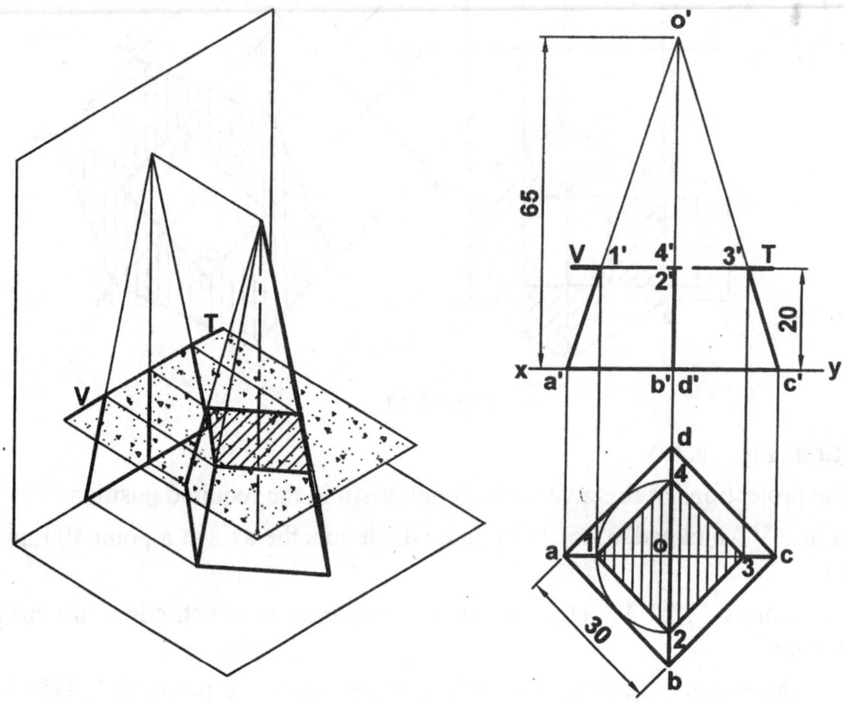

Fig. 10.30

Construction (Fig. 10.30):

1. Draw the projections of a square pyramid in the required position.
2. In the front view, draw a line VT parallel to and 20 mm above xy line.
3. Mark the points 1', 2', 3' and 4' at which edges are cut by VT.
4. Project points 1' and 3' on the edge oa and oc at 1 and 3 respectively.

5. Points 2' and 4' cannot be projected directly. For this, with o as centre and radius equal to o1 (or o3), draw arcs on ob and od at 2 and 4 respectively.
6. Join 1-2-3-4 in a proper sequence and hatch the area 1234.

Problem 10.22 (Fig. 10.31): *A pentagonal pyramid having edge of base 25mm and axis 65mm long is resting on its base on the H.P. and edge of base parallel to the V.P. It is cut by a horizontal section plane passing through a point on the axis 30 mm from the apex. Draw its front view and sectional top view.*

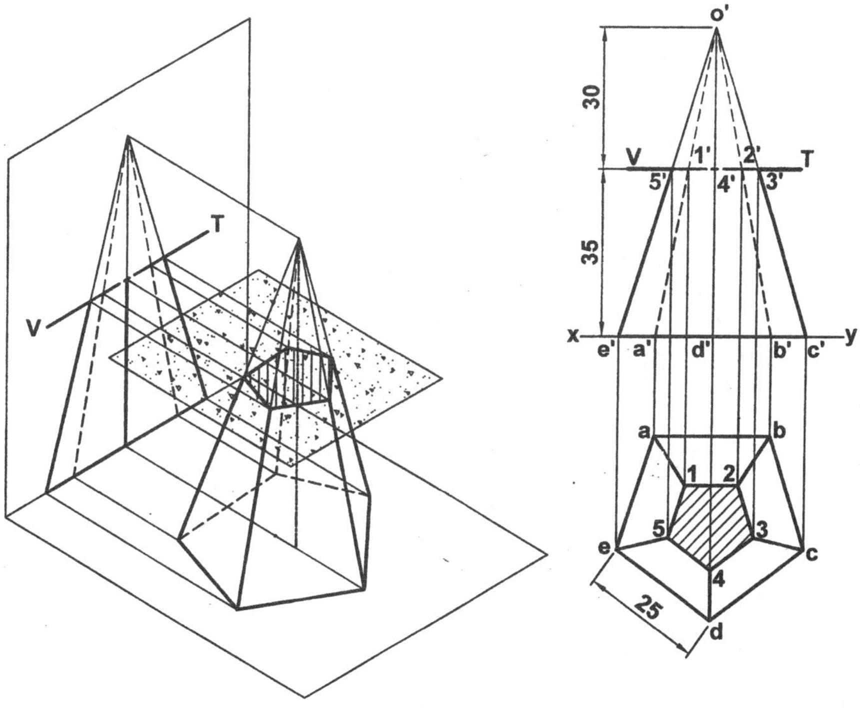

Fig. 10.31

Construction (Fig. 10.31):

1. Draw the projection of given pentagonal pyramid in the required position.
2. Draw a line VT parallel to xy line and 35 mm above the base.
3. Mark the points 1', 2', 3'....etc., which are cut by section plane.
4. Project these points on the corresponding edges in the top view.
5. Join 1, 2, 3....etc., in a correct sequence and hatch the pentagon 12345.

Problem 10.23 (Fig. 10.32): *A pentagonal pyramid having edge of base 25 mm and axis is 65 mm long lies on one of its triangular faces on the H.P. and axis parallel to the V.P.*

It is cut a horizontal section plane passing through a point on the axis 10 mm away from the base. Draw front view and sectional top view.

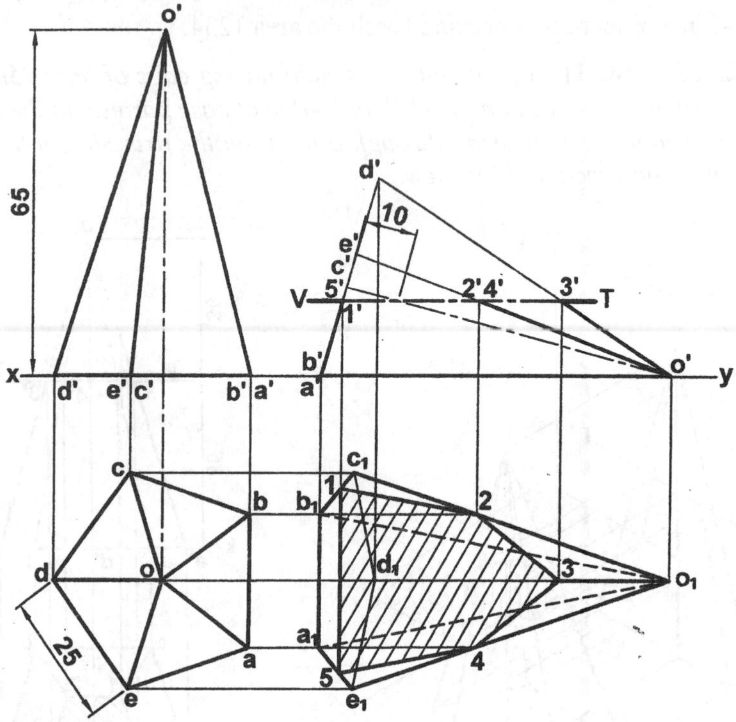

Fig. 10.32

Construction (Fig. 10.32):

1. Draw the projection of pentagonal pyramid in the required position.
2. Draw a line VT parallel to xy line which passes through a point on the axis 10 mm from the base.
3. Mark the points 1', 2', 3', 4' and 5' at which edges are cut by section plane.
4. Project these points on the corresponding edges in the top view at 1, 2, 3....etc., respectively.
5. Join 1-2-3-4-5 and hatch the portion of figure in the top view.

10.5.2 Section Plane Parallel to the V.P.

Problem 10.24 (Fig. 10.33): *A square pyramid edge of base 30 mm and axis 65 mm long is resting on its base on the H.P. with edges of base equally inclined to the V.P. It is cut by a section plane parallel to the V.P. and 12 mm in front of the axis. Draw top view and sectional front view.*

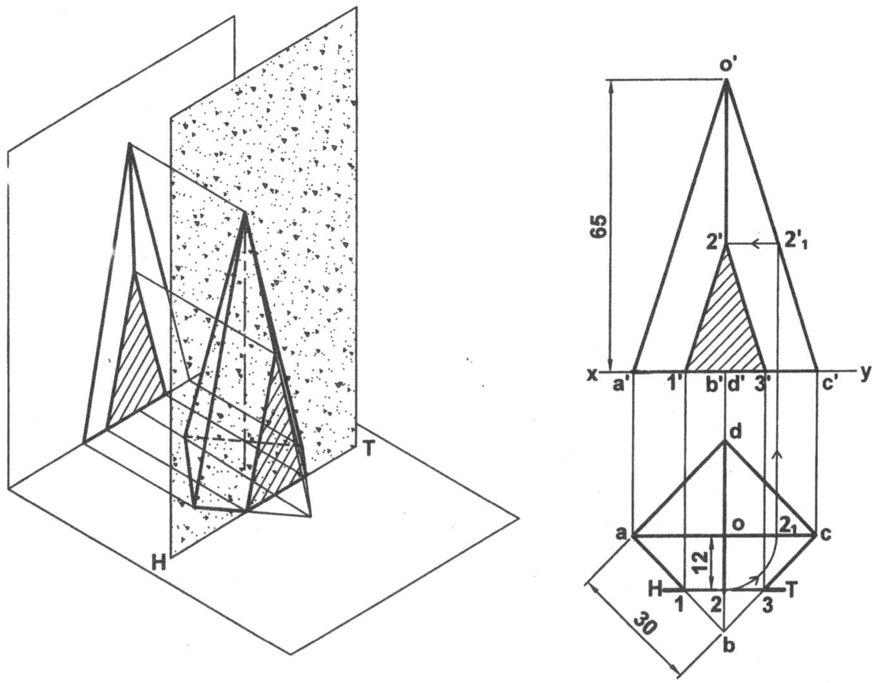

Fig. 10.33

Construction (Fig. 10.33):

1. Draw the top view and front view of given square pyramid.
2. In the top view, draw a line HT parallel to the xy line and 12 mm away from o (top view of the axis).
3. Mark points 1, 2 and 3 at which edge are cut by HT.
4. Project points 1 and 3 on the corresponding edges at 1' and 3' respectively in the front view.
5. Project point 2 at 2' in the front view as explained below.
   (a) With o as centre and radius equal to o2, draw an arc to cut oc at 2_1.
   (b) Through 2_1, draw a vertical projector to intersect o'c' at $2'_1$.
   (c) Through $2'_1$, draw a horizontal line to cut edge o'b' at 2'.
6. Join 1'-2'-3' and hatch the area enclosed by triangle 1'2'3'.

Problem 10.25 (Fig. 10.34): *A pentagonal pyramid having side of base 25 mm and axis 65 mm long is resting on its base on the H.P. with a side of base parallel to the V.P. and nearer to the V.P. It is cut by a section plane parallel to the V.P. and 12 mm away from the axis nearer to the observer. Draw its top view and sectional front view.*

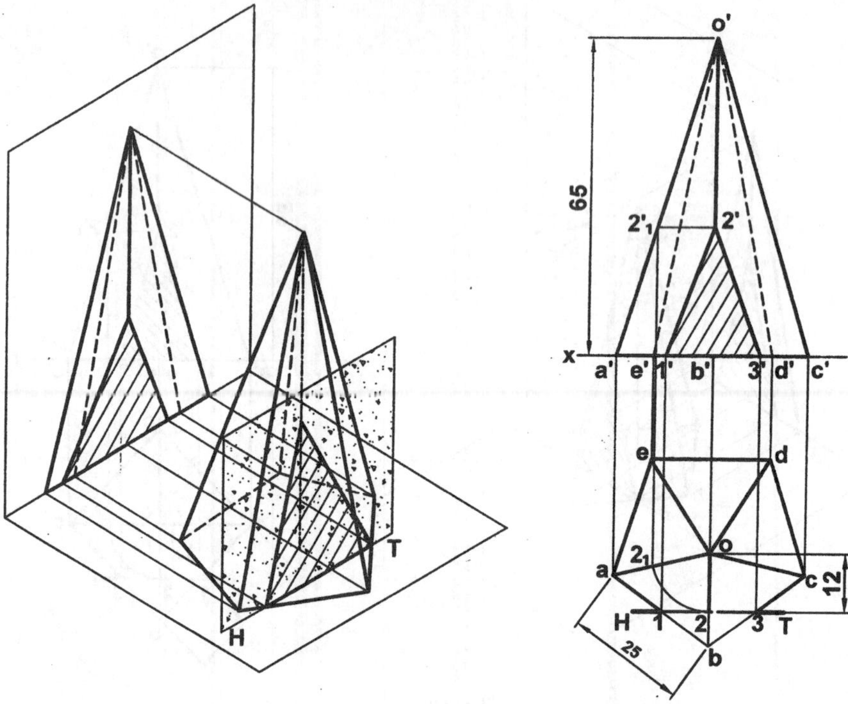

Fig. 10.34

Construction (Fig. 10.34):

1. Draw the projections of given pentagonal pyramid in the required position.
2. Since, cutting plane is parallel to V.P. its top view is a line parallel to xy. So, draw a line HT parallel to xy line and 12 mm away from the axis.
3. Name the points 1, 2 and 3 at which edges are cut.
4. Project points 1 and 3 on the corresponding edges in the front view.
5. Point 2 cannot be projected directly on edge o'b'. For this, with o as centre and radius equal to o2, draw an arc to meet the edge ao at 2_1. Project 2_1 to $2_1'$ on slant edge o'a'. From $2_1'$ draw a horizontal line to intersect o'b' at 2'.
6. Join 1'-2'-3' and hatch the triangle 1'2'3'.

Problem 10.26 (Fig. 10.35): *A triangular pyramid having edge of base 30 mm and axis 65 mm long is lying on one of its triangular faces on the H.P. with axis parallel to the V.P. It is cut by a sectional plane parallel to the V.P. and 10 mm away from the axis nearer to the observer. Draw its sectional front view and top view.*

Construction (Fig. 10.35):

1. Draw the projections of given triangular pyramid in the required position.
2. Draw a line HT parallel to xy line and 10 mm away from the axis.
3. Mark points 1, 2 and 3 at which edges are cut by section plane. Project these points on the corresponding edges in the front view at 1', 2' and 3' respectively.
4. Join 1'-2'-3' and hatch the triangle 1'2'3'.

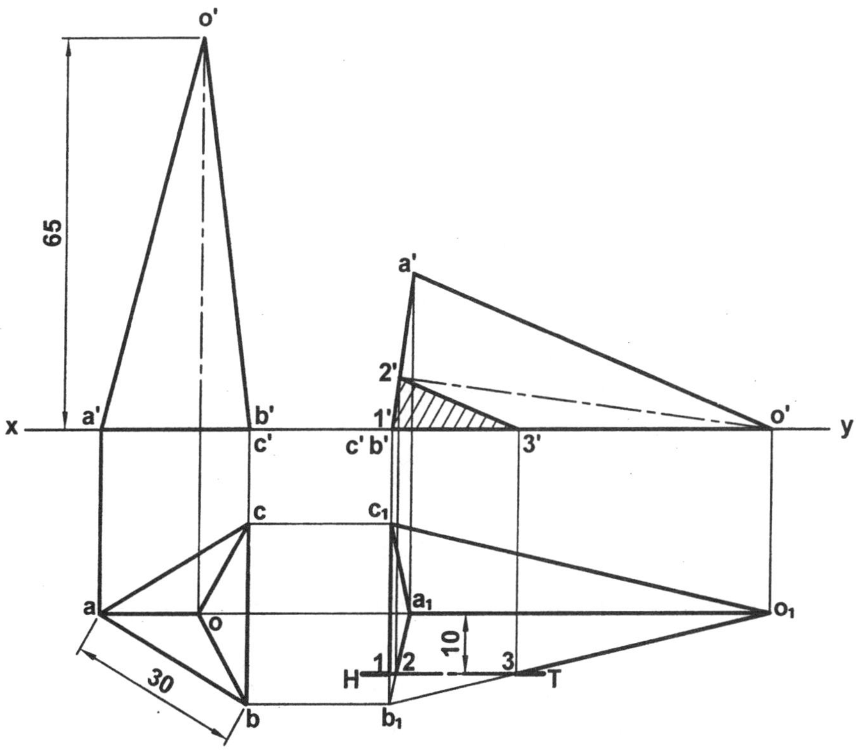

Fig. 10.35

Problem 10.27 (Fig. 10.36): *A pentagonal pyramid side of base 30 mm and axis is 65 mm long rests on its base on H.P. such that one of the edges of the base is perpendicular to V.P. A section plane perpendicular to H.P. and parallel to the V.P. cuts the pyramid at a distance of 15 mm from the corner of the base nearer to the observer. Draw its top view and sectional front view.*

Construction (Fig. 10.36):

Draw the projections of given pentagonal pyramid and obtain the sectional front view as explained in problem 10.25.

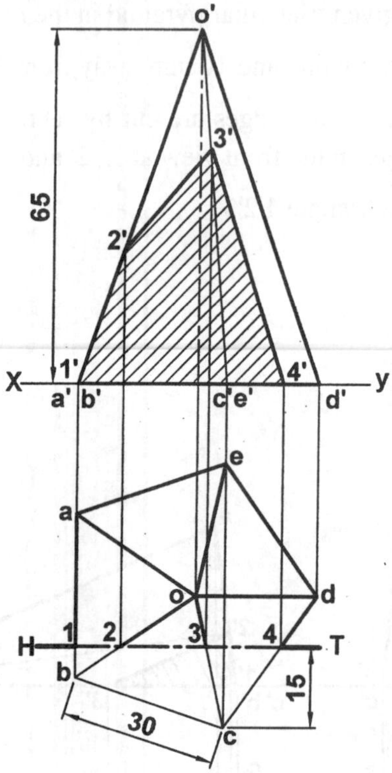

Fig. 10.36

10.5.3 Section Plane Perpendicular to V.P. and Inclined to H.P. (A.I.P)

Problem 10.28 (Fig. 10.37): *A square pyramid having edge of base 30 mm and axis 65 mm long is resting on its base on the H.P. with edges of base equally inclined to the V.P. It is cut by a section plane perpendicular to the V.P. inclined at 60° to the H.P. and passing through a point on the axis at 30 mm from the base. Draw front view section, top view section, side view and true shape of the section.*

Construction (Fig. 10.37):

1. Draw the projections of given square pyramid.
2. Draw line VT inclined at 60° to xy and cuts the axis 30 mm above the base.
3. Mark the points 1', 2', 3'......etc., at which edges are cut by section plane.
4. Project 1', 3' and 5' on the corresponding edges in the top view.
5. Points 2' and 4' cannot be projected directly. For this, draw a horizontal line through 2' meeting the slant edge at $2_1'$. Project $2_1'$ to 2' on the edge oa in the top view.

6. With o as centre and radius equal to $o2_1$, draw arcs cutting ob and od at 2 and 4 respectively. Join 1-2-3-4-5 and hatch the figure as shown.
7. Project the sectional side view. Draw true shape of the section as explained in Problem10.15.

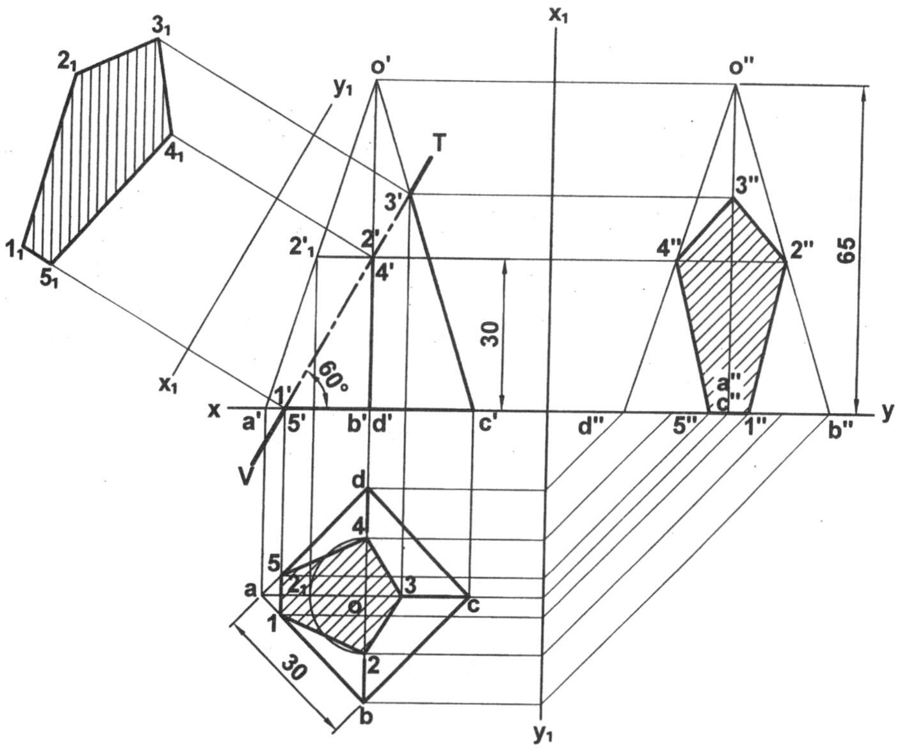

Fig. 10.37

Problem 10.29 (Fig.10.38): *A square pyramid base 40 mm side and axis 70 mm long has its base on the H.P. and all the edges of the base are equally inclined to the V.P. It is cut by a section plane perpendicular to the V.P. and inclined at 45° to H.P. and bisecting the axis. Draw the sectional top view and true shape of the section.*

Construction (Fig. 10.38):

1. Draw the projections of given square pyramid in the required position.
2. In the front view draw a line VT inclined at 45° to xy and bisecting the axis.
3. Mark the points 1', 2', 3'....etc., at which edges are cut by section plane.
4. Project these points on the corresponding edges in the top view. Join 1-2-3-4-1 and hatch the figure 1234.

5. To draw true shape of the section, through 1', 2', 3'......etc., draw projectors perpendicular to VT and on these mark 1_1, 2_1, 3_1......etc., keeping their distances from VT equal to their corresponding distances from xy in the top view.
6. Join 1_1-2_1-3_1-4_1-1_1 and hatch the figure obtained which is the required true shape of the section.

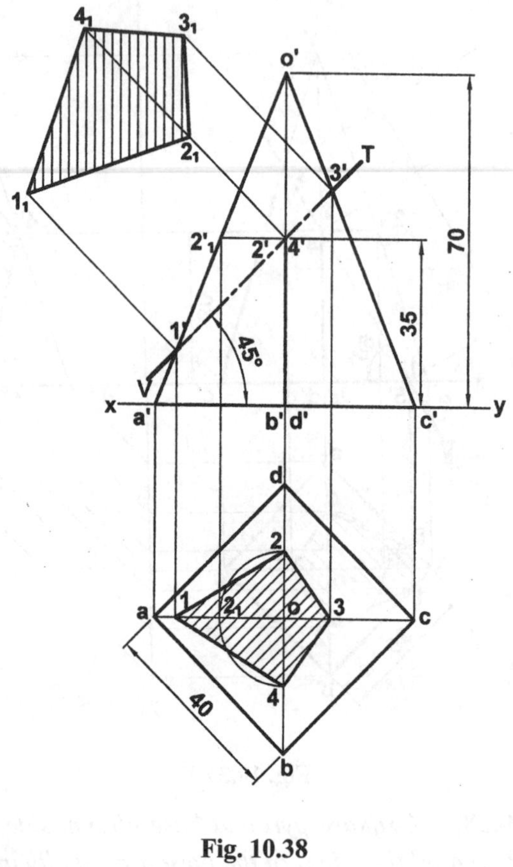

Fig. 10.38

Problem 10.30 (Fig. 10.39): *A pentagonal pyramid having sides of base 30 mm and axis is 65 mm long, is resting on its base on the H.P. with one of its base side parallel to the V.P. It is cut by a section plane perpendicular to the V.P. and inclined at 60° to the H.P. and bisecting the axis. Draw its front view, sectional top view and true shape of the section.*

Construction (Fig. 10.39):

1. Draw the projections of given pentagonal pyramid in the required position.
2. In the front view, draw a line VT inclined at 60° to xy line and bisecting the axis.

3. Mark the points 1', 2', 3'....etc., at which the edges are cut by section plane. Project these points on the corresponding edges in the top view.
4. Point 2 cannot be projected directly. For this, draw a horizontal line through 2' to meet the slant edge o'c' at $2_1'$. Project $2_1'$ on oc in the top view. Now with o as centre and radius equal to $o2_1$ draw an arc cutting ob at 2.
5. Join 1, 2, 3....etc., in the correct sequence and hatch the figure 123456.
6. Obtain true shape of the section as explained in problem 10.29.

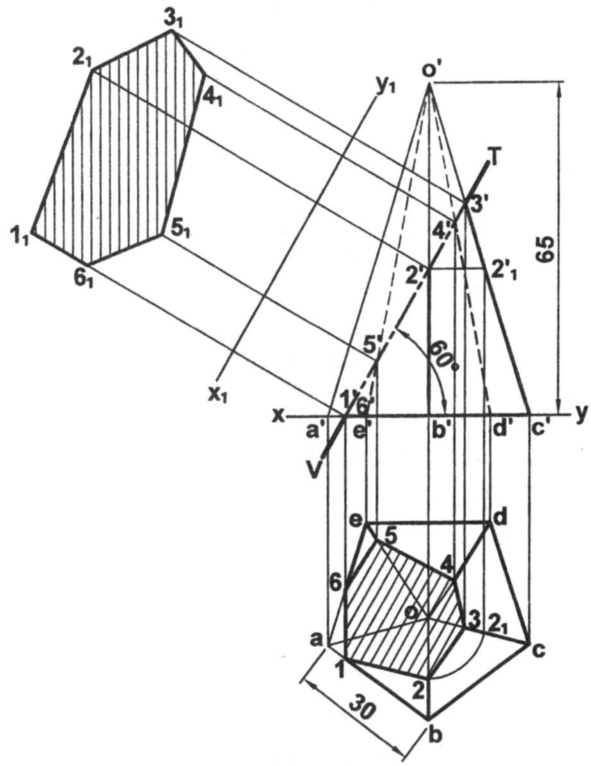

Fig. 10.39

Problem 10.31 (Fig. 10.40): *A hexagonal pyramid base 30 mm side and axis 65 mm long is resting on its base on the H.P. with two edges parallel to the V.P. It is cut by a section plane perpendicular to the V.P. and inclined at 45° to the H.P. and intersecting the axis at a point 25 mm above the base. Draw the front view, sectional top view, sectional side view and true shape of the section.*

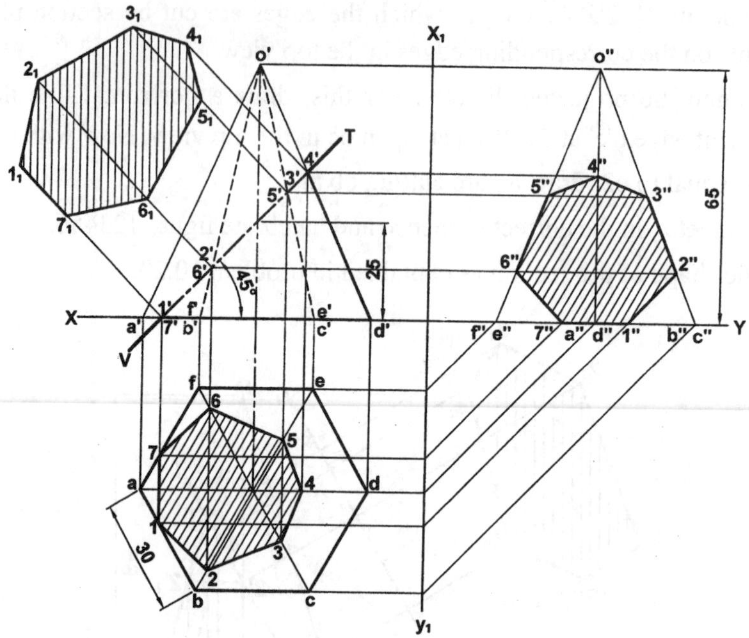

Fig. 10.40

Construction (Fig. 10.40):

1. Draw the projections of given hexagonal pyramid in the required position.
2. In the front view, draw a line VT inclined at 45° to xy line and cuts the axis at a point 25 mm above the base.
3. Obtain sectional top view and true shape of the section as explained in problem 10.30.
4. Draw a reference line x_1y_1 perpendicular to xy to represent auxiliary vertical plane.
5. Project the side view with the help of front view and top view.
6. Through 2', 3', 4', 5' and 6', draw horizontal projectors to intersect corresponding edges at 2", 3", 4", 5" and 6" respectively.
7. Project 1' and 7' on the sides ab and af in the top view respectively.
8. Through 1 and 7 draw horizontal lines to meet x_1y_1 then turn them at 45° to x_1y_1 to meet xy at 1" and 7".
9. Join 1", 2", 3".....etc., in a correct sequence and hatch the enclosed figure. This is the required side view.

10.5.4 Section Plane Perpendicular to the H.P. and Inclined to V.P. (A.V.P.)

Problem 10.32 (Fig. 10.41): *A square pyramid having edge of base 30 mm and axis 65 mm long is resting on its base on the H.P. with its edges of base equally inclined to the V.P. It is cut by a section plane perpendicular to the H.P. and inclined at 30° to the V.P. and 10 mm away from the axis. Draw sectional front view and top view.*

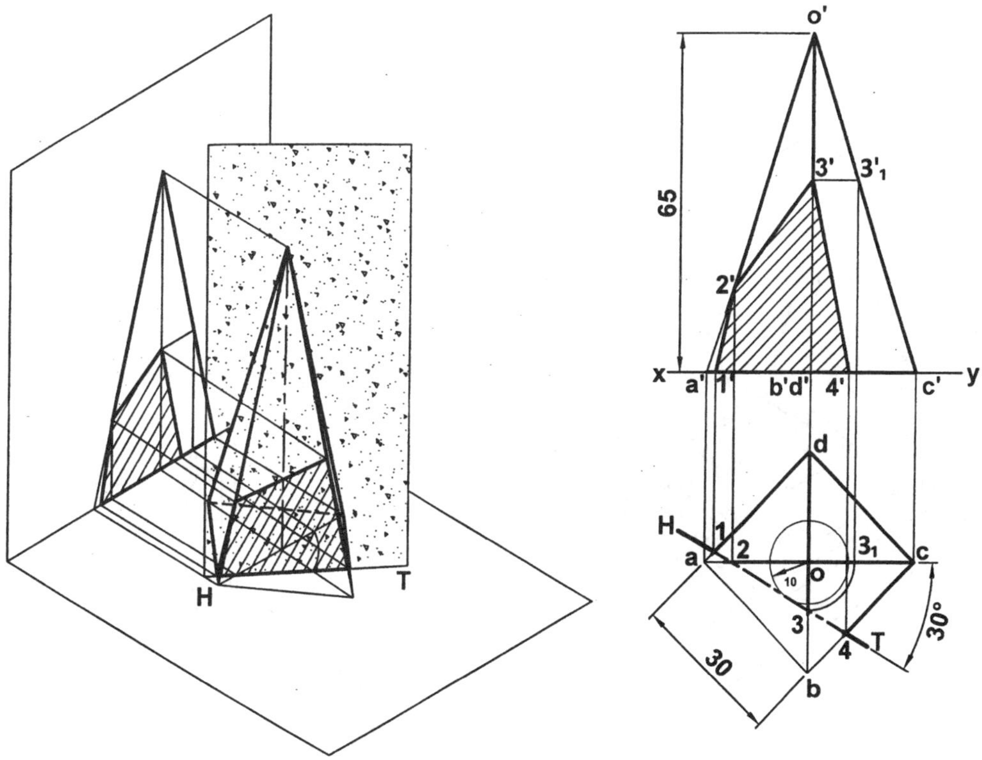

Fig. 10.41

Construction (Fig. 10.41):

1. Draw the projections of given square pyramid in the required position.
2. With o as centre and radius equal to 10 mm, draw a circle. Now, draw HT inclined at 30° to xy and tangential to the circle.
3. Mark points 1, 2, 3 and 4 at which edges are cut by the section plane.
4. Project them on the corresponding edges in the front view.
5. Join 1', 2', 3' and 4' in a proper sequence by straight lines.
6. Hatch the area 1'2'3'4'.

Problem 10.33 (Fig. 10.42): *A pentagonal pyramid, having edge of its base 25 mm and axis 65 mm long, is resting on its base on the H.P. with an edge of the base parallel to the V.P. which is also nearer to the V.P. It is cut by an A.V.P. inclined at 45° to the V.P. and 10 mm away from its axis. Draw its sectional front view, top view and true shape of its section.*

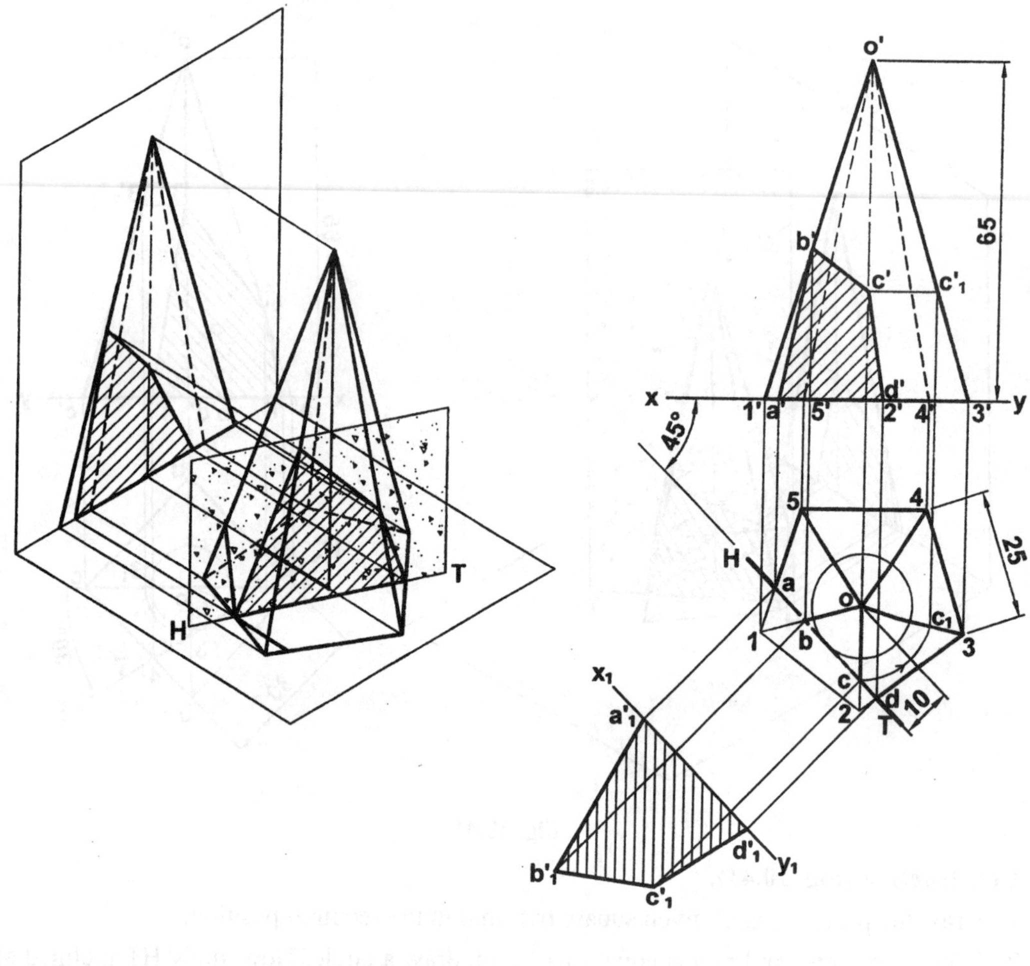

Fig. 10.42

Construction (Fig. 10.42):

Front view and Top view

1. Draw top view and front view of given pentagonal pyramid.

Section Plane (HT)

2. In the top view, with o as centre and radius equal to 10 mm, draw a circle. Draw a line HT tangential to the circle and inclined at 45° to xy line.

Sectional Front View

3. Mark the points a, b, c and d at which edges are cut by section plane.
4. Project a, b and c on the corresponding edges at a', b' and c' respectively in the front view.
5. Point c cannot be projected directly. For this, with o as centre and radius equal to oc, draw an arc to meet o3 at c_1. Project c_1 on slant edge o'3' at c_1'. Through c_1' draw a horizontal line to meet o'd' at c'. Join a', b', c'.....etc., to obtain close figure a'b'c'd'a' and hatch it.

True Shape of the Section

6. Draw a new reference line x_1y_1. Project the true shape on it as explained in problem 10.30.

Problem 10.34 (Fig. 10.43): *A pentagonal pyramid base 30 mm side and axis 70 mm long is lying on one of its triangular faces on the H.P. such that its axis remain parallel to the V.P. It is cut by an A.V.P. bisecting its axis and inclined at 30° to the V.P. and removing the portion nearer to the observer. Draw its sectional front view, top view and true shape of the section.*

Construction (Fig. 10.43):

Front View and Top View

1. Draw the projections of given pentagonal pyramid in the required position.

Section Plane (HT)

2. Draw a line HT inclined at 30° to xy line and bisecting the axis.

Sectional Front view

3. Mark the points 1, 2, 3...etc., at which edges/sides are cut by section plane. Project these points on the corresponding edges in the front view at 1', 2', 3'....etc., Join 1', 2', 3'....etc., in a proper order and hatch the figure enclosed.

True Shape of the Section

4. Draw a new reference line x_1y_1 and project the true shape of the section as explained in problem 10.29.

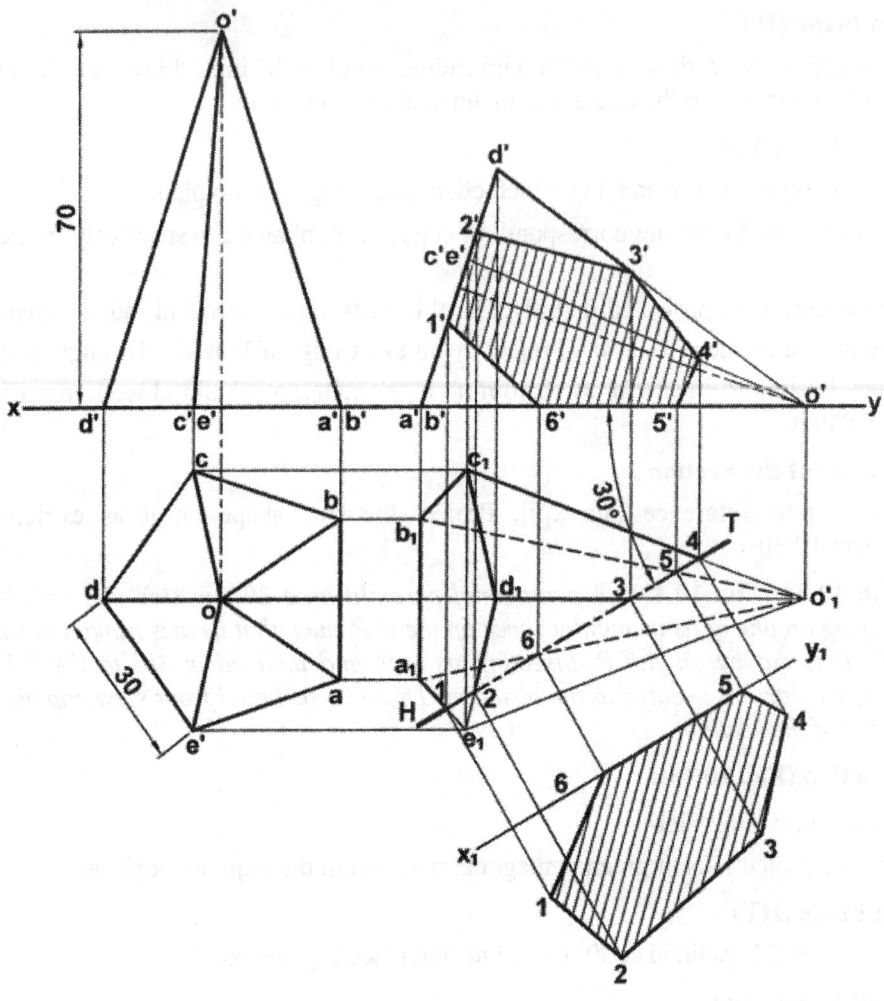

Fig. 10.43

Problem 10.35 (Fig. 10.44): *A hexagonal pyramid having base with 25 mm side and axis 65 mm long is resting on its base corner on the H.P. with its axis parallel to both H.P. and V.P. It is cut by an A.V.P. whose H.T. is inclined at 30° to the V.P. and cuts the axis at a point 25 mm from the base. If the portion of the solid containing the apex is assumed to be retained, draw its sectional front view, top view and true shape of the section.*

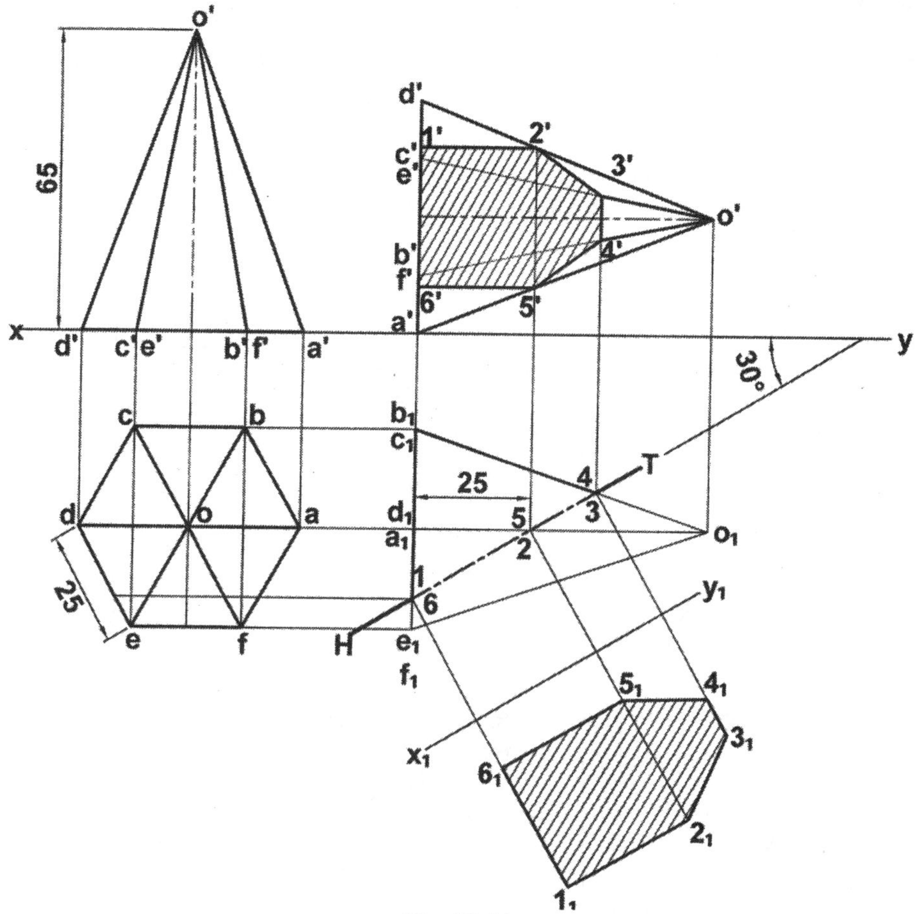

Fig. 10.44

Construction (Fig. 10.44):

Front View and Top View

1. Draw the projection of hexagonal pyramid in the required position.

Section Plane (HT)

2. Draw HT inclined at 30° to xy and bisecting the axis in the top view.

Sectional Front View and True Shape of Section

3. Obtain the sectional front view and true shape of the section as explained in problem 10.34.

10.5.5 Section Plane Perpendicular to both H.P. and V.P.

Problem 10.36 (Fig. 10.45): *A hexagonal pyramid having side of the base 25 mm and axis 65 mm long is resting on an edge of its base on the H.P. such that its axis inclined at 45° to the H.P. and parallel to the V.P. It is cut by a section plane whose V.T. and H.T. are perpendicular to the xy line and passes through the edge on which the solid is resting. Draw front view, top view and sectional side view.*

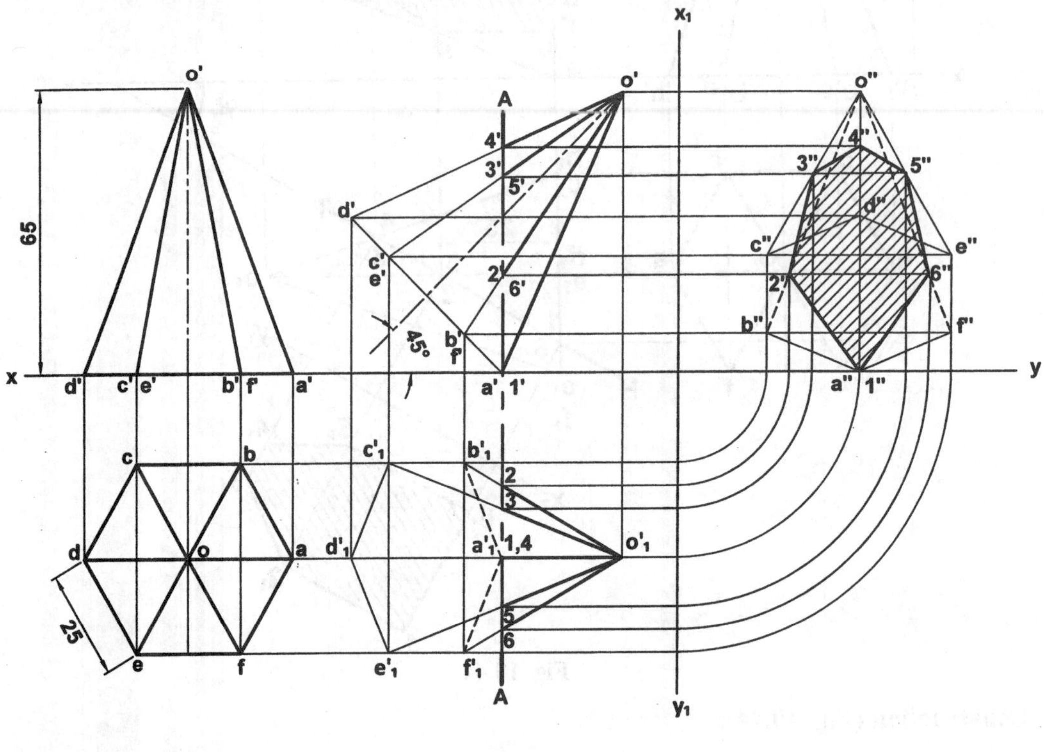

Fig. 10.45

Construction (Fig. 10.45):

In this problem, the section plane is perpendicular to both the H.P. and V.P. Therefore, section will be seen in the side view only.

Projections

1. Draw front view, top view and side view of given hexagonal pyramid.

Section Plane (HT and VT)

2. Draw a common HT-VT vertical line AA passing through the base corner a'.

Sectional Side View

3. Mark points 1, 2, 3...etc., at which line AA cuts various edges in the top view.
4. Similarly mark points 1', 2', 3'...etc., at which line AA cuts the various edges in the front view.
5. Project these points in the side view to obtain 1", 2", 3"...etc., respectively.
6. Join 1"-2"-3"-4"-5"-6" in a proper order and hatch the enclosed area.

10.6 Sections of Cones

Sections of cones are illustrated with the help of several problems for following positions of section plane.

(a) Section plane parallel to the H.P.

(b) Section plane parallel to the V.P.

(c) Section plane perpendicular to the V.P. and inclined to H.P.(A.I.P.)

(d) Section plane perpendicular to the H.P. and inclined to the V.P.(A.V.P.)

(e) Section plane perpendicular to both H.P. and V.P.

Sections of cones are obtained in the same way as explained in section 10.5 for pyramids. Since cone does not have edges, it is required to draw a number of generators on its surface. These are imaginary lines joining apex and points on the circumference of base circle. The generators are used (instead of edges in case of pyramids) for obtaining various points where section plane intersects.

10.6.1 Section Plane Parallel to the H.P.

Problem 10.37 (Fig. 10.46): *A cone having base diameter 40 mm and axis 70 mm long is resting on its base on the H.P. It is cut by a section plane at a distance of 35 mm from the base and parallel to the H.P. Draw its front view and sectional top view.*

Construction (Fig. 10.46):

Front View and Top View

1. Draw the front view and top view the cone in the given position.

Section Plane (VT)

2. Draw a line VT 35 mm above xy line in the front view.

Sectional Top View

3. Since section plane is parallel to H.P., the section will be a circle. Mark the points a' at which generator o'1' is cut by section plane. Similarly mark b' at which generator o'3' is cut by section plane.
4. Project a' and b' to a and b respectively in the top view.
5. With o as centre and radius equal to oa, draw a circle and hatch it.

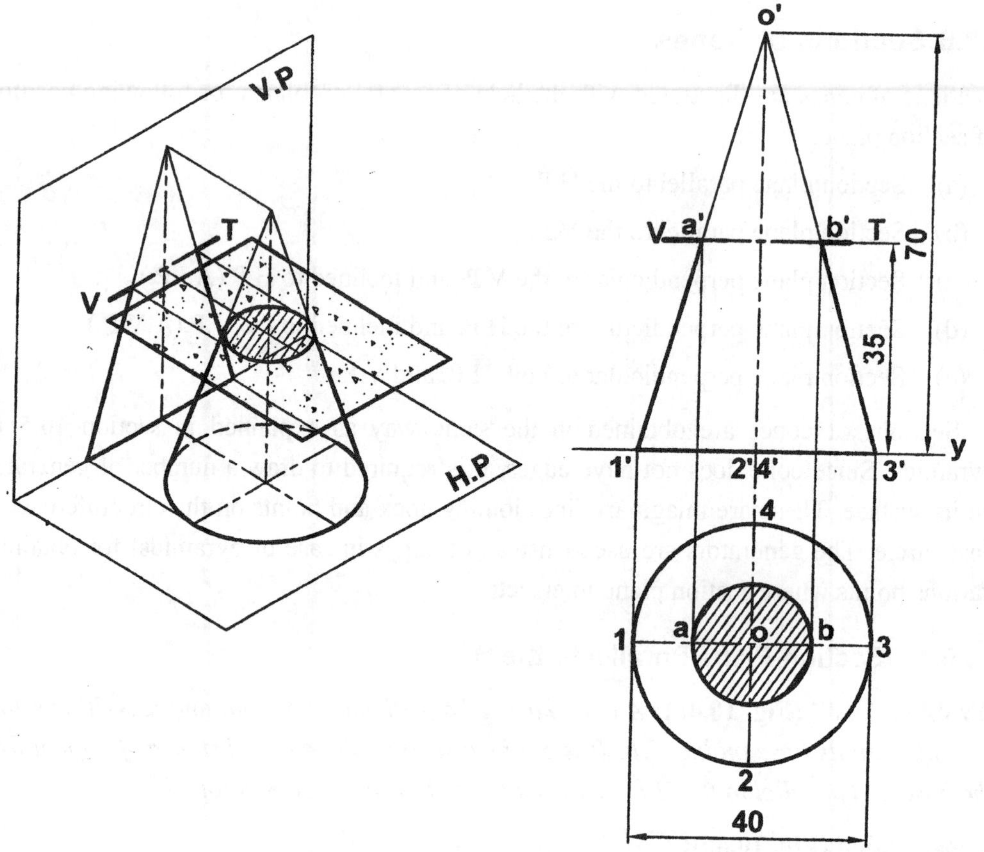

Fig. 10.46

Problem 10.38 (Fig. 10.47): *A cone with base diameter 40 mm and height 70 mm is lying on one of its generators on the H.P. in such a way that its axis is parallel to the V.P. It is cut by a section plane parallel to and 10 mm above the H.P. Draw its sectional top view and front view.*

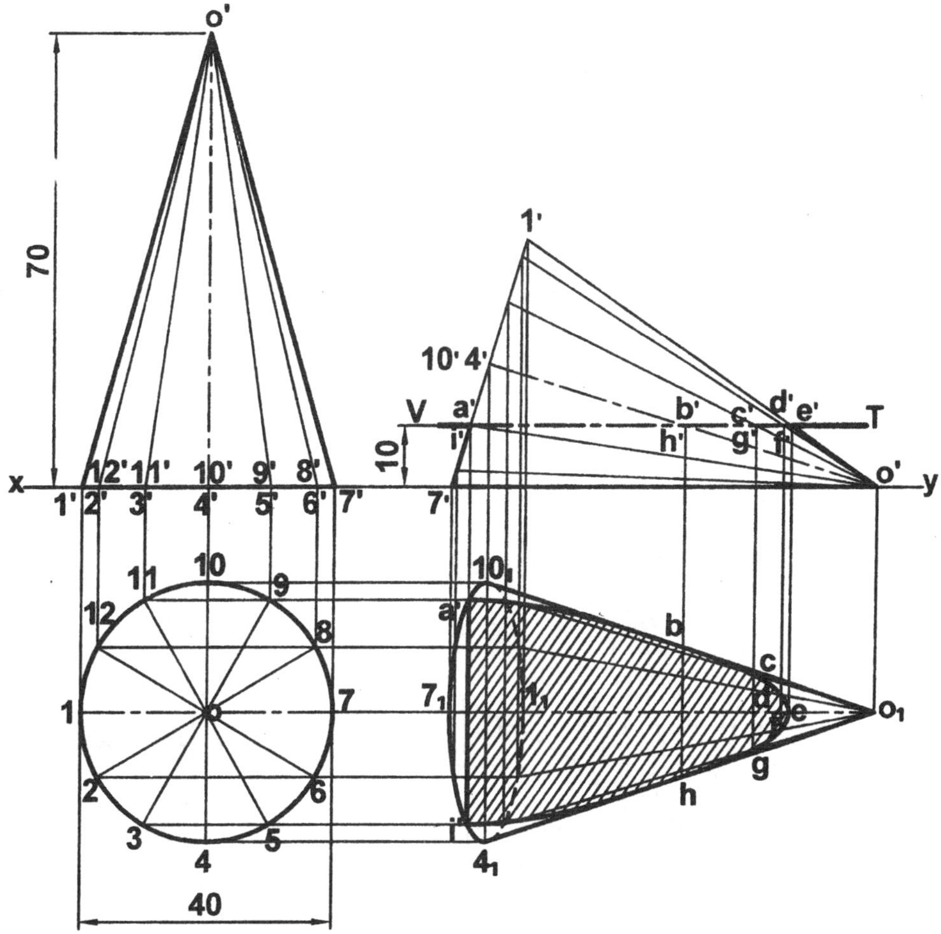

Fig. 10.47

Construction (Fig. 10.47):

Front View and Top View

1. Draw front view and top view of given cone in the given position.

Section Plane (VT)

2. Draw a line VT parallel to and 10 mm above xy line.

Sectional Top View

3. In the front view mark the points a', b', c'...etc., at which generators are cut by section plane.

4. Project these points on the corresponding generators in the top view.
5. Join a, b, c...etc., to obtain closed curve and hatch it. This is required sectional top view.

10.6.2 Section of the Cone by a Plane Parallel to the V.P.

Problem 10.39 (Fig. 10.48): *A cone having 40 mm base diameter and 65 mm long axis is resting on its base on the H.P. A section plane parallel to the V.P. and 15 mm away from the axis nearer to the observer, cuts the cone. Draw sectional front view and top view.*

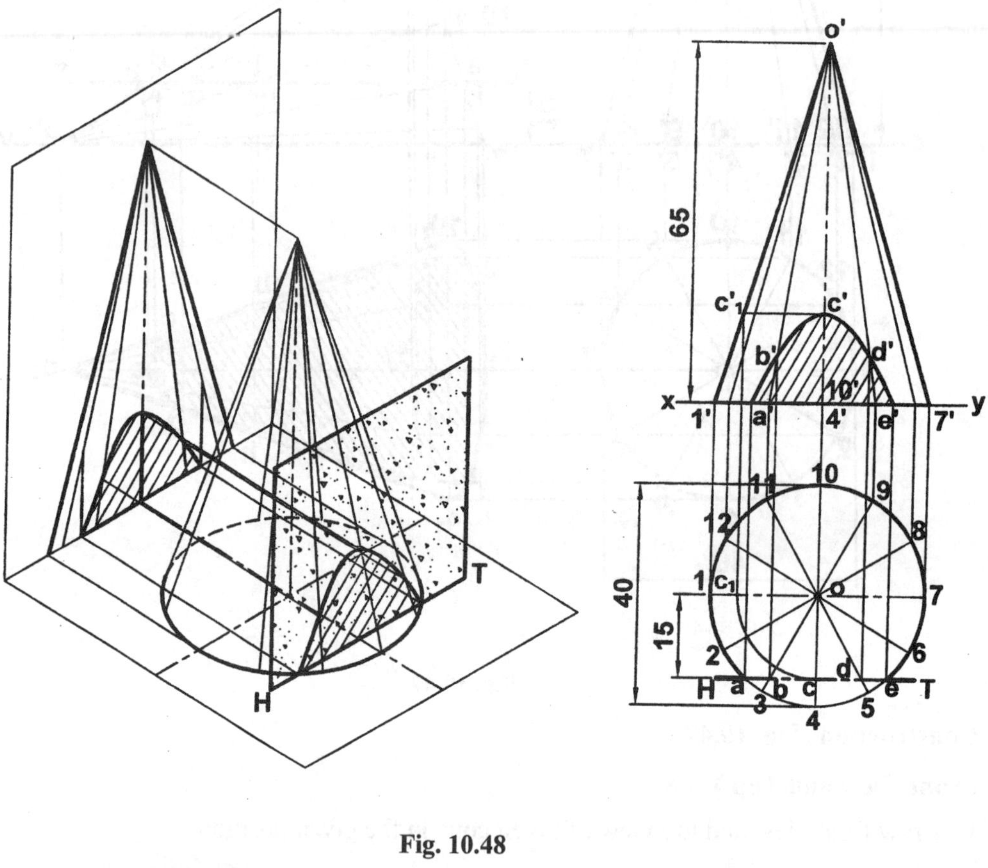

Fig. 10.48

Construction (Fig. 10.48):

Front View and Top View

1. Draw the front view and top view of cone in the given position.

Section Plane (HT)

2. Draw a line HT parallel to xy line and 15 mm away from the axis.

Sectional Front View

3. Mark points a, b, c... etc., at which generators are cut by section plane.
4. Project these points on the corresponding generators in the front view.
5. Join a', b', c'...etc., to obtain a curve. Hatch the enclosed area. This is the required sectional front view. Since section is parallel to V.P. this also represents true shape of the section.

Problem 10.40 (Fig. 10.49): *A cone with base diameter 40 mm and axis 70 mm long is lying on one of its generators on the H.P. with its axis parallel to the V.P. It is cut by a section plane parallel to the V.P. and 10 mm away from the axis. Draw its sectional front view and top view.*

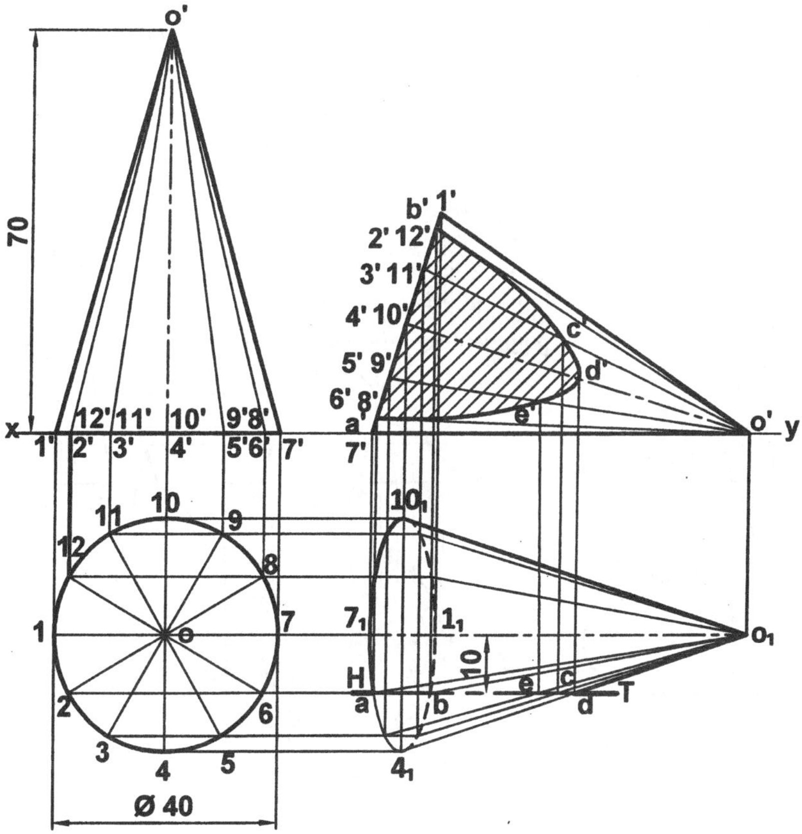

Fig. 10.49

Construction (Fig. 10.49):

Front View and Top View

1. Draw front view and top view of given cone in the required position.

Section Plane (HT)

2. In the top view draw a line HT parallel to and 10 mm away from the axis.

Sectional Front View

3. Mark points a, b, c...etc., at which generators are cut by HT.
4. Project these points at a', b', c'...etc., respectively on the corresponding generators in the front view.
5. Join a', b', c'...etc., by a smooth curve.
6. Hatch the area enclosed by curve a'b'c'd'e'.

10.6.3 Section Plane Perpendicular to the V.P. and Inclined to H.P. (A.I.P.)

Problem 10.41 (Fig. 10.50): *A cone base diameter 50 mm and axis 70 mm long is resting on its base on the H.P. It is cut by a section plane perpendicular to the V.P. and inclined at 45° to the H.P. and cutting the axis at a point 35 mm above the base. Draw its front view, sectional top view and true shape of the section.*

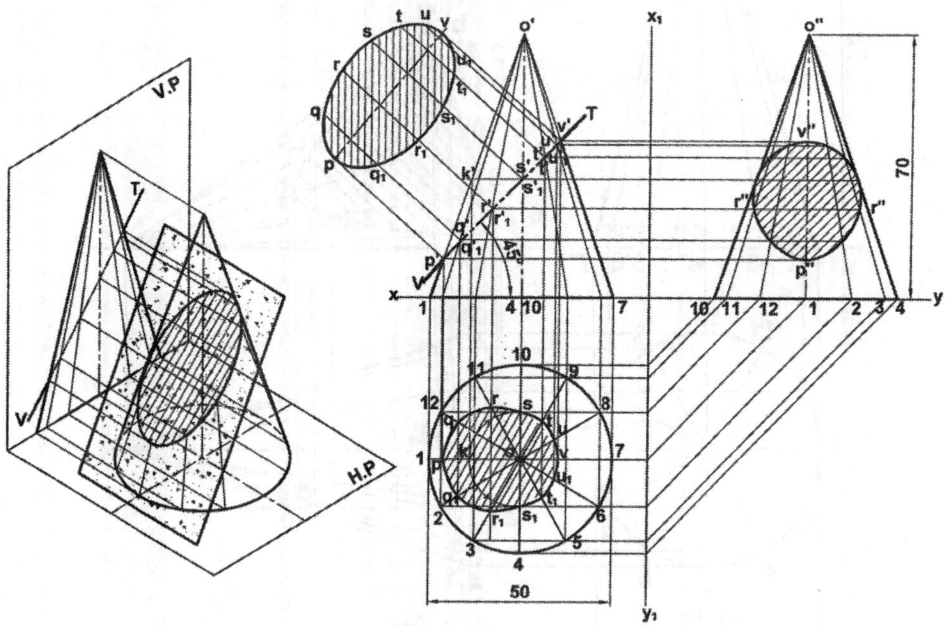

Fig. 10.50

Construction (Fig. 10.50):

Front View, Top View and Side View

1. Draw front view, top view and side view of the cone in the given position.

Section Plane (VT)

2. Draw a line VT inclined at 45° to xy line and bisecting the axis.

Sectional Top View

3. Mark points p', q', r'...etc., at which generators are cut by section plane.
4. Project these points on the corresponding generators in the top view.
5. Since, points s' and s cannot be projected directly, draw a horizontal line through s_1' to meet extreme left generator at k' then project it down to meet op at k. Now with o as centre and radius equal to ok, draw arcs to meet the generator through o' at s and s_1. Join p, q, r...etc., by a smooth curve and hatch the enclose area.

Sectional Side View

6. To draw sectional side view, project p', q', r'...etc., horizontally to meet corresponding generators at p", q", r"...etc. Join these points to obtain smooth curve. Hatch the area. This is required sectional side view.

Problem 10.42 (Fig. 10.51): *A cone, base diameter 40 mm and axis 65 mm long is resting on its base on the H.P. It is cut by a section plane perpendicular to the V.P. and parallel to and 10 mm away from one of its end generators. Draw its front view, sectional top view and true shape of the section.*

Construction (Fig. 10.51):

Front View and Top View

1. Draw front view and top view of the cone in given position.

Section Plane (VT)

2. Draw a line VT parallel to and 10 mm away from extreme generator o'1'.

Sectional Top View

3. In the front view, mark points p', q', r'.....etc., at which generators are cut by the section plane.
4. Project these points on the corresponding generators in the top view. Project points q' and q_1' on the corresponding generators as explained in problem 10.41. Join p, q, r...etc., by smooth curve and hatch the enclosed figure.

True Shape of the Section

5. Draw true shape of the section as explained in problem 10.29.

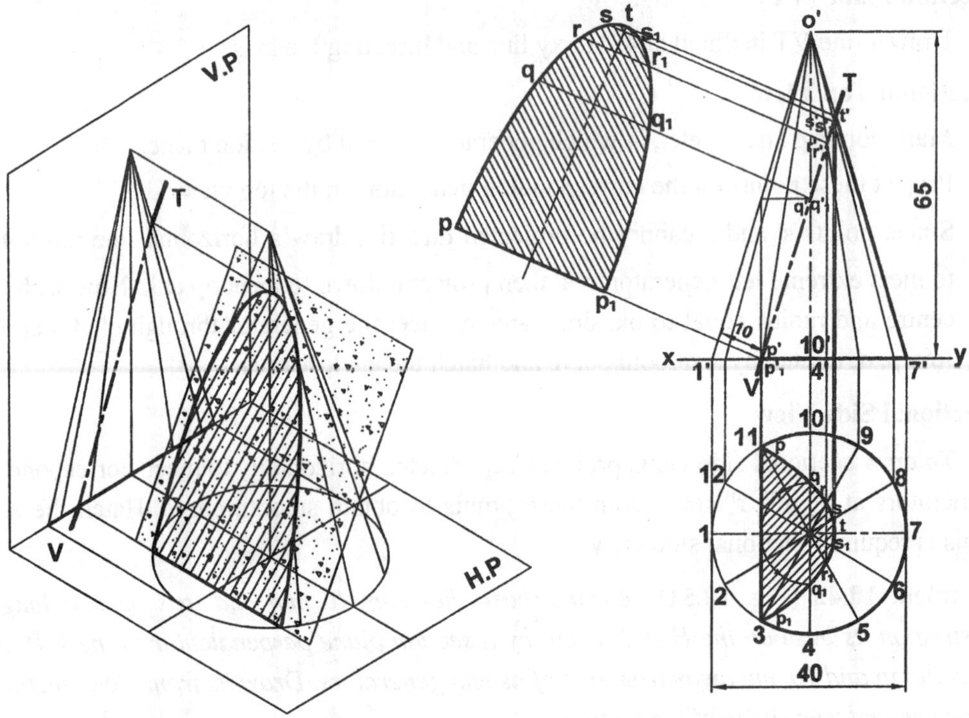

Fig. 10.51

10.6.4 Section Plane Perpendicular to H.P. and Inclined to the V.P. (A.V.P.)

Problem 10.43 (Fig. 10.52): *A cone having base diameter 50 mm and axis 75 mm long is resting on its base on the H.P. It is cut by a section plane inclined at 60° to the V.P. and perpendicular to the H.P. in such a way that section plane is 10 mm away from the axis. Draw its sectional front view and true shape of the section.*

Construction (Fig. 10.52):

Front View and Top View

1. Draw front view and top view of given cone in the required position.

Section Plane

2. Draw a line HT inclined at 60° to xy and 10 mm away from the axis.

Sectional Front View

3. Mark the points a, b, c...etc., at which generators are cut by section plane.

4. Project these points on the corresponding generators in the front view at a', b', c'...etc., respectively.
5. Join these points by a smooth curve and hatch the figure.

True Shape of the Section

6. Draw a new reference line x_1y_1 parallel to the HT. Through a, b, c...etc., draw projectors perpendicular to x_1y_1, and on these generators mark a_1', b_1', c_1'... etc., such that their distances from x_1y_1 are equal to their corresponding distances from xy in the front view. Join these points by smooth curve and hatch the figure as shown.

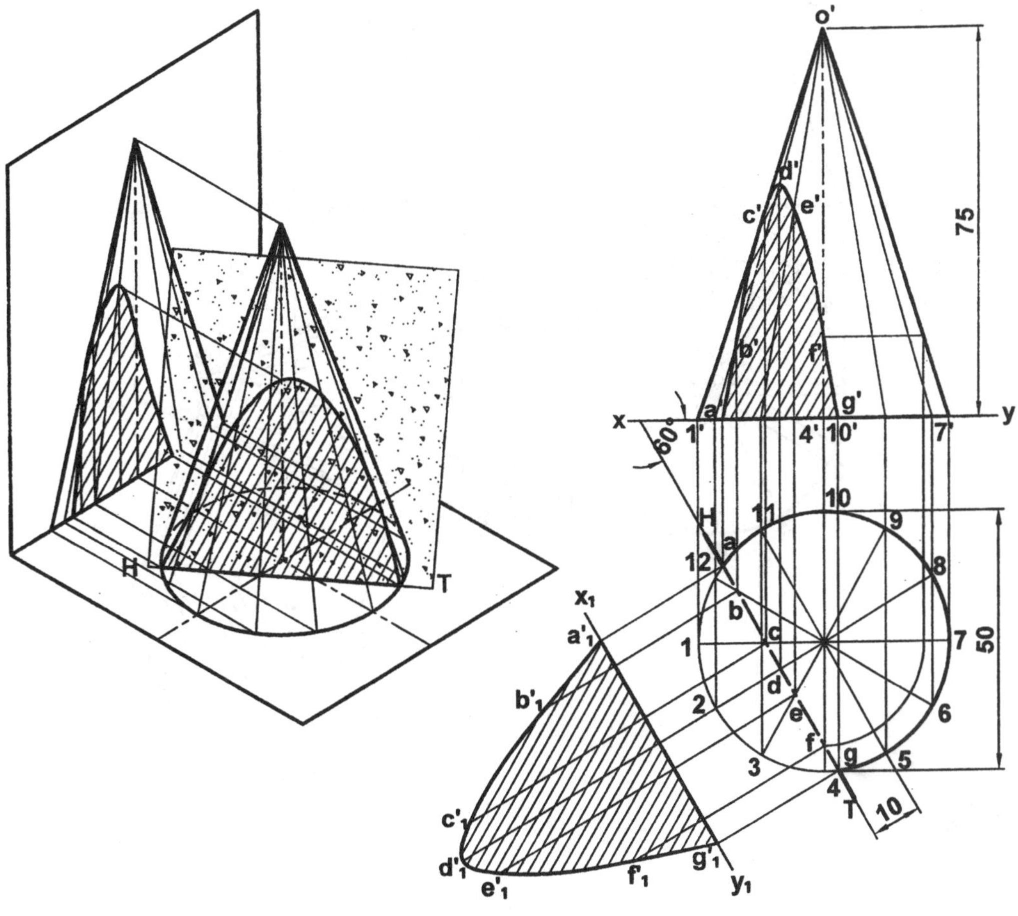

Fig. 10.52

Problem 10.44 (Fig. 10.53): *A cone with base diameter 50 mm and axis 65 mm long is lying on one of its generators on the H.P. with its axis parallel to the V.P. A vertical section plane parallel to the generator which is tangent to the ellipse (for the base) in the top view cuts the cone bisecting the axis and removing the portion containing the apex. Draw sectional front view, top view and true shape of the section.*

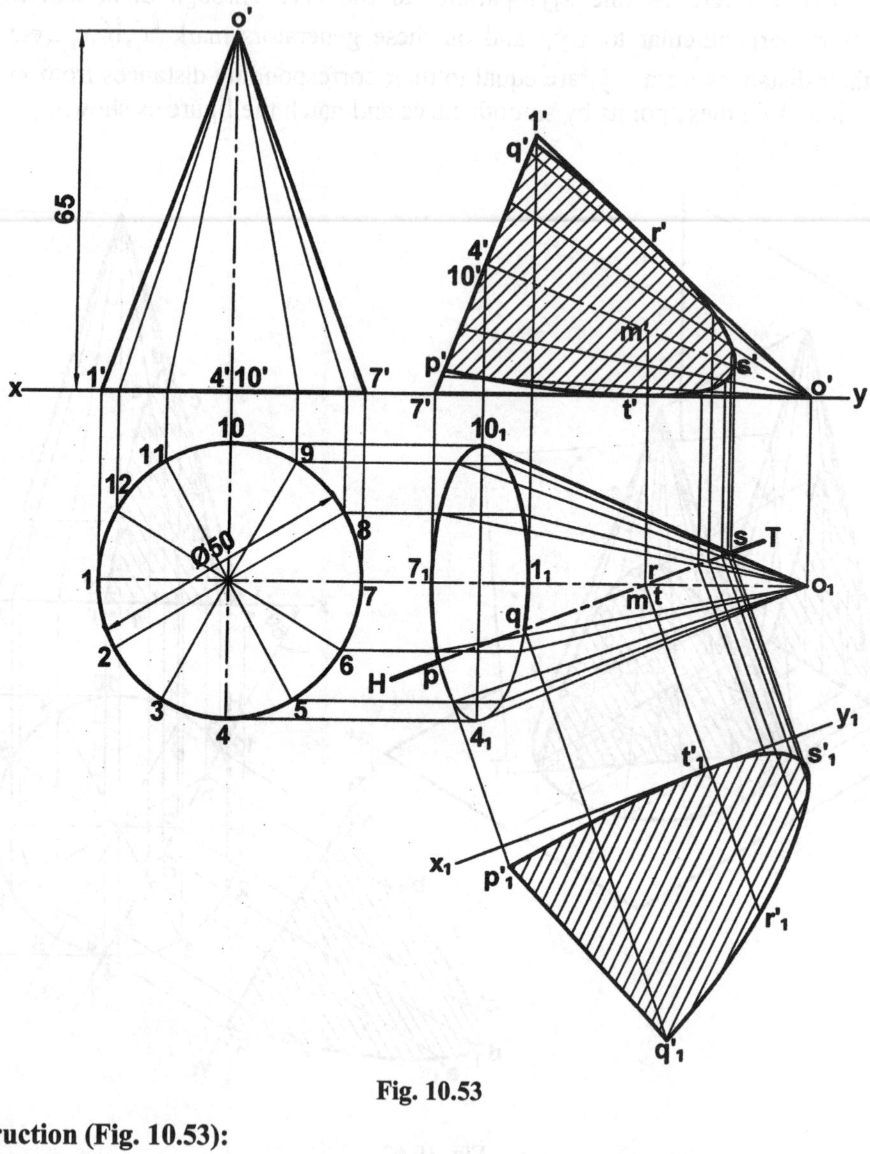

Fig. 10.53

Construction (Fig. 10.53):

Front View and Top View

1. Draw the projections of given cone in the required position.

Section Plane

2. In the front view, mark m' as midpoint of axis and project it to m in the top view. Draw a line HT passing through m (in the top view) and parallel to the generator $o_1 4_1$.

Sectional Front View

3. Mark p, q, r, s and t at which generators/points on base curve, are cut by the section plane.
4. Project these points at p', q', r' and s' respectively on the corresponding generators in the top view.
5. Join p', q', r', s' and t' and hatch the figure.

True Shape of the Section

6. Draw a new reference line x_1y_1. Through p, q, r, s and t, draw projectors perpendicular to x_1y_1 and on them mark $p_1', q_1', r_1' s_1'$...etc., such that their distances from x_1y_1 are equal to their corresponding distances from xy.

10.6.5 Section Plane Perpendicular to the H.P. and V.P.

Problem 10.45 (Fig. 10.54): *A cone having base diameter 40 mm and axis 65 mm long is resting on its base on the H.P. It is cut by a section plane perpendicular to the both H.P. and the V.P. and 10 mm away from the axis. Draw its elevation, plan and sectional side view.*

Construction (Fig. 10.54):

In this problem, the section plane is perpendicular to both H.P. and V.P. so neither front view nor top view shows sectional view. But section will be seen in side view which is obtained as follows:

Front View, Top View and Side View

1. Draw projections of the given cone in the required position.

Section Plane (VT & HT)

2. Draw a line representing VT and HT perpendicular to xy and 10 mm away from the axis. Mark the points p', q'...etc., at which generators are cut by section plane.

Sectional Side View

3. Through p', q', r'....etc., draw horizontal projectors to meet the corresponding generators at p", q", r"...etc., respectively in the side view.
4. Join p", q", r"...etc., by smooth curve. Hatch the figure enclosed by the curve.

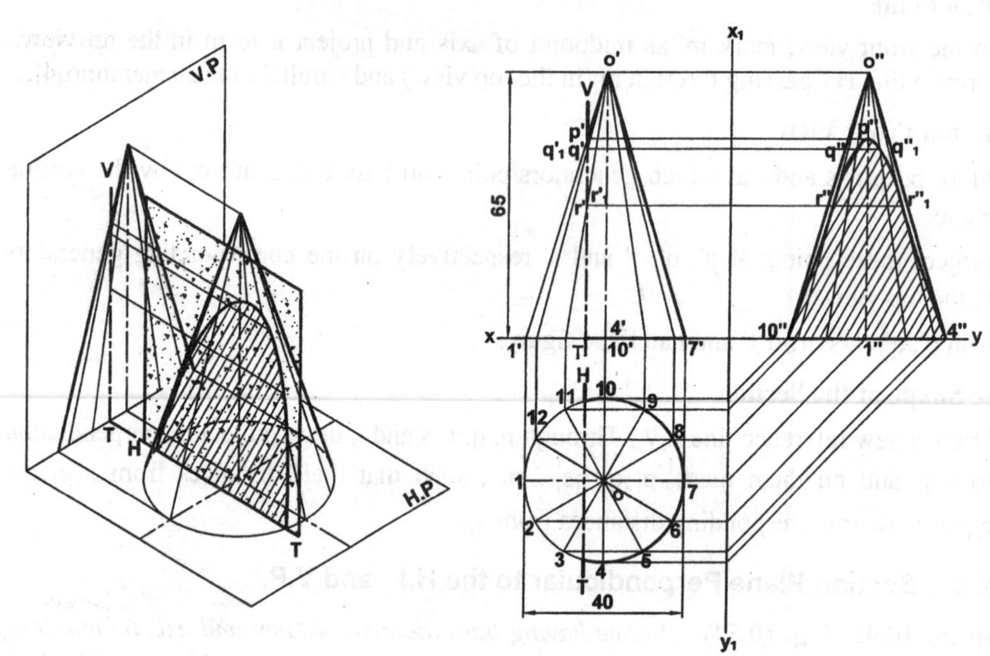

Fig. 10.54

10.7 Sections of Cylinder

In this section, method of obtaining section of cylinder has been explained with the help of some typical problems for following positions of section plane.

(a) Section plane parallel to the H.P.

(b) Section plane parallel to the V.P.

(c) Section plane perpendicular to the V.P. and inclined to the H.P.(A.I.P)

(d) Section plane perpendicular to the H.P. and inclined to the V.P.(A.V.P.)

10.7.1 Section Plane Parallel to the H.P.

Problem 10.46 (Fig. 10.55): *A cylinder having base diameter of 40 mm and axis 65 mm long is resting on its base on the H.P. It is cut by a section plane parallel to and 45 mm above the H.P. Draw its front view and sectional top view.*

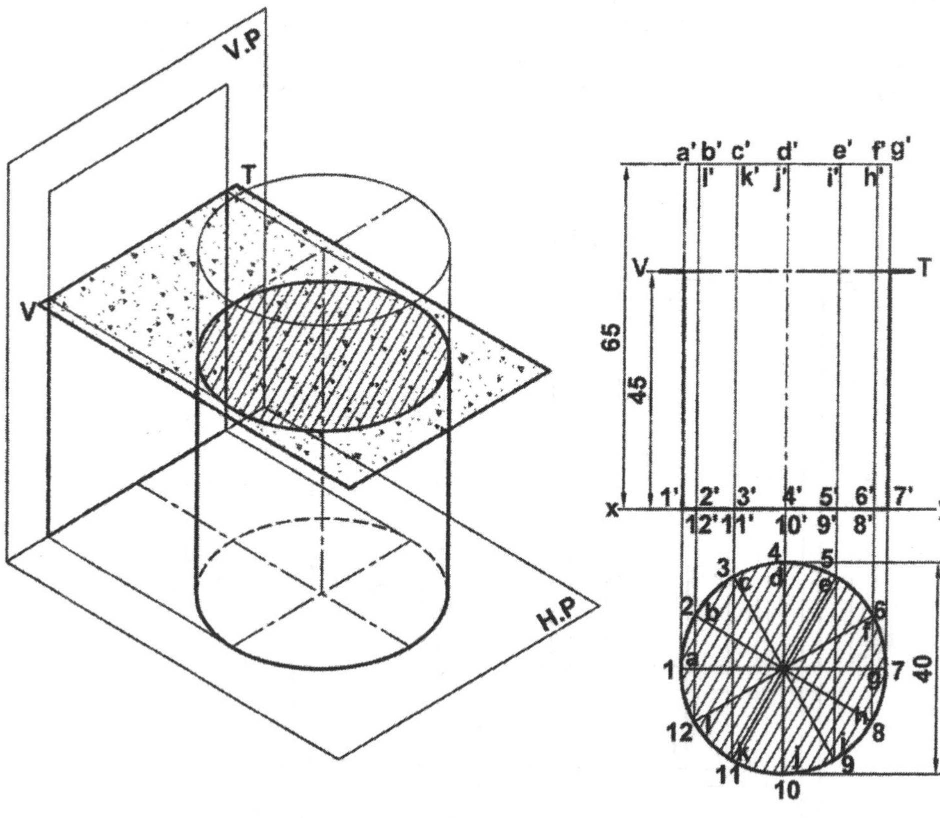

Fig. 10.55

Construction (Fig. 10.55):

In this problem cutting plane is parallel to H.P. therefore, it will cut all the generators of the cylinder. The sectional top view will also represent true shape of the section.

Front View and Top View

1. Draw top view and front view of given cylinder in the required position.

Section Plane (VT)

2. Draw a line VT parallel to xy and 45 mm above the base to represent section plane.

Sectional Top View

3. Since, all the generators are cut by the section plane, hatch the base circle which represents sectional top view.

Problem 10.47 (Fig. 10.56): *A hollow cylinder, having outside and inside diameters 50 and 30 mm respectively and axis 70 mm long is resting on a point on its base circle such that its base is inclined at 30° to the H.P. and axis is parallel to the V.P. It is cut by a horizontal section plane bisecting its axis. Draw its front view and sectional top view.*

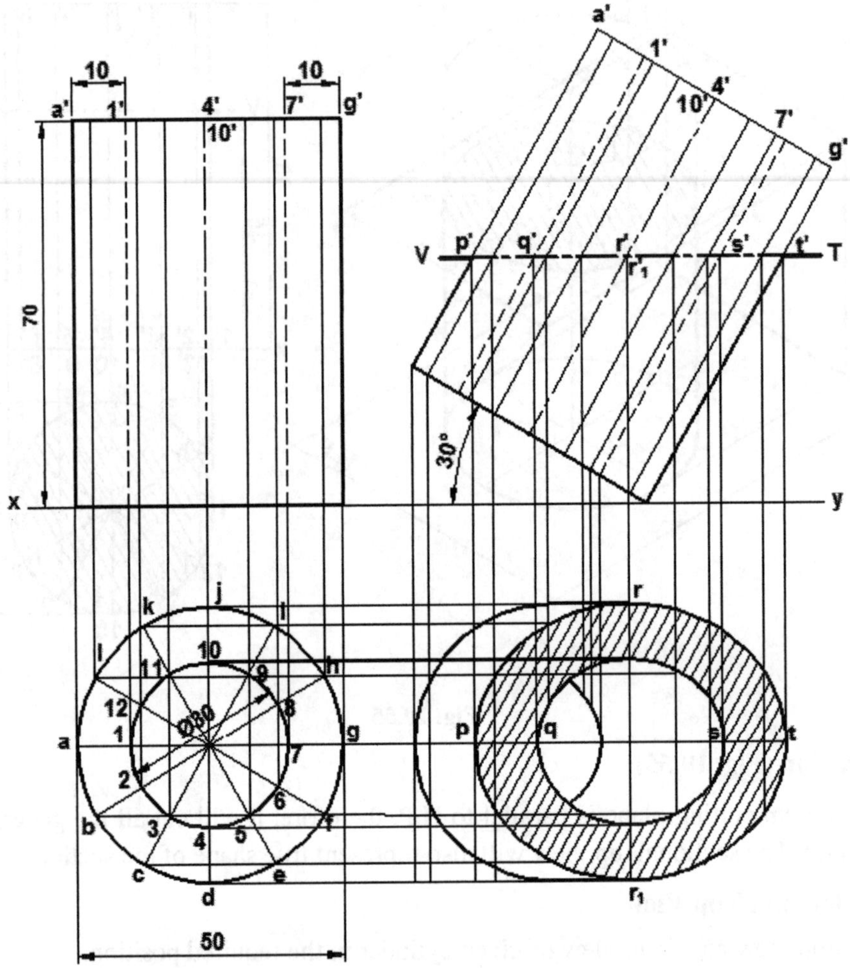

Fig. 10.56

Construction (Fig. 10.56):

Front View and Top View

Stage I

1. Draw two concentric circles of diameter 50 and 30 mm below xy line. Divide them in twelve equal parts. Name the division points 1, 2, 3...etc., and a, b, c...etc. Project these points upward.
2. Project the front view above xy and draw its generators as shown.

Stage II

3. Tilt the front view such that its base makes an angle of 30° with the xy line.
4. Project the final top view.

Section Plane (VT)

5. Draw a line VT parallel to xy and bisecting the axis of the cylinder.

Sectional Top View

6. Mark the points p', q', r'...etc., at which generators are cut by section plane.
7. Project these points on the corresponding generators in the top view and complete the sectional top view as shown in the figure.

10.7.2 Section Plane Parallel to the V.P.

Problem 10.48 (Fig. 10.57): *A cylinder having 40 mm base diameter and 60 mm long axis, is resting on its base on the H.P. It is cut by a section plane parallel to the V.P. and 10 mm away from the axis. Draw sectional front view and top view.*

Construction (Fig. 10.57):

Front View and Top View

1. Draw front view and top view of the cylinder in given position. Draw its generators.

Section Plane (HT)

2. Draw a line HT parallel to xy and 10 mm away from axis.

Sectional Front View

3. Mark the points p and q at which base circle is cut by section plane. Project p to p' and p_1' and q to q' and q_1' in the front view.
4. Join p' to p_1' and q' to q_1' and hatch the rectangle $p'p_1'q_1'q'$. This is the required sectional front view.

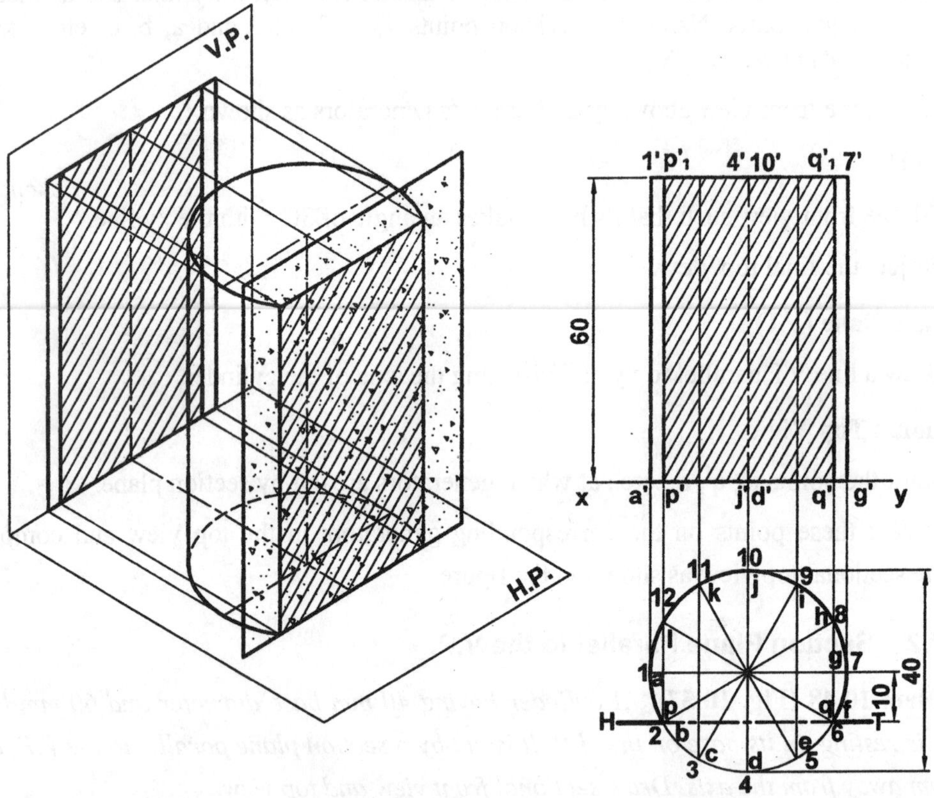

Fig. 10.57

10.7.3 Section Plane Perpendicular to the V.P. and Inclined to the H.P. (A.I.P)

Problem 10.49 (Fig. 10.58): *A cylinder having 40 mm base diameter and 65 mm height is resting on its base on the H.P. It is cut by a section plane perpendicular to the V.P. and inclined at 45° to the H.P. in such a way that it bisects the axis. Draw its front view, sectional top view, sectional side view and true shape of the section.*

Construction (Fig. 10.58):

Projections

1. Draw top view, front view and side view of given cylinder in the required positions.

Section Plane

2. Draw a line VT inclined at 45° to xy line and bisecting the axis in the front view.

Sectional Top View

3. In the front view mark the points p', q', r'...etc., at which generators are cut by the section plane.
4. Since all the generators of cylinders are cut by the section plane, a circle is seen in the top view. So hatch the base circle completely which represents sectional top view.

Sectional Side View

5. Through p', q', r'...etc., draw horizontal projectors to meet corresponding generators in the side view at p", q", r"...etc. Join these points to obtain a circle of 40 mm diameter.

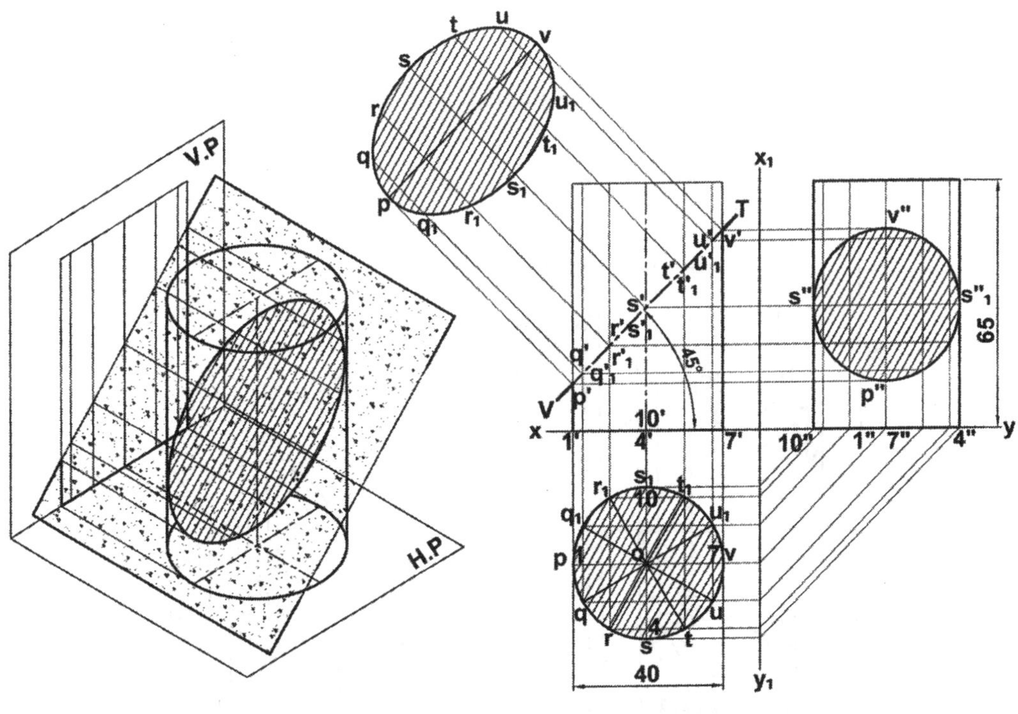

Fig. 10.58

Problem 10.50 (Fig. 10.59): *A cylinder with 40 mm base diameter and 65 mm long axis is resting on its base on the H.P. It is cut by a section plane perpendicular to V.P. whose V.T cuts the extreme left generator (in the front view) at a point 20 mm above the base and is inclined at 60° to the H.P. Draw its front view, sectional top view and true shape of the section.*

Construction (Fig. 10.59):

Front View and Top View

1. Draw top view (circle of 40 mm diameter) and front view of the given cylinder in the required position.

Section Plane (VT)

2. Draw a line VT inclined at 60° to xy line and passing through a point on the extreme left generator and 20 mm above the base.

Sectional Top View

3. Mark the points p', q', r'...etc., at which generators/points on base circular are cut by section plane.
4. Project these points on the corresponding generators in the top view.
5. Join t-t_1 and hatch the portion of circle towards left of t-t_1.

True Shape of the Section

6. Obtain true shape of the section as explained in problem 10.44.

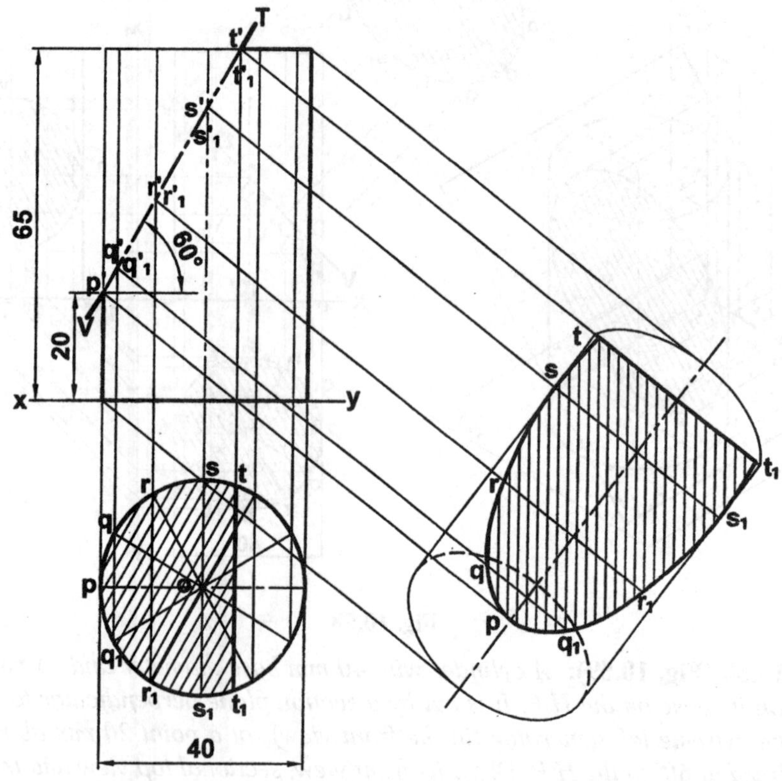

Fig. 10.59

Problem 10.51 (Fig. 10.60): *A cylinder resting on the ground is cut by a section plane to give the true shape of the section as an ellipse of minor axis 40 mm and major axis 70 mm. If this is the maximum possible length of the major axis, determine the diameter of the cylinder, the height of the cylinder and the inclination of section plane with H.P. Also draw front view, sectional top view and sectional side view.*

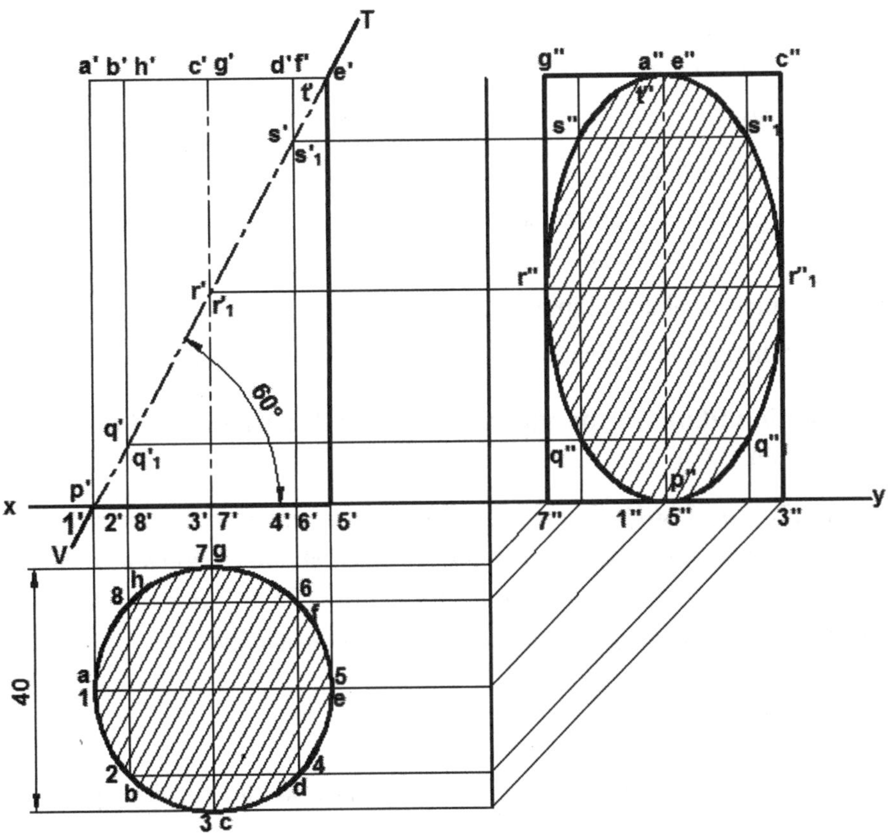

Fig. 10.60

Construction (Fig. 10.60):

Under the given conditions, the length of the minor axis of the ellipse will be equal to diameter of the cylinder (i.e., ϕ 40 mm) so begin with top view.

Sectional Top View

1. Draw a circle of 40 mm diameter in the top view. Divide it into eight equal parts. Name the division points and hatch the complete circle.

Section Plane and Front View

2. Project the extreme generator points say 1 and 5 to 1' and 5' on the xy.
3. Through 1' and 5' draw vertical lines of any convenient length.
4. With 1' as centre and radius equal to 70 mm (equal to major axis of ellipse in true shape) draw an arc cutting vertical line through 5' at e'. Complete the rectangle representing front view. Draw a line passing through 1' and e' which is the required VT. Draw its generators.

Sectional Side View

5. Project the side view and draw generators by projecting division points of base circle.
6. Through p', q', r'.....etc., draw horizontal projectors to meet corresponding generators at p", q", r"......etc., in the side view.
7. Join these points to obtain ellipse.
8. Height of cylinder and inclination of section plane can be obtained by measurement.

10.7.4 Section Plane Perpendicular to the H.P. and Inclined to the V.P. (A.V.P.)

Problem 10.52 (Fig. 10.61): *A cylinder having base diameter 40 mm and axis 60 mm long is resting on its base the H.P. It is cut by a section plane perpendicular to the H.P. and inclined at 60° to the V.P. in such a way that its H.T is 5 mm away from the axis. Draw sectional front view, top view and true shape of the section.*

Construction (Fig. 10.61):

Front View and Top View

1. Draw top view and front view of given cylinder in the required position. Draw generators of the cylinder.

Section Plane (HT)

2. Draw a line HT inclined at 60° to xy line and 5 mm away from the axis.

Sectional Front View

3. Mark the points p, q, r...etc., and p_1, q_1, r_1...etc., at which generators are cut by section plane.
4. Project points p, p_1, q and q_1 at p', p_1', q' and q_1' respectively in the front view. Draw the area enclosed by these points by dark lines and hatch it.

True Shape of the Section

5. Draw a new reference line x_1y_1 and project the true shape of the section on it.

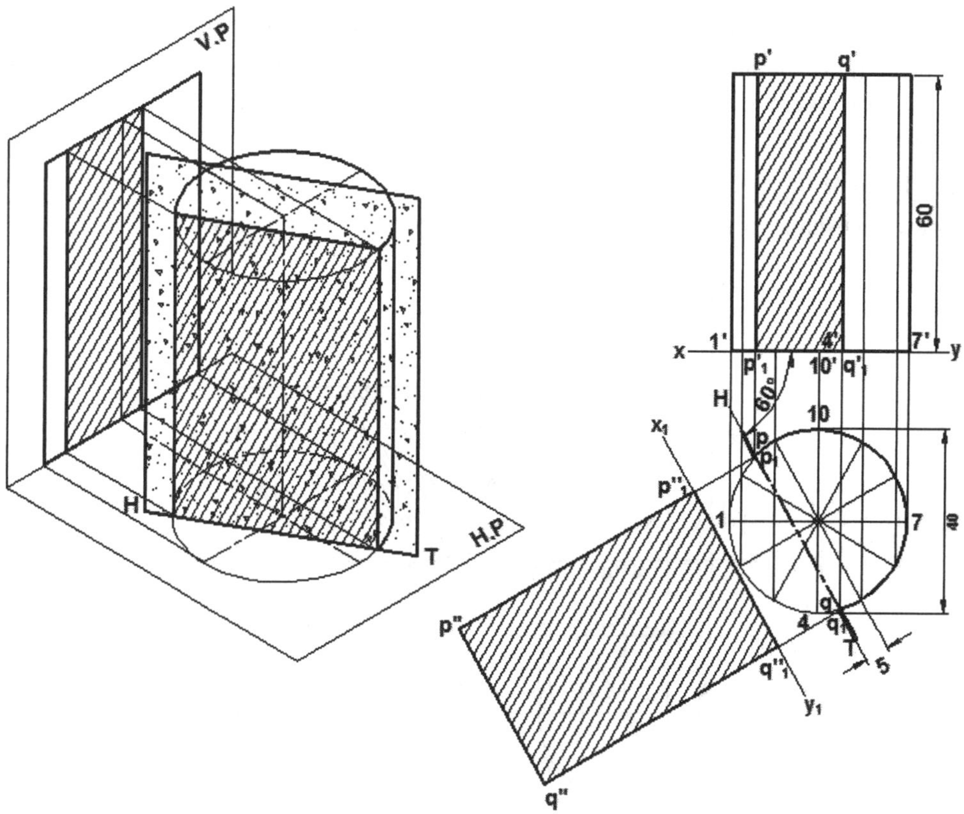

Fig. 10.61

Problem 10.53 (Fig. 10.62): *A cylinder, 50 mm base diameter and 70 mm long, is lying on one of its generator on the H.P. in such a way that its axis is parallel to both H.P. and V.P. It is cut by a vertical section plane inclined at 30° to the V.P. such that axis is cut at a point 20 mm from one of its end. Draw sectional front view, top view and true shape of the section.*

Construction (Fig. 10.62):

Front View and Top View

1. Draw front view and top view of given cylinder in the required position.

Section Plane (HT)

2. Draw a line HT inclined at 30° to xy and passing through a point on the axis 20 mm away from a base.

Sectional Front View

3. Mark the points p, q, r...etc., at which the generators are cut by section plane.
4. Project these points on the corresponding generators at p', q', r'...etc., in the front view. Join them in a correct order by a smooth curve. Hatch the area enclosed by curve.

True Shape of the Section

5. Draw true shape of the section as explained in previous problems.

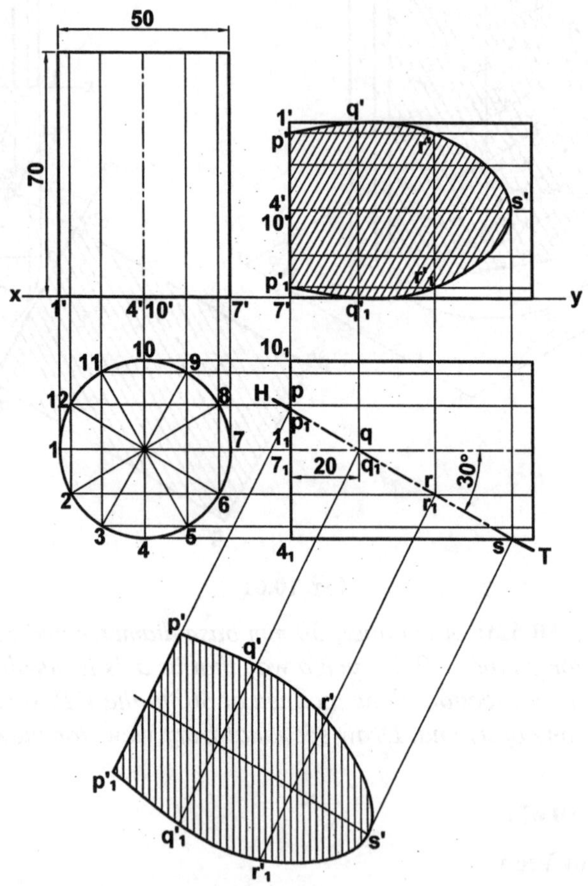

Fig. 10.62

Problem 10.54 (Fig. 10.63): *A cylinder, base diameter 40 mm and axis 65 mm long, has a square hole of 20 mm side cut through it such that the axis of the hole coincides with that of the cylinder. The cylinder is lying on the H.P. with the axis perpendicular to the V.P. and the faces of the holes are equally inclined to the H.P. It is cut by a vertical section plane inclined at 60° to the V.P. in two equal halves. Draw sectional front view and true shape of the section.*

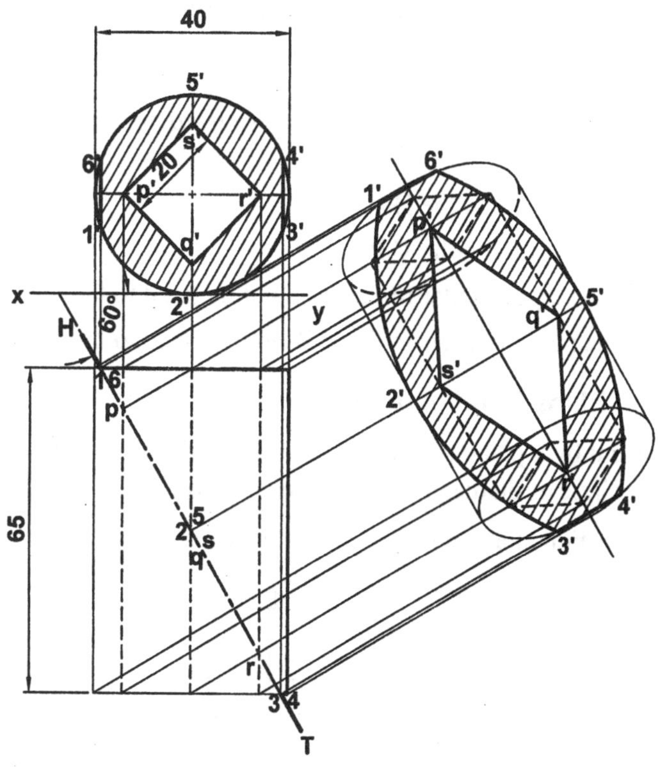

Fig. 10.63

Construction (Fig. 10.63):

Front View and Top View

1. Draw the front view and top view of the given cylinder having square hole in the given position.

Section Plane (HT)

2. Draw a line HT inclined at 60° to xy line and bisecting the axis.

Sectional Front View

3. Mark the points 1, 2, 3...etc., at which generators are cut by section plane in the top view.

4. Project points 1 and 6 to 1' and 6' respectively in the front view. Similarly project points 3 and 4 to 3' and 4' respectively in the front view.
5. Join 1'-6' and 3'-4'. Hatch the area enclosed by 1'-2'-3'-4'-5'-6'-1' leaving the square p'q'r's'.

True Shape of the Section

6. Project the true shape of the section as explained earlier.

10.8 Sections of Spheres

Methods of obtaining sections of spheres are discussed for following positions of the section plane.

(i) Section plane parallel to the H.P.

(ii) Section plane parallel to the V.P.

(iii) Section plane perpendicular to the V.P. and inclined to the H.P.

(iv) Section plane perpendicular to the H.P. and inclined to the V.P.

10.8.1 Section Plane Parallel to the H.P.

When a sphere is cut by a plane parallel to the H.P. as shown in Fig. 10.64 the true shape of the section is a circle of diameter equal to the chord p'p'.

The width of the section at any point say q' will be equal to the chord qq in the top view.

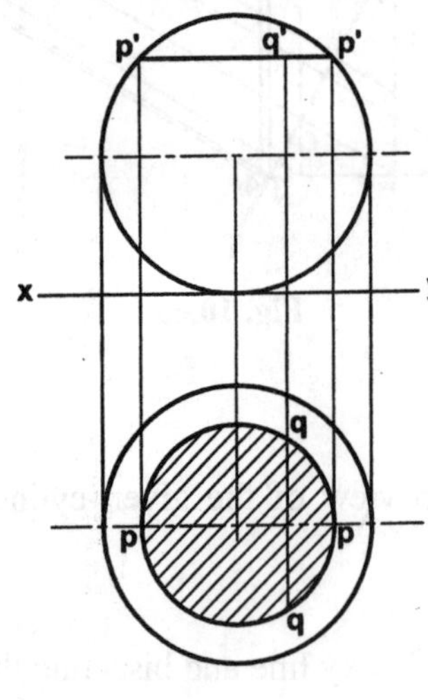

Fig. 10.64

10.8.2 Section Plane Parallel to the V.P.

When a section plane parallel to the V.P., cuts the sphere the true shape of the section is a circle of diameter equal to length of rr as shown in Fig.10.65 and the width of the section at any point say s is equal to the length of chord s's'.

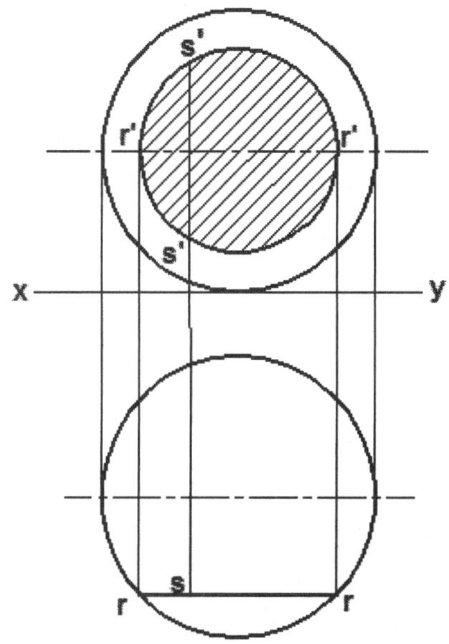

Fig. 10.65

10.8.3 Section Plane Perpendicular to the V.P. and Inclined to the H.P.

Problem 10.55 (Fig. 10.66): *A sphere of 45 mm diameter is resting on the H.P. It is cut by a section plane perpendicular to the V.P. and inclined at 45° to the H.P. and 10 mm away from its centre. Draw the sectional top view and true shape of the section.*

Construction (Fig. 10.66):

Front View and Top View

1. Draw front view and top view of the given sphere of 45 mm diameter.

Section Plane (VT)

2. Draw a circle of 15 mm diameter in the centre of bigger circle.
3. Draw a line VT inclined at 45° to xy and tangential to smaller circle.

True Shape of the Section

4. Mark the point 1' and 7' at which circle is cut by section plane.
5. Through 1' and 7', draw projectors perpendicular to VT and on them mark 1_1 and 7_1 such that the line 1_1-7_1 is parallel to VT.
6. With mid point of 1_1-7_1 as centre and radius equal to half of 1_1-7_1, draw a circle and hatch the area enclosed. Divide it into twelve equal parts. This is the required true shape of the section.

Sectional Top View

7. Through 1_1, 2_1, 3_1....etc., draw projectors perpendicular to x_1y_1 intersecting the VT at 1', 2', 3' etc., respectively. Through these points, draw vertical projectors and on them, mark point 1, 2, 3...etc., such that their distances from xy are equal to their corresponding distances from x_1y_1.
8. Join 1, 2, 3....etc., by smooth curve. Hatch the area enclosed, which is required sectional top view.

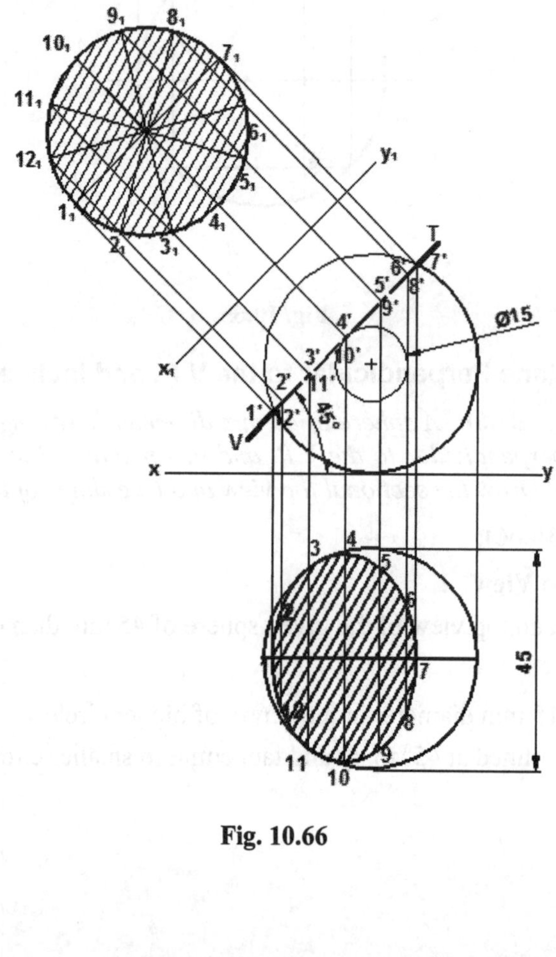

Fig. 10.66

10.8.4 Section Plane Perpendicular to the H.P. and Inclined to the V.P.

Problem 10.56 (Fig. 10.67): *A sphere of 45 mm diameter is resting on the H.P. It is cut by a plane perpendicular to the H.P. and inclined at 45° to the V.P. which passes through a point 10 mm away from the centre. Draw sectional front view, top view and true shape of the section.*

Construction (Fig. 10.67):

1. Draw front view and top view of the sphere in the required position.
2. In the top view draw line VT inclined at 45° and 10 mm away from the centre.
3. Project the true shape of the section (as centre) on a new reference line x_1y_1.
4. Finally obtain the sectional front view using the procedure explained in the problem 10.55.

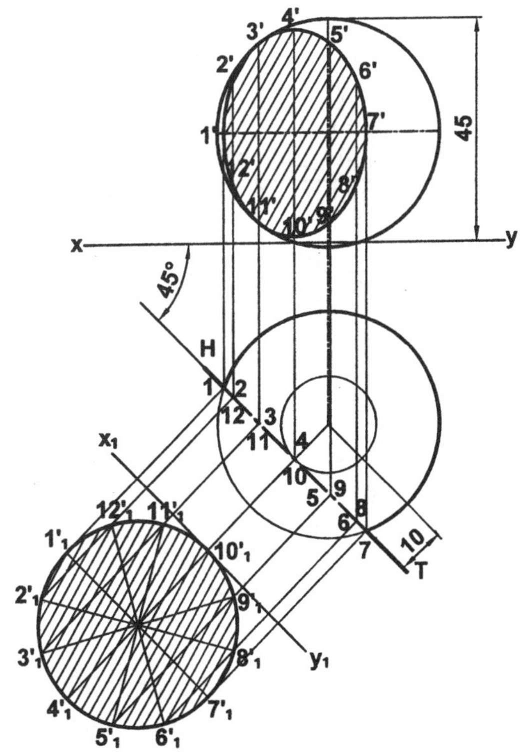

Fig. 10.67

Additional Problems

Problem 10.57 (Fig. 10.68): *A cone having 50 mm base diameter and 75 mm long is resting on its base on the H.P. An auxiliary inclined plane (A.I.P) cuts the cone in such a way that true shape of the section is an isosceles triangle of 40 mm base. Draw front view, sectional top view and true shape of the section.*

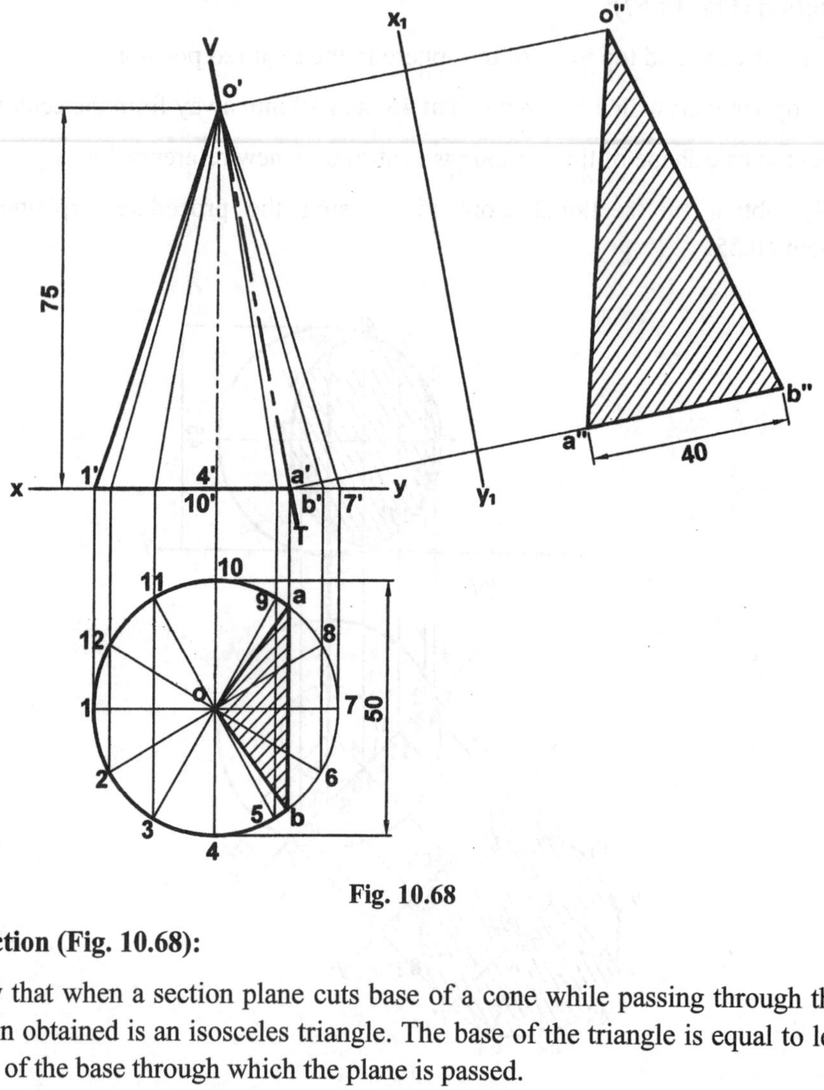

Fig. 10.68

Construction (Fig. 10.68):

We know that when a section plane cuts base of a cone while passing through the apex, the section obtained is an isosceles triangle. The base of the triangle is equal to length of the chord of the base through which the plane is passed.

1. Draw the top view and front view of given cone. Draw its generators in front view and top view.

2. In the top view (on base circle) draw a vertical chord ab whose length is equal to 40 mm (base of the required triangle). Join o to a and o to b. Hatch the triangle oab.
3. Project points a and b vertically on xy line as a' and b' respectively.
4. Join a' to o' and produce at both ends. This line represents VT of section plane.
5. Draw a new reference line x_1y_1 parallel to VT and project true shape of the section on it.

Problem 10.58 (Fig. 10.69): *A tetrahedron of 50 mm long edge is resting on one of its faces on the H.P. with an edge perpendicular to V.P. It is cut by a section plane perpendicular to V.P. and inclined to H.P. such that the true shape of the section is an isosceles triangle of 40 mm base and 30 mm altitude. Draw its front view, sectional top view and true shape of the section. Also determine the angle of inclination of section plane with the H.P.*

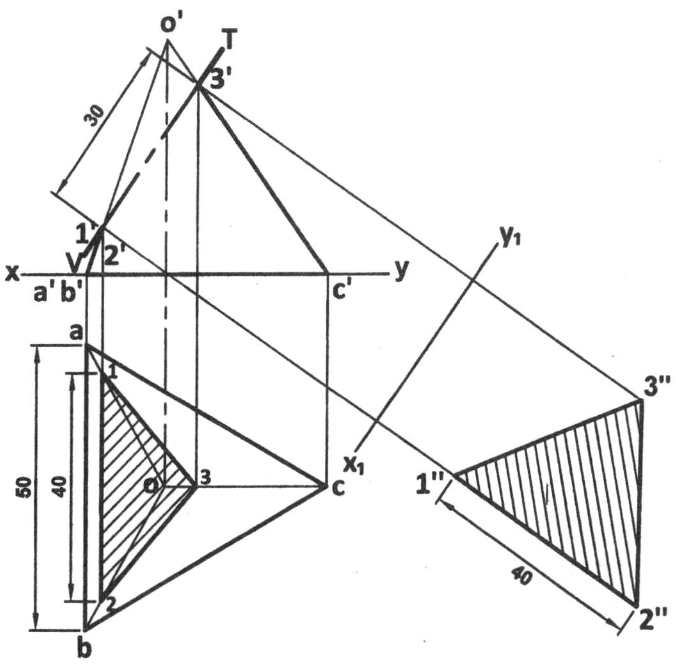

Fig. 10.69

Construction (Fig. 10.69):

Top View

1. Draw an equilateral triangle abc of 50 mm long edge below xy line. Locate its centre o and join it with corners a_1b_1.

Front View

2. Draw the axis through o and extend it above xy line.
3. Since edge oc is parallel to V.P. its front view represents true length. So, with c' as centre and radius equal to 50 mm (true length of edge) draw an arc on the axis at o'.
4. Complete the front view as shown.

Sectional Top View

5. On edges oa and ob mark points 1 and 2 respectively such that line 12 is equal to 40 mm.
6. Project these points on the corresponding edges in the front view at 1' and 2' respectively.
7. Mark point 3' on o'c' such that 1'3' = 30 mm. Project this point on the corresponding edge in the top view at 3. Hatch the area enclosed by triangle 123. The line through 1' and 3' is the required VT.

True Shape of the Section

8. Draw a new reference line x_1y_1 parallel to VT and project true shape of the section on it.

Problem 10.59 (Fig. 10.70): *A cylinder of 60 mm diameter is resting on its base on the H.P. It is cut by an auxiliary vertical plane (AVP) inclined at 30° to the V.P. such that the true shape of the section is a rectangle of 40 mm × 75 mm sides. Draw sectional front view, top view and true shape of the section.*

Construction (Fig. 10.70):

1. Draw a circle of 60 mm diameter below xy line to represent top view of the cylinder.
2. Project the front view considering height of cylinder equal to 75 mm. (since, the required section has length and width equal to 75 mm and 40 mm respectively).
3. Draw a line H.T. inclined at 30° to xy line such that the chord length is equal to 40 mm in the top view.
4. Mark points p, q, r and s at which the section plane (HT) cuts the circle.
5. Project these points on the bottom and top base circles.
6. Join p'-q'-r'-s'. Hatch the area p'q'r's'. Thus view above xy line represent sectional front view.
7. Draw a new reference line x_1y_1 parallel to HT and project the true shape of the section on it. Thus rectangle p"q"r"s" represents true shape of the section.

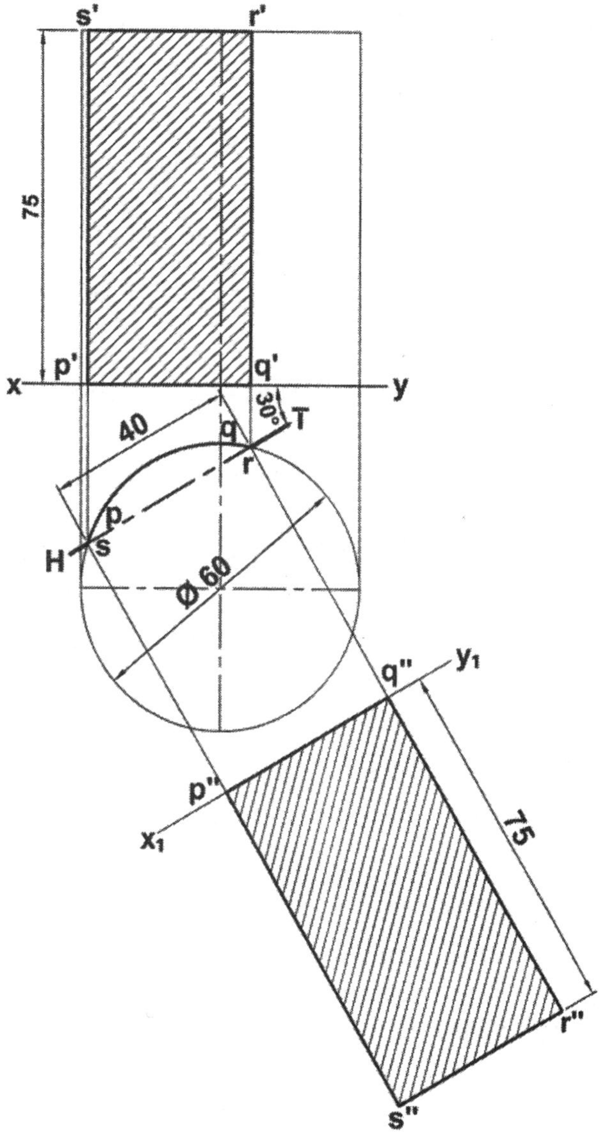

Fig. 10.70

10.9 Intersection of Cylinders

When one solid penetrates into another solid, line or curve which are formed are known as intersection of surfaces. Since, lateral surfaces of cylinders are curved the line of intersection between them, will also be curved.

The intersection of surfaces is required for the following engineering applications.

(i) Sheet metal work i.e., for fabrication of cylindrical ducts.

(ii) Boiler fittings e.g., tubes, shells etc.

(iii) Pipe fittings for various applications.

(iv) Construction of aircrafts components such as wings, etc.

(v) Construction of tanks with piping etc.

In case of a cylinder penetrating into another cylinder nature of curve of intersection depends on the size and relative position of their axes. There can be following typical cases.

10.9.1 Intersecting Cylinders have different Diameters and their Axes Intersect at Right Angle

In this case, the line of intersection will be a smooth curve. A convenient number of points (say 12) are required to obtain the line of intersection. The method of obtaining line/curve of intersection can be understood with the help of following examples.

Problem 10.60 (Fig. 10.71): *A cylinder of 50 mm diameter and 70 mm height is resting on its base on the H.P. It is completely penetrated by a horizontal cylinder of 36 mm diameter and 70 mm length such that their axes bisect each other at right angles. Draw its projections and curve of intersection.*

Construction (Fig. 10.71 & Fig. 10.72):

This problem can be solved by following two methods.

1. Line method
2. Cutting Plane method

1. **Line Method:** In this method, a number of lines (generators) are drawn on the surface of penetrating cylinder in the front view. Corresponding lines are also drawn on the surface of penetrating cylinder in the top view. In the top view these lines intersect the circumference of the circle (top view of vertical cylinder) at various points. Corresponding positions of these points are also marked in the front view. Finally these points are joined by a smooth curve which is the required curve of intersection or interpenetration. Consider the following steps (Fig. 10.71):

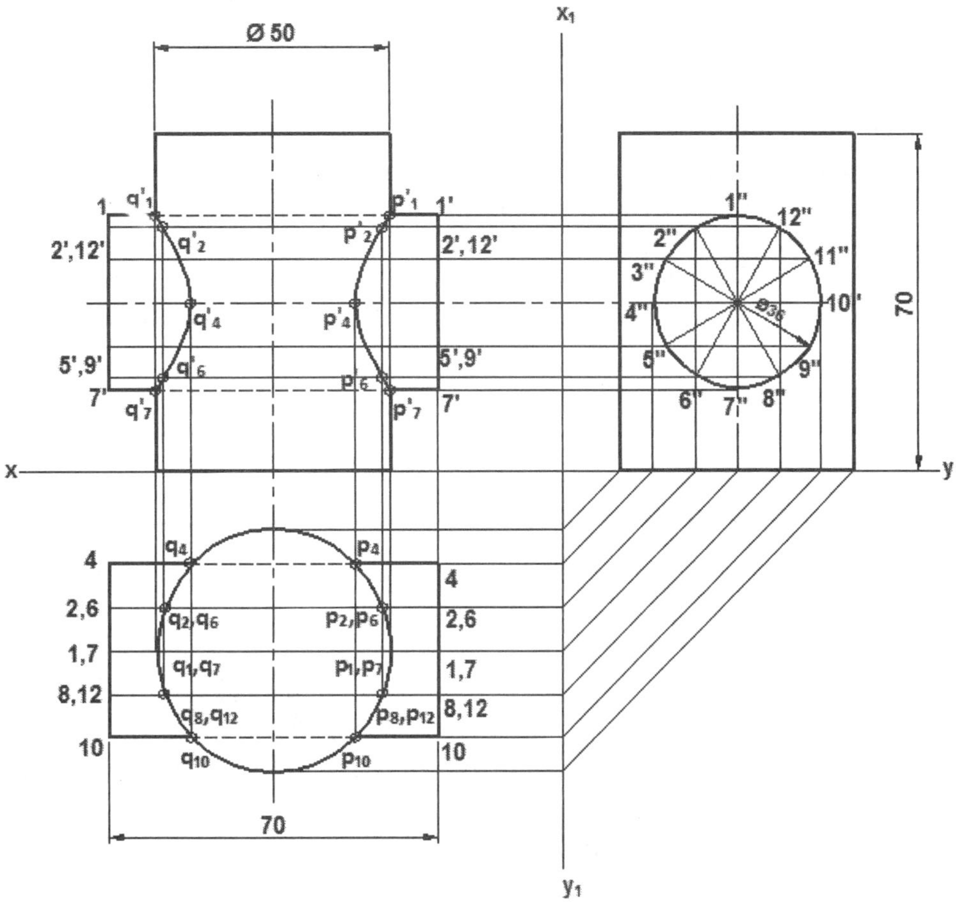

Fig. 10.71

1. Draw front view, top view and side view of given intersecting cylinders.
2. In the side view, divide the circle into twelve parts. Mark the division points as 1", 2", 3"...etc., and project them horizontally to obtain generators 1'-1', 2'-2', 3'-3', ...etc., on the surface of horizontal cylinder in the front view.
3. Through 1", 2", 3"...etc., (in the side view), draw vertical projectors to meet xy and then turn them at 45° to intersect x_1y_1 and then make them horizontal in the top view. Mark them 1-1, 2-2, 3-3...etc.
4. In the top view, mark the points of intersection of horizontal lines with the circle as p_1, p_2, p_3, ... etc., on one side of the axis and q_1, q_2, q_3... etc., on other side of the axis.

5. Project these points on the corresponding lines in the front view as p_1', p_2', p_3'... etc., and q_1', q_2', q_3'... etc. These points are called as key points.
6. Join these points by a smooth curve on both sides of the axis. These curves are required curves of intersection.

2. **Cutting Plane Method:** In this method, a series of horizontal cutting planes passing through the lines on the horizontal cylinder are assumed. These sections of horizontal cylinder are rectangles of different width while sections of vertical cylinder are circles of same diameter. The points at which sides of a rectangle intersects the circle are marked. These points lie on the curve of intersection. Consider the following steps (Fig. 10.72)

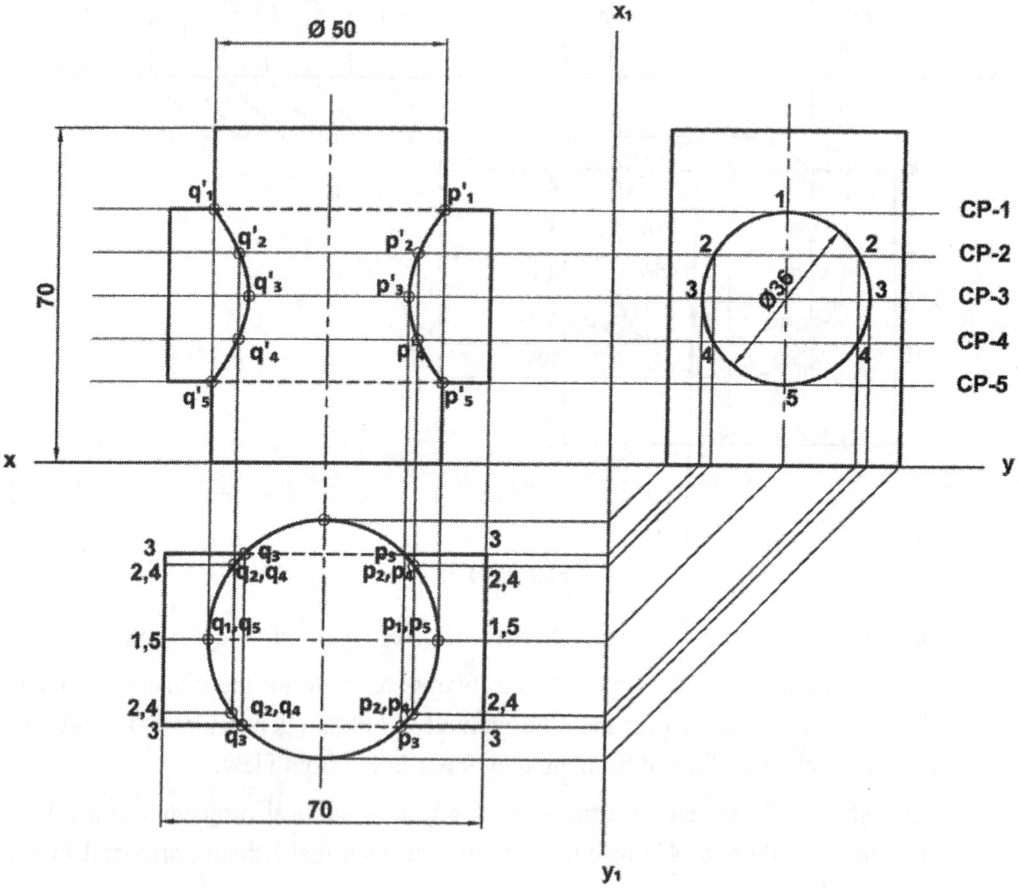

Fig. 10.72

1. Draw front view, top view and side view of intersecting cylinders.
2. Draw a number of horizontal lines CP-1, CP-2, CP-3 .... etc., representing cutting planes.
3. Mark the points at which sides of the rectangles intersect the circle as 2-2, 3-3 4-4 etc.
4. Project these points to p_1, p_2, p_3 ....etc., on one side of the axis and q_1, q_2, q_3 etc., on other side of the axis in the top view.
5. Project p_1, p_2, p_3, etc., to p_1', p_2', p_3'... etc., in the front view.
6. Join these points in a correct sequence by smooth curves. These curves are the required curves of intersection.

10.9.2 Intersecting Cylinders have same Diameters and their Axes Intersect at Right Angles

In this case, the line of intersection will be two straight lines. Four key points on the circumference of horizontal cylinder are sufficient to draw the lines of intersection.

Problem 10.61 (Fig. 10.73): *A cylinder of 50 mm diameter and 80 mm height is resting on its base on the H.P. It is completely penetrated by a horizontal cylinder of same diameter and same length such that their axes bisect each other at right angles. Draw its projections and lines of intersection of two cylinders.*

Construction (Fig. 10.73):

1. Draw front view, top view and side view of given intersecting cylinders.
2. In the side view mark the key points 1", 2", 3"...etc., on the circumference of penetrating cylinder.
3. Through points 1", 2", 3" and 4", draw horizontal lines 1'-1', 2'-2', 3'-3' and 4'-4' respectively.
4. Draw corresponding horizontal lines in the top view also.
5. Mark the points of intersection of these lines with the circle in the top view as p_1, p_2, p_3.... etc., on one side of the axis and q_1, q_2, q_3,.... etc., on other side of the axis.
6. Project these points on the corresponding lines in the front view and join them in a correct sequence by straight lines.

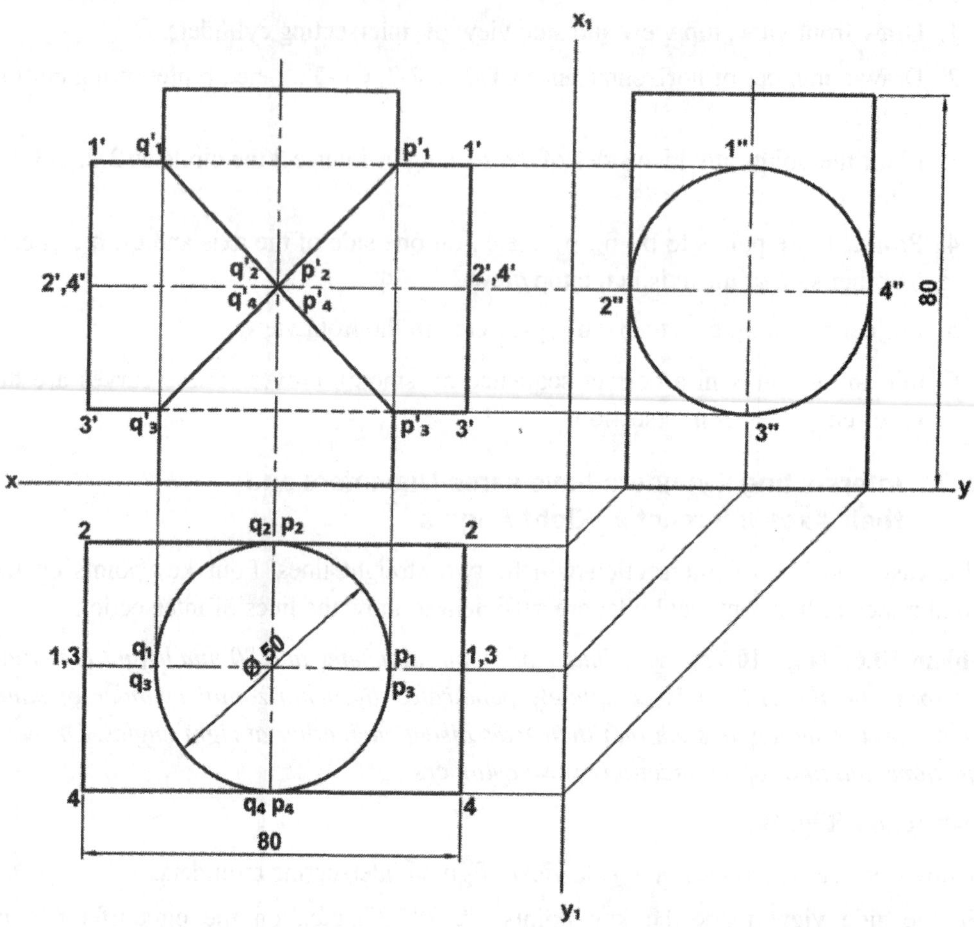

Fig. 10.73

10.9.3 Intersecting Cylinders have different Diameters and their Axes do not Intersect each other

In this case, the intersection of curves are obtained in the same way as explained in Problem 10.60.

Problem 10.62 (Fig. 10.74): *A cylinder of 50 mm diameter and 70 mm height is resting on its base on the H.P. It is completely penetrated by another horizontal cylinder of 30 mm diameter and 70 mm length such that axis of horizontal cylinder is 6 mm away from the axis of vertical cylinder. Draw its projections showing curves of intersection.*

Construction (Fig. 10.74):

1. Draw front view, top view and side view of given intersecting cylinders such that the axis of penetrating cylinder is 6 mm away from the axis of vertical cylinder.
2. In the side view mark the key points 1", 2", 3"... etc., on the circumference of penetrating cylinder. Project them in the front view and the top view.
3. In the top view, draw horizontal lines 1-1, 2-2, 3-3 ... etc., and in the front view, draw horizontal lines 1'-1', 2'-2'...etc.
4. Mark p_1, p_2, p_3 etc., on one side of the axis and q_1, q_2, q_3 etc., on other side of the axis in the top view at which the horizontal lines meet the circle.
5. Project p_1, p_2, p_3 ... etc., to, p'_1, p'_2, p'_3... etc., respectively and q_1, q_2, q_3 ...etc., to q'_1, q'_2, q'_3... etc., respectively in the front view.
6. Join these points by smooth curves which are required curves of intersection.

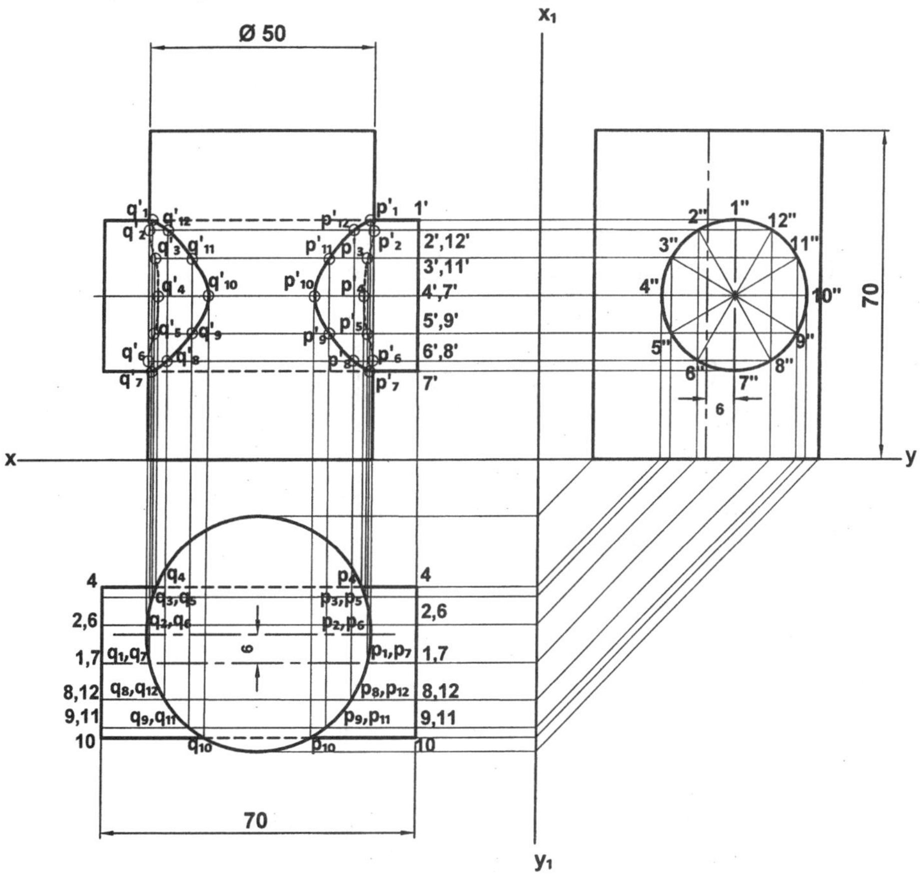

Fig. 10.74

Problem 10.63 (Fig. 10.75): *A cylinder of 50 mm diameter and 80 mm height is resting on its base on the H.P. It is completely penetrated by another cylinder of same size. The axis of penetrating cylinder is parallel to both H.P. and V.P. and is 10 mm away from the axis of vertical cylinder. Draw the projections of the solids showing curves of intersection.*

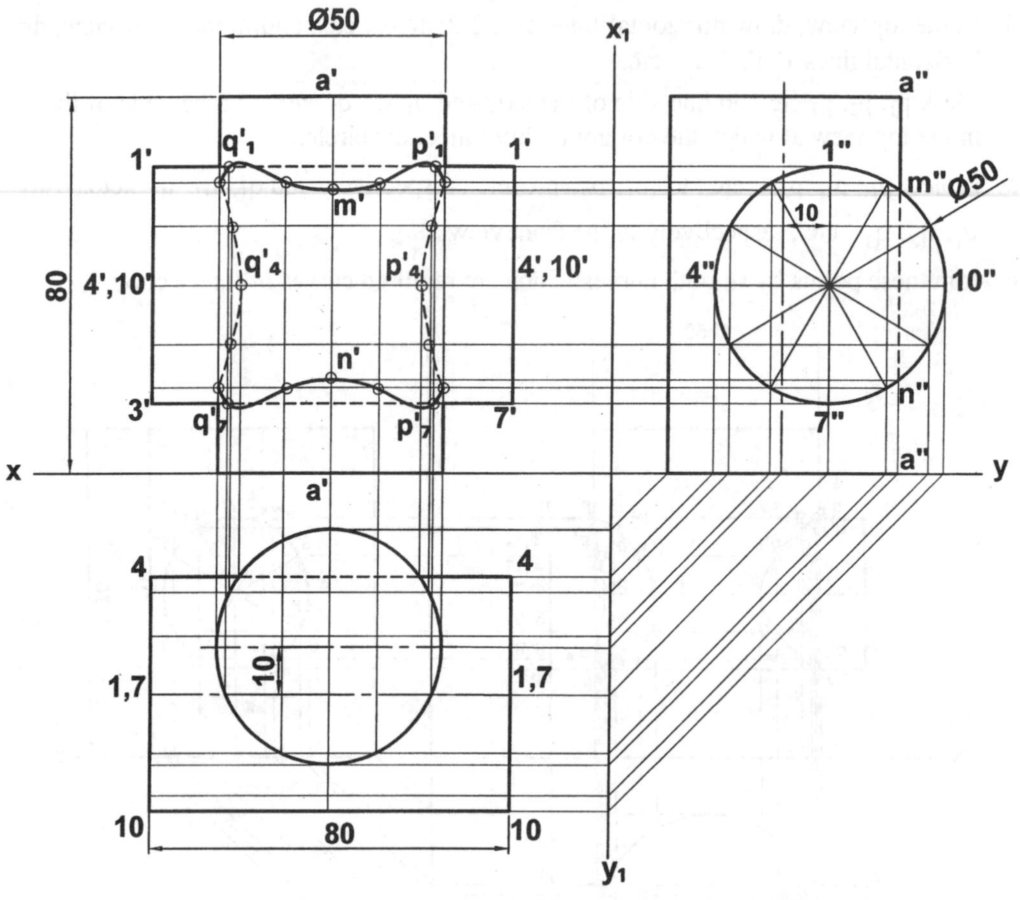

Fig. 10.75

Construction (Fig. 10.75):

1. Draw front view, top view and side view of given intersecting cylinders such that the axis of penetrating cylinder is 10 mm away from the axis of vertical cylinder.
2. In the side view mark the key points 1", 2", 3"...etc., on the circumference of penetrating cylinder. Project them in the front view and the top view.
3. Draw horizontal lines 1-1, 2-2, 3-3... etc., in the top view and horizontal lines 1'-1', 2'-2', 3'-3' ... etc., in the front view.

4. Mark two additional key points m'' and n'' on a''a''.
5. Project m'' and n'' on the corresponding lines at m' and n' respectively in the front view. Mark key points p_1', p_2', p_3'... etc., and q_1', q_2', q_3'... etc., as explained earlier.
6. Join these points by smooth curves.

Problem 10.64 (Fig. 10.76): *A cylinder of 50 mm diameter and 70 mm height is resting on its base on the H.P. It is penetrated by another cylinder of 30 mm diameter and 90 mm long axis such that the axis of penetrating cylinder is parallel to V.P., bisects the axis of the vertical cylinder and inclined at 60° to the axis of vertical cylinder. Draw its projections showing curves of intersection.*

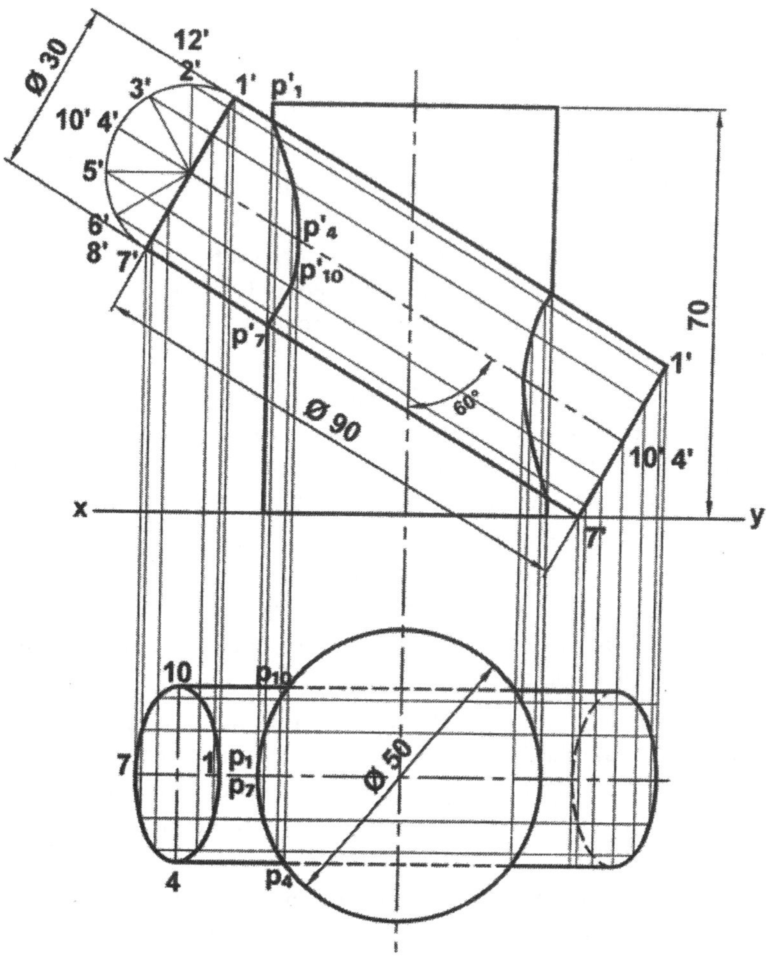

Fig. 10.76

Construction (Fig. 10.76):

1. Draw the front view and top view of intersecting cylinders in the required positions.
2. Draw a number of generators on the surface of inclined cylinder.
3. Project them in the top view.
4. Mark the key points p_1, p_2, p_3...etc., at which generators intersect the circle in the top view.
5. Project these points on the corresponding generators at p_1', p_2', p_3'... etc., respectively.
6. Join these points by smooth curves. Similarly curve of intersection may be obtained on the right side of the vertical axis.

Exercises - (a)

1. A cube having side of base 40 mm, is resting on one of its base side on the H.P. with a face containing that edge is inclined at 45° to the H.P. It is cut by a horizontal section plane 25 mm above the H.P. Draw its front view and sectional top view.
2. A pentagonal pyramid base side 30 mm and axis 60 mm long, resting on its base on the H.P. with an edge of base inclined at 30° to V.P. It is cut by a horizontal section plane at a distance of 30 mm above the base. Draw its front view and sectional top view.
3. A hexagonal prism, side of base 30 mm and axis 65 mm long, is resting on one of its base corners on the H.P. with a rectangular face parallel to the V.P. The axis is inclined at 60° to the H.P. It is cut by a horizontal section plane bisecting the axis. Draw its sectional top view and true shape of the section.
4. A hexagonal pyramid, side of base 30 mm and axis 65 mm long is resting on one of its corners on the H.P. with the longer edge containing that corner inclined at 60° to the H.P. and axis parallel to the V.P. It is cut by a horizontal section plane in such a way that it divides the axis in two equal halves. Draw the front view, sectional top view and true shape of the section.
5. A square pyramid, base 40 mm side and axis 70 mm long, is resting on the H.P. on one of its triangular faces, the top view of the axis making an angle of 30° with the V.P. It is cut by a horizontal section plane such that the V.T. of which intersects the axis at a point 10 mm above the base. Draw its front view and sectional top view.
6. A hexagonal prism, side of base 30 mm and axis 70 mm long, is resting on a corner of its base on the H.P. with its axis inclined at 45° to the H.P. and parallel to the V.P. It is cut by a horizontal section plane such that it divides the prism in two equal halves. Draw its front view and sectional top view.

7. Hexagonal prism, side of base 25 mm and axis 65 mm long, is lying on its rectangular face on the H.P. with its axis parallel to the V.P. It is cut by a section plane, the H.T. of which makes an angle of 45° with the xy line and bisects the axis. Draw the sectional front view and true shape of the section.
8. A triangular pyramid, side of base 30 mm and axis 60 mm long, is lying on one of its triangular faces on the H.P. A section plane parallel to V.P. and perpendicular to H.P.cuts the pyramid at a distance of 15 mm from the axis. Draw the sectional front view and top view.
9. A hexagonal prism, side of base 25 mm and axis 60 mm long, is resting on its base on the H.P. It is cut by a section plane parallel to the V.P. and 12 mm in front of the axis. Draw its top view and sectional front view.
10. A cube of 40 mm sides is resting on its base on the H.P. with a vertical face inclined at 30° to the V.P. A section plane parallel to V.P. and perpendicular to H.P. cuts the cube at a distance of 12 mm from the axis. Draw its top view and sectional front view.
11. A cone, base diameter 60 mm and axis 70 mm long, is lying on its generator on the H.P. with its axis parallel to the V.P. A section plane parallel to V.P. and perpendicular to the H.P. cuts the cone at a distance of 10 mm away from the axis. Draw its sectional front view.
12. A square prism, side of base 30 mm and axis 70 mm long, is resting on its base on the H.P. with a vertical face parallel to the V.P. It is cut by a section plane perpendicular to H.P. and inclined at 30° to V.P. and passing through the axis. Obtain its sectional front view and true shape of the section.
13. A pentagonal pyramid, side of base 30 mm and axis 70 mm long, has its base on the H.P. and an edge of base is inclined at 30° to V.P. It is cut by a section plane perpendicular to the H.P. and inclined at 30° to the V.P. at a distance of 10 mm from the axis. Draw its sectional front view and true shape of the section.
14. A right circular cone, base diameter 40 mm and height 65 mm, is resting on its base on the H.P. It is cut by a section plane, the H.T. of which makes an angle of 30° with the V.P. and 10 mm away from the top view of the axis. Draw its sectional front view and true shape of the section.
15. A cylinder, with a 50 mm base diameter and axis 70 mm long is resting on its base on the H.P. It is cut by a section plane, the HT of which makes an angle of 45° with the reference line and passes through a point at a distance of 15 mm from the axis. Draw its sectional front view, top view and true shape of the section.
16. A cylinder with a base diameter of 50 mm and axis 75 mm long, is lying on a generator with its axis parallel to the V.P. It is cut by a section plane perpendicular to the H.P. and inclined at 30° to the V.P. such that it divides the axis in two halves.

17. A pentagonal pyramid, side of base 30 mm and height 70 mm, is lying on one of its triangular faces on the H.P. with its axis parallel to V.P. A section plane inclined at 30° to the V.P. and perpendicular to the H.P. cuts the pyramid and passes through a point 20 mm from the apex along the axis. Draw the sectional front view and true shape of the section.
18. A pentagonal prism, side of base 40 mm and axis 70 mm long has a rectangular face on the H.P. with its axis parallel to the V.P. It is cut by a vertical section plane, the H.T. of which makes an angle of 30° with the reference line and bisects the axis. Draw its sectional front view, top view and true shape of the section.
19. A hollow square prism of 6 mm thickness, 35 mm external edge of base and axis 65 mm long, is resting on its base on the H.P. with its vertical face inclined at 30° to the V.P. It is cut by a section plane perpendicular to V.P. and inclined at 45° to the H.P. and passing through a point distant 10 mm below top end of the axis. Draw its sectional top view and true shape of the section.
20. A hollow cylinder, outer diameter 50 mm, inner diameter 40 mm and axis 70 mm long is resting on its base on the H.P. It is cut by a section plane perpendicular to V.P. and inclined at 30° to the H.P. and passing through a point on the axis 15 mm from its top end. Draw its sectional top view, sectional side view and true shape of the section.
21. A hexagonal prism, side of base 30 mm and axis 65 mm long, is resting on one of its rectangular faces on the H.P. with its axis inclined at 30° to the V.P. A section plane inclined at 45° to the H.P. and perpendicular to the V.P. cuts the prism such that it passes through a point on the axis at a distance of 30 mm from one end nearer to V.P. Draw the sectional top view and true shape of the section.
22. A cone, base diameter 50 mm and axis 70 mm long, is resting on its base on the H.P. It is cut by a section plane perpendicular to V.P. and inclined at 40° to H.P. and bisecting the axis. Draw its sectional top view and true shape of the section.
23. A cone having diameter of base 40 mm and axis 70 mm long, is lying on one of its generator on the H.P. with its axis parallel to the V.P. It is cut by a section plane perpendicular to the V.P. and inclined at 30° to the H.P. bisecting the axis. Draw the sectional top view and true shape of the section.
24. A sphere of 55 mm diameter is resting on the H.P. It is cut by a section plane perpendicular to the V.P. and inclined at 45° to H.P. and 10 mm away from the centre of the sphere. Draw the sectional top view and true shape of the section.
25. A pentagonal pyramid, side of base 30 mm and axis 70 mm long is resting on its base on the H.P. with an edge of base parallel to the V.P. It is cut by a section plane perpendicular to V.P. and inclined at 45° to the H.P. and bisecting the axis. Draw its sectional top view and true shape of the section.

26. A sphere having 60 mm diameter, is resting on the H.P. It is cut by a section plane, perpendicular to the V.P. and inclined at 45° to the H.P. such that the true shape of the section is a circle of 50 mm diameter. Draw its front view, sectional front view and sectional side view.

27. A hexagonal pyramid, side of base 30 mm and axis 70 mm long is resting on a triangular face on the H.P. with its axis parallel to the H.P. It is cut by a section plane perpendicular to V.P. and inclined at 45° to the H.P. and bisecting the axis. Draw the sectional top view and true shape of the section.

28. A hexagonal pyramid, side of base 25 mm and axis 60 mm long is resting on its base on the H.P. with a base side parallel to V.P. A section plane perpendicular to the H.P. and inclined at 45° to the V.P. cuts the pyramid at a distance of 10 mm from the axis. Draw its sectional front view and true shape of the section.

29. A square pyramid, side of base 30 mm axis 70 mm long, is resting on its base on the H.P. with a base side parallel to the V.P. It is cut by a section plane perpendicular to the V.P. and inclined at 45° to H.P. such that its axis is divided into two halves. Draw its sectional top view, sectional side view and true shape of the section.

30. A pentagonal pyramid, side of base 25 mm and axis 65 mm long is resting on its base on the H.P. with an edge of base perpendicular to V.P. It is cut by a section plane perpendicular to V.P. and inclined at 45° to the H.P. and cuts the axis at a point 20 mm above the base. Draw the sectional top view, sectional side view and true shape of the section.

31. A cone having base diameter 50 mm and axis 70 mm long, is resting on its base on the H.P. It is cut by a section plane, the HT of which makes an angle of 30° with the xy line and 12 mm away from the axis. Draw its sectional front view and true shape of the section.

32. A pentagonal pyramid side of base 30 mm and axis 65 mm long is resting on its base on the H.P. with an edge of base parallel to the V.P. A horizontal section plane cuts the pyramid at distance of 35 mm from the base. Draw its sectional top view.

33. A cone having base diameter 50 mm and axis 70 mm long is resting on its base on the H.P. it is cut by a section plane perpendicular to V.P. and inclined at 45° to H.P. which bisects the axis. Draw its sectional top view and true shape of the section.

34. A cone having diameter of base 50 mm and axis 70 mm long is lying on its generator on the H.P. with its axis parallel to the V.P. It is cut by a section plane, the HT of which makes an angle of 45° with the reference line and bisecting the

axis such that the apex is removed. Draw its sectional front view and true shape of the section.

35. A cylinder, diameter of base 60 mm and axis 80 mm long, is lying in one of its generators on the H.P. with its axis inclined at 30° to the V.P. It is cut by a vertical section plane in such a way that the true shape of the section is an ellipse of major axis 70 mm long. Draw its sectional front view and true shape of the section.

36. A cube having 60 mm long edges is resting on its base on the H.P. with its vertical faces equally inclined to the V.P. It is cut by a section plane perpendicular to V.P. such that the true shape of the section is a regular hexagon. Determine the angle of inclination of cutting plane with H.P. and draw its sectional top view and true shape of the section.

Exercises - (b)

1. A vertical cylinder of 70 mm diameter is completely penetrated by another cylinder of 54 mm diameter. Their axes bisect each other at right angles. If the axis of penetrating cylinder is parallel to the V.P., draw projections of the arrangement showing curves of intersection.
2. Two cylinders having base diameter 40 mm and height 80 mm, penetrate each other such that the axes bisect each other at right angle. Draw the curves of penetration.
3. A vertical cylinder having base diameter 75 mm, is penetrated by another cylinder of 50 mm diameter, the axis of which is parallel to both the H.P. and V.P. The two axes are 10 mm apart. Draw the projections showing curves of intersection.
4. A cylinder of 50 mm diameter and 60 mm height is standing on its base on the H.P. It is penetrated by a horizontal cylinder of 35 mm diameter and 70 mm height such that their axes bisect each other at right angles and are parallel to the V.P. Draw the projections of arrangement showing curves of intersection.

CHAPTER 11

Development of Surfaces

11.1 Introduction

When an object covered or wrapped by thin sheet of paper, or metal is unwrapped and covering is laid on a flat plane, the flattened out cover (paper or sheet) is known as development of surfaces of the object. Fig 11.1 shows a square prism covered with paper being opened out. Flattened out paper is the development of square prism which consists of four similar rectangles and two squares.

The knowledge of development of surfaces is very much essential for manufacturing of various objects such as automobiles, ships, aircrafts, boxes of sheet metals, packaging, boilers, hopper, funnels, chimney etc. Before manufacturing the sheet metal objects, it is required to mark the dimensions and cut the sheet in required size and shape. After cutting the sheet, fabrication process is completed. Prior knowledge of development helps to minimize manufacturing time, cost and labour work. Various objects may have plane surfaces singly curved surface or doubly curved surfaces. Based on nature of surface, the suitable method for development is adopted.

11.2 Methods of Development

Following are principle methods of development.

1. **Parallel Line Method:** This method is employed for the development of objects such as prisms and cylinders in which stretch-out principle is used. The line equal to the perimeter or circumference of the object is known as stretch-out line. In this method, two parallel lines (stretch out lines) are drawn and faces of the prism are marked between stretch-out lines to represent development of lateral part of prism. In case of cylinder, the stretch out line is equal to circumference (π D) of base of the cylinder. Fig. 11.1 shows development of a square prism using this method.

2. **Radial Line Method:** Radial line method is used for the development of cones and pyramids. In this method, the true length of the slant edge or generator is used as radius. An arc is drawn using this radius and edges or generators are marked on this arc to represents development of pyramid or cone.

3. **Triangular Method:** This method is adopted for development of transition pieces. In this method, a surface is divided into a number of triangles which are transformed into development.

4. **Approximate Method:** The approximate method is used to develop objects of double curved surfaces such as sphere, parabolic, ellipsoid etc.

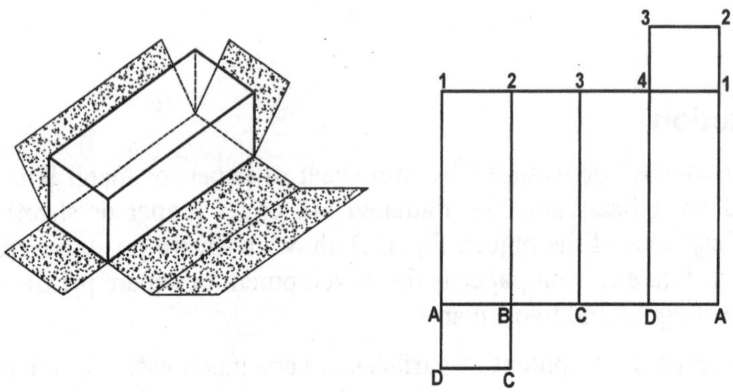

Fig. 11.1

11.3 Development of Cube and Prism

The development of the surface of cube comprises of six equal squares. Cubes are developed using parallel line method. Two parallel lines (stretch out lines) are drawn first. Four faces are marked between the parallel lines to represent development of lateral part of the cube.

Development of prism are drawn by parallel line method. Given front view and top view of the prism are drawn first. Then two parallel stretch out lines are drawn whose length is equal to the perimeter of its base. On these parallel lines, faces of the prism are marked, which shows development of lateral part of the prism.

For truncated prism, point of intersection of cutting plane and the vertical edges, are marked, then horizontal lines through these points are drawn to meet the corresponding edges. Points obtained on the edges (drawn between stretch out lines) are joined to get final development.

Points to Remember

- Every line on the development must represent the true length of the corresponding edge.
- For a closed object, first and last corner or edge is given the same name.
- Usually lateral surface of the prism are developed. The ends may be omitted unless it is required.

Problem 11.1 (Fig.11.2): *Draw the development of a triangular prism having 30 mm edge of base and axis 65 mm long.*

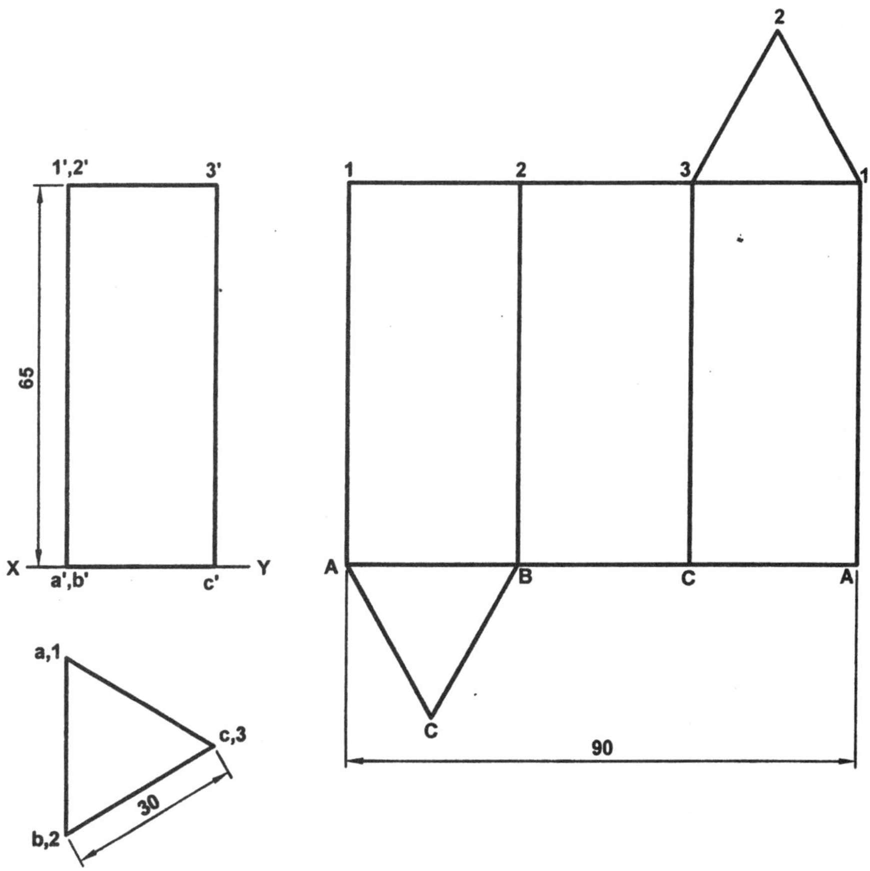

Fig. 11.2

Construction (Fig. 11.2):

1. Draw front view and top view of a triangular prism as shown in the figure.
2. Draw stretch out lines A-A and 1-1 equal to the perimeter of base.

3. Divide A-A and 1-1 into three equal parts. Name the intermediates divisions as shown.
4. Draw lines A-1, B-2, C-3 and again A-1.
5. Attach triangles ABC and 123 as top and bottom bases.

Problem 11.2 (Fig. 11.3): *Draw the development of the lateral surface of the part B of a triangular prism having 30 mm edge of base and 65 mm long axis as shown in Fig. 11.3.*

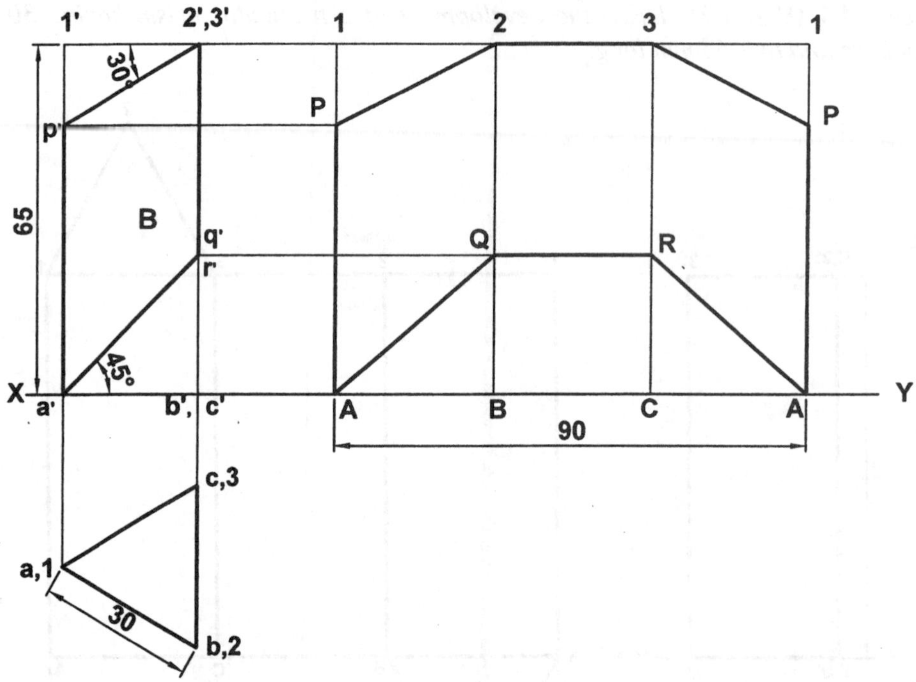

Fig. 11.3

Construction (Fig. 11.3):

1. Draw front view and top view of a triangular prism of given dimensions. Name all the corners.
2. Draw the cutting planes in the front view.
3. In the front view mark cutting points p', q' and r' on the vertical edges.
4. Draw stretch out lines A-A and 1-1 equal to 90 mm (perimeter of base), in line with the front view and divide each of them in three equal parts. Name the intermediate division as shown in the figure.
5. Draw lines A-1, B-2, C-3.

6. Draw horizontal lines through p', q' and r' to intersects the vertical edges at P, Q and R respectively.
7. Keeping the lines for removed part fainter, complete the development.

Problem 11.3 (Fig. 11.4): *Draw the development of the surface of the part A of a cube of 40 mm sides, the front view of which is shown in Fig. 11.4.*

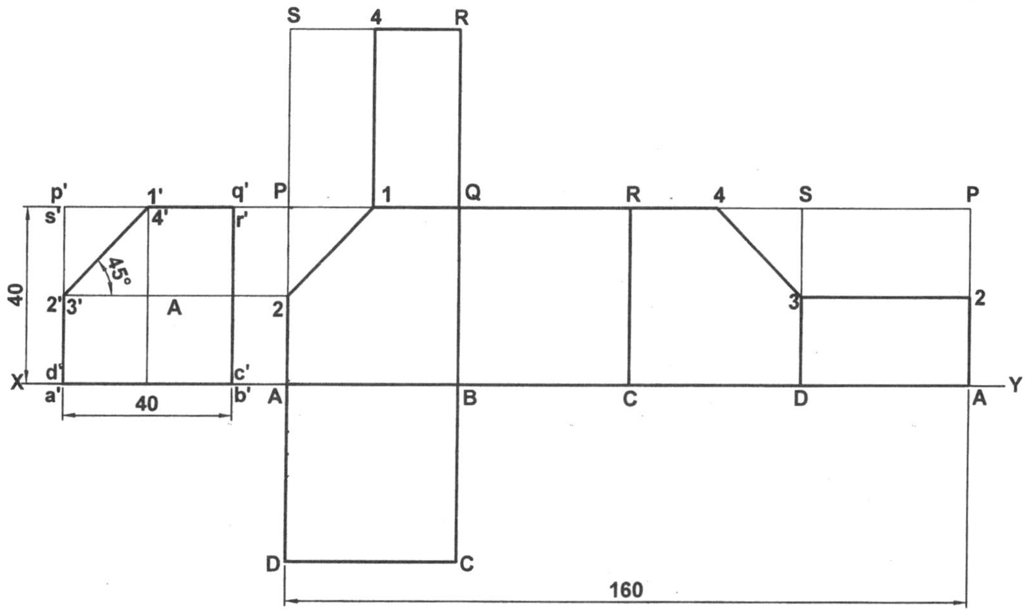

Fig. 11.4

Construction (Fig. 11.4):

1. Draw front view of a cube having 40 mm sides and draw a cutting plane line 1'2' or 3'4'.
2. Draw stretch out lines A-A and P-P. Divide these lines into four equal parts. Draw vertical lines AP, BQ, CR and DS. Attach two faces.
3. Mark points 1, 2, 3 and 4 on the edges such that, P1 = p'1', P2 = p'2', D3 = d'3' and R4 = r'4'.
4. Join 1-2, 1-4, 3-4, 2-3. Keeping the lines for the removed portion thin and fainter, complete the development.

Problem 11.4 (Fig. 11.5): *A cube of base edge 30 mm is resting on the H.P. with its all faces equally inclined to the V.P. A section plane perpendicular to the V.P. and inclined* *to H.P. passes through the diagonally opposite corner. Draw its projections and development of surfaces.*

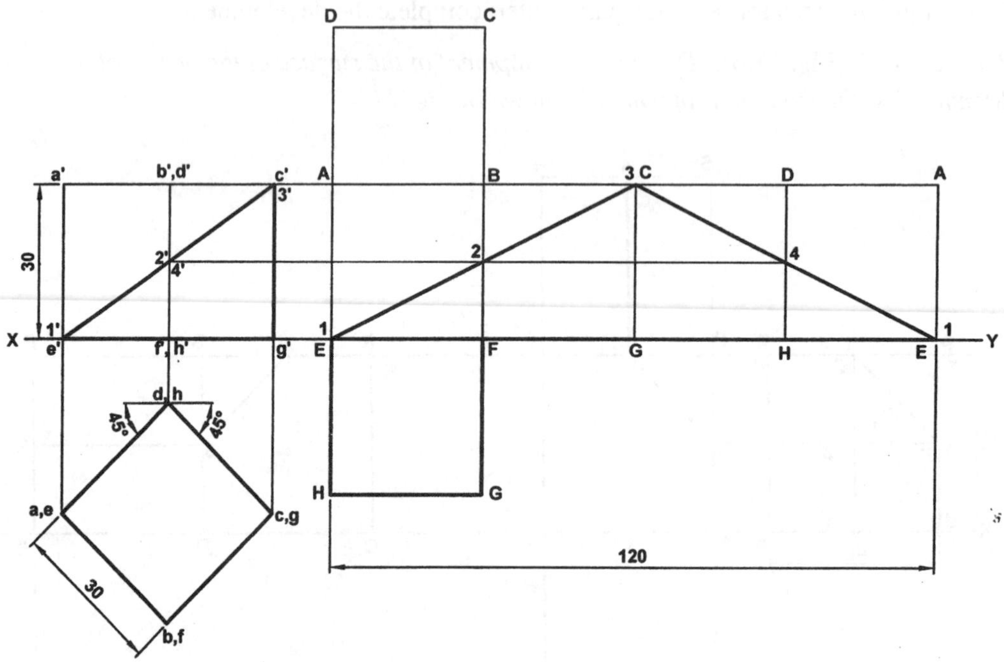

Fig. 11.5

Construction (Fig. 11.5):

1. Draw front view and top view of a cube of 30 mm sides with its vertical faces equally inclined to V.P.
2. Draw the cutting plane (V.T.) and mark its points of intersection with the vertical edges as 1', 2', 3' and 4'.
3. Draw stretch out lines A-A and E-E each equal to perimeter of base (i.e., 120 mm) and divide these lines into four equal parts. Name the intermediate divisions as shown. Through A, B, C...etc., draw vertical lines to meet at E, F, G....etc., respectively.
4. Attach two square ABCD and EFGH thus completing development of cube considering it to be whole.
5. Through 1', 2', 3' and 4', draw horizontal lines to cut the corresponding vertical edges at 1, 2, 3, and 4 respectively.
6. Join these points and complete the development keeping the removed portion thin and fainter.

Problem 11.5 (Fig.11.6): *Draw the development of a square prism of 30 mm base edge and 70 mm long axis.*

Construction (Fig. 11.6):

1. Draw front view and top view of square prism of given dimensions with a side of base parallel to the V.P.
2. Draw stretch out lines A-A and 1-1 in line with front view having length equal to perimeter of base (i.e., 120 mm) and divide them into four equal parts. Name intermediate divisions as shown.
3. Join A-1, B-2, C-3, D-4 and A-1. Draw two squares ABCD and 1234 to represent top and bottom faces.

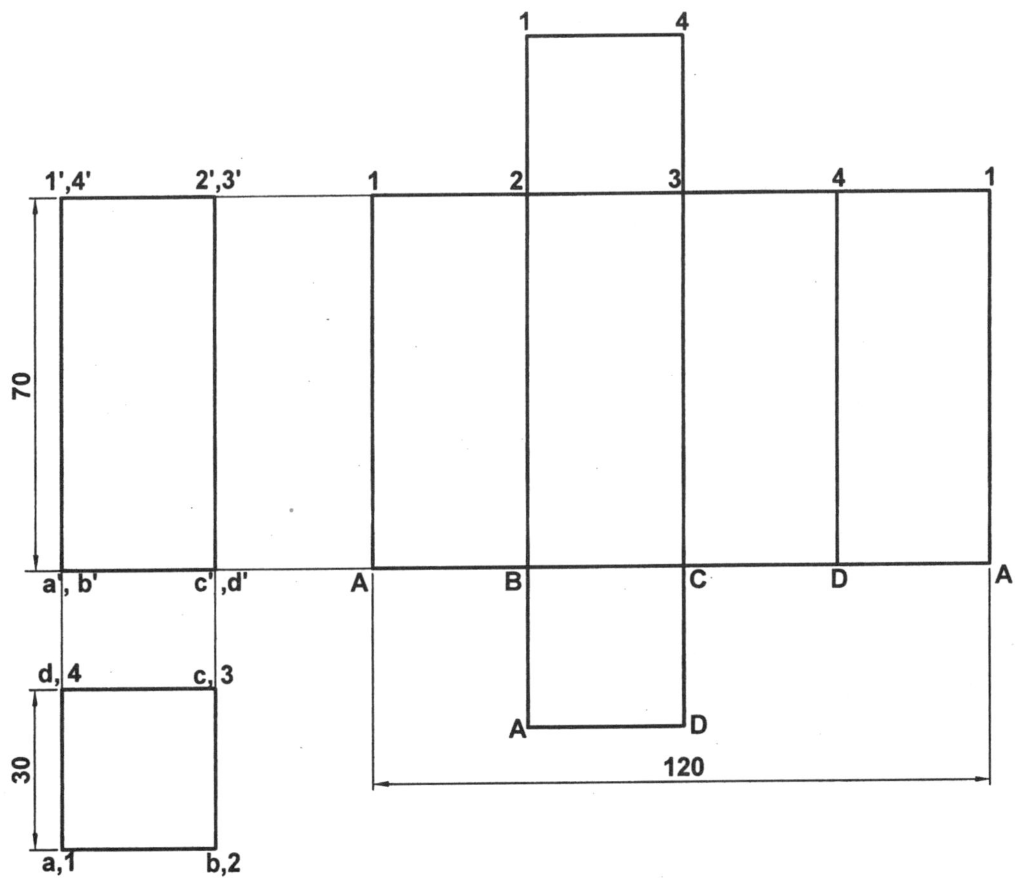

Fig. 11.6

Problem 11.6 (Fig. 11.7): *A square prism of 40 mm base edge and 65 mm height stands on one of its faces on the H.P. with a vertical face making 45° angle with the V.P. A horizontal hole of 20 mm diameter is drilled centrally through the prism such that the hole passes through the opposite vertical face of the cube. Draw its development.*

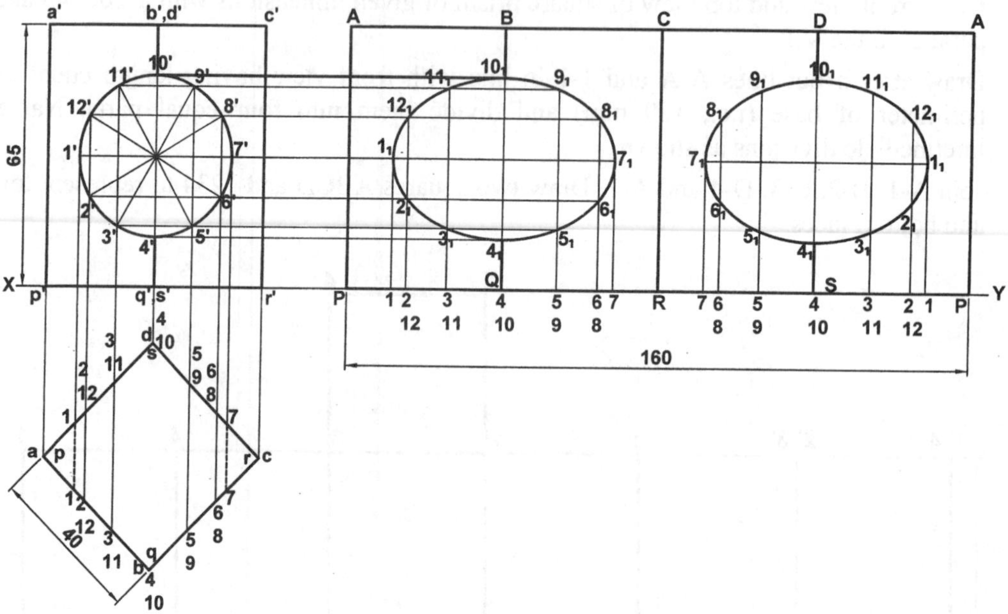

Fig. 11.7

Construction (Fig. 11.7):

1. Draw a square of 40 mm side as top view.
2. Project the front view from the top view. Draw a circle of 40 mm diameter in the front view such that centre of circle lies on the midpoint of the axis. Divide the circle into twelve equal parts.
3. Draw stretch out lines A-A and P-P, 160 mm long (equal to perimeter of base).
4. Divide P-P and A-A into four equal parts and join A-P, B-Q, C-R, D-S and A-P.
5. Through 1', 2', 3' ...etc., draw vertical projections to meet the base edge at 1, 2, 3...etc., respectively.
6. Mark points 1, 2, 3.......etc., on the stretch out line such that p1 = P1, p2 = P2, p3 = P3 etc.
7. Draw horizontal lines through 1', 2', 3'...etc., and vertical lines through 1, 2, 3...etc.
8. Mark 1_1, 2_1, 3_1...etc., at the intersection of corresponding projectors and join them to obtain smooth curves.

Problem 11.7 (Fig.11.8): *A square prism of 30 mm base edge and 75 mm long axis is resting on its base on the H.P. with its vertical faces equally inclined with the V.P. A cylindrical hole of 30 mm diameter is drilled such that its axis is perpendicular to the V.P. and 10 mm away from the left edge. Draw the development of lateral surface of the prism.*

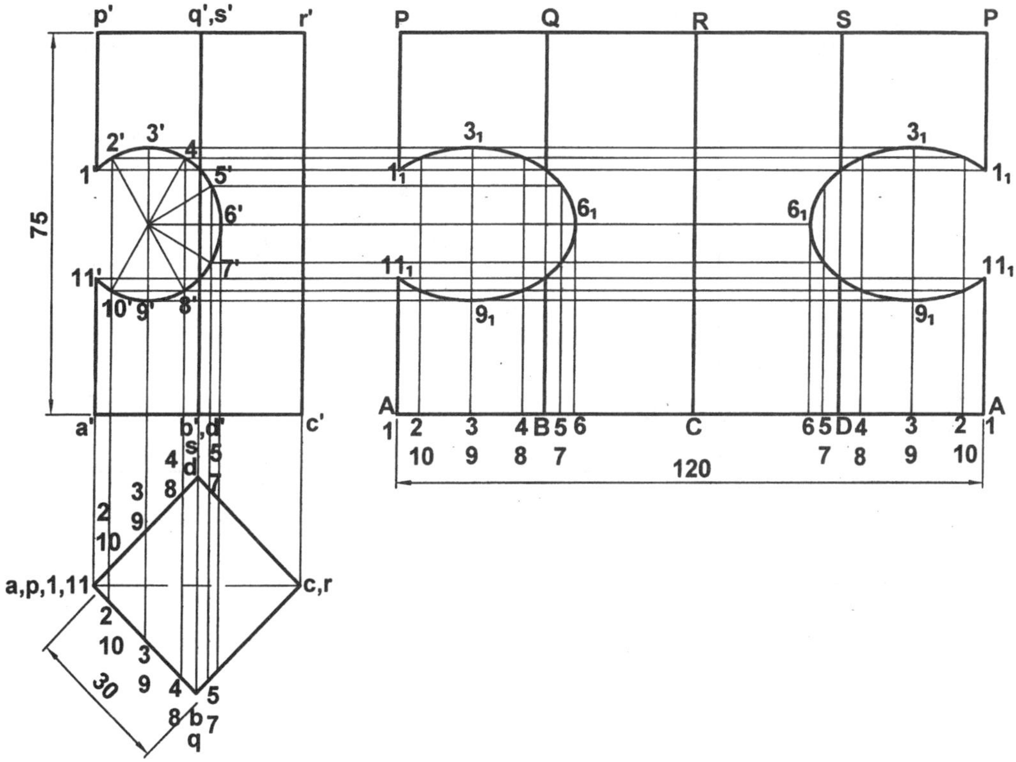

Fig. 11.8

Construction (Fig. 11.8):

1. Draw the front view and top view of a given square prism. Draw a circle 30 mm diameter in the centre of front view such that its centre is 10 mm away from the left edge.
2. Mark points 1', 2', 3'...etc., on the circumference of circle and draw vertical lines through these points to intersect edge of base at 1, 2, 3...etc.
3. Transfer points 1, 2, 3...etc., on the stretch out line A-A such that a2 = A2, a3 = A3...etc.
4. Through 1', 2', 3'...etc., draw horizontal lines and through 1, 2, 3...etc., draw vertical lines.

5. At the intersection of corresponding horizontal and vertical projectors, obtain 1_1, 2_1, 3_1...etc., and join them by a smooth curves.
6. Complete the development as shown.

Problem 11.8 (Fig. 11.9): *A pentagonal prism having a base with a 30 mm side and 72 mm long axis, is resting on its base on the H.P. with one of its faces parallel to the V.P. It is cut by a plane perpendicular to the V.P. and inclined at 45° to the H.P. and passes through the mid- point of the axis. Draw the development of lateral surface of the truncated pentagonal prism.*

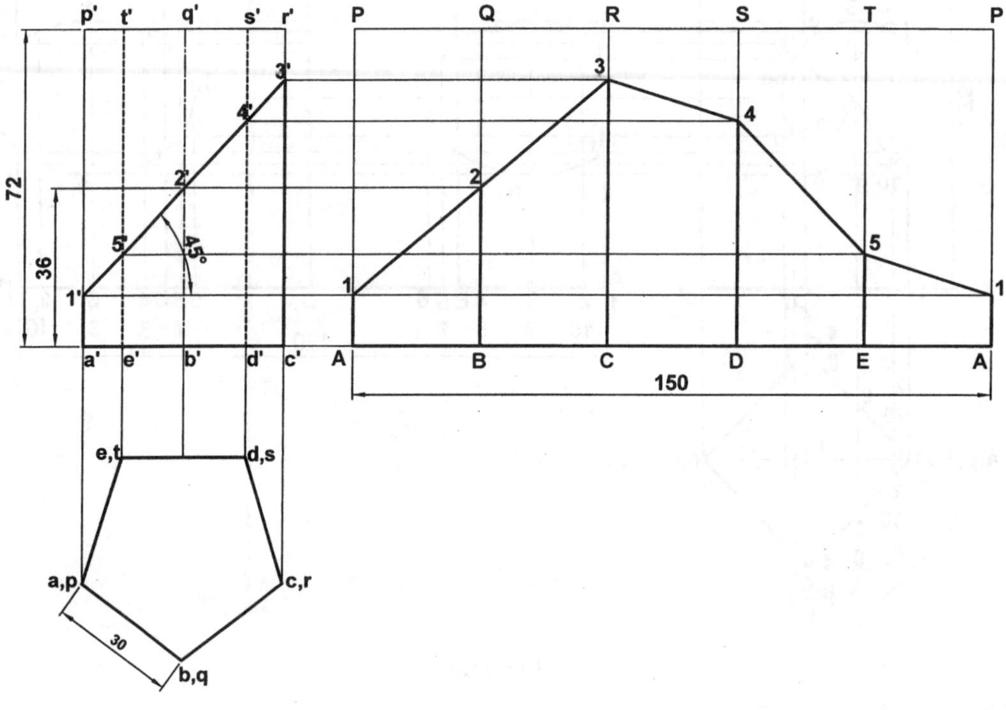

Fig. 11.9

Construction (Fig. 11.9):

1. Draw front view and top view of the pentagonal prism of given dimensions.
2. Draw cutting plane V.T. inclined at 45° to xy line in the front view.
3. Name the cutting edges as 1', 2', 3', ...etc.
4. Draw stretch out lines A-A and P-P, 150 mm long in line with front view and divide them into five equal parts.
5. Draw vertical lines A-P, B-Q, C-R, D-S, E-T and A-P.

6. Draw horizontal lines through 1', 2', 3'...etc., to cut the corresponding edges at 1, 2, 3.....etc., respectively.
7. Join 1-2, 2-3, 3-4...etc., and keeping the removed portion fainter, complete the development.

Problem 11.9 (Fig. 11.10): *Draw the development of the lateral surface of truncated pentagonal prism having 25 mm base edge and 70 mm long axis, resting on its base on the H.P. with an edge of the base perpendicular to the V.P. A cylindrical hole of 35 mm diameter is cut as shown in Fig.11.10.*

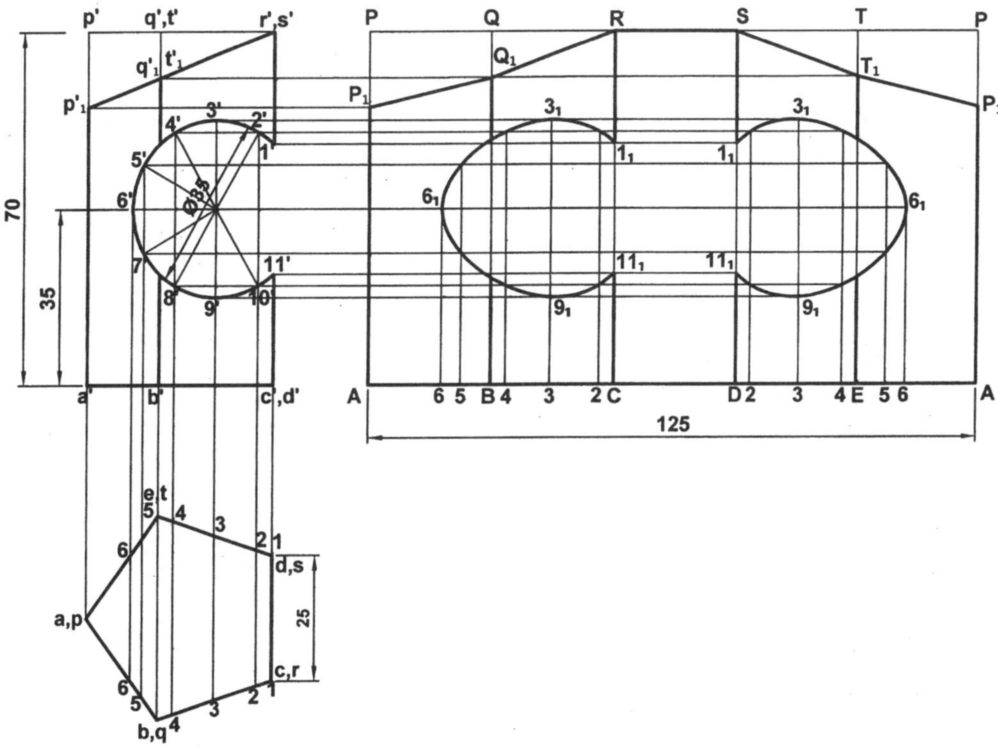

Fig. 11.10

Construction (Fig. 11.10):

1. Draw the front view and top view of pentagonal prism of given dimensions.
2. Draw a circle of 35 mm diameter keeping its centre 35 mm above the base.
3. Draw the cutting plane V.T.
4. Draw development of pentagonal prism assuming it to be whole.
5. Draw horizontal lines through p_1' and q_1' to intersect corresponding lines at P_1, Q_1, T_1 and P_1. Join P_1-Q_1, Q_1-R, R-S, S-T_1 and T_1-P_1.

6. Mark some points 1', 2', 3'... etc., on the circumference of circle (in the front view).
7. Project these points in the top view and transfer them to stretch out line A-A.
8. Draw the horizontal lines through 1', 2', 3' ... etc., to meet corresponding vertical lines at 1_1, 2_1, 3_1 ... etc., respectively. Join 1_1, 2_1, 3_1 ... etc., by a smooth curve.
9. Complete the development keeping removed portion fainter.

Problem 11.10 (Fig. 11.11): *A hexagonal prism having side of base 25 mm and axis 60 mm long is resting on its base on the H.P. with one rectangular face parallel to the V.P. It is cut by a section plane perpendicular to the V.P. and inclined at 45° to the H.P. which passes through the right corner of the top face of the prism. Draw its projections and develop the lateral surface of the truncated prism.*

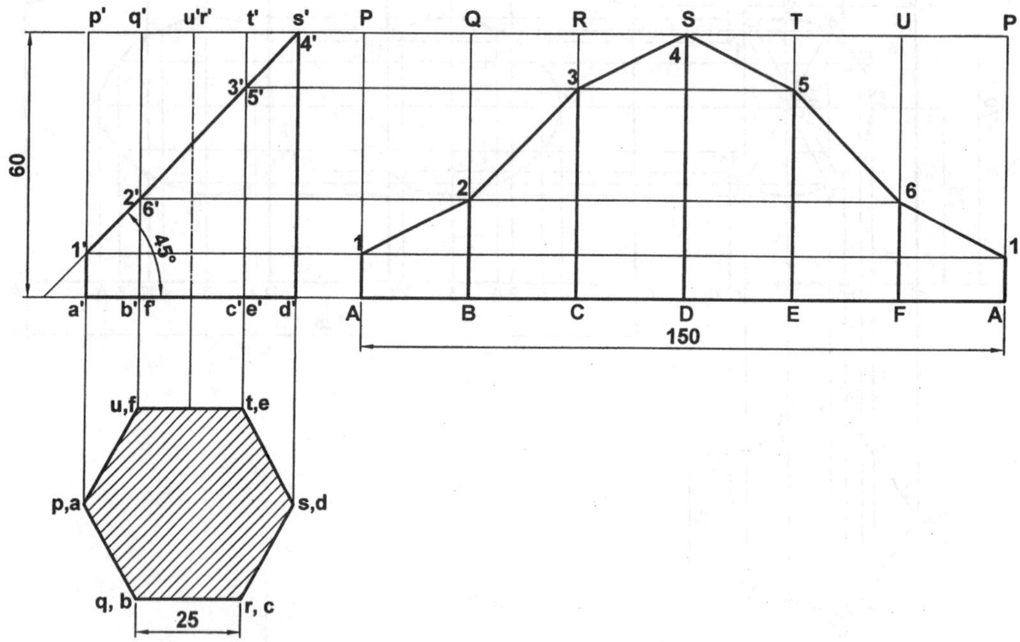

Fig. 11.11

Construction (Fig. 11.11):

1. Draw the front view and top view of hexagonal prism of given dimensions. In the front view, draw cutting plane VT and mark its points of intersection with vertical edges as 1', 2', 3'... etc.
2. Develop the lateral surface of the prism assuming it to be whole.
3. Draw horizontal lines through 1', 2', 3' ... etc., to intersect corresponding lines representing edges at 1, 2, 3 ... etc., respectively.
4. Join 1-2, 2-3, 3-4, 4-5, 5-6 and 6-1 and complete the development as shown.

Problem 11.11 (Fig. 11.12): *A hexagonal prism of base side 20 mm and axis 60 mm long is resting on its base on the H.P. with one of its rectangular faces perpendicular to the V.P. It is cut by a section plane perpendicular to the V.P. and inclined at 45° to the H.P. and which cuts the axis at a point 10 mm below the top end. Draw the development of lateral surface of the truncated prism.*

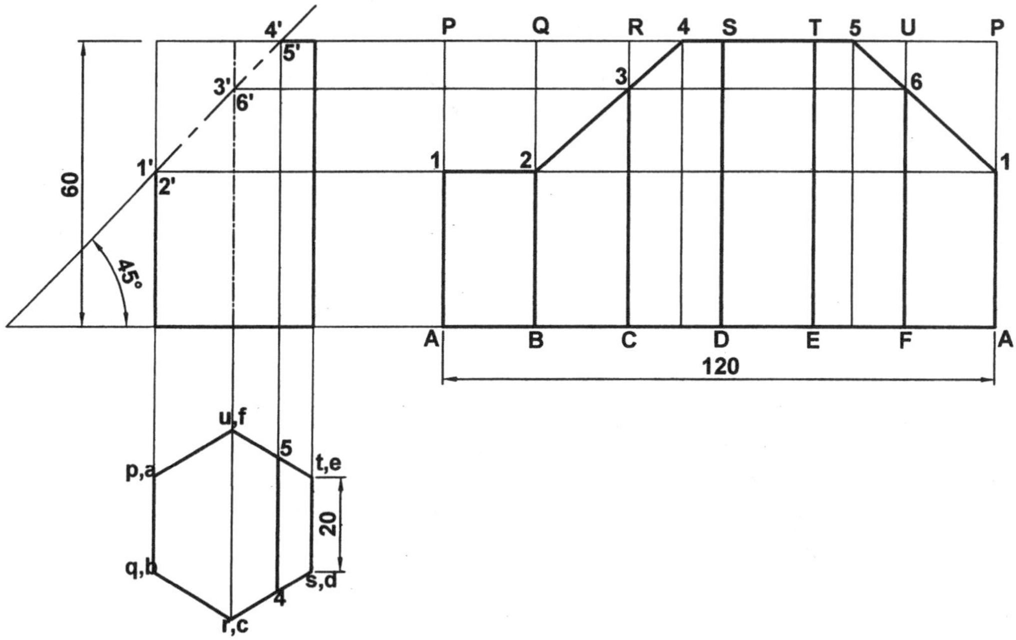

Fig. 11.12

Construction (Fig. 11.12):

1. Draw a hexagon of 20 mm side as top view and project the front view. Draw cutting plane VT and mark its point of intersection with vertical edges as 1', 2', 3'...etc.
2. Develop the lateral surface of prism assuming it to be whole using parallel line method.
3. Project points 4' and 5' on the top view. Mark points 4 and 5 on the stretch out line P-P such that r4 = R4 and u5 = U5.
4. Draw horizontal lines through 1', 2', 3'...etc., to cut the corresponding lines representing vertical edges at 1, 2, 3...etc., respectively.
5. Join 1-2, 2-3, 3-4, 4-5, 5-6 and 6-1.
6. Complete the development as shown.

Problem 11.12 (Fig. 11.13): *Draw the development of the lateral surface of part A of a hexagonal prism as shown in Fig. 11.13.*

Construction (Fig. 11.13):

1. Draw front view and top view of given hexagonal prism and show cutting plane VT in the front view and mark the points of intersection of cutting plane with the vertical edges as 1', 2', 3'...etc.
2. Draw stretch out lines A-A and P-P equal to the perimeter of base (i.e., 120 mm) and develop its lateral surface assuming the prism to be whole.
3. Divide A-A and P-P into six equal parts. Name the intermediate divisions of A-A as B, C, D, E and F. Similarly name the intermediate divisions of P-P as Q, R, S, T and U.
4. Draw vertical lines BQ, CR, DS ...etc.
5. Draw horizontal lines through 1', 2', 3'...etc., to intersects the corresponding vertical lines at 1, 2, 3...etc., respectively.
6. Join 1-2, 2-3, 3-4, 4-5, 5-6, 6-7, 8-9, 9-10, and 10-1.
7. Complete the development as shown.

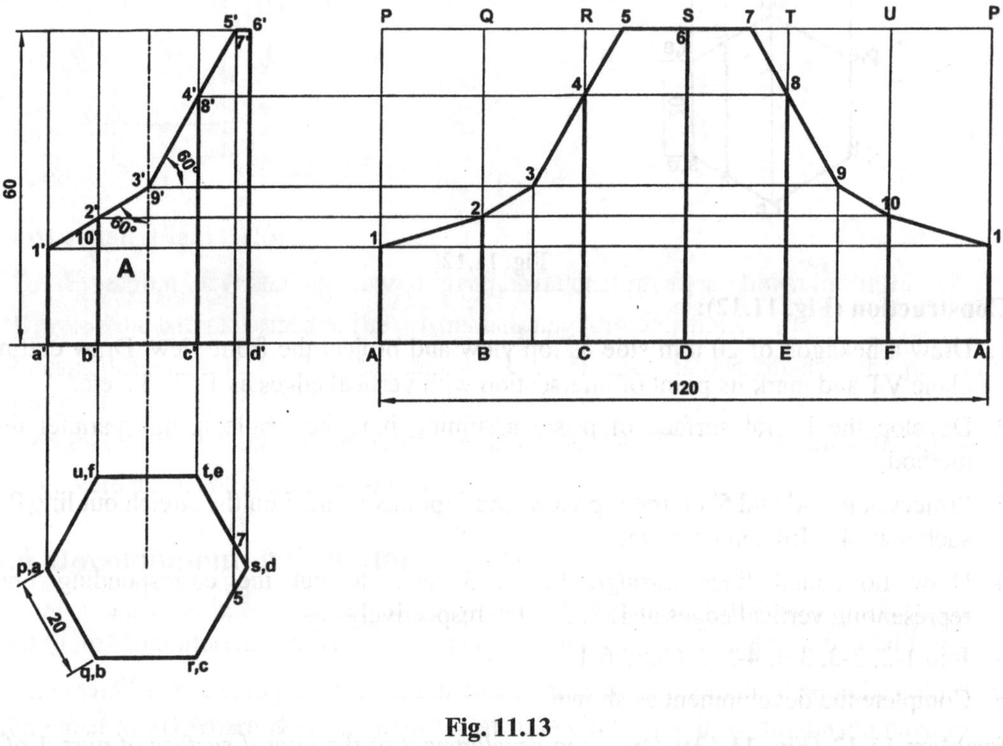

Fig. 11.13

Problem 11.13 (Fig.11.14): *Develop the lateral surface of a right regular hexagonal prism having 25 mm edge of base and 70 mm long axis resting on its base on H.P. with a* *base side perpendicular to the V.P. A horizontal cylindrical hole of 30 mm diameter is drilled centrally whose axis is perpendicular to the V.P.*

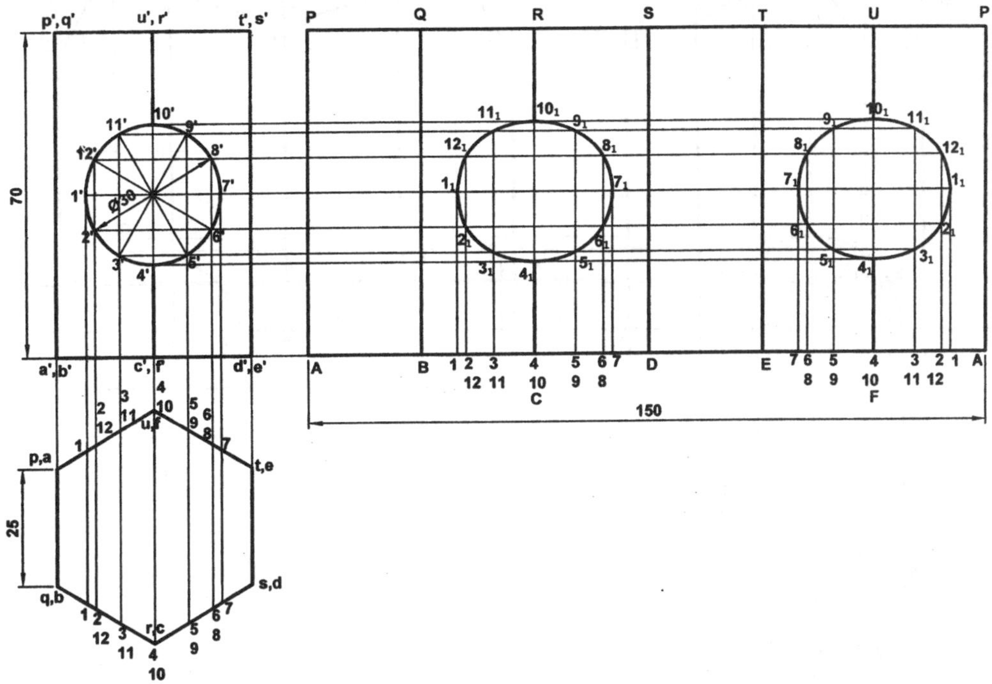

Fig. 11.14

Construction (Fig. 11.14):

1. Draw front view and top view of the given hexagonal prism. Draw a circle of 30 mm diameter such that its centre lies in the middle of the axis.
2. Divide the circle into 12 equal parts and mark 1',2',3'...etc., on its circumference.
3. Project these points on the base sides as 1, 2, 3...etc., respectively.
4. Develop the lateral surface of the prism as in the usual manner.
5. Draw horizontal projectors through points1', 2', 3'...etc.
6. Mark points 1, 2, 3...etc., on the stretch out lines A-A such that b1= B1, b2 = B2...etc.
7. Through point 1, draw a vertical line to intersect corresponding horizontal line through 1' at 1_1.
8. Similarly obtain points 2_1, 3_1, 4_1...etc. Join them by smooth curve and complete the development.

Problem 11.14 (Fig.11.15): *Draw the development of a lateral surface of truncated hexagonal prism having 25 mm base edge and 80 mm long axis resting on its base on the H.P. with an edge of the base parallel to the V.P.*

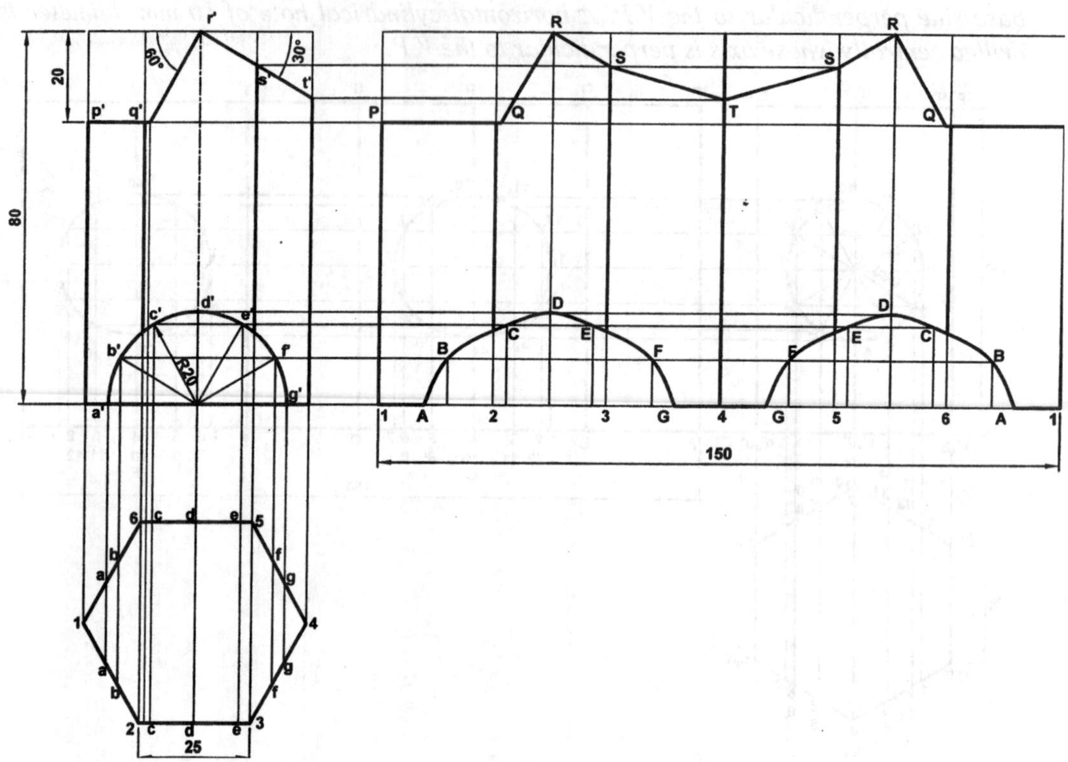

Fig. 11.15

Construction (Fig. 11.15):

1. Draw the front view and top view of given hexagonal prism as shown in Fig.11.15.
2. Develop the lateral surface of the prism assuming it to be whole.
3. Obtain the points P, Q, R... etc., and A, B, C, D...etc., as described in the previous problem.
4. Join P-Q, Q-R, R-S etc., by straight lines and A-B, B-C, C-D etc., by smooth curves.
5. Complete the development as shown.

11.4 Development of Cylinder

The development of lateral surface of a cylinder is a rectangle. The length of rectangle is equal to the circumference of base circle and breadth is equal to height of the cylinder.

The cylinder is developed using parallel line method. The length of stretch out line is taken equal to πD where D is the diameter of base circle. To draw the development of a cylinder, first of all its generators are drawn which are imaginary lines on the circumference. These generators are used instead of edges as in case of prism.

Problem 11.15 (Fig. 11.16): *A cylinder having 40 mm diameter of base and 60 mm long axis is resting on its base on the H.P. It is cut by a section plane perpendicular to the V.P. and inclined at 45° to the H.P. The section plane is passing through the top end of an extreme generator. Draw the development of lateral surface of the truncated cylinder.*

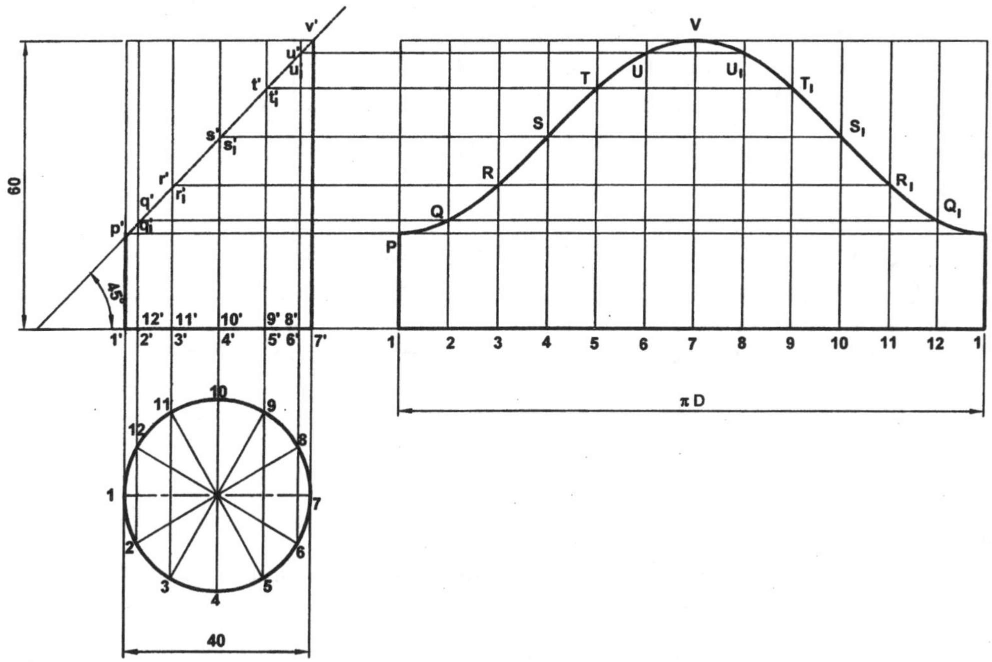

Fig. 11.16

Construction (Fig. 11.16):

1. Draw a circle of 40 mm diameter as top view and divide it into twelve equal parts. Mark division points as 1, 2, 3...etc.
2. Project the division points to the front view and draw generators.
3. Draw cutting plane VT in the front view such that it is inclined at 45° to the xy line and passing through its top end.
4. Mark the points p', q', r'...etc., at which the generators are cut.
5. Draw the stretch out line 1-1 equal to circumference of base circle (i.e., πD) and divide it into twelve equal parts and name them 1, 2, 3...etc.
6. Complete the rectangle and draw vertical lines through 1, 2, 3, 4...etc.
7. Draw horizontal lines through p', q', r'...etc., to cut the corresponding lines at P, Q, R...etc.
8. Join these points by a smooth curve. Complete the development as shown.

Problem 11.16 (Fig. 11.17): *Draw the development of the lateral surface of a truncated cylinder as shown in Fig.11.17.*

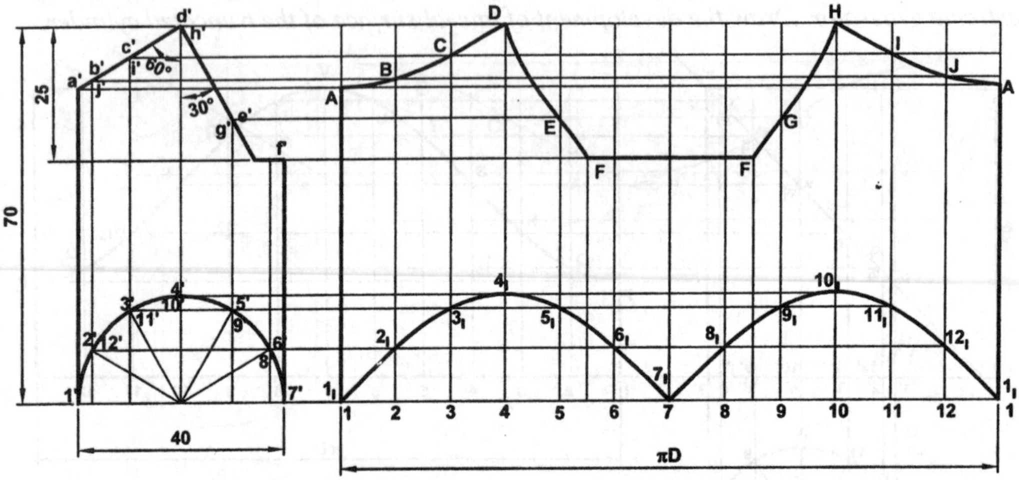

Fig. 11.17

Construction (Fig. 11.17):

1. Draw front view as given in Fig.11.17 and divide the semi-circle into six equal parts. Name the divisions as shown.
2. Draw stretch out line 1-1 equal to πD and divide it into twelve equal parts. Name the division points as 1, 2, 3, 4...etc.
3. Develop the lateral surface of the cylinder assuming it to be whole.
4. Draw a number of generators through 1', 2', 3'...etc., in the front view.
5. Draw vertical lines through 1, 2, 3, 4....etc., representing generators.
6. Draw horizontal lines through a', b', c'...etc., intersecting the corresponding generators at A, B, C...etc., respectively. Join these points.
7. Similarly, draw horizontal lines through 1', 2', 3'...etc., to obtain points 1_1, 2_1, 3_1...etc. Join these points by smooth curve and complete the development as shown.

Problem 11.17 (Fig. 11.18): *Draw the development of the lateral surface of the truncated cylinder as shown in Fig.11.18.*

Construction (Fig. 11.18):

1. Draw circle of 50 mm diameter as top view and divide it into twelve equal parts. Project these points to the front view and draw generators through 1', 2', 3'...etc.
2. In the front view, draw cutting plane VT and mark its point of intersection with generators as a', b', c'...etc.

3. Draw stretch out line 1-1 equal to the circumference of base circle (i.e., πD) and divide it into twelve equal parts. Complete the rectangle.
4. Draw horizontal projectors through a', b', c'...etc., to intersect corresponding lines at A, B, C...etc., respectively.
5. Join A-B, B-C, C-D...etc., and complete the development as shown.

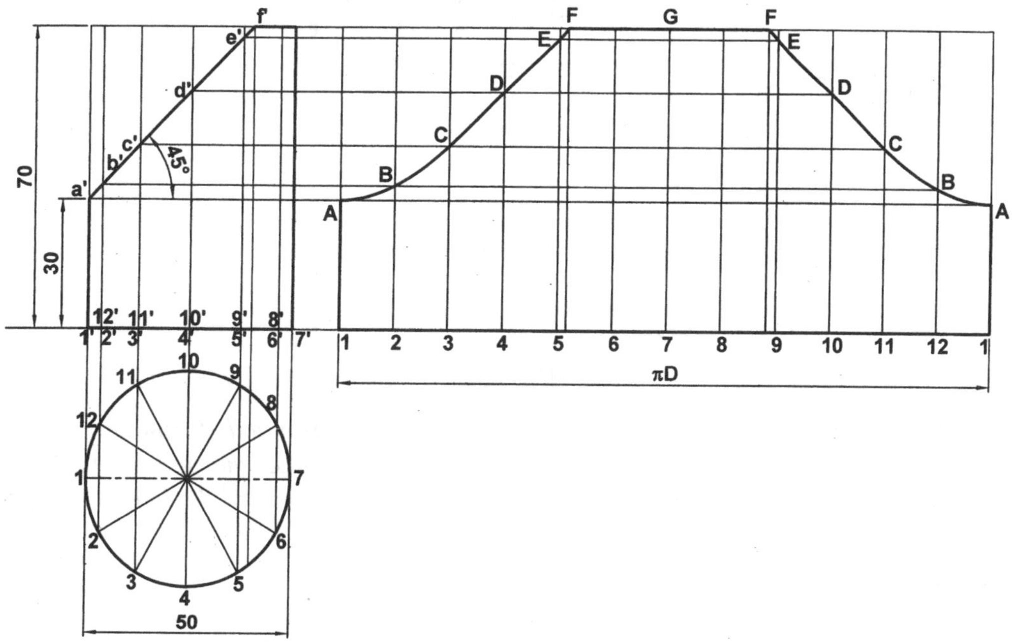

Fig. 11.18

Problem 11.18 (Fig. 11.19): *A vertical chimney of 50 mm diameter is mounted over a slopping roof at an angle of 30° with the horizontal. The shortest portion over the roof is 35 mm. Determine the shape of metallic sheet required for fabrication of the chimney.*

Construction (Fig. 11.19):

1. Draw front view and top view of given cylindrical chimney.
2. Divide the circle (top view) into eight equal parts and project these points to the front view and draw generators.
3. Complete the development as explained in problem 11.17.

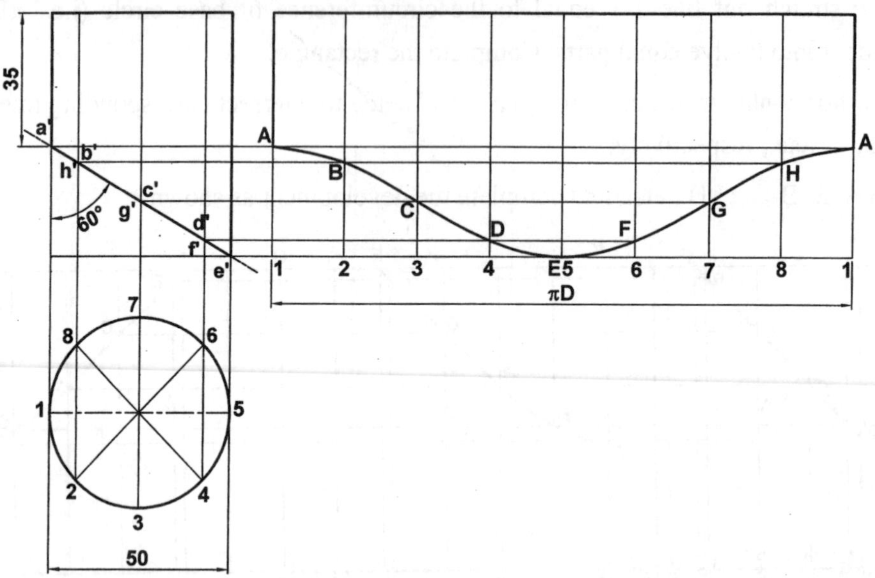

Fig. 11.19

Problem 11.19 (Fig. 11.20): *A cylinder of base diameter 40 mm and 70 mm long axis is resting on its base on the H.P. The cylinder has a centrally made square hole of 20 mm side whose axis is perpendicular to the V.P. Draw the development of the lateral surface of cylinder.*

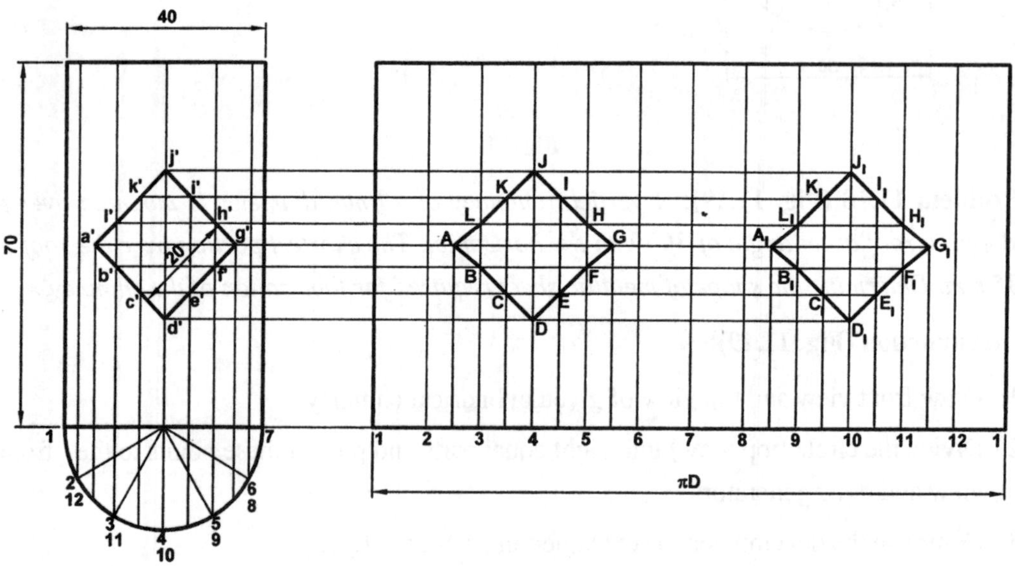

Fig. 11.20

Construction (Fig. 11.20):

1. Draw the front view with a square hole in the centre of cylinder. Draw generators of the cylinder as shown in Fig. 11.20.
2. Name the points where the generators are cut by edges of the hole as a', b', c', d'...etc.
3. Project a' and g' to the top view and transfer the same on the stretch out line 1-1.
4. Draw horizontal projectors through a', b', c'...etc., cutting the corresponding edges at A, B, C...etc. Join AB, BC, CD......etc.
5. Similarly obtain A_1, B_1, C_1...etc. Join A_1-B_1, B_1-C_1, C_1-D_1...etc., and complete the development as shown.

Problem 11.20 (Fig. 11.21): *A cylinder of base diameter 50 mm and axis 70 mm long is resting on its base on the H.P. It is cut by three planes as shown in fig.11.21. Draw the development of lateral surface of the cylinder.*

Construction (Fig.11.21):

1. Draw the front view as shown in Fig.11.21.
2. Develop the lateral surface of the cylinder assuming it to be whole.
3. Obtain P, Q, R, S...etc., and A, B, C...etc., as explained in previous problems.
4. Join these points and complete the development keeping the removed portion fainter.

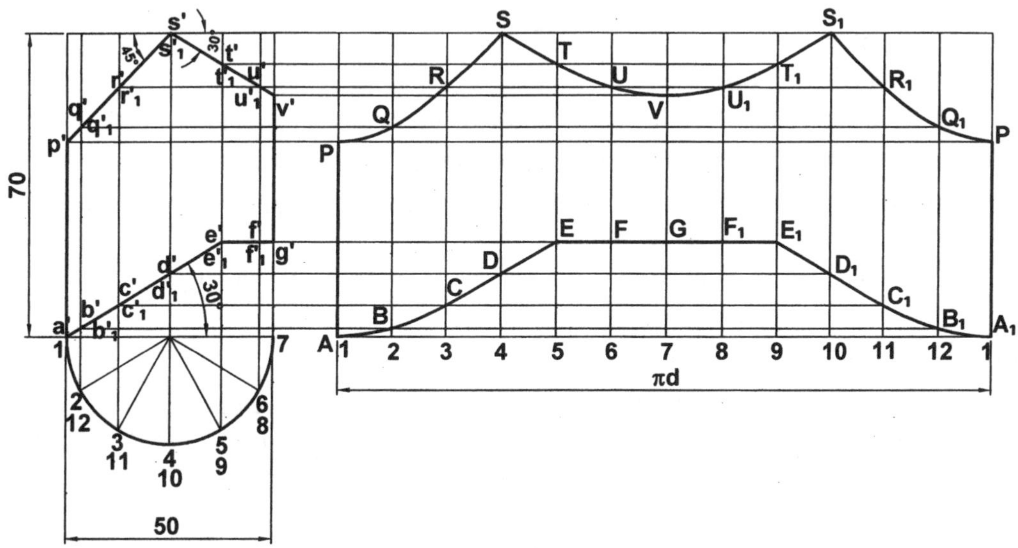

Fig. 11.21

Problem 11.21 (Fig.11.22): *A right cylinder, diameter of base 40 mm and height 65 mm is cut by two section planes as shown in Fig.11.22. Draw development of lateral surface of the truncated cylinder.*

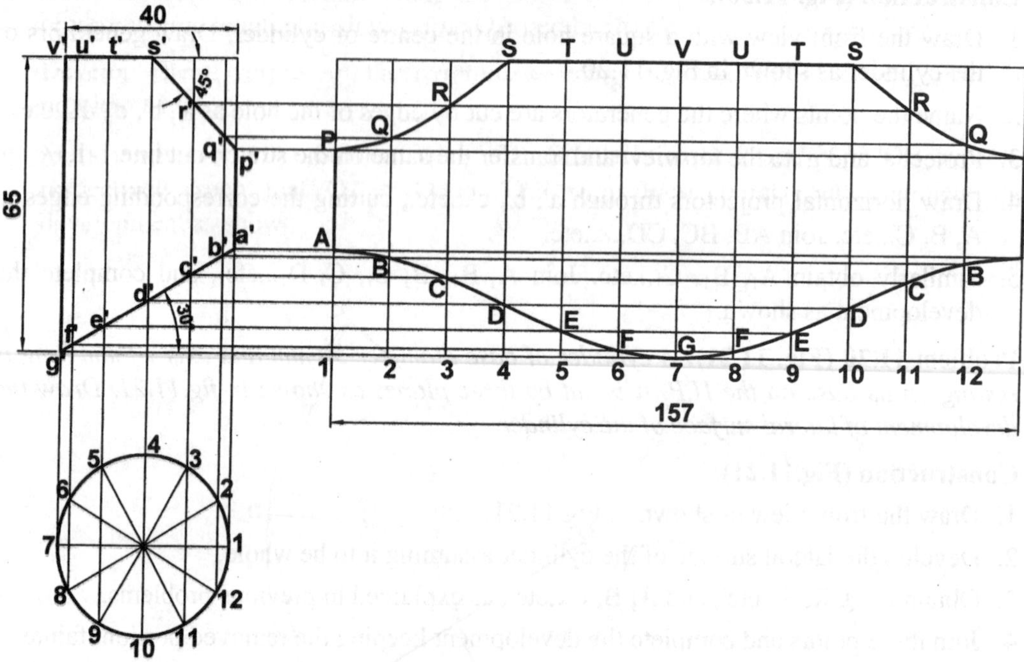

Fig. 11.22

Construction (Fig. 11.22):

1. Draw front view and top view of the truncated cylinder as shown in Fig.11.22.
2. Develop the lateral surface of the cylinder assuming it to be whole.
3. Obtain the point P, Q, R...etc., and A, B, C...etc., as explained in the previous problems. Join these points by smooth curves.
4. Complete the development keeping the removed portion fainter and lighter.

11.5 Development of Pyramid

Since, a pyramid has a number of isosceles triangular faces, development of its lateral surface consists of the number of isosceles triangles connected side by side. The base of each triangle is equal to side of base and side of each triangle is equal to the true length of slant edge of pyramid.

Procedure of developing lateral surface of a pyramid is explained as follows:
(Refer Fig. 11.23)

1. In the given position of square pyramid since, no slant edge is parallel to V.P., none of the line representing slant edge will give true length. Therefore, with O as centre and

radius equal to ob, draw an arc meeting the horizontal line through o, at b_1. Project b_1 to b_1' and join $o'b_1'$ which is the true length of slant edge.

2. Draw a line OA parallel to $o'b_1'$.
3. Take any point O, as centre and radius equal to the true length of slant edge (OA) of the pyramid, draw an arc A-A.
4. With radius equal to true length of side of base, (say AB) step-off divisions equal to the number of base edges.
5. Join these division points in correct sequence. The figure thus obtained, is the development of lateral surface of a pyramid.

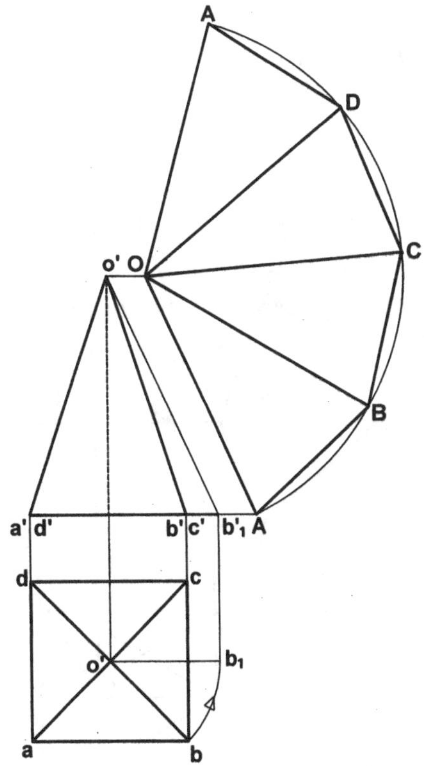

Fig. 11.23

Problem 11.22 (Fig. 11.24): *A triangular pyramid having base 40 mm and axis 60 mm long is resting on its base on the H.P. with a base side perpendicular to the V.P. It is cut by an auxiliary inclined plane as shown in Fig.11.23. Draw development of lateral surface of the truncated pyramid.*

Construction (Fig. 11.24):

1. Draw the front view and top view of given triangular pyramid as shown in Fig.11.24.

2. Since, the top view of slant edge oa is parallel to the V.P., its front view o'a' represents true length. So, draw a line OA parallel to o'a'.
3. Develop lateral surface of the pyramid assuming it to be whole, as explained in problem 11.21.
4. Mark point 1 on OA such that o1 = o'1'. Similarly mark 2 and 3 on OB and OC respectively such that O2 = O3 = O'2'. Join these points and complete the development as shown.

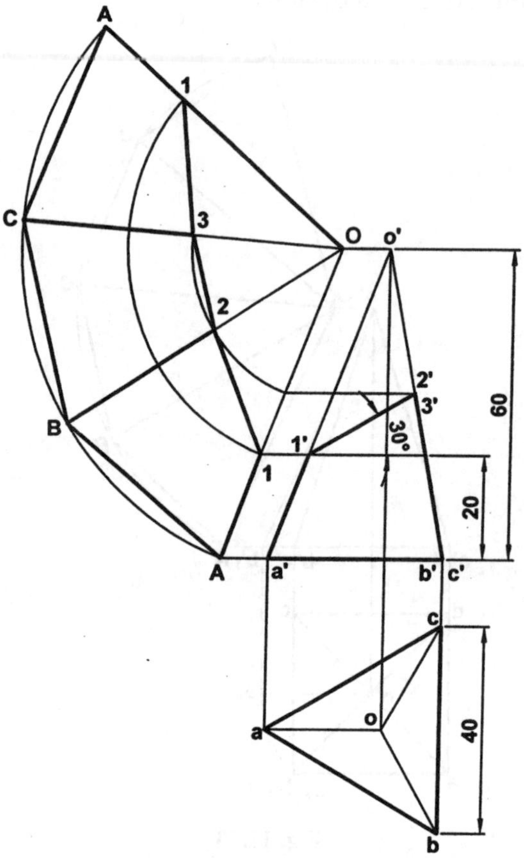

Fig. 11.24

Problem 11.23 (Fig.11.25): *A square pyramid having side of base 30 mm and axis 60 mm long is resting on its base on the H.P. with a base side parallel to the V.P. It is cut by a section plane perpendicular to the V.P. and inclined at 45° to the H.P. which passes through the mid-point of the axis. Draw development of the lateral surface of the cut solid.*

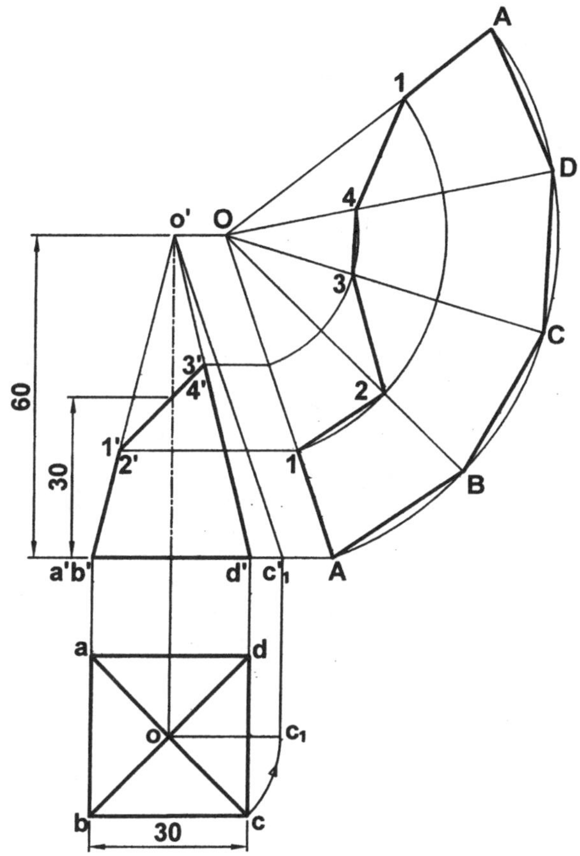

Fig. 11.25

Construction (Fig.11.25):

1. Draw the front view and top view of given square pyramid. With o as centre and radius equal to oc draw an arc meeting c_1 such that oc_1 is parallel to the V.P. project c_1 to c_1'.

2. Draw OA parallel to $o'c_1'$ and draw development of lateral surface of the pyramid assuming it to be whole.

3. Draw horizontal line through 1' to meet OA at 1.

4. With O as centre and radius equal to o'1' draw an arc cutting OA and OB at 1 and 2 respectively.

5. Similarly obtain point 3 and 4.

6. Join 1-2, 2-3, 3-4, 4-1 in correct sequence and complete the development as shown.

Problem 11.24 (Fig. 11.26): *A frustum of a square pyramid base 40 mm, top 20 mm side and height 50 mm is resting on its base on the H.P. with a side of base parallel to the V.P. Draw the development of its lateral surface. Also draw the projections of the frustum indicating the line joining the mid-point of a top edge of one face with the mid-point of the bottom edge of the opposite face by the shortest distance.*

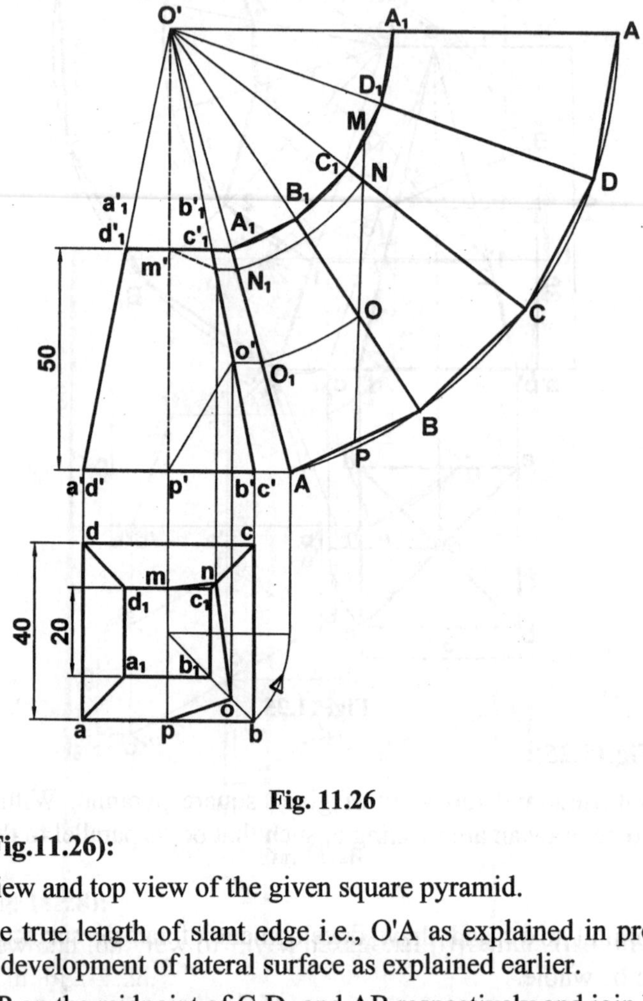

Fig. 11.26

Construction (Fig.11.26):

1. Draw front view and top view of the given square pyramid.
2. Determine the true length of slant edge i.e., O'A as explained in problem 11.23 and complete the development of lateral surface as explained earlier.
3. Mark M and P on the midpoint of C_1D_1 and AB respectively and join M-P.
4. Mark the points N and O on O'C and O'B where the line MP cuts the slant edges.
5. Transfer points M, N, O and P on the corresponding edges in the front view and top view.
6. In the top view. Join m-n-o-p by straight lines.

Problem 11.25 (Fig. 11.27): *A pentagonal pyramid base 25 mm side and axis 65 mm long is resting on its base on the H.P. with a side of base perpendicular to the V.P. It is cut by a section plane perpendicular to the V.P. and inclined at 60° to the H.P. and passes through the axis at a point 35 mm from the apex. Draw the development of lateral surface of the truncated pentagonal pyramid.*

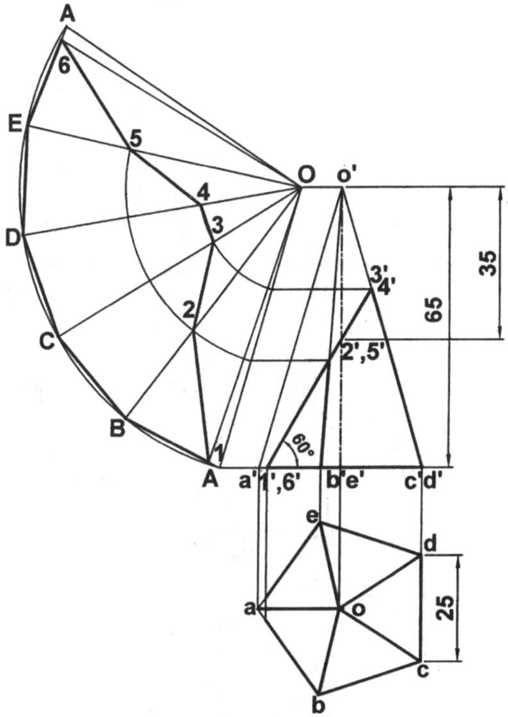

Fig. 11.27

Construction (Fig. 11.27):

1. Draw front view and top view of given pentagonal pyramid and draw section plane V.T. as shown in the figure. Mark 1', 2', 3'......etc., at which slant edges are cut by cutting plane.
2. Since, o'a' gives true length of slant edge, draw line OA parallel to o'a'.
3. With O as centre and radius equal to OA draw as arc A-A. With radius equal to side of base, step-off five divisions and mark them A, B, C, D...etc.
4. Join A, B, C...etc., by straight lines and also join A, B, C...etc., with O.
5. Mark points 1, 2, 3...etc., on the slant edges as explained in problem 11.22.
6. Join 1, 2, 3......etc., and complete the development as shown.

Problem 11.26 (Fig. 11.28): *A hexagonal pyramid having base side 25 mm and axis 60 mm long is resting on its base on the H.P. with a side perpendicular to the V.P. It is cut by two section planes as shown in Fig. 11.28. Draw the development of lateral surface of the truncated hexagonal pyramid.*

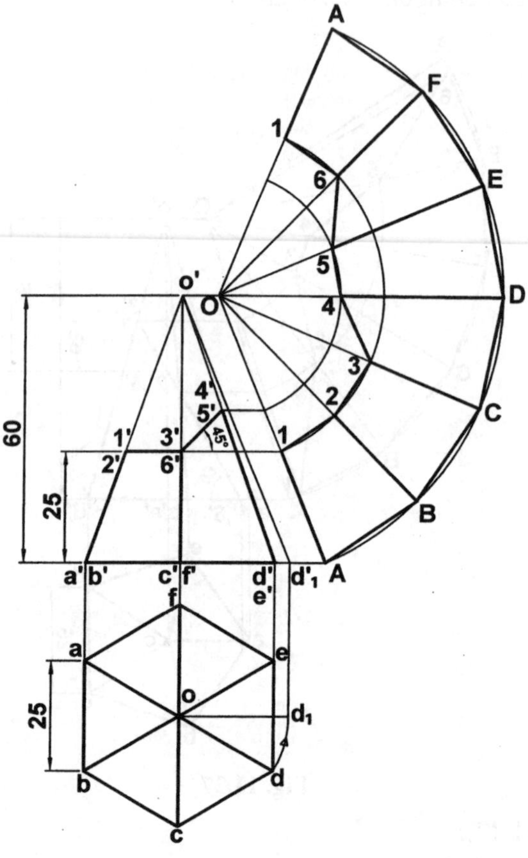

Fig. 11.28

Construction (Fig. 11.28):

1. Draw front view and top view of given hexagonal pyramid. Also draw cutting plane V.T. as shown in Fig. 11.28.
2. With o as centre and radius equal to od turn od to od_1. Project d_1 to d_1'. Join o' with d_1'. Thus $o'd_1'$ represents true length of slant edge.
3. Draw OA parallel to $o'd_1'$ and draw development of lateral surface of pyramid assuming it to be uncut.
4. Mark points 1, 2, 3...etc., on the slant edges as explained in problem 11.22.
5. Join 1, 2, 3...etc., and complete the development keeping removed portion fainter.

Problem 11.27 (Fig.11.29): *A hexagonal pyramid having base side 25 mm and axis 60 mm long is resting on its base on the H.P. with a side perpendicular to the V.P. It is cut by a section plane perpendicular to the V.P. and inclined to the H.P. such that it passes through the mid-point of the axis and bottom edge which is perpendicular to the V.P. Draw the development of lateral surface of the truncated hexagonal pyramid.*

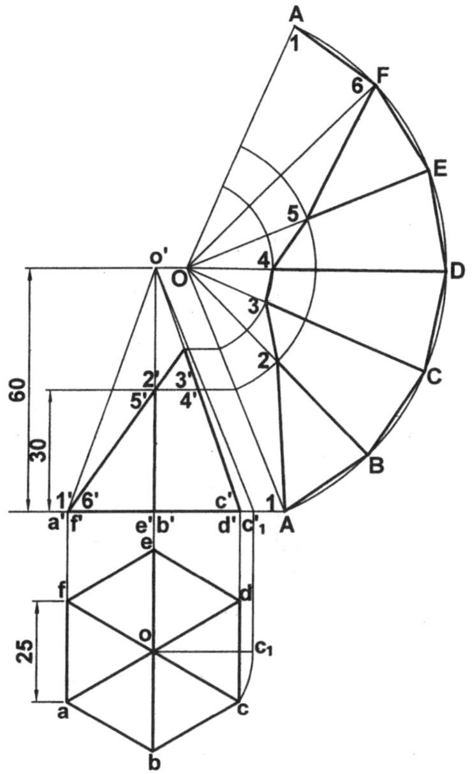

Fig. 11.29

Construction (Fig.11.29):

1. Draw front view and top view of a given hexagonal pyramid. Also draw cutting plane V.T. as shown in Fig. 11.29.
2. With o as centre and radius equal to oc, turn c to c_1. Project c_1 to c_1'. Thus $o'c_1'$ represents true length of slant edge.
3. Draw OA parallel to $o'c_1'$ and draw development of its lateral surface assuming it to be un-cut.
4. Mark points 1, 2, 3...etc., on OA, OB, OC...etc., respectively as explained in the problem 11.22.
5. Join 1, 2, 3... etc., and complete the development as shown in the Fig. 11.29.

11.6 Development of Cone

A cone is developed using radial line method. The development of its curved surface is a sector of circle whose radius is equal to slant height of the cone. The sector subtends an angle θ which can be calculated as follows.

$$\theta = \frac{r}{s} \times 360$$

where, r = radius of base circle and s = slant height of the cone.

The procedure to develop curved surface of a cone is explained with the help of following examples.

Problem 11.28 (Fig. 11.30): *A cone having base diameter 60 mm and axis 80 mm long is resting on its base on the H.P. Draw the development of the lateral surface of the cone.*

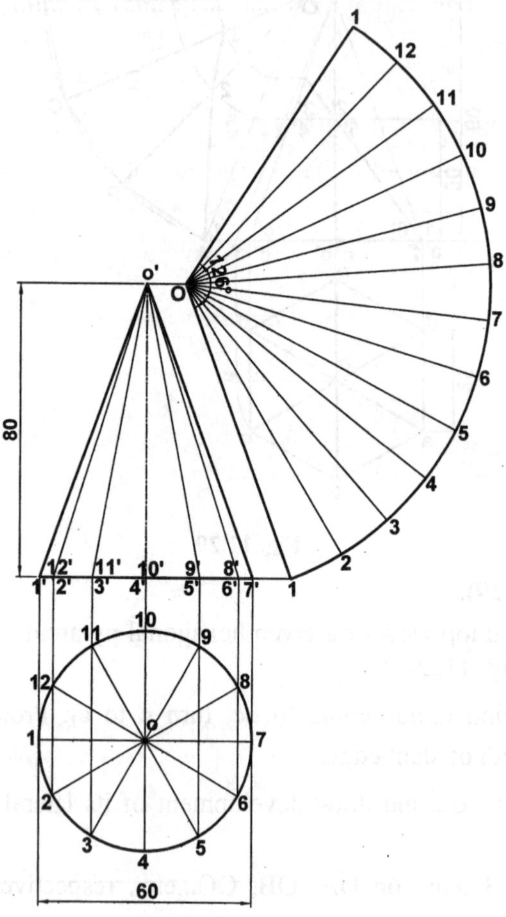

Fig. 11.30

For developing a cone, first of all it is required to calculate θ as follows:

Let, r = radius of base circle

s = slant height

θ = subtended angle

h = height

Then,

$$s = \sqrt{r^2 + h^2} = \sqrt{30^2 + 80^2} = 85.44 \text{ mm}$$

$$\theta = \frac{r}{s} \times 360 = \frac{30}{85.44} \times 360 = 126.4^\circ \text{ (Approx)}$$

Construction (Fig. 11.30):

1. Draw front view and top view of a given cone. In the top view, divide the base circle into twelve equal parts and name the divisions points as 1, 2, 3... etc.
2. Project these points on the front view and name them as 1', 2', 3'... etc., join them with o'.
3. Since the extreme generator in the front view shows true length, draw a line o1 parallel to o'7'.
4. With O as centre and radius equal to O1, draw a sector of circle which subtends an angle of θ (i.e., 126.4°).
5. Divide the sector into twelve equal parts and draw generators O1, O2, O3.... etc.

Problem 11.29 (Fig. 11.31): *A cone of base diameter 50 mm and axis 60 mm long is resting on its base on the H.P. It is cut by a plane parallel to the H.P. and passing through the axis at a distance of 15 mm from the apex. Draw the development of lateral surface of truncated cone.*

Construction (Fig. 11.31):

1. Draw front view and top view of the cone.
2. Develop its curved surface as explained in problem 11.28.
3. Draw cutting plane V.T. and mark its point of intersection with generators. Name them a', b', c'...etc.
4. Since, VT cuts all the generators at the equal distance from the apex, draw an arc A-A with O as centre and radius equal to OA.
5. Complete the development as shown.

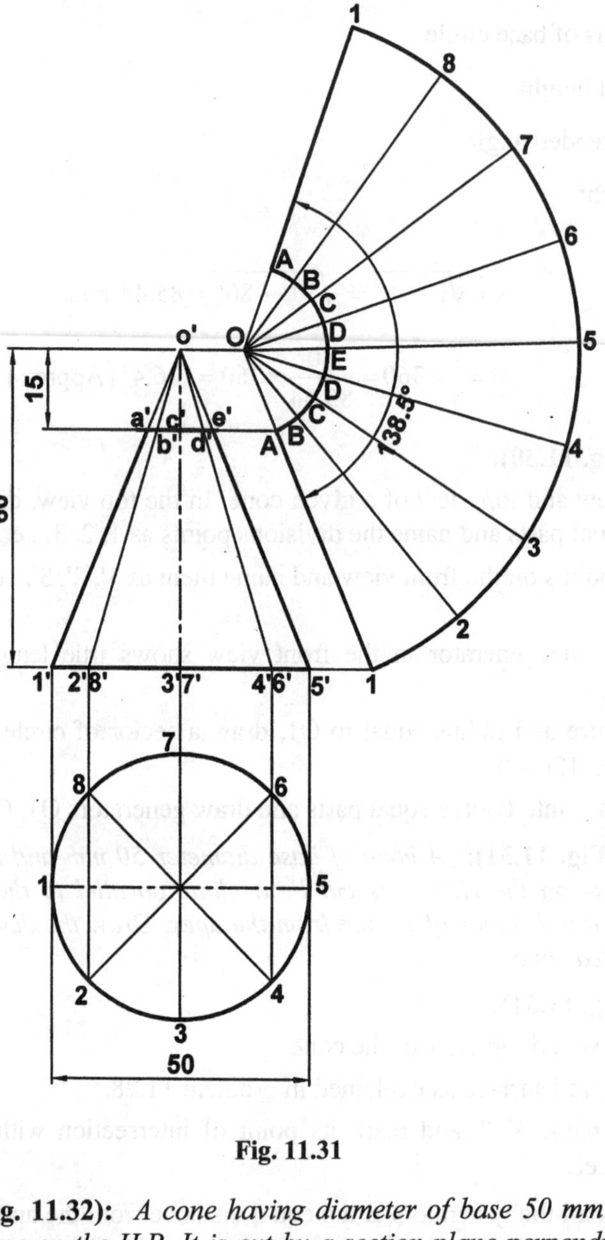

Fig. 11.31

Problem 11.30 (Fig. 11.32): *A cone having diameter of base 50 mm and axis 80 mm long rests on its base on the H.P. It is cut by a section plane perpendicular to the V.P. and inclined at 45° to the H.P. and bisects the axis. Draw the development of lateral surface of the truncated cone.*

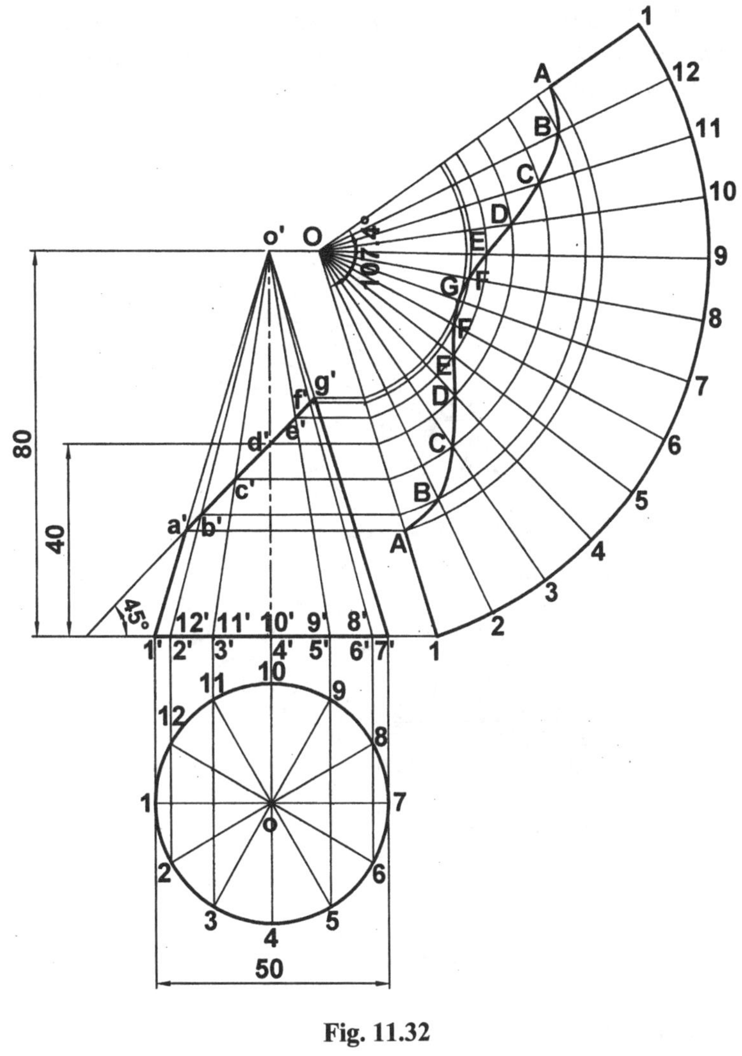

Fig. 11.32

Construction (Fig. 11.32):

1. Draw front view and top view of the given cone.
2. Draw cutting plane VT inclined at 45° to the xy line.
3. Develop the curved surface of the cone as explained in problem 11.28.
4. Locate the point of intersection of cutting plane VT with the generators as a', b', c'...etc.
5. Mark points A, B, C, D...etc., on the slant edges such that OA = o'a', OB = o'b', OC = o'c'... etc.
6. Join A, B, C.... etc., in the correct sequences.

Problem 11.31 (Fig.11.33): *A cone having base diameter 50 mm and axis 70 mm is resting on its base on the H.P. It is cut by a section plane perpendicular to both H.P. and V.P. which is 10 mm away from the axis. Draw the development of lateral surface of cut cone.*

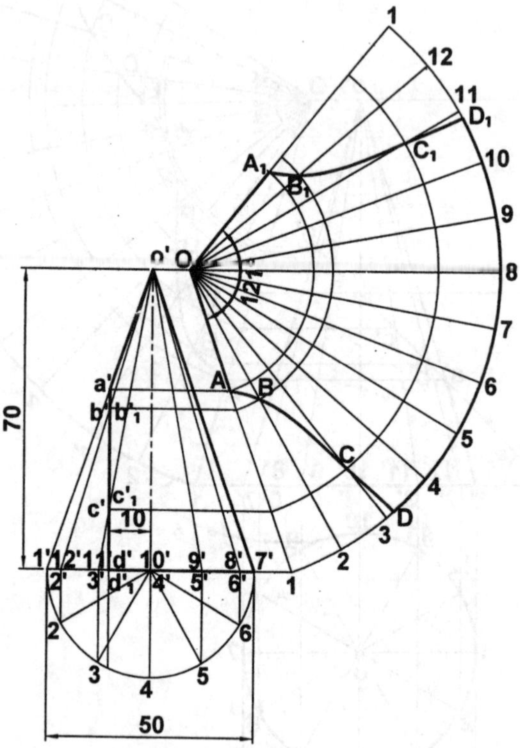

Fig. 11.33

Construction (Fig.11.33):

1. Draw front view of the cone and cutting plane VT, 10 mm away from the axis.
2. Draw a semi circle at the base of front view and divide it into six equal parts. Name them as 1, 2, 3...etc., and project these points to the base line as 1', 2', 3'...etc., respectively.
3. Draw its generators as o'1', O'2', O'3'...etc.
4. Draw line O1 parallel to o'7' and draw a sector of circle with o as centre and radius equal to O1 and having subtended angle $\theta = 121^\circ$.
5. Mark points A, B, C...etc., on the generators O1, O2, O3... etc., respectively as explained in problem 11.28 and complete development as shown.

Problem 11.32 (Fig. 11.34): *Draw the development of the lateral surface of truncated cone as shown in Fig. 11.34.*

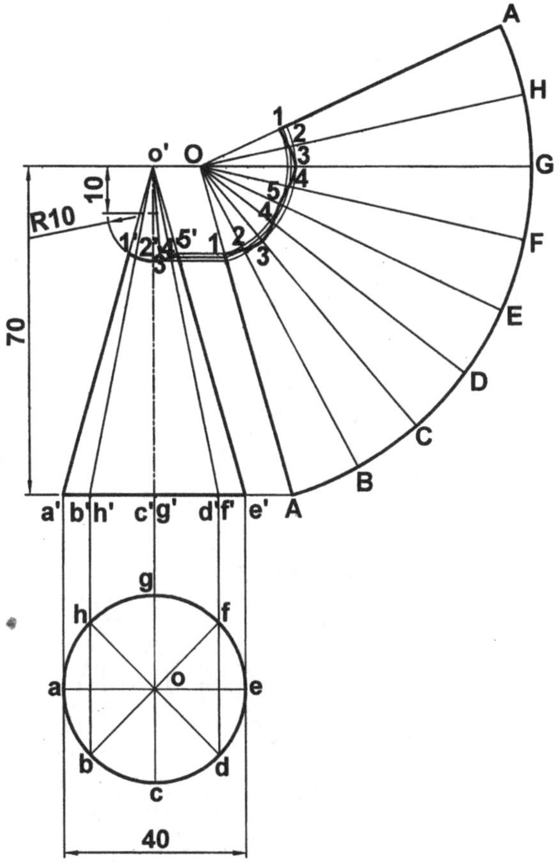

Fig. 11.34

Construction (Fig. 11.34):

1. Draw front view and top view of the truncated cone as shown in Fig.11.34.
2. Divide the circle (top view) in eight equal parts and project division points to the front view. Join o'-a', o'-b', o'-c'.... etc., as generators.
3. Draw development of curved surface of un-cut cone (sector of circle with OA as radius and subtended angle $\theta = 99^{o}$)
4. Divide the sector into eight equal parts and join O-A, O-B, O-C... etc.
5. Mark points 1, 2, 3... etc., on OA, OB, OC... etc., as explained in problem 11.28.
6. Join these points and complete the development as shown.

Problem 11.33 (Fig.11.35): *A cone having base diameter 50 mm and axis 70 mm long is resting on its base on the H.P. Draw the projections of the cone and show on it the shortest path traced by a point P on the circumference of the base which moves around it and reaches the same point.*

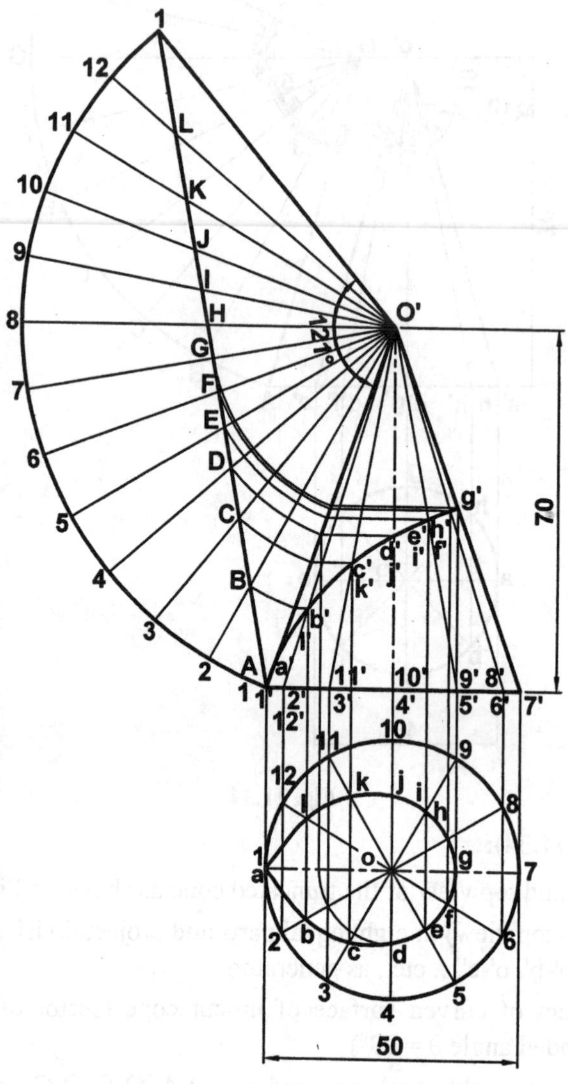

Fig. 11.35

Construction (Fig.11.35):

1. Draw front view and top view of the given cone and draw its generators.

2. Draw the development of its curved surface which is a sector of circle having radius equal to slant edge and subtended angle 121°.
3. In the development, draw a straight line A-1 to represent shortest path of a point P on the base circle.
4. Mark the points A, B, C... etc., at the intersection of line A-1 with the generators.
5. With O' as centre transfer these points on O'1' and draw horizontal lines through these points to intersect corresponding generators at a', b', c'... etc.
6. Join a', b', c'... etc., by a smooth curve.
7. Draw vertical lines through a', b', c'... etc., to intersect the corresponding generator in the top view at a, b, c... etc., respectively.
8. Join a, b, c... etc., by a smooth curve which shows the required shortest path.

11.7 Development of Sphere

Since, a sphere consists of double curved surface, it can only be developed using approximate methods as explained below.

1. **Zone Method:** A portion of sphere between two parallel planes which are perpendicular to the axis, is called as zone. In this method, the sphere is assumed to be cut into a number of convenient zones. Each zone is considered as frustum of cone except the top most zone which is a cone of small altitude. Each zone is developed separately.
2. **Lune Method:** The portion between two planes which contain the axis of the sphere is known as a Lune.

   In this method, a sphere is assumed to be divided into twelve Lunes. Each lune is developed separately whose length is equal to one half the circumference of the sphere and width is equal to maximum distance between two planes.

   Following examples illustrate development of spheres using above mentioned methods.

Problem 11.34 (Fig. 11.36): *Draw development of a sphere of 48 mm diameter using zone method.*

Construction (Fig. 11.36):

1. Draw the front view and top view of the sphere as shown in Fig. 11.36 (a).
2. Divide upper half of the sphere into a number of zones say four.
3. Zone-1 is in the form of a small cone of low altitude, which can be developed in the usual manner.
4. Zones 2, 3 and 4 are in the form of cone frusta which can be developed separately. For example, to develop zone-2 (Fig. 11.36(b)) which is frustum of a cone, mark its apex at O_2' where lines joining 1'-2' intersect.

5. With O_2' as centre and radius equal to $O_2'1'$, draw an arc 1'-1' subtending an angle of θ which can be calculated as explained in problem 11.33.
6. With O_2' as centre and radius equal to $O_2'2'$, draw another arc 2'-2' subtending an angle θ.
7. Circular strip may be divided into twelve parts.
8. Similarly, Zone-3 (Fig. 11.36 (c)) and zone-4 can also be developed.

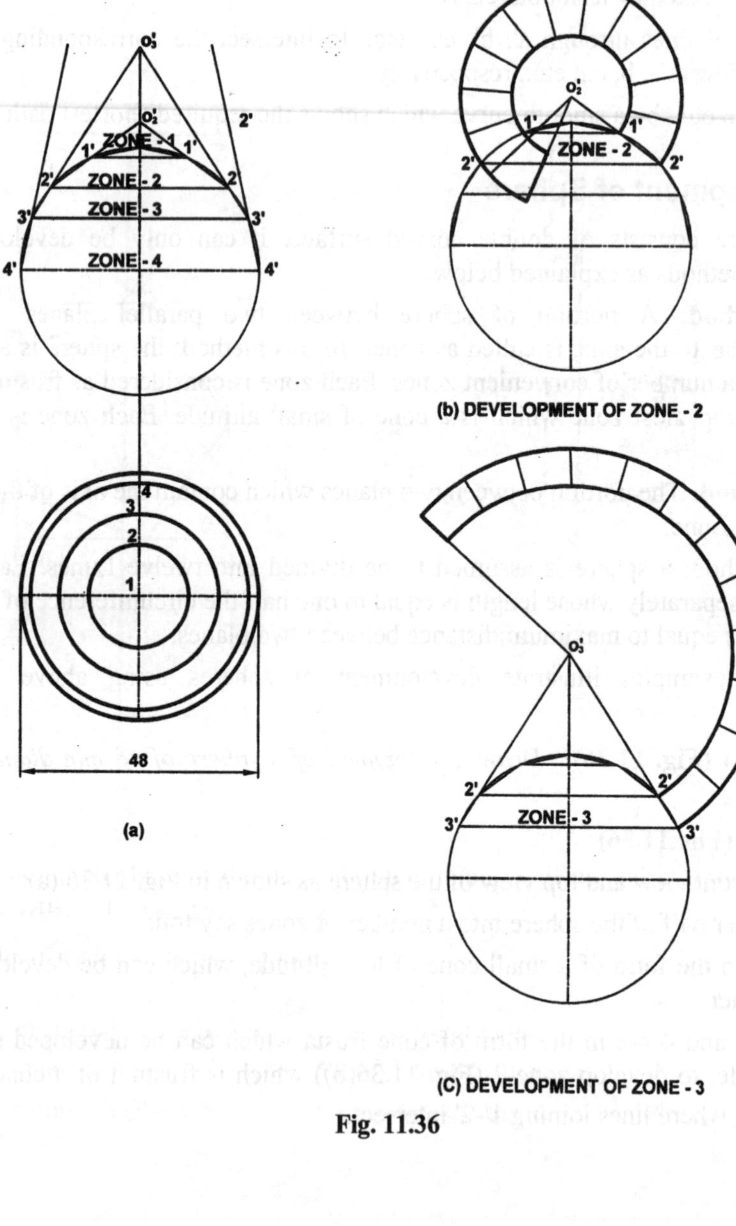

Fig. 11.36

Problem 11.35 (Fig.11.37): *Draw development of a sphere of 50 mm diameter using lune method.*

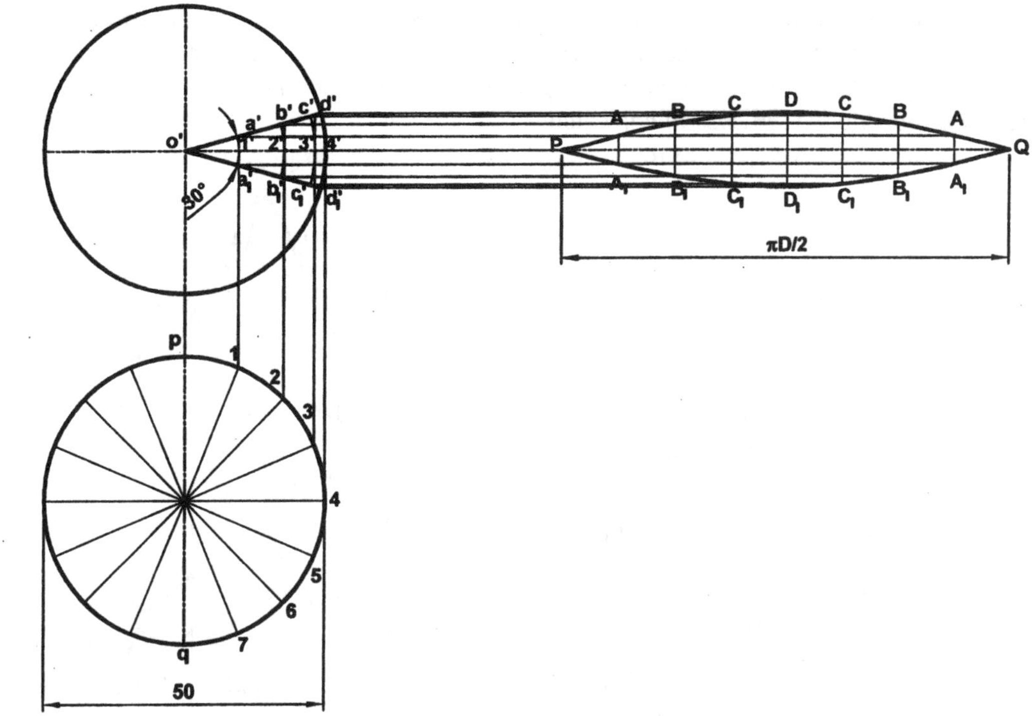

Fig. 11.37

Construction (Fig. 11.37):

1. Draw front view and top view of given sphere.
2. Considering the sphere to be divided into lunes, one of which is drawn as $o'd'd'_1$ which subtends an angle of 30° at the centre.
3. In the top view, divide the circle into sixteen divisions. Show the division points as 1, 2, 3 ... etc. Project these points to the front view as 1', 2', 3' ... etc.
4. With o' as centre and radius equal to o'1', draw an arc cutting o'd' at a' and $o'd'_1$ at a'_1. Similarly obtain b' and b'_1, c' and c'_1 ... etc.
5. Draw a line PQ equal to length of arc pq (i.e., $\pi D/2$). Divide it into eight equal parts and draw vertical lines through these division points.
6. Draw horizontal lines through a' and a'_1 to intersect vertical line through first division point at A and A_1 respectively. Similarly obtain B and B_1, C and C_1, ...etc. Join these points by smooth curve which is required development of this lune. Similarly other lunes can also be developed.

11.8 Additional Problems

Problem 11.36 (Fig. 11.38): *A hollow right circular cone made of paper is opened out whose development appears as a semi-circle of 40 mm radius. Inside this semi-circle, a full circle of the largest possible size is drawn in ink and the paper is folded back so as to form a cone. Draw the front view and top view of the cone assuming it to be in simple position showing the ink lines in each view.*

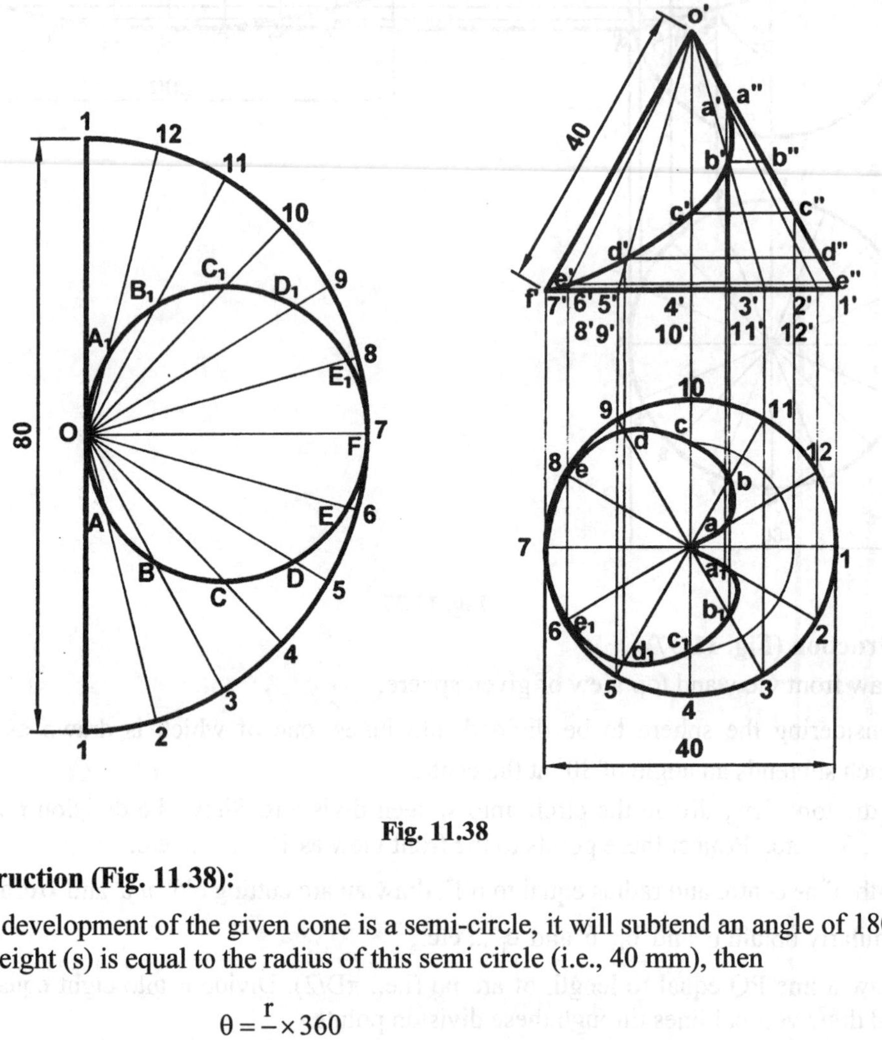

Fig. 11.38

Construction (Fig. 11.38):

Since, development of the given cone is a semi-circle, it will subtend an angle of 180° and slant height (s) is equal to the radius of this semi circle (i.e., 40 mm), then

$$\theta = \frac{r}{s} \times 360$$

$$180 = \frac{r}{40} \times 360 \text{ which gives } r = 20 \text{ mm}$$

1. Draw the front view and top view of a cone of base radius 20 mm and slant height 40 mm.
2. Divide the base circle into twelve equal parts and name the division points as 1,2.3... etc.
3. Project these points in the front view and draw its generators.
4. Draw a semi circle of 40 mm radius to represent the development of cone. Divide it into twelve equal parts. Name the division points as 1, 2, 3....etc., and draw radial lines O1, O2, O3... etc.
5. With O7 as diameter equal to 40 mm, draw a circle inside the semicircle which intersects the radial lines at A, B, C, A_1, B_1, C_1 ...etc.
6. In the front view, mark points a", b" c".... etc such that o'a" = OA, o'b" = OB, o'c" = OC ...etc. Draw horizontal lines through a", b" c"...etc., respectively intersecting corresponding generators at a', b', c'....etc. Draw a smooth curve through these points.
7. Project a', b', c' ...etc., to intersect corresponding generators at a, b, c...etc. Join these points by a smooth curve.

Problem 11.37 (Fig. 11.39): *The projections of a solid, composed of a truncated half cylinder and a cut half prism are shown in Fig.11.39(a). Draw the development of its lateral surface.*

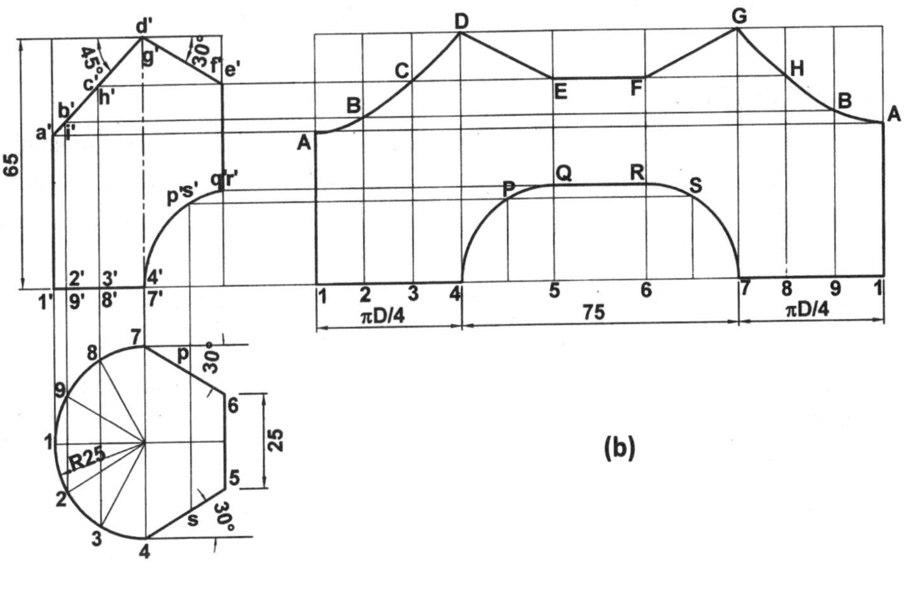

Fig. 11.39

Construction (Fig.11.39):

1. Draw the front view and top view of combined solid as shown in Fig. 11.39(a).
2. Draw stretch out line 1-1 equal to sum of half the circumference of the circle and half the perimeter of hexagonal prism.
3. Draw the development first assuming it to be un-cut.
4. Divide the semi circle in the top view into six equal parts and project the division points 1, 2, 3...etc., to the front view as 1', 2', 3'...etc. Draw the generators of half cylinder in the front view and draw vertical lines representing generators in the development also.
5. Draw line ps in the top view. Project these points to the quarter circle at p', s'.
6. Mark the points of intersection of cutting plane VT with the generators and vertical edges as a', b', c'...etc.
7. Draw horizontal lines through a', b', c'...etc., to cut the corresponding generators/edges at A, B, C...etc., respectively. Join these points as shown in the figure.
8. Similarly mark P, Q, R and S and draw a smooth curve 4-P-Q-R-S-7 and complete the development as shown

Problem 11.38 (Fig. 11.40): *A solid is made of half pyramid and half cone whose front view and top view are shown in Fig.11.40 (a). Draw the development of the lateral surface of the cut solid.*

Construction (Fig. 11.40):

1. Draw the top view and front view of the combined solid. Divide the base semi circle into six equal parts. Name the division points as 1, 2, 3,...etc., and project these points in the front view as 1', 2', 3'...etc., respectively. Also draw its generators through these points.
2. Draw O1 parallel to o'4'. With O as centre and radius equal to O1, draw an arc whose length is equal to circumference of base semi circle plus perimeter of half hexagon.
3. Draw cutting plane VT and mark its points of intersection with generators/slant edges as p', q', a', b', c'...etc.
4. Draw horizontal lines through p', q', a', b' etc., to intersect O1 and obtain points P, Q, A, B, C...etc., where the arcs cut the corresponding generators/edges.
5. Join these points and complete the development as shown in Fig. 11.40 (b).

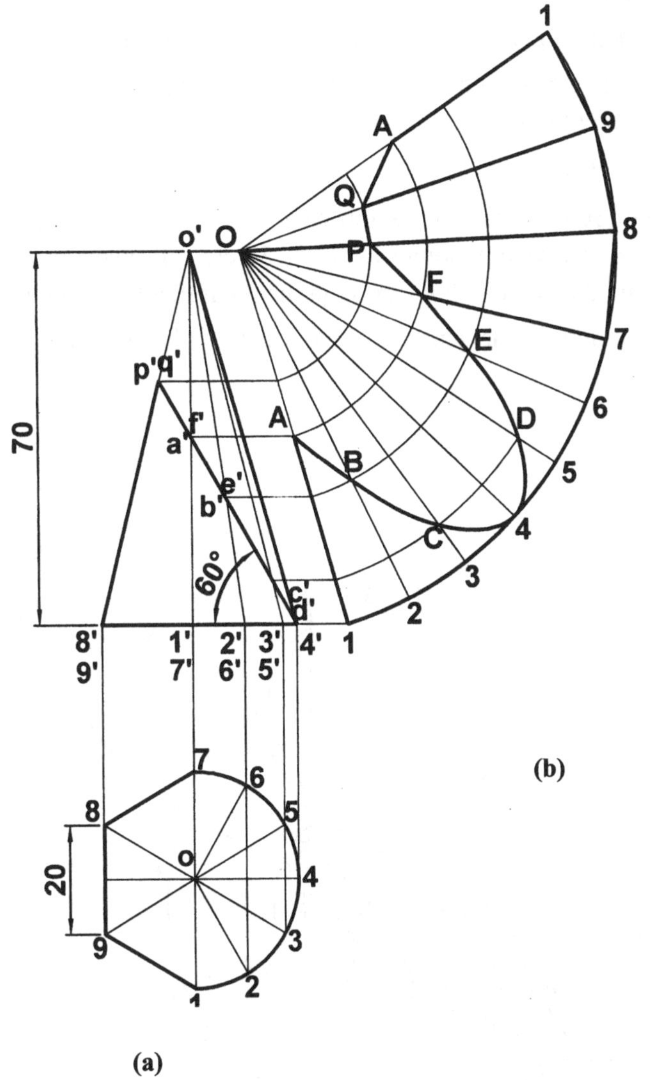

Fig. 11.40

Problem 11.39 (Fig. 11.41): *Draw development of the lateral surface of a funnel shown in Fig.11.41(a).*

Construction (Fig. 11.41):

The given funnel can be considered in two parts. Part A is truncated cone and part B is in the form of cylinder cut at one end.

Development of part A:

1. Draw development of given cone considering it to be uncut.
2. Through point of intersection of generators with cutting plane, draw horizontal lines to cut O1 at different points. Obtain the curve in usual manner.

Development of part B:

1. Draw a circle of 30 mm diameter as top view of cylindrical part. Divide it into twelve equal parts. Name division points as 1, 2, 3...etc. Project division points on the cutting plane and then extend lines through them parallel to stretch out line.
2. Obtain various points on the generators and join them by smooth curve.

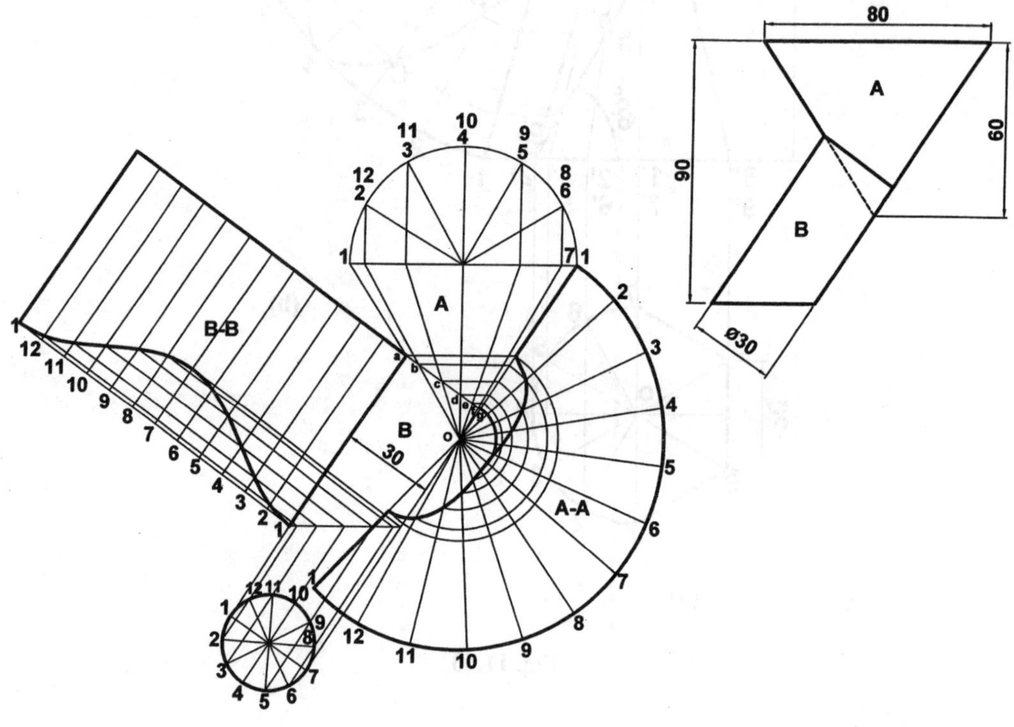

Fig. 11.41

Problem 11.40 (Fig.11.42): *A funnel shown in Fig. 11.42(a) is required to be fabricated using a metallic sheet. Obtain the shape of sheet metal required for its fabrication.*

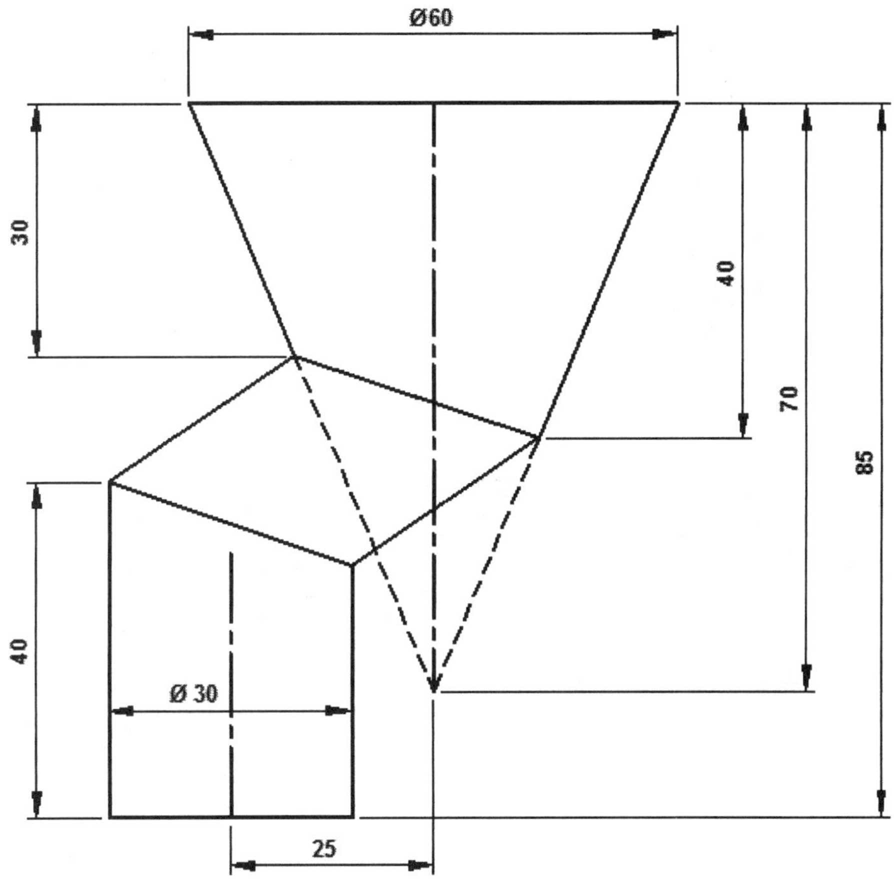

Fig. 11.42(a)

Construction (Fig. 11.42(b)):

1. Draw the front view of a given funnel.
2. Divide the front view in three parts A, B and C.
3. Part A is a truncated cone. Draw a number of generators on it and draw the development as shown (A-A).
4. Part B is in the form of a truncated cylinder at both ends. Draw its development as shown (B-B).
5. Part C is in the form of a truncated cylinder at one end. Draw its development as shown (C-C).

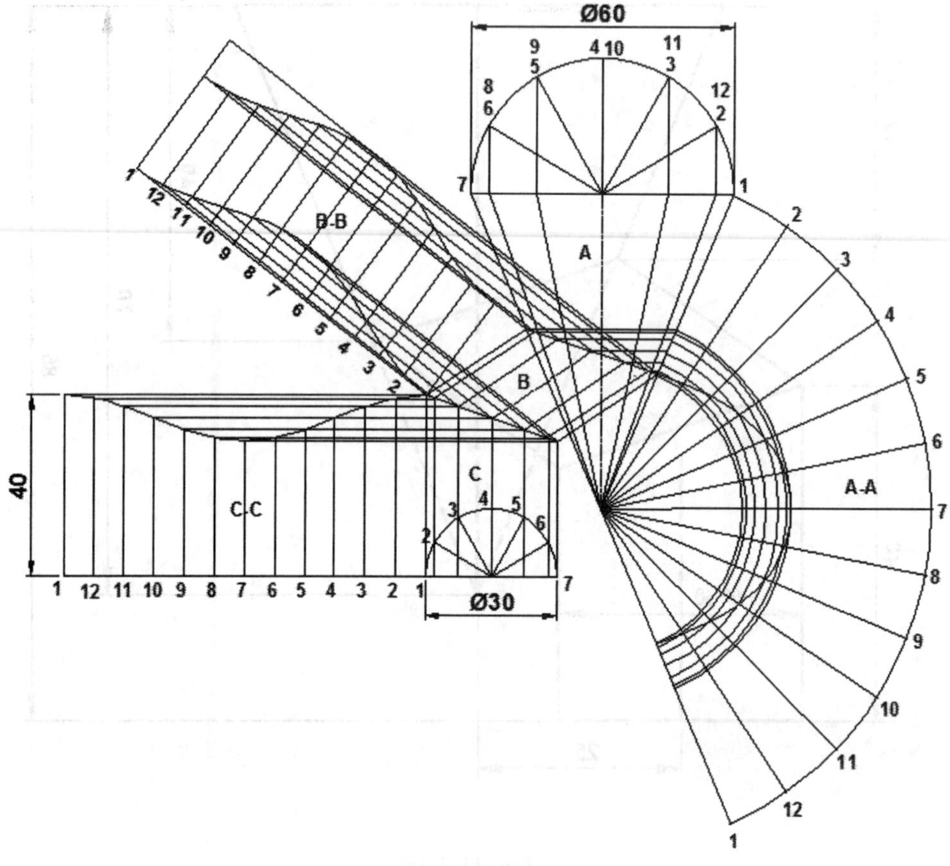

Fig. 11.42(b)

Problem 11.41 (Fig. 11.43): *A solid is made up of frusta of a cone and a frusta of hexagonal pyramid as shown in Fig. 11.43(a). Draw the development of its lateral surface.*

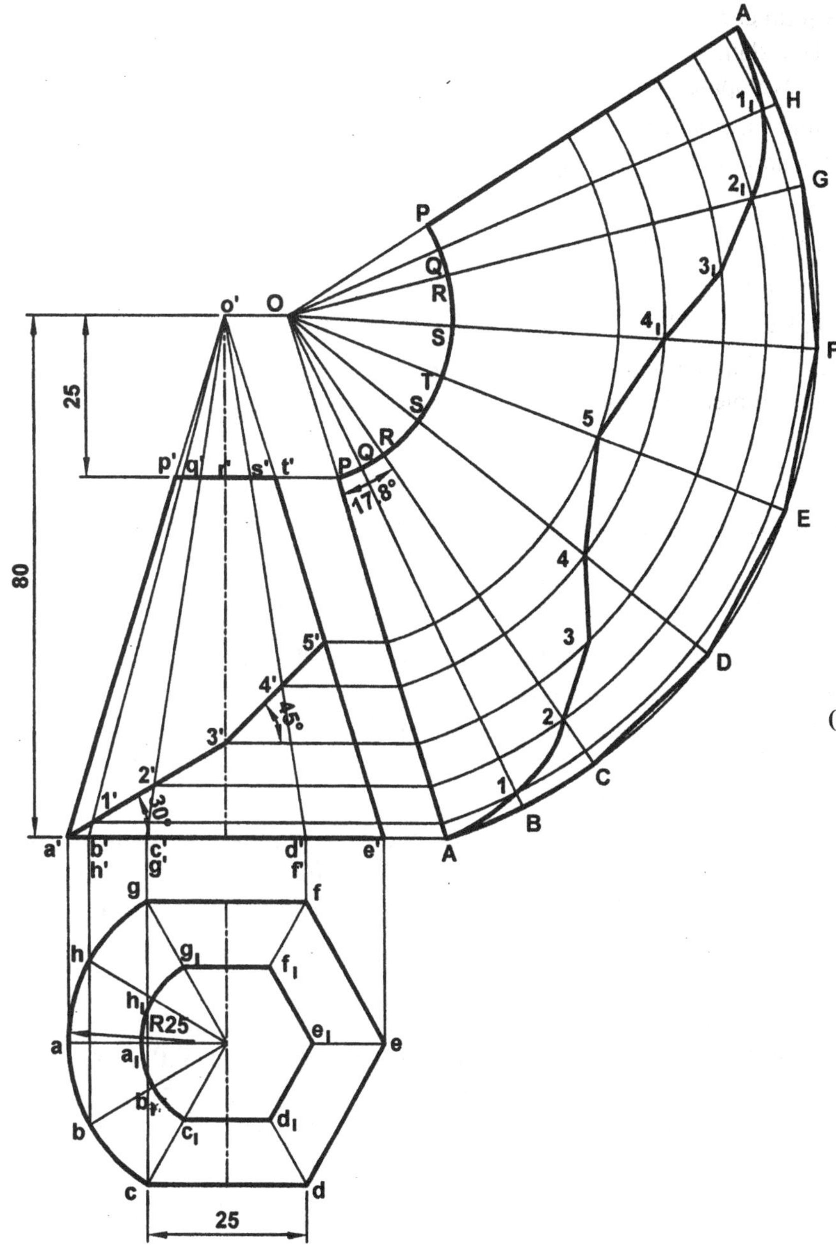
A
H
G
F
E
D
C
B
A
P
Q
R
S
T
S
R
Q
P
O
o'
1_1
2_1
3_1
4_1
5
4
3
2
1
17
8°
25
80
p' q' r' s' t'
5'
4'
3'
2'
1'
45°
30°
a' b' c' d' e'
h' g' f'
g f
h
g_1
f_1
h_1
R25
a
a_1
e_1
e
b_1
b
c_1
d_1
c d
25

(a)

(b)

Fig. 11.43

Construction (Fig. 11.43):

The given solid consists of a portion of frusta of cone and a portion of frusta of hexagonal pyramid. The slant height of conical portion is equal to the slant edge of hexagonal portion. Its development is drawn as follows:

1. Draw front view and top view of the solid as shown in Fig. 11.43 (a). Draw cutting plane VT and mark its points of intersection with generators and slant edges as 1', 2', 3' ... etc.
2. Draw OA parallel to o'e'. With O as centre and radius equal to OA and OP, draw arcs.
3. On the outer arc, step off divisions as follows,

   AB = ab, BC = bc, CD = DE = EF = FG = Side of base.
4. Join these points with O. Mark 1,2,3 ... etc. as explained in previous problems. Join them and complete the development as shown.

Exercises

1. Draw the development of the lateral surface a cube of 50 mm side.
2. Draw the development of lateral surface of a triangular prism having side of base 30 mm and axis 80 mm long when a vertical face is kept parallel to the V.P.
3. A square prism, side of base 30 mm and axis 70 mm long, is resting on its base on the H.P. with its vertical faces equally inclined with the V.P. It is cut by a section plane inclined at 45° to the H.P. and perpendicular to the V.P. in such a way that the prism is divided into two equal halves. Develop the lateral surface of the truncated prism.
4. A square prism, side of base 40 mm and axis 70 mm long, is resting on its base on the H.P. with its vertical faces equally inclined to the V.P. A circular hole of 40 mm diameter is drilled centrally through the prism in such a way that the axis of hole is perpendicular to the solid axis. Draw development of lateral surface of the prism.
5. A pentagonal prism, side of base 25 mm and axis 60 mm long, is resting on its base on the H.P. with a vertical face perpendicular to the V.P. It is cut by a plane inclined at 45° to the H.P. and perpendicular to the V.P. which passes through a point on the axis at a distance of 15 mm from the top end. Draw the development of the lateral surface of the prism.
6. A hexagonal prism, side of base 25 mm and axis 65 mm long is resting on its base on the H.P. with a vertical face parallel to the V.P. It is cut by a section plane perpendicular to the V.P. and inclined at 60° to the H.P. such that it divides the axis in two halves. Draw development of lateral surface of truncated prism.

7. Develop the lateral surface of a right regular hexagonal prism (side of base 30 mm and height 70 mm) kept vertically with a base side perpendicular to the V.P. and having a cylindrical hole of 40 mm diameter drilled centrally with the axis of hole being perpendicular to the V.P.

8. A triangular pyramid of 25 mm side of base and 60 mm height, is resting on its base on the H.P. with a side of base parallel to V.P. A section plane making an angle of 60° with the H.P. and perpendicular to V.P., cuts the axis at a height of 35 mm from the base. Draw development of truncated pyramid.

9. A frustum of a square pyramid has its base 40 mm side, top 25 mm side and height 70 mm. Draw the development of its lateral surface.

10. A square pyramid, side of base 40 mm and axis 70 mm long, is resting on its base on the H.P. with its sides of base equally inclined to the V.P. A horizontal section plane cuts the pyramid at a height of 40 mm from the base. Develop the lateral surface of the pyramid.

11. A pentagonal pyramid having side of base 30 mm and height 65 mm is resting on its base on the H.P. with a side of base perpendicular to the V.P. It is cut by a section plane inclined at 60° to the H.P. and bisect the axis. Draw the development of lateral surface of remaining portion of the pyramid.

12. A hexagonal pyramid, side of base 30 mm and axis 65 mm long, is resting on its base on the H.P. with two of its base edges perpendicular to the V.P. An auxiliary vertical plane whose HT is inclined at 60° to the V.P. cuts the pyramid through a point 10 mm away from the axis. Draw its projections and develop the lateral surface of the retained portion of the pyramid.

13. A cylinder having diameter of base 40 mm and axis 60 mm long, is resting on its base on the H.P. It is cut by a section plane perpendicular to the V.P. and inclined at 45° to the H.P. which passes through the top end of an extreme generator of the cylinder. Draw development of the lateral surface of the cut cylinder.

14. A cylinder base diameter 40 mm and axis 75 mm long is resting on its base on the H.P. A square hole of 30 mm side is made centrally through the cylinder such that axis of the hole is perpendicular to the axis of the solid. Draw the development of lateral surface of the cylinder.

15. A cone having diameter of base 50 mm and axis 75 mm long, is resting on its base on the H.P. It is cut by a section plane perpendicular to both H.P. and V.P. and 10 mm away from the axis. Draw the development of lateral surface of retained portion of the cone.

16. A cone having diameter of base 60 mm and axis 80 mm long, is resting on its base on the H.P. A section plane perpendicular to the V.P. and inclined to the H.P., is passing through a point on the axis 25 mm from the apex and cuts the

cone parallel to an extreme generator. Draw the development of lateral surface of the retained solid.

17. Draw the development of surface of a sphere having 70 mm diameter by zone method.

18. Draw the development of surface of a sphere having 65 mm diameter by Lune method.

19. A funnel consists of a conical part and a cylindrical pipe as shown in Fig.11.44. Draw development of lateral surfaces.

20. Draw the development of the lateral surface of the part P of a square prism having side of base 30 mm and axis 70 mm long whose front view and top view are shown in Fig. 11.45.

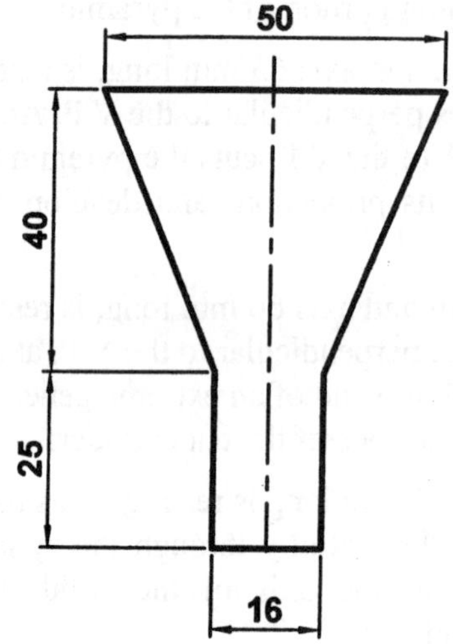

Fig. 11.44

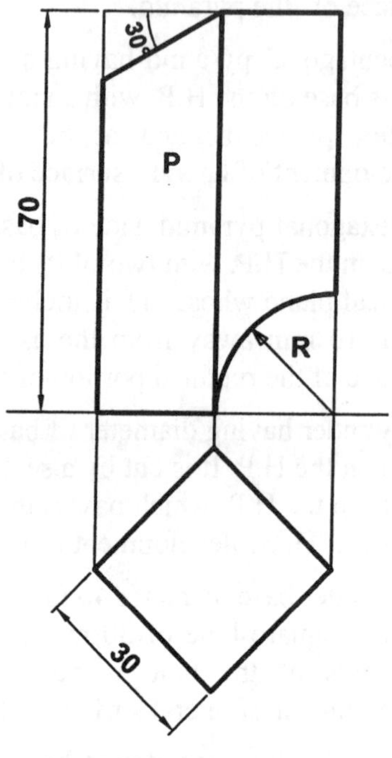

Fig. 11.45

21. Draw the development of lateral surface of a truncated cylinder as shown in Fig. 11.46.

22. A right circular cylinder having diameter of base 50 mm and height 70 mm is truncated from its two ends by two different section planes as shown in Fig. 11.47. Develop the lateral surface of the truncated cylinder.

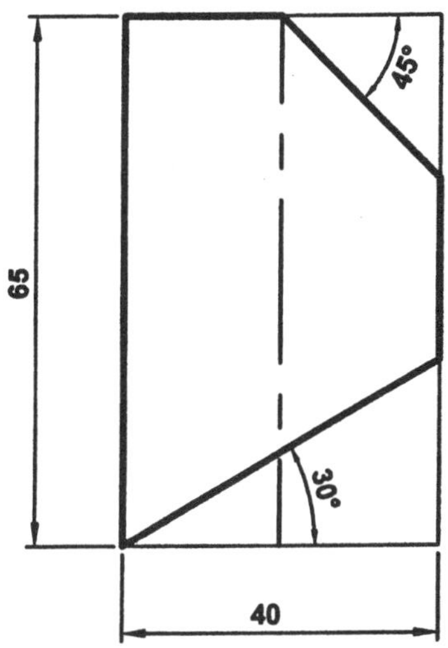

Fig. 11.46

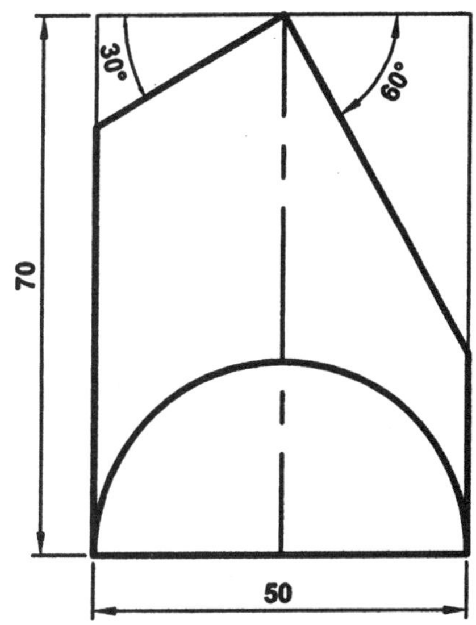

Fig. 11.47

CHAPTER 12

Isometric Projections

12.1 Introduction

Isometric projection is a kind of pictorial projection in which all three dimensions (i.e., length, breadth and height) are shown in one view. It offers an additional advantage that, the actual sizes can be measured directly from it. An isometric drawing of an object is useful to understand objects of complex nature. In engineering practice, many complex objects are required to be designed and drawn before actual manufacturing. Sometimes orthographic projections alone do not bring its clear understanding but when they are presented with their isometric views, it becomes more easier to understand their designs.

To understand the concept of isometric projection consider a cube resting on one of its corner on the H.P. with its solid diagonal perpendicular to the V.P. Projections of the cube in the above mentioned position are drawn in the following three stages as shown in Fig.12.1

Stage 1

Draw the front view and top view assuming the cube to be kept on its base on the H.P. with the vertical faces equally inclined to the V.P.

In this position, the solid diagonal AG is parallel to the V.P. and inclined to the H.P.

Stage 2

Tilt the front view about corner c' so that the line a'g' becomes parallel to the xy line. Project the second top view from the tilted front view. Now solid diagonal ag is parallel to both the places.

Stage 3

Rotate the second top view in such a way that the top view of solid diagonal (a_1g_1) is perpendicular to the xy line. Now project the third front view. This front view is an isometric projection of the cube.

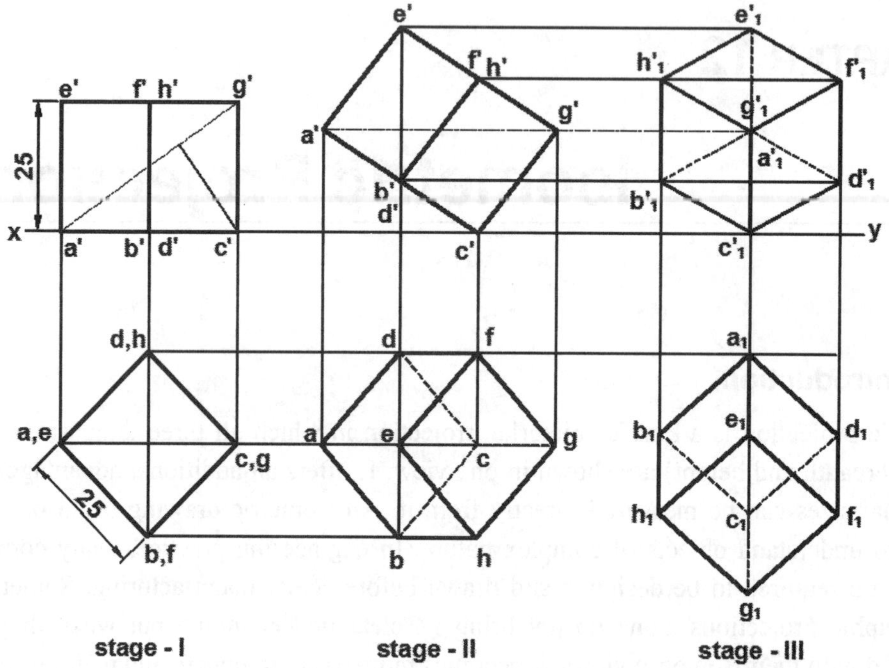

Fig. 12.1

So from this example, it is clear that, if a cube is placed on one of its corner on the ground with a solid diagonal perpendicular to the V.P., the front view is the isometric projection of the cube.

Consider the front view of the cube in above position with the corner designated in capital letters as shown in Fig. 12.2. Following points can be observed.

1. All the faces of the cube are equally inclined to the V.P. which are similar and equal rhombuses.
2. The three lines AC, AH and AF meeting at point A, represent three edges of the cube, which are:
   (i) Equally inclined to the V.P.
   (ii) Equally foreshortened
   (iii) Equally inclined to each other i.e., at 120°

The line AC is vertical whereas other lines AH and AF are inclined at 30° with the horizontal.

3. All other lines representing the edges of the cube are parallel and equal to one or the other of the lines AC, AH and AF.
4. The diagonal HF of the top face is parallel to the V.P. and hence remains equal to its true length.

12.2 Important Terms

1. **Isometric Axes:** Three lines AC, AH and AF meets at a point C and making an angle of 120° with each other are known as isometric axes.
2. **Isometric Lines:** Lines parallel to the isometric axes such as BF, DH, EA, BC, CD etc., are called as isometric lines.
3. **Isometric Planes:** Planes representing faces of the cube are termed as isometric planes. However, planes parallel to these planes are also known as isometric planes. For example HAFE, ACBF, ACDH... etc., are isometric planes.

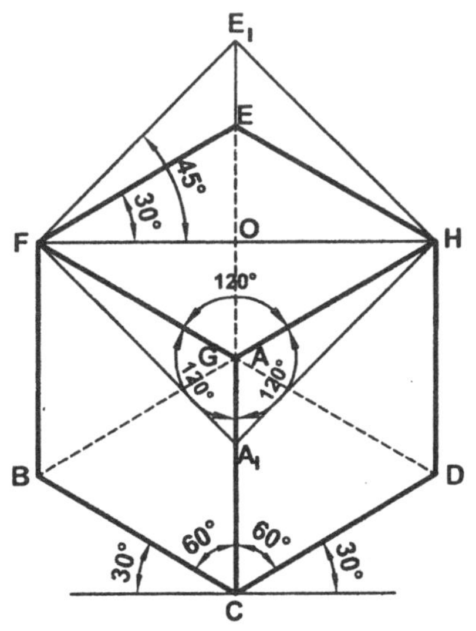

Fig. 12.2

Important Note: *It is clear from the above discussion that the given horizontal line in orthographic view becomes a line inclined at 30° to horizontal in isometric view where as vertical line in orthographic view, remains vertical in the isometric view.*

12.3 Isometric Scale

We have seen in previous sections that since, all the edges of the cube are equally foreshortened, the square faces are seen rhombuses. For example the rhombus HEFA (refer Fig. 12.2) is an isometric projection of the top square face of the cube where HF is true length of the diagonal. Around diagonal HF, construct a square E_1 H A_1 F then HE_1 represents true length of HE.

In triangle EHO,
$$\frac{HE}{HO} = \frac{1}{\cos 30^\circ} = \frac{2}{\sqrt{3}} \quad \text{..... (1)}$$

In triangle E_1HO,
$$\frac{HE_1}{HO} = \frac{1}{\cos 45^\circ} = \frac{\sqrt{2}}{1} \quad \text{..... (2)}$$

Dividing (1) by (2) we get,
$$\frac{HE}{HE_1} = \frac{2}{\sqrt{3}} \times \frac{1}{\sqrt{2}} \times \frac{\sqrt{2}}{\sqrt{3}} = 0.815$$

or
$$\frac{\text{Isometric Length}}{\text{True Length}} = \frac{HE}{HE_1} = 0.815$$

It shows that the isometric projection is reduced in the ratio of $\sqrt{2}:\sqrt{3}$ or we can say that the isometric length is 0.815 of the true length. Hence, for drawing isometric projection of any object, it is required to convert true length into isometric length. For ease of conversion, an isometric scale can be constructed and used conveniently. The procedure of construction of an isometric scale is explained as follows:

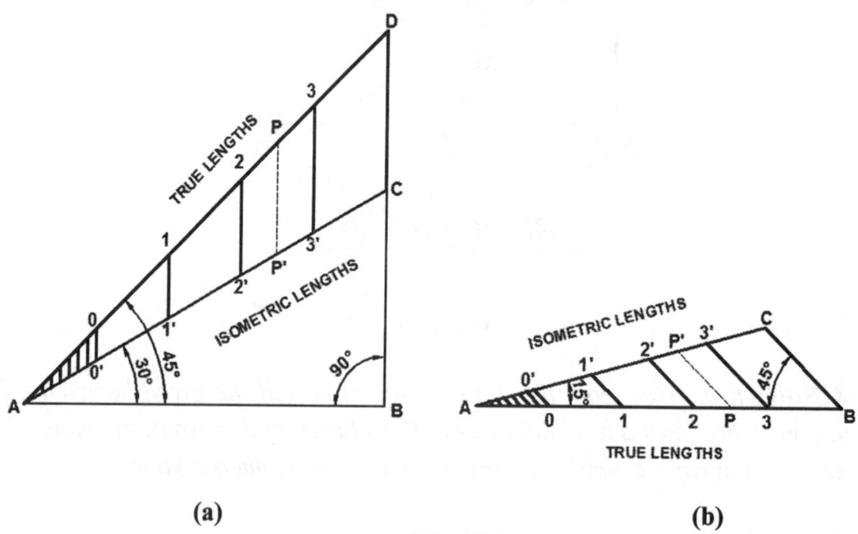

Fig. 12.3

Method I (Fig. 12.3(a)):

1. Draw a horizontal line AB of any length.
2. Through A, draw another lines AC and AD inclined at 30° and 45° to the horizontal respectively.
3. Mark a number of divisions on line of true length (on the AD).
4. From each division points, draw vertical lines meeting AC at respective points. Thus division obtain on line AC give lengths on isometric scale. For example, to convert AP into isometric length, mark point P on AD at a given distance. Then through P, draw a vertical line meeting AC at P'. Thus AP' gives the isometric length of AP.

Method II (Fig. 12.3(b)):

1. Draw a horizontal line AB of any convenient length.
2. At A, draw another line AC inclined at 15° to the AB.
3. At B, draw line BC inclined at 45° to the AB.
4. Mark a number of division points 0, 1, 2, 3... etc., on AB. Through 0, 1, 2, 3....etc., draw lines parallel to BC and meeting the respective division points on AC.
5. In order to convert a true length AP into isometric length, first mark P on the AB at the given distance. Then through P draw a line parallel to BC and meeting AC on P'. Thus AP' is the required isometric length.

12.4 Isometric Projections Vs Isometric Views

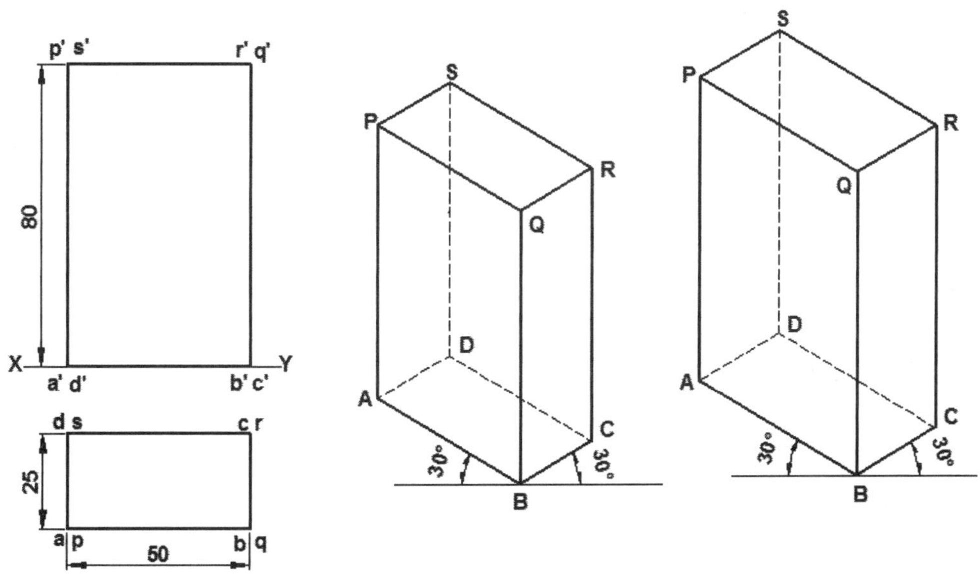

(a) Orthographic Projection (b) Isometric Projection (c) Isometric View

Fig. 12.4

The basic differences between isometric projection and isometric view are tabulated in Table 12.1 and shown in Fig. 12.4

Table 12.1

S. No	Isometric View	Isometric Projection
1.	True lengths are used to prepare the drawing.	Isometric lengths are used to prepare the drawing.
2.	It is larger than isometric projection.	It is shorter than isometric view.
3.	The dimensions can be measured directly from the drawing.	The dimension cannot be measured directly but conversion is required.

12.5 Important Tips for Preparation of Isometric Drawing

Following tips are important for preparing isometric projections or views of an object. (Refer Fig.12.4)

1. The vertical lines on the object will remain vertical in isometric projection.
2. The horizontal lines on the object will be shown by line inclined at 30° to the horizontal in isometric projections or views.
3. The lines parallel to the isometric axes (isometric lines) are equally foreshortened in isometric projection.
4. It can be observed that the lines which are not parallel to the isometric axes, are not reduced according to the fixed standard ratio. Such lines are known as non isometric lines. Therefore, for measurement purpose, only isometric lines or axes are used. For drawing non-isometric lines, positions of their ends are first located on the isometric planes and then they are joined.
5. For dimensioning, true lengths of the edges are shown on the isometric view or projections.
6. All the dimension lines or extension lines must be drawn parallel to the isometric lines.
7. The hidden lines are usually avoided unless they are very essential.

12.6 Isometric Drawing of Plane Figures

Before drawing the isometric view of a plane it should be checked carefully whether given plane represents its front view or top view. There may be following two cases.

1. If front view of a plane is given, then surface is assumed to be vertical. So in isometric view vertical edges will be drawn vertical, while horizontal edges will be drawn inclined at 30° to the horizontal.

2. If top view of a plane is given, then surface is assumed to be horizontal and therefore all the edges will be horizontal. So in isometric view all the edges will be drawn inclined at 30° to horizontal.

Problem 12.1 (Fig. 12.5): *Draw the isometric view of a triangular plane*

(a) *Whose front view is shown in Fig. 12.5(a)*

(b) *Whose top view is shown in Fig. 12.5(a)*

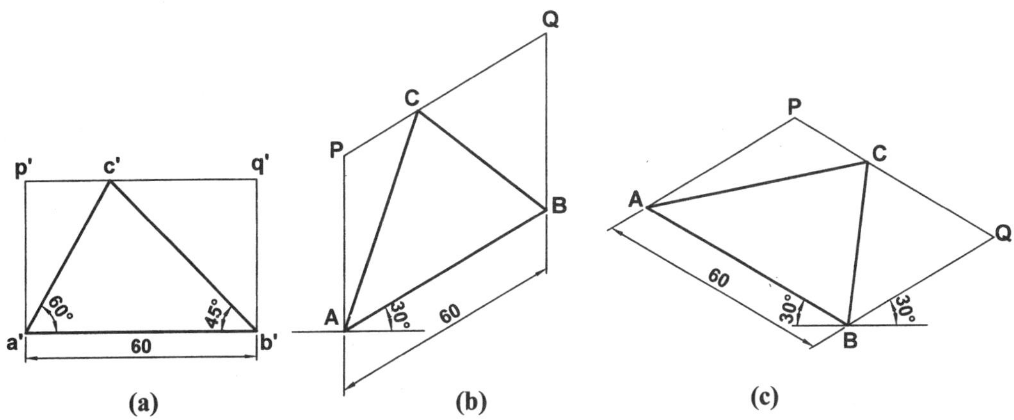

Fig. 12.5

Construction (Fig. 12.5):

(a) When Fig. 12.5(a) shows its front view:

Since, line AC and BC are non-isometric lines, these lines cannot be drawn directly. So enclose the triangle in a rectangle a'b'c'd'.

1. Since surface of the triangle is vertical and its base AB is horizontal. So draw a line AB 60 mm long and inclined at 30° to the horizontal as shown in Fig. 12.5(b)
2. At A, draw a vertical line AP and complete the rhombus ABQP. Mark a point C on PQ such that p'c' = PC.
3. Join AB, BC and CA. The triangle ABC is the required isometric view.

(b) When Fig. 12.5(a) shows its top view:

1. In this case surface of the triangle a'b'c' is horizontal. So at B, draw two lines BA and BQ inclined at 30° to the horizontal such that AB = a'b' and BQ = b'q'. Complete the Rhombus ABQP as shown in Fig.12.5(c).
2. Mark a point C on PQ such that p'c' = PC.
3. Join AB, BC and CA to obtain triangle ABC. The triangle ABC is the required isometric view.

Problem 12.2 (Fig. 12.6): *Draw the isometric view of a square plane of the 40 mm side when,*

(a) *Fig.12.6(a) represents its front view.*

(b) *Fig.12.6(a) represents its top view.*

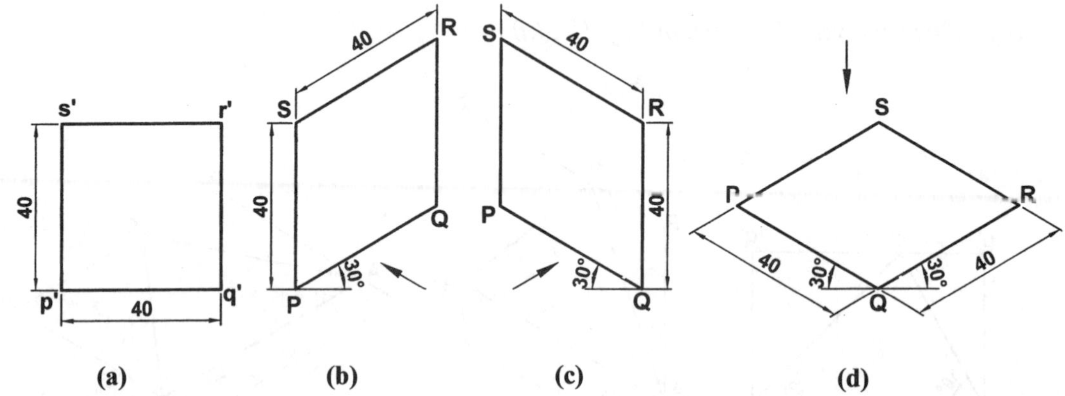

Fig. 12.6

Construction (Fig. 12.6):

(a) When Fig. 12.6(a) represents front view:

Refer Fig.12.6 (b) and (c)

1. Since, front view is a square and its surface is vertical, then vertical edge will be shown by vertical line and horizontal edge will be shown by line inclined at 30° to the Horizontal.
2. So at any point P, draw a line PQ inclined at 30° to the horizontal and a vertical line PS equal to 40 mm. Complete the rhombus PQRS as shown in Fig.12.6(b). The view can also be drawn towards other slopping axis as shown in Fig.12.6(c).

(b) When Fig. 12.6(a) represents top view:

Refer Fig. 12.6(d)

1. As the top view is a square, the surface of the square is horizontal so all the edges of the square will be shown by inclined lines at 30° to the horizontal. At a point Q, draw lines QP and QR, 40 mm long and inclined at 30° to the horizontal.
2. Complete the rhombus PQRS which is the required isometric view.

Problem 12.3 (Fig. 12.7): *The top view of a rectangular plane having length 50 mm and breadth 25 mm, is shown in Fig.12.7(a). Draw its isometric view.*

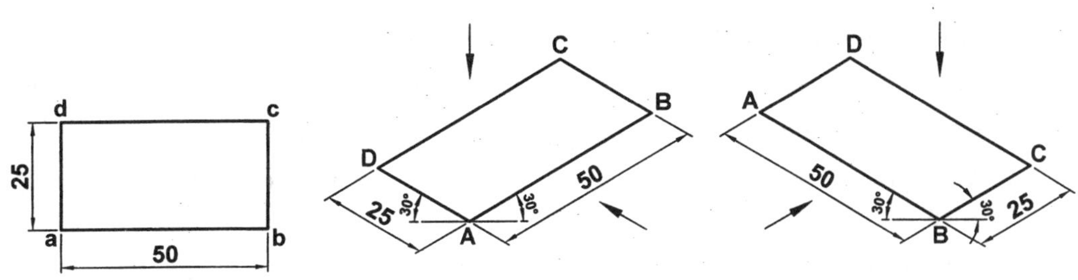

(a) (b) (c)

Fig. 12.7

Construction (Fig. 12.7):

1. Since, the given figure shows the top view of a rectangle, its surface is horizontal and therefore, all the edges will be shown inclined at 30° to the horizontal. So at any point A draw two lines AB and AD inclined at 30° to the horizontal such that AB = ab and AD = ad.
2. Complete the rhombus ABCD, which is the required isometric view
3. The view can also be drawn taking longer side towards left sloping axis as shown in Fig.12.7(c)

Problem 12.4 (Fig.12.8): *Draw the isometric views of a quadrilateral when,*

(a) Fig.12.8(a) shows its front view.

(b) Fig.12.8(a) shows its top view.

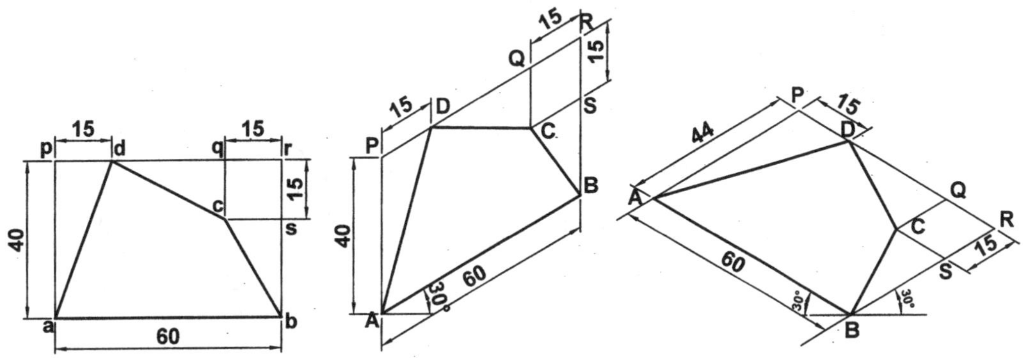

(a) (b) (c)

Fig. 12.8

Construction (Fig. 12.8):

(a) When the given figure represents its front view:

(Refer Fig.12.8(b))

1. Enclose the given quadrilateral in a rectangle aprb.
2. Through c, draw lines parallel to ab and br to mark the point s and q respectively.
3. Draw the isometric view of rectangle abrp as shown in Fig.12.8(b).
4. Mark point D and Q on PR such that pd = PD and qr = QR. Mark point S on BR such that rs = RS.
5. Through Q, draw a line parallel to BR and through S draw a line parallel to PR intersecting at C.
6. Join AB, BC, CD and DA. The quadrilateral ABCD is required isometric view.

(b) When the given figure represents its top view:

(Refer Fig.12.8(c))

1. At any point B, draw two lines BA and BR inclined at 30° to horizontal such that ab = AB and br = BR.
2. Complete the rhombus ABRP and obtain point C and D as discussed above.
3. Join AB, BC, CD and DA. The quadrilateral ABCD is required isometric view.

Problem 12.5 (Fig.12.9): *Draw the isometric view of a pentagonal plane when:*

(a) Fig.12.9(a) represents its front view.

(b) Fig.12.9(a) represents its top view.

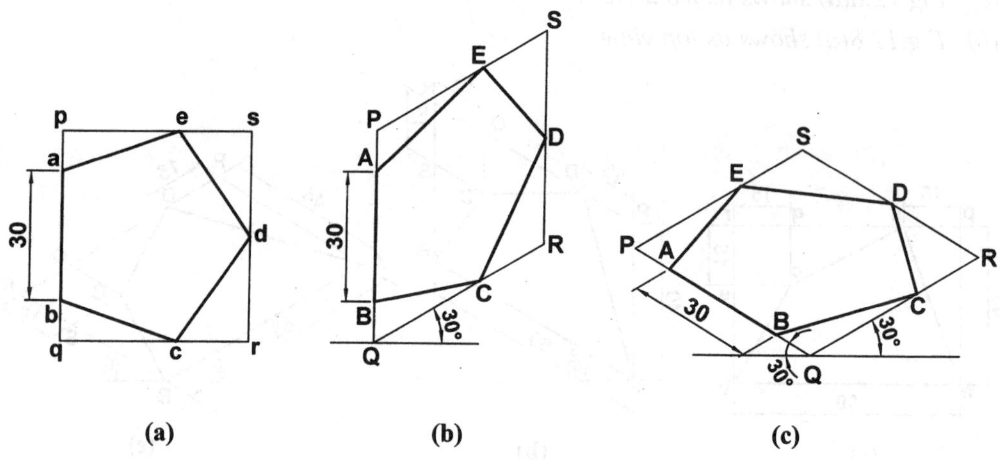

Fig.12.9

Construction (Fig.12.9):

(a) When the given figure represents its front view:

1. Enclose the pentagon in a rectangle pqrs.

2. At any point say Q, draw a line QR inclined at 30° to the horizontal and a vertical line QP such that qr = QR and pq = PQ as shown in Fig. 12.9(b)
3. Complete the rhombus PQRS. Mark the point A, B, C, D and E in such a way that pa = PA, bq = BQ, qc = QC, rd = RD and pe = PE.
4. Join AB, BC, CD, DE and EA. The figure obtained is the required isometric view.

(b) When the given figure represents its top view:

1. At any point say Q (Fig.12.9(c)) draw lines QR and QP inclined at 30° to the horizontal such that pq = PQ and qr = QR.
2. Complete the rhombus PQRS and obtain the points A, B, C, D and E as explained earlier.
3. Join AB, BC, CD, DE and EA by straight lines. The figure obtained is the required isometric view.

Problem 12.6 (Fig.12.10): *Draw isometric view of a hexagonal plane shown in Fig. 12.10(a) when*

(a) It represents front view and

(b) It represents top view.

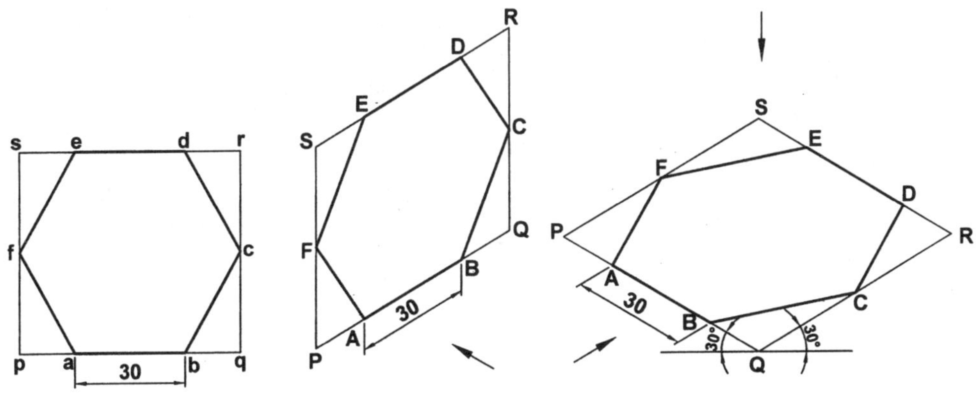

(a) (b) (c)

Fig.12.10

Construction (Fig.12.10):

(a) When given figure represents front view:

1. Enclose the hexagon in a rectangle pqrs.

2. At any point say P, draw a line PQ inclined at 30° to the horizontal and a vertical line PS such that pq = PQ and ps = PS as shown in Fig.12.10(b)
3. Complete the rhombus PQRS and mark points A, B, C, D, E and F such that pa = PA, qb = QB, qc = QC etc.
4. Join AB, BC, CD, DE, EF and FA by straight lines, which is the required isometric view.

(b) When the given figure represents its top view:

1. At a point say Q (Fig.12.10(c)), draw lines QR and QP inclined at 30° to the horizontal such that qr = QR and qp = QP.
2. Complete the rhombus PQRS and mark points A, B, C, D, E and F on it.
3. Join AB, BC, CD, DE, EF and FA by straight lines. The figure obtained is the required isometric view.

Problem 12.7 (Fig.12.11): *A circular plane having 60 mm diameter shown in Fig. 12.11(a). Draw its isometric view using method of points when:*

(a) It represents front view.

(b) It represents top view.

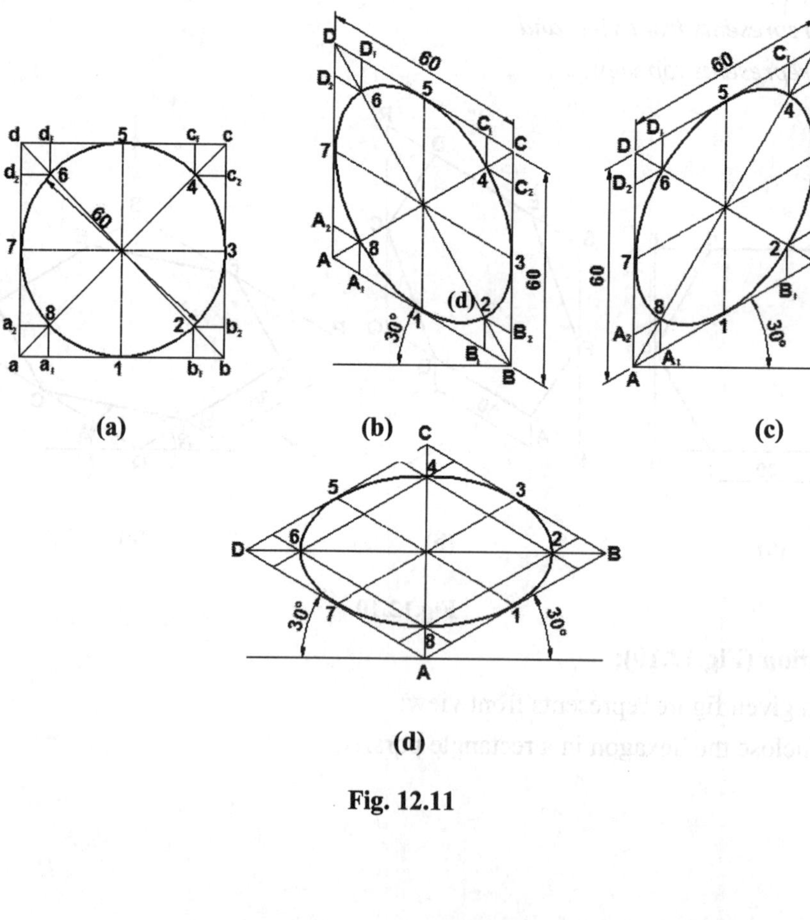

Fig. 12.11

Construction (Fig.12.11):

Orthographic view (see Fig.12.11(a)):

1. Draw a circle of 60 mm diameter and enclose it in a square abcd as shown in Fig.12.11(a). Mark mid points of its sides as 1, 3, 5 and 7.
2. Draw its diagonal ac and bd intersecting the circle at point 2, 4, 6 and 8.

Isometric view (see Fig.12.11(b) and (c)):

(a) When Fig.12.11(a) represents front view:

1. Draw the isometric view of the square abcd as shown in fig.12.11(b) or (c) and mark mid points of the rhombus ABCD as 1, 3, 5 and 7.
2. Draw diagonals AC and BD. Mark points 2, 4, 6 and 8 on AC and BD such that $2B_1 = 2b_1$ and $2B_2 = 2b_2$, $8A_1 = 8a_1$ and $8A_2 = 8a_2$....etc.
3. Join 1, 2...etc., to obtain isometric view.

(b) When Fig.12.11(a) represents top view:

1. Draw rhombus ABCD which represents isometric view of square abcd. (Refer Fig.12.11 (d))
2. Mark mid points of sides as 1, 3, 5 and 7.
3. Draw diagonals AC and BD. Mark points 2, 4, 6 and 8 on the diagonals as described earlier.

Problem 12.8 (Fig.12.12): *Draw the isometric view of a circle of 60 mm diameter as shown in Fig. 12.12(a) using four centre method when:*

(a) Fig.12.12(a) represents front view.

(b) Fig.12.12(a) represents top view.

Construction (Fig.12.12(b) and (c)):

(a) Fig.12.12(a) represents front view:

1. Draw rhombus ABCD representing isometric view of square abcd.
2. Mark midpoints of sides of rhombus as 1, 2, 3 and 4.
3. Draw its longer diagonal BD.
4. Join the ends of the minor diagonals (A and C) to the midpoints of their opposite sides. Mark centres P and Q as shown.
5. Draw the ellipse using four centres P, Q, A and C which is the required isometric view. Alternatively the isometric view can be drawn on the other sloping axis using the procedure explained above as shown in Fig.12.12(c).

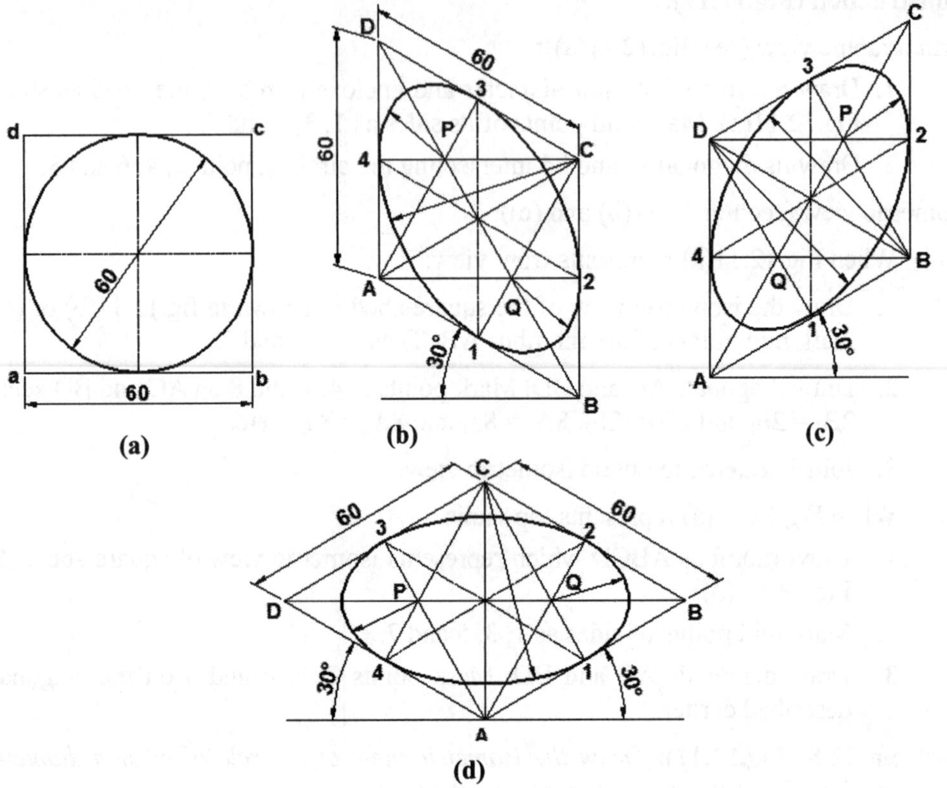

Fig. 12.12

(b) Fig.12.12(a) represents top view:

Refer Fig.12.12(d)

1. Draw rhombus ABCD to represent isometric view of the square abcd.
2. Mark midpoints of the sides as 1, 2, 3 and 4.
3. Draw major diagonal BD.
4. Join A and C (ends of the minor diagonals) to the midpoints of the opposite sides and obtain centres P and Q as shown.
5. Draw an ellipse using above four centres (P, Q, A and C) which is the required isometric view.

Problem 12.9 (Fig.12.13): *Draw the isometric view of a semi-circular plane of 50 mm diameter when*

(a) *It is kept parallel to the vertical plane.*

(b) *It is kept on the horizontal plane.*

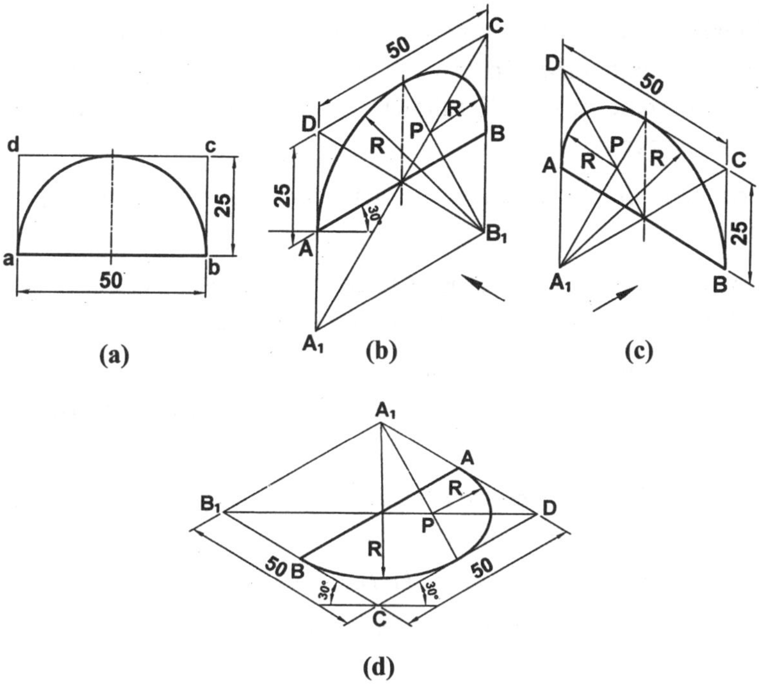

Fig.12.13

Construction (Fig.12.13):

(a) When the given plane is kept parallel to the V.P.

1. Enclose the given semi-circle in a rectangle abcd and draw its isometric view ABCD as shown in Fig.12.13(b).
2. Complete the rhombus A_1B_1CD and obtain point P as explained in problem 12.8.
3. Using two centres P and B_1 (A_1 in case of Fig.12.12(c)) draw the half ellipse which is the required isometric view.

(b) When the given plane is kept on the H.P:

Refer Fig.12.13(d)

1. Assuming square plane draw the isometric view of it as rhombus A_1B_1CD and obtain centre P as explained in problem 12.7.
2. Using centres P and A_1 draw half ellipse and join A to B.
3. This is required isometric view.

12.7 Isometric Drawing of Solids

Containing isometric lines some solid such as cube, square prism, rectangular prism etc., have right angles at their corner. In drawing, edges of such solids are represented by the lines parallel to the isometric axes. For drawing isometric drawings of any solid following points must be kept in mind:

1. Isometric view should be drawn in such a way that its maximum possible details are visible.
2. At every corner of the solid, at least three lines representing edges must converge. Out of these, at least two lines must be for visible edges.
3. Hidden lines representing the invisible edges, are usually avoided. But it is advisable to check up every corner so that no line representing visible edges is left out.

Problem 12.10 (Fig.12.14): *Draw the isometric view of a square prism having side of its base, 30 mm and axis 60 mm long, when*

(a) It is resting on its square base on the H.P.

(b) It is resting on its rectangular face on the H.P.

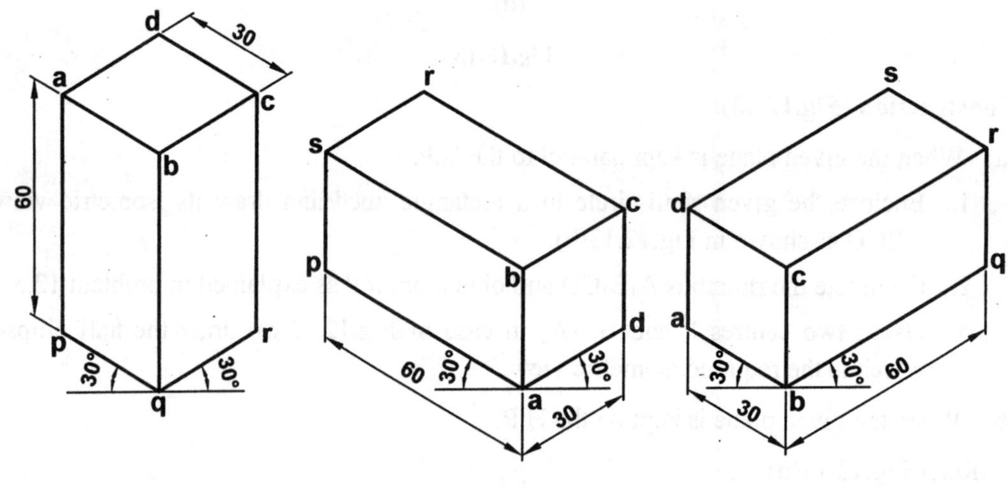

Fig. 12.14

Construction (Fig.12.14):

(a) When it is resting on its square base on the H.P. (Refer Fig.12.14(a)):

1. Since, square base is horizontal, it will be seen as rhombus in isometric view. So draw rhombus abcd to represent isometric view of the top face as shown in Fig.12.14(a).
2. Draw vertical lines a-p, b-q and c-r 60 mm long. Since, edge d-s will not be visible, it should not be drawn.
3. Join p to q and q to r thus completing the required isometric view of the square prism.

Construction may also be initiated from point q, and then proceeding upward.

(b) When it is resting on a rectangular face. (Refer Fig.12.14(b) and (c)):

In this case ends will be vertical

1. Draw isometric view of the end abcd (Fig.12.14(b))
2. Draw line ap, 60 mm long inclined at 30° the horizontal. Through point d and c draw lines ds and cr equal to the length of 60 mm and parallel to ap.
3. Join p-s and s-r to get the required isometric view.
4. The length can also be drawn on right isometric axis as shown in Fig.12.14(c).

12.8 Isometric Drawing of the Solid containing Non-isometric Edges

Many objects have edges which are not parallel to isometric axes. In the isometric view, such edges will be shown by non isometric lines. The isometric view of such objects are drawn by using one of the following methods.

(i) Box method

(ii) Offset or co-ordinate method

(i) Box Method: This method is used when the inclined lines (non isometric lines) and their ends lie in the isometric plane. For drawing isometric view of such objects, first of all the given object is assumed to be enclose in a rectangular box. Then isometric view of the box is drawn. The ends of inclined lines are then located be measuring their distances from or on the outlines of the box.

Following problem has been solved using box method.

Problem 12.11 (Fig.12.15): *Draw the isometric view of a pentagonal prism of 25 mm base side and axis 65 mm long resting on the H.P. with one of the base edge perpendicular to the H.P.*

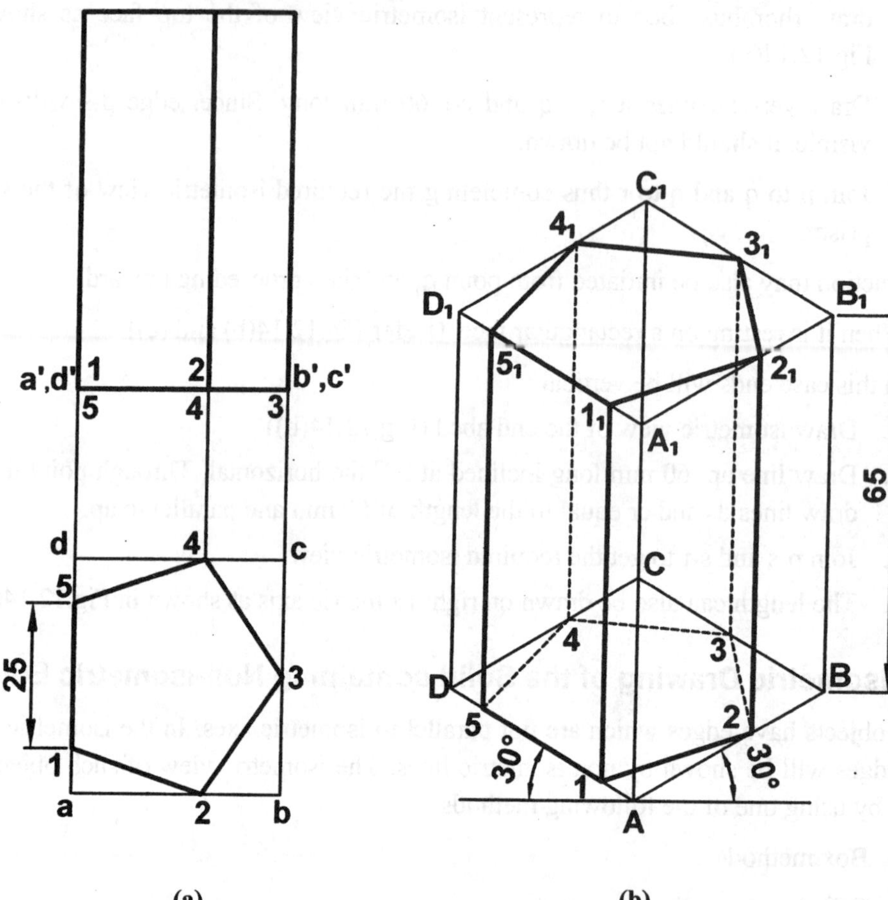

Fig.12.15

Construction (Fig. 12.15):

1. Draw the top view12345 and enclose it in a rectangle abcd. Project its front view as shown in Fig. 12.15 (a).
2. Draw the isometric view of rectangle abcd (Fig. 12.15 (b)). The rhombus ABCD represents the base of the box. Through points A, B, C and D draw vertical lines AA_1, BB_1, CC_1 and DD_1 equal to 65 mm.
3. Join A_1-B_1-C_1-D_1 to get isometric view of the box.
4. Mark points 1, 2, 3, 4 and 5 on the bottom face and 1_1, 2_1, 3_1, 4_1 and 5_1 on the top face such that A1 = a1, A2 = a2...etc., and join all the corners by straight lines.
5. Join line 1-1_1, 2-2_1, 3-3_1, 4-4_1 and 5-5_1 as vertical edges.
6. The figure obtained is the required isometric view as shown in Fig.12.15(b).

(ii) Offset or Co-ordinate Method: This method is used to draw the isometric view of the objects in which neither non isometric lines nor their ends lie in the isometric planes.

To locate any point, perpendicular or offset is drawn on the edge from it. Then by measuring and transferring the distance, the required point is located in isometric drawing for example for locating ends of the axis in Fig.12.16, first an offset or perpendicular op is drawn on any edge say ab then a line OP = op is drawn parallel to BC (Fig.12.16(b)). Thus centre O can be located in isometric drawing. Similarly any other point can be located.

Problem 12.12 (Fig.12.16): *Draw an isometric view of a hexagonal prism having side of base 25 mm and axis 65 mm long resting on its base on the H.P. with an edge of its base parallel to the V.P.*

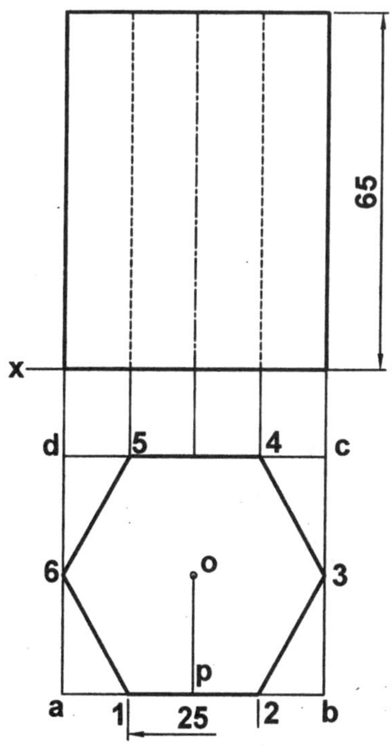

(a)

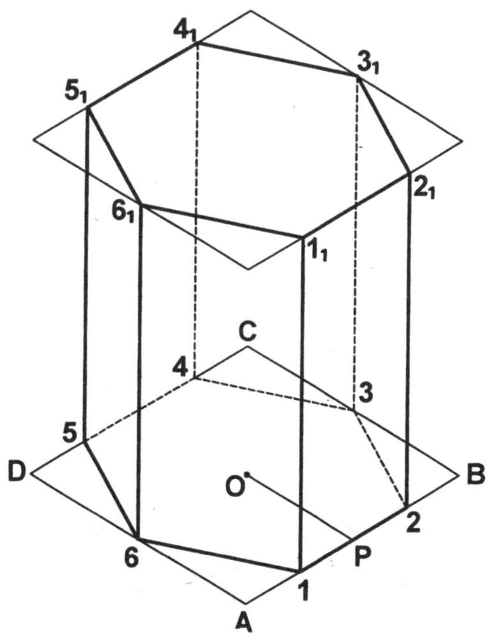

(b)

Fig.12.16

Construction (Fig. 12.16):

1. Draw a regular hexagon of 25 mm base to represent top view of the hexagonal prism (Fig. 12.16 (a)). Project the top view to draw its front view.
2. Enclose the top view in a rectangle abcd. Mark the corner of hexagon as 1, 2...etc.

3. Draw the isometric view of the rectangle abcd as ABCD, as shown in Fig.12.16(b).
4. Locate the corner 1, 2, 3, 4, 5 and 6 such that A1 = a1, B2 = b2...etc.
5. Through points 1, 2, 3...etc., draw vertical lines 11_1, 22_1, 33_1...66_1. Join point 1_1, 2_1, 3_1... 6_1.
6. The figure obtained is isometric view of a hexagonal prism.

Problem 12.13 (Fig.12.17): *Draw the isometric view of a square pyramid having side of its base 30 mm and axis 70 mm long which is resting on its base on the H.P. with an edge of its base perpendicular to the V.P.*

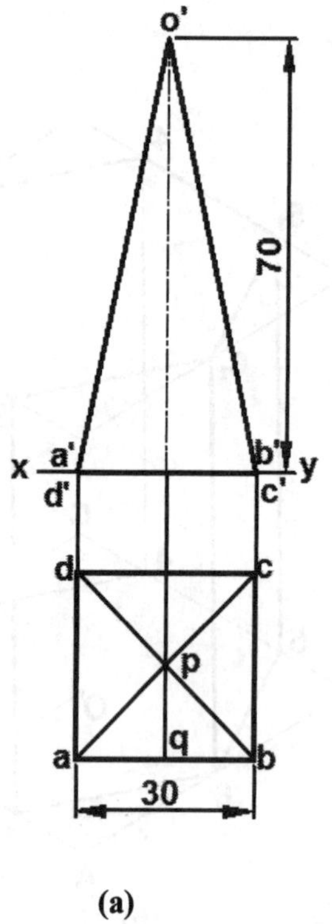

(a)

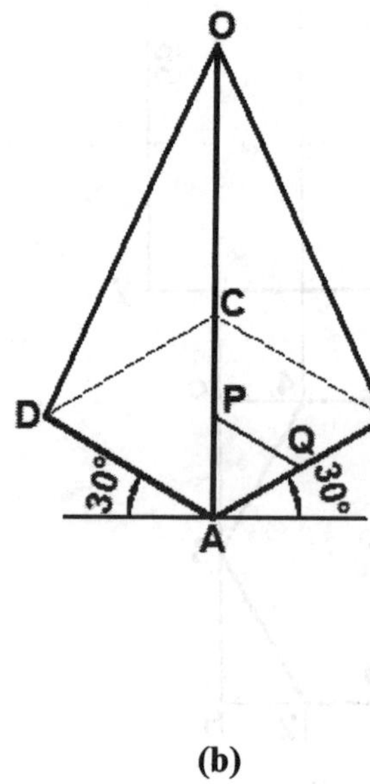

(b)

Fig.12.17

Construction (Fig. 12.17):

1. Draw a square abcd to represents the top view of the square pyramid. Join a to c and b to d. Project the top view to obtain the front view as shown in Fig.12.17(a). From point p, draw a perpendicular pq on the edge ab.
2. Draw the rhombus ABCD as isometric view of the base. (Refer Fig. 12.17 (b)).

3. Mark point Q on the AB such that AQ = aq and draw a line PQ = pq and parallel to AD.
4. Through P draw a vertical axis PO 70 mm long.
5. Join all the corner points A, B, C and D to O.
6. The dotted lines may be erased.
7. The figure obtained is the required isometric view.

Problem 12.14 (Fig.12.18): *Draw the isometric view of pentagonal pyramid having side of the base 30 mm and axis 70 mm long whose projections are shown in Fig. 12.18(a) when*

(a) Its axis is vertical.

(b) Its axis is horizontal.

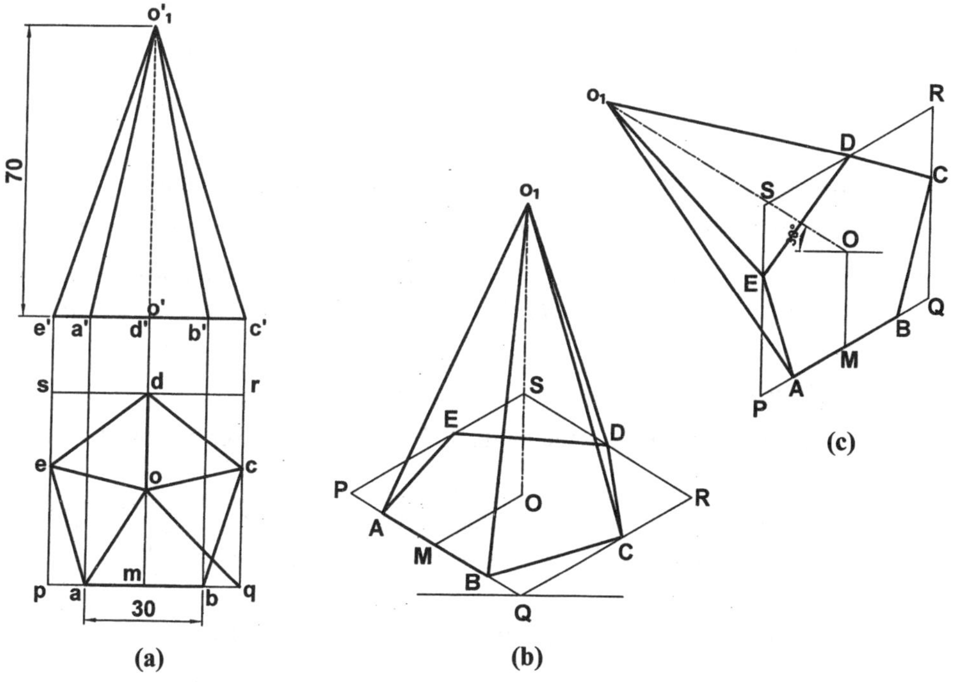

Fig.12.18

Construction (Fig.12.18):

(a) When its axis is vertical:

1. Enclose the base (top view) in a rectangle pqrs and draw an offset om on the line pq as shown in Fig. 12.18(a).

2. Draw the isometric view of the rectangle and locate the corners A, B, C, D and E of the base in it (Fig. 12.18(b)).
3. Mark a point M on the line PQ such that PM = pm. From M, draw a line MO whose length is equal to mo and parallel to QR.
4. Draw a vertical axis OO_1 equal to 70 mm.
5. Join O_1 with corners A, B, C, and D.
6. The edge O_1E is not visible so it is not required to be shown.
7. The visible edges may be darkened thus completing the isometric view of the pyramid.

(b) When the axis is horizontal:

1. In this case, the base will be in vertical plane. So drawn the isometric view of the rectangle pqrs assuming it to be vertical, as shown in Fig.12.18(c).
2. Locate the corners of the base of pyramid as A, B, C, D and E.
3. Mark M on the line PQ such that PM = pm and draw a vertical line MO equal to mo.
4. Draw axis OO_1 70 mm long and inclined at 30° to the horizontal.
5. Join O_1 with A, B, C, D and E, thus completing isometric view of the pyramid.

Problem 12.15 (Fig.12.19): *Draw the isometric view of a hexagonal pyramid having side of base 25 mm and axis 60 mm long which is resting on its base on the H.P. with a side of base parallel to the V.P.*

Construction (Fig.12.19):

1. Enclose the top view in a rectangle 1234. Draw as offset om on the line bc as shown in Fig.12.19(a).
2. Draw the isometric view of the rectangle 1234 as shown in Fig. 12.19(b). Locate the corner A, B, C and F on it. The invisible corners D and E need not to be shown.
3. Mark a point M on the line 23 such that BM = bm. Draw a line OM parallel to 12 and equal to the length of om.
4. From O_1 draw a vertical axis OO_1 equal to 60 mm.
5. Join O_1 with the corners A, B, C and F thus completing the required isometric view.

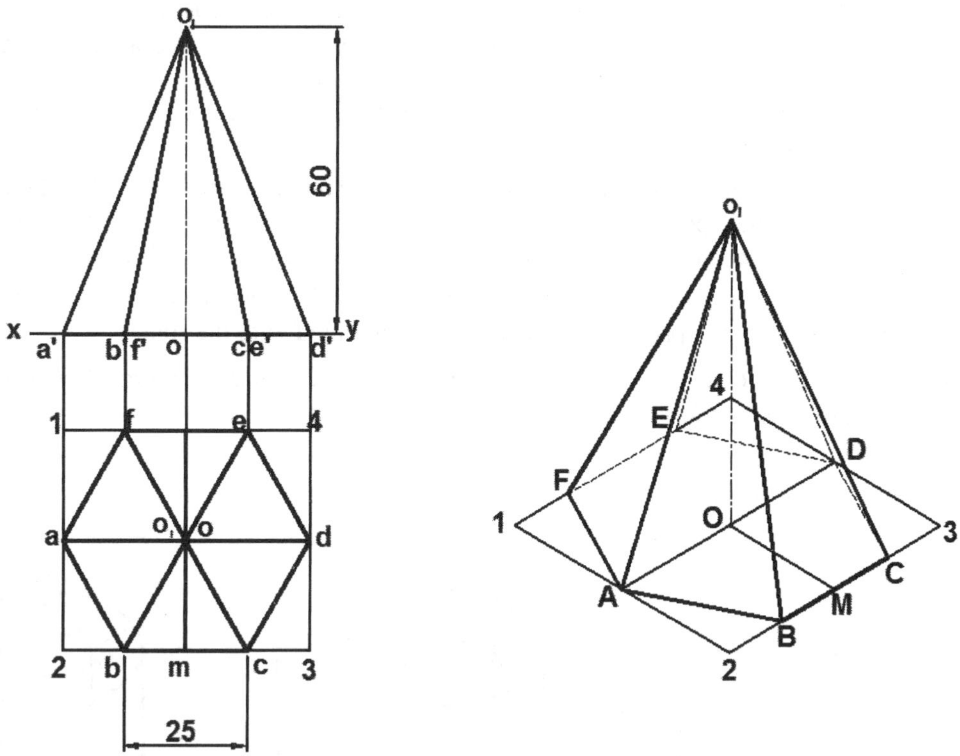

(a) (b)

Fig. 12.19

12.9 Isometric Drawing of Solids of Revolution

Solid such as cylinder, cone and sphere are known as solids of revolution. Both cylinders and cones have circular base whose isometric view appears as an ellipse.

The isometric view of a cylinder can be drawn by following methods.

1. **Method 1 (Box Method):** The cylinder is assumed to be enclosed in a square prism, which has two rhombuses (top and bottom). Ellipse is drawn in each rhombus using four centre method. Two ellipses are joined by two common tangents to get the final isometric view of a cylinder. The detailed procedure is explained in problem12.16.
2. **Method 2:** In this method, one rhombus is drawn for upper end of the prism and an ellipse is drawn in it using four centre method. From these four centres, vertical lines of length equal to the length of the axis, are drawn. The lower ends of these lines act as four centres for lower ellipse. Upper and lower ellipse are connected by to common tangents to get the final isometric view of the cylinder.

For drawing isometric view of a cone, the base is enclosed in a square. Then, isometric view of the square is drawn as rhombus. As ellipse is drawn inside this rhombus using method explained as above. A vertical axis from the centre of the ellipse is drawn. Two tangents are drawn from the apex to get the final isometric view of a cone.

The isometric projection of a sphere is a circle whose diameter is equal to the true diameter of the sphere. The distance of the centre of the sphere from its point of contact with the ground should be equal to the isometric radius of the sphere.

So here it is important to note that isometric scale must invariably be used to draw the isometric projections of any solid combined with the sphere.

Problem 12.16 (Fig.12.20): *Draw the isometric view of a cylinder having 50 mm base diameter and axis 70 mm long when the*

(a) *Axis is vertical.*

(b) *Axis is horizontal.*

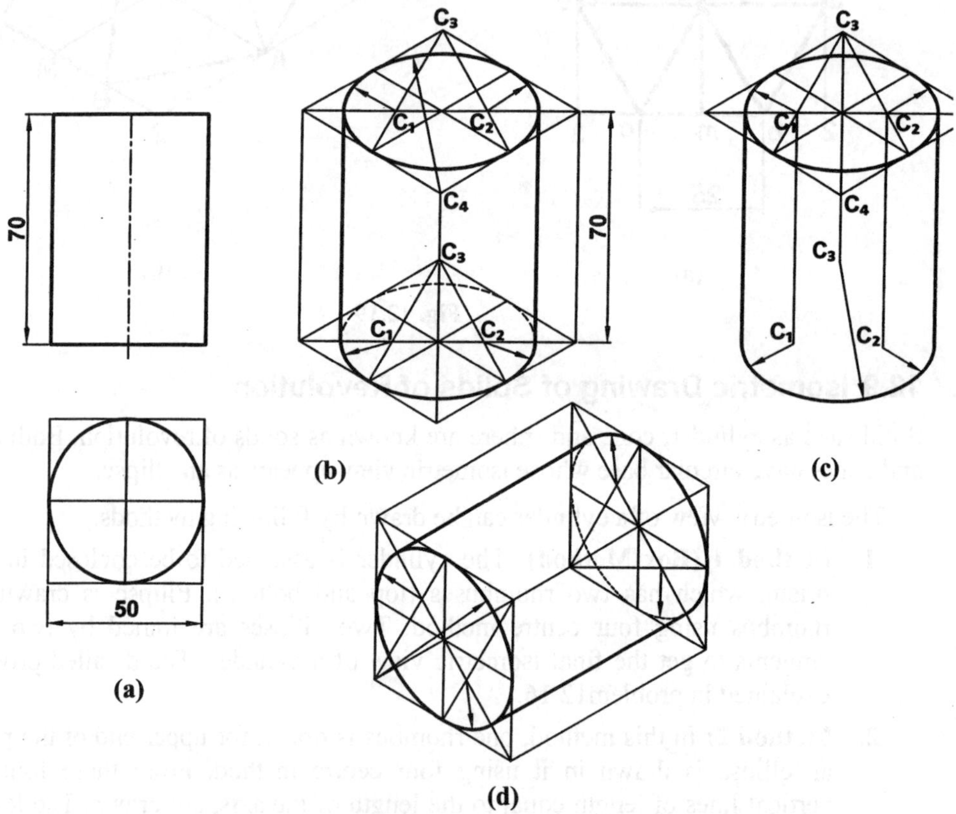

Fig.12.20

Construction (Fig.12.20):

(a) When the axis is vertical:

Method 1 (Fig. 12.20(b))

1. Assuming given cylinder to be enclosed in a square prism, draw the isometric view of the square prism as shown in Fig.12.20(b).
2. In top and bottom rhombuses, draw the ellipse using four centre method. The four centres C_1, C_2, C_3 and C_4 are obtained first as described earlier. These centres are used to draw different parts of the ellipse.
3. Draw two common tangents to the ellipse to get the required isometric view of the cylinder.

Method 2: (Fig. 12.20(c))

1. Draw the rhombus for the top end of the prism. Obtain for centres C_1, C_2, C_3 and C_4 as described earlier. Draw the ellipse using these centres.
2. Through centres C_1, C_2, C_3 and C_4 draw vertical lines of length equal to the length of the axis.(i.e., 70 mm).
3. The lower ends of these lines are used as centres for drawing ellipse for the lower part of the cylinder.

(b) When the axis is horizontal (Fig.12.20(d)):

1. Enclose the cylinder in a square prism.
2. Draw the isometric view of the square prism in horizontal position.
3. Draw the ellipse in two rhombuses using four centre method.
4. Draw common tangents to these ellipses to obtained final isometric view of the cylinder.

Problem 12.17 (Fig.12.21): *Draw the isometric view of a cone having diameter of its base 40 mm and axis 65 mm long when the base is (a) in the H.P. and (b) in the V.P.*

Construction (Fig.12.21):

(a) When the base is in H.P. (Fig.12.21(b)):

1. Enclose the base circle in a square.
2. Draw the isometric view of the square and draw an ellipse in it using four centre method.
3. Locate the position of apex O_1 using offset method.
4. Draw tangents to the ellipse from the apex O_1.
5. Erase the dotted part of ellipse.

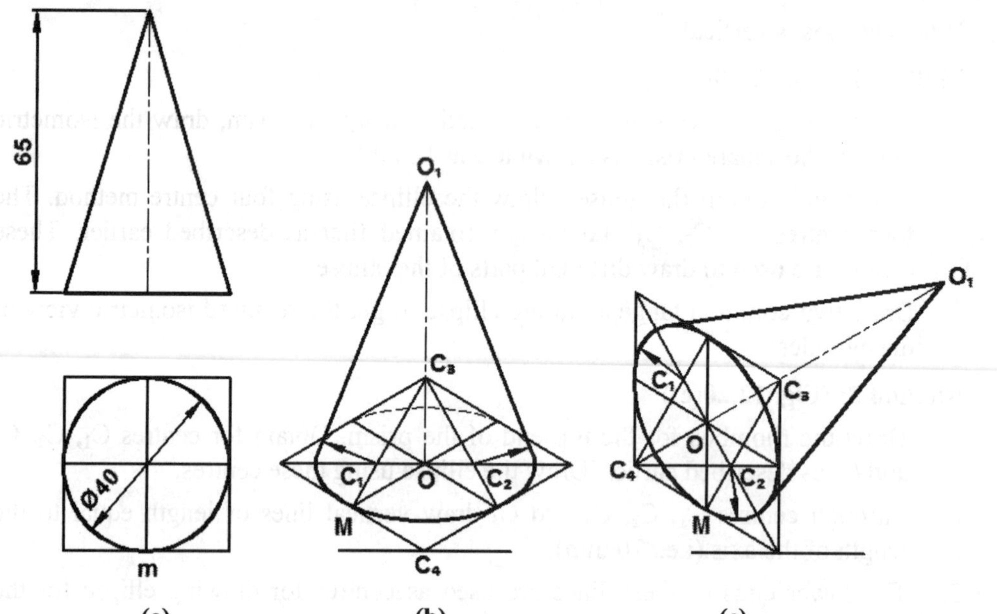

Fig.12.21

(b) When the base is in the V.P. (Fig.12.21(c)):

1. Assuming base to be vertical, enclose the base circle in a square and draw its isometric view as shown in Fig.12.21(c). Mark centre O and draw axis OO_1 65 mm long and inclined at 30° to horizontal.
2. Draw an ellipse inside the rhombus using four centre method.
3. Draw common tangents to the ellipse from the apex.

12.10 Isometric Drawing of Frustums

For drawing isometric drawing of frustums of solid such as pyramids and cones, box method is more suitable. The frustums are assumed to be enclosed in a rectangular box. Then top and bottom ends are drawn inside the rhombuses as explained earlier. Final view is obtained by connecting top and bottom ends by lines or tangents. Detailed procedure is explained in problems 12.18 and 12.19 as follows:

Problem 12.18 (Fig.12.22): *Draw the isometric view of a frustum of hexagonal pyramid having side of the base 30 mm, side of the top 20 mm and axis 70 mm long. The solid is resting on its base on the H.P. with its two edges parallel to the V.P.*

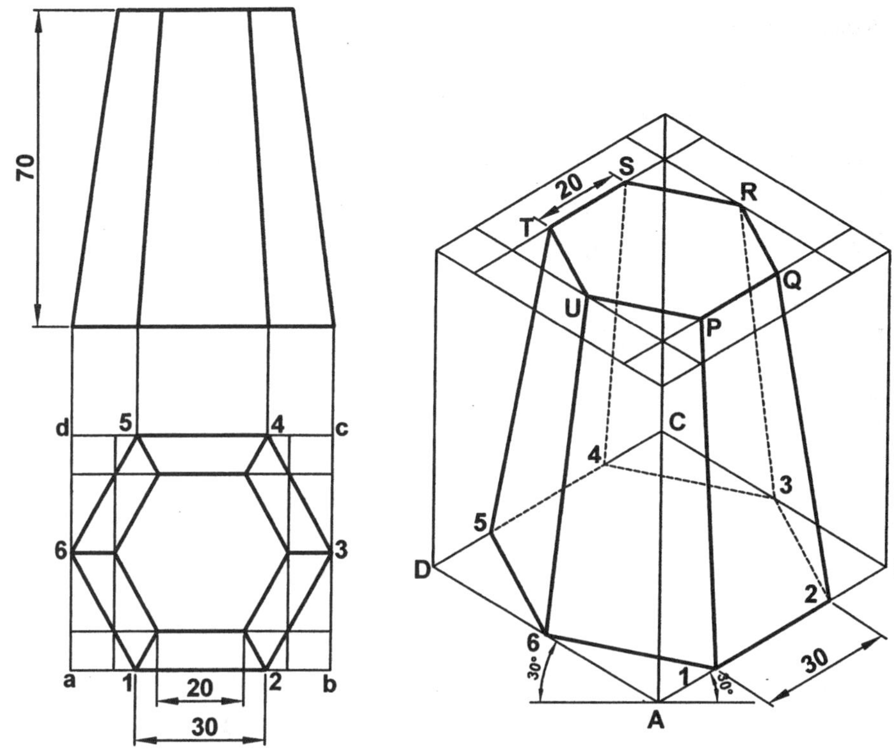

(a) (b)

Fig. 12.22

Construction (Fig.12.22):

1. Draw the front view and top view of frustum of hexagonal pyramid as shown in Fig. 12.22(a).
2. Enclose the base in rectangular box.
3. Draw the isometric view of a rectangular prism.
4. Locate the corners 1, 2, 3, 4, 5 and 6 of the base such that A1 = a1, B1 = b1.. etc.
5. Similarly locate the corner of top end as P, Q, R, S, T and U.
6. Join 1-P, 2-Q, 3-R, 4-S, 5-T and 6-U.
7. Erase the dotted lines in final isometric view.

Problem 12.19 (Fig.12.23): *Draw the isometric view of the frustum of a cone with 50 mm base diameter, 30 mm top diameter and axis 65 mm long resting on its base on the H.P.*

Construction (Fig. 12.23):

1. Assuming the frustum to be enclosed in a square prism and draw the isometric view of the square prism having side of base 50 mm and axis 65 mm long, as shown in Fig.12.23(b).
2. Draw an ellipse inside the bottom rhombus ABDC using four centre method.
3. Draw the rhombus $A_2B_2C_2D_2$ to represent isometric view of square of 30 mm base side and draw an ellipse inside it using four centre method.
4. Draw two common tangents to ellipses.
5. Erase the inner half of (dotted line) the ellipse to get final isometric view.

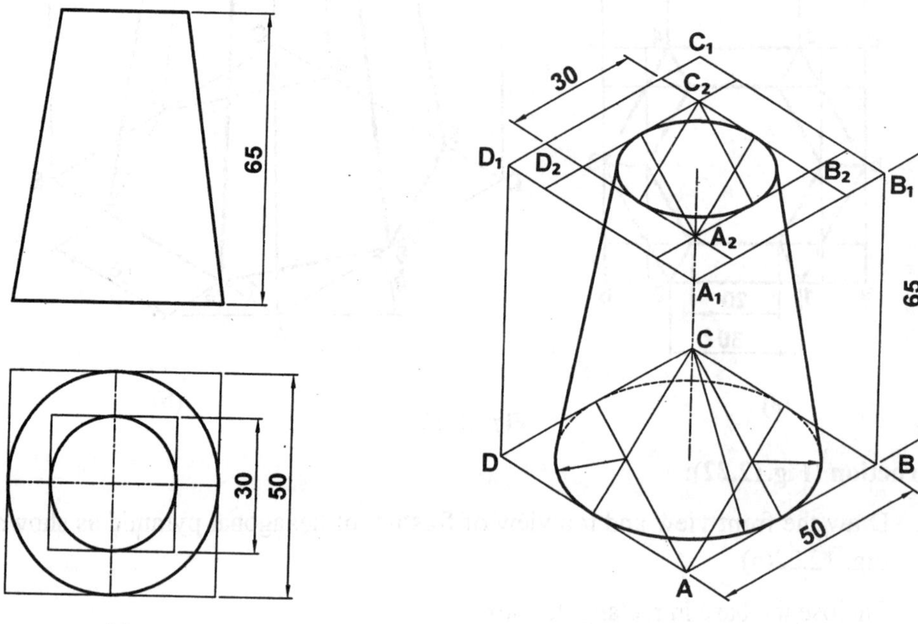

(a) (b)

Fig.12.23

Problem 12.20 (Fig.12.24): *A spherical ball of 50 mm diameter is resting centrally on the top of square prism of 70 mm base side and 25 mm thickness. Draw the isometric projection of the arrangement.*

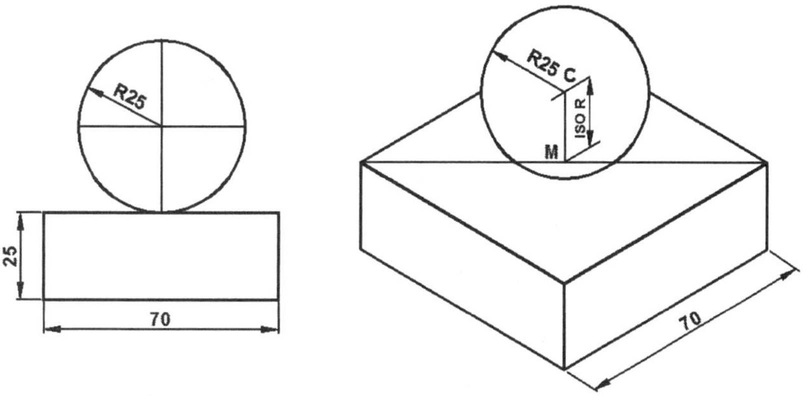

(a) (b)

Fig.12.24

Construction (Fig.12.24):

1. In this case isometric projection (with isometric scale) will be drawn because the arrangement consists of a square. So draw an isometric projection of a square prism using isometric length.
2. Locate the centre M on the top surface of square prism. (Fig. 12.24(b)).
3. Through M draw a vertical line MC such that MC = isometric radius (R_{iso}).

   Where $R_{iso} = 25 \times 0.815 = 20.34$ (approx)
4. With C as centre and radius equal to 25 mm, draw a circle. This is the required isometric view.

Problem 12.21 (Fig.12.25): *A cube of 50 mm side rests centrally on a square block of 70 mm side and 25 mm thickness in such a way that edges of two solids are parallel to each other. Draw the isometric view of the arrangement.*

Construction (Fig.12.25):

1. Draw the isometric view of a square prism of 70 mm base side and 25 mm thickness.
2. Locate the centre C of upper rhombus ABCD.
3. Draw another rhombus $A_2B_2C_2D_2$ of side 50 mm with the same centre C.
4. With $A_2B_2C_2D_2$ as base, draw the isometric view of a cube of 50 mm side.
5. Erase invisible sides (shown by dotted lines) to get final isometric view of combined solid.

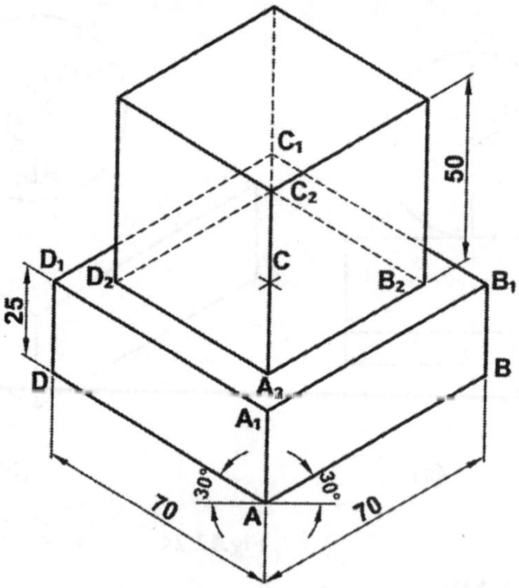

Fig.12.25

Problem 12.22 (Fig.12.26): *A sphere of 40 mm diameter is resting centrally on the top of a cube of 40 mm side. Draw the isometric view of the combined solid.*

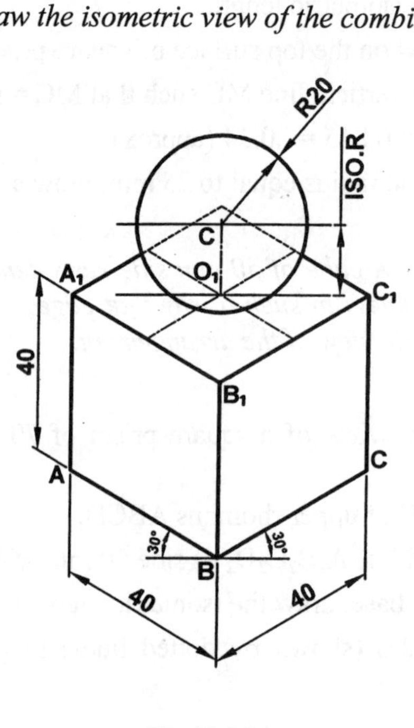

Fig.12.26

Construction (Fig. 12.26):

1. Draw the isometric view of a cube of 40 mm side.
2. Locate the centre O_1 of the upper rhombus $A_1B_1C_1D_1$.
3. Draw a vertical line through O_1 and locate the centre C such that O_1C = isometric radius of the sphere.
4. With C as centre and radius equal to true radius R (20 mm), draw a circle.

Problem 12.23 (Fig. 12.27): *A sphere of 50 mm diameter is placed centrally on the top of a frustum of square pyramid. The base of the frustum is a square of 50 mm side, its top is square of 30 mm side and its height is 60 mm. Draw the isometric view of the arrangement.*

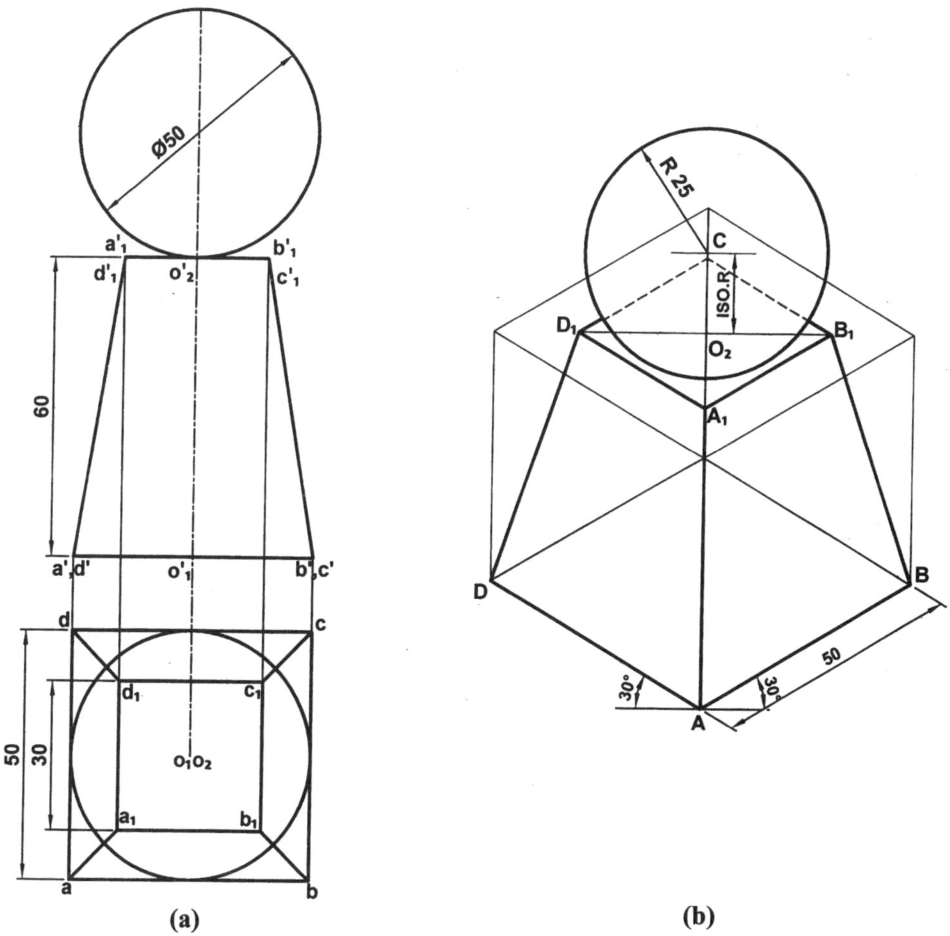

Fig.12.27

Construction (Fig. 12.27):

1. Draw the front view and top view of the arrangement as shown in Fig.12.27(a).
2. Assuming the frustum to be enclosed in a square prism draw the isometric projection of the square prism with isometric dimensions as illustrated in Fig. 12.27(b)
3. In the upper rhombus, locate points A_1, B_1, C_1 and D_1 as described earlier representing corners of top square.
4. Join A-A_1, B-B_1, C-C_1 and D-D_1.
5. Locate the centre O_2 of rhombus $A_1B_1C_1D_1$ and through it draw a vertical line O_2C such that O_2C = isometric radius.
6. With C as centre and radius equal to true radius of the sphere, draw a circle.

Problem 12.24 (Fig. 12.28): *Draw the isometric projection of a spherical ball of 40 mm diameter resting centrally on the top of a pentagonal prism of base 30 mm and height 50 mm.*

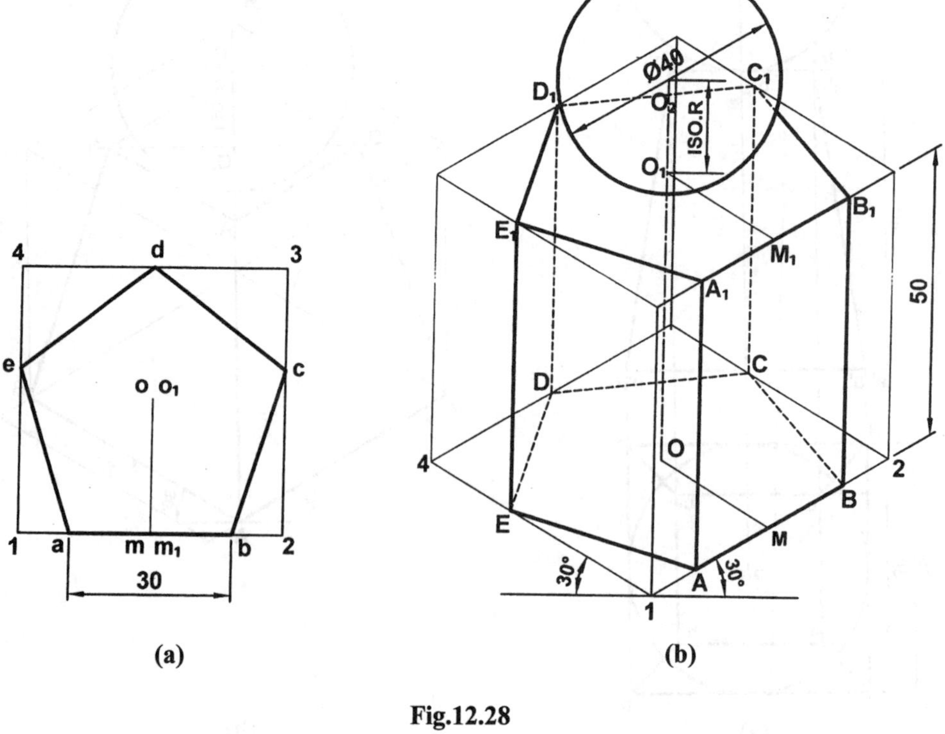

Fig.12.28

Construction (Fig. 12.28):

1. Draw the top view of a pentagonal prism of 30 mm side and enclose it in a rectangle 1234, as shown in Fig.12.28 (a).
2. Draw the isometric projection of the rectangular prism and locate the corners A, B, C, D and E on the lower rhombus. Similarly locate the corner A_1, B_1, C_1, D_1 and E_1 on the top rhombus. (Refer Fig. 12.28 (b)). Locate centres O and O_1.
3. Join A-A_1, B-B_1, C-C_1, D-D_1 and E-E_1.
4. At O_1 draw a vertical line O_1O_2 such that O_1O_2= isometric radius.
5. With O_2 as centre and radius equal to 20 mm draw a circle and complete the required isometric projection.

Problem 12.25 (Fig. 12.29): *A hexagonal prism having side of base 30 mm height 60 mm has a hemispherical top touching all the sides. Draw the isometric projection of the arrangement.*

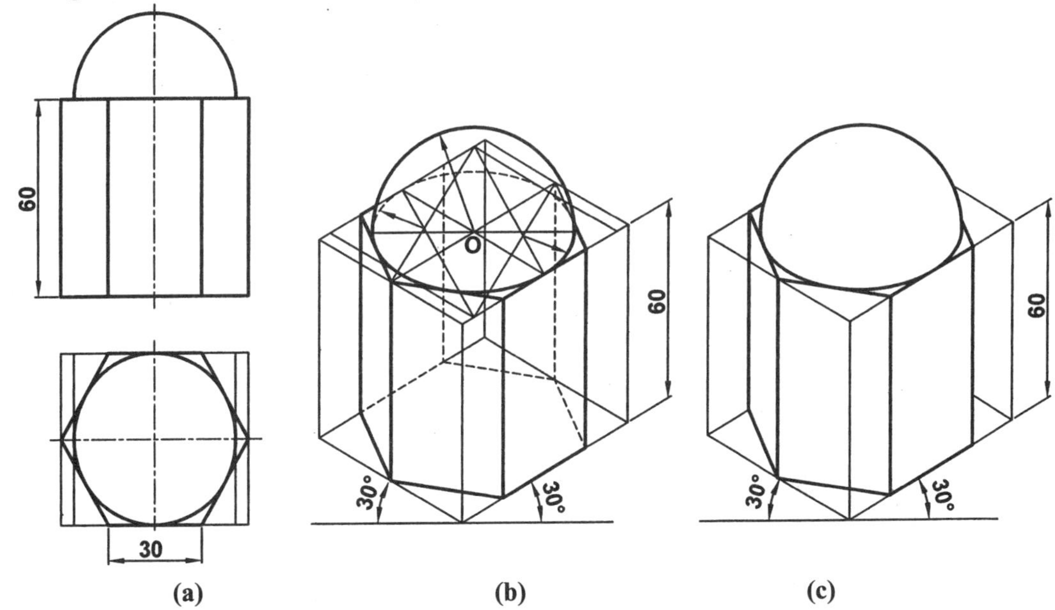

Fig.12.29

Construction (Fig. 12.29):

1. Draw the front view and top view of the arrangement as shown in Fig. 12.29(a). Enclose the hexagonal base in a rectangle.
2. Enclose the hexagonal prism in a rectangular prism and draw the isometric projection of the hexagonal prism.

3. Inside the top rhombus, draw an ellipse using four centre method.
4. Locate the centre O inside the top rhombus.
5. Draw a semicircle with O as centre and complete, the isometric projection as shown in Fig. 12.29(c).

Problem 12.26 (Fig. 12.30): *A square funnel is made of very thin sheet whose orthographic projections are shown in Fig. 12.30(a). Draw its isometric projection.*

Construction: (Fig. 12.30(b))

1. Assuming the funnel to be enclosed in a square prism, draw the isometric projection of the square prism using isometric dimensions (when all the lengths are shortened to isometric lengths).
2. Mark the rhombuses ABCD, $A_1B_1C_1D_1$, $A_2B_2C_2D_2$ and $A_3B_3C_3D_3$.
3. Draw rhombus EFGH inside the rhombus $A_2B_2C_2D_2$ centrally.
4. Draw rhombus $E_1F_1G_1H_1$ inside the rhombus $A_3B_3C_3D_3$ centrally.
5. Join the corners and complete the isometric view as shown in Fig.12.30(b).

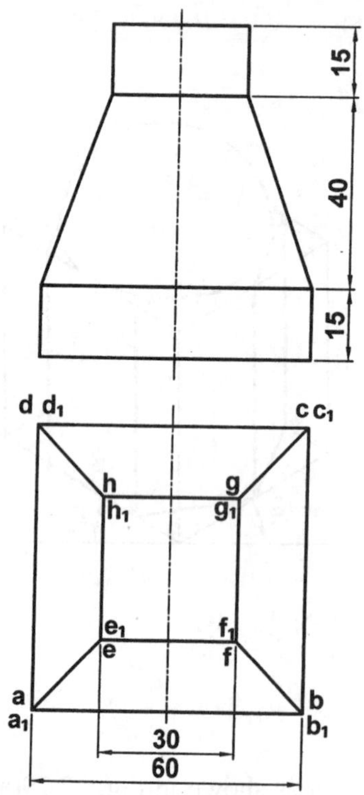

(a)

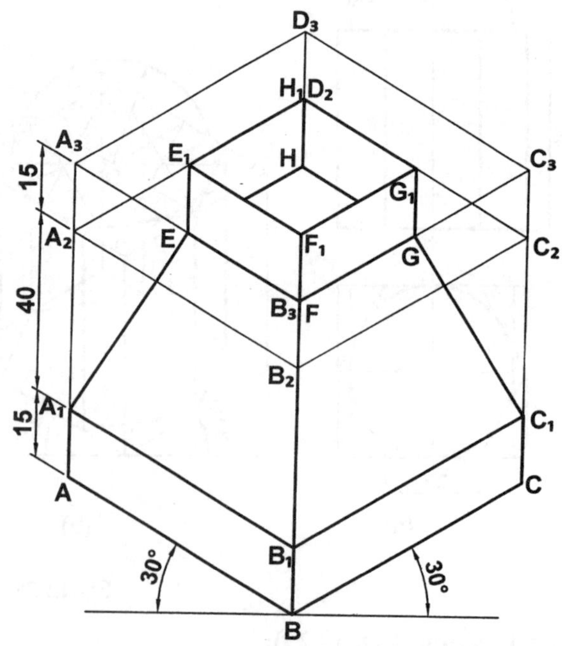

(b)

Fig.12.30

Problem 12.27 (Fig. 12.31): *Draw the isometric projection of a paper weight consisting of the frustum of a cone, cylinder and a cut sphere whose front view is shown in Fig. 12.31(a).*

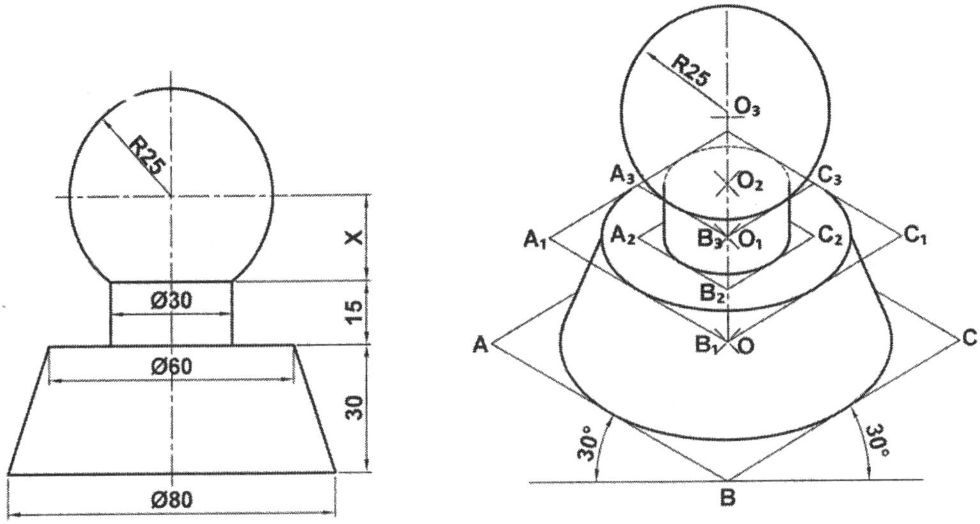

(a) Fig. 12.31 (b)

Construction (Fig. 12.31(b)):

In this problem all the lengths are to be shortened to the isometric length, except for the radius of circle representing sphere.

Following steps are taken.

1. Draw rhombus ABCD, of side equal to isometric diameter of base and locate its centre O. Locate O_1 such that OO_1 = isometric length of 30 mm.
2. Draw rhombus $A_1B_1C_1D_1$ of required side with centre O_1. Inside it, draw a small rhombus $A_2B_2C_2D_2$.
3. Locate O_2 on the vertical axis such that O_1O_2 = isometric length of 15 mm.
4. With centre O_2, draw a rhombus $A_3B_3C_3D_3$.
5. Mark O_3 such that O_2O_3 = isometric length of x.
6. Draw ellipses inside the four rhombuses.
7. With O_3 as centre and radius equal to 25 mm draw a circle.
8. Erase the invisible lines.

   Note: *Distance x can be determined graphically from the front view.*

Problem 12.28 (Fig. 12.32): *Three solids frustum of square pyramid, frustum of cone and a cut sphere are kept one above the other, with their axes in a straight line as shown in Fig. 12.32(a).*

Draw the isometric view of the arrangement.

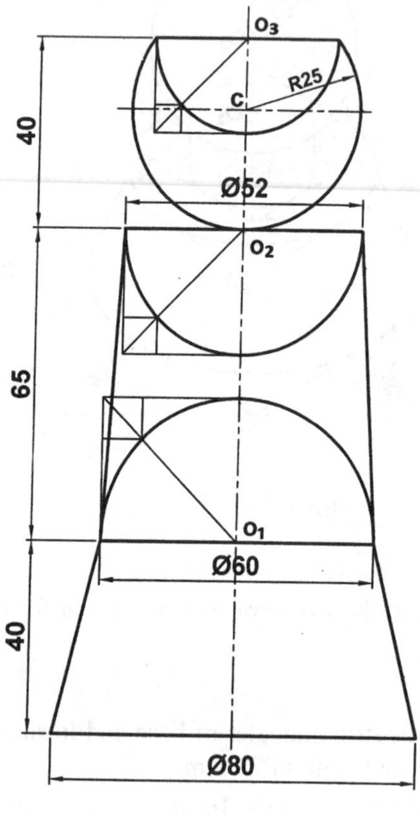

(a)

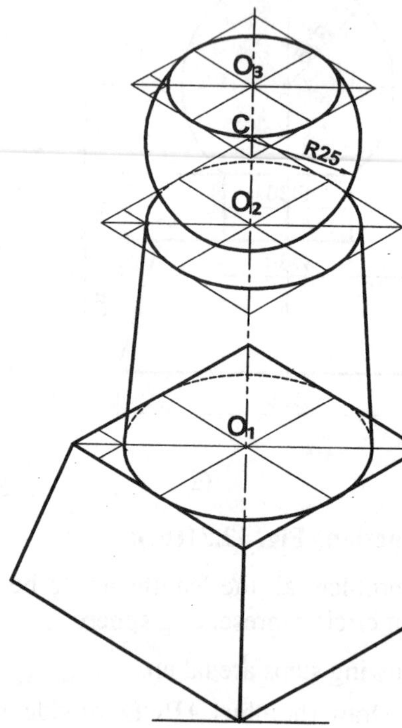

(b)

Fig. 12.32

Construction (Fig.12.32):

In this problem, all dimensions except for radius of sphere must be shortened to the isometric length. Combined isometric view of frustum of square pyramid and frustum of cone is drawn in this usual manner.

The centre C of the sphere is obtained such that the length O_2C is equal to its isometric radius.

i.e., $O_2C = R_{iso} = 25 \times 0.816 = 20.4$ mm

Then circle for the sphere is drawn with the true radius (i.e., 25 mm). Three ellipses are drawn inside three rhombuses constructed around the points O_1, O_2 and O_3 on the vertical axis.

Problem 12.29 (Fig. 12.33): *The orthographic projections of an object is shown in Fig. 12.33(a). Draw its isometric view.*

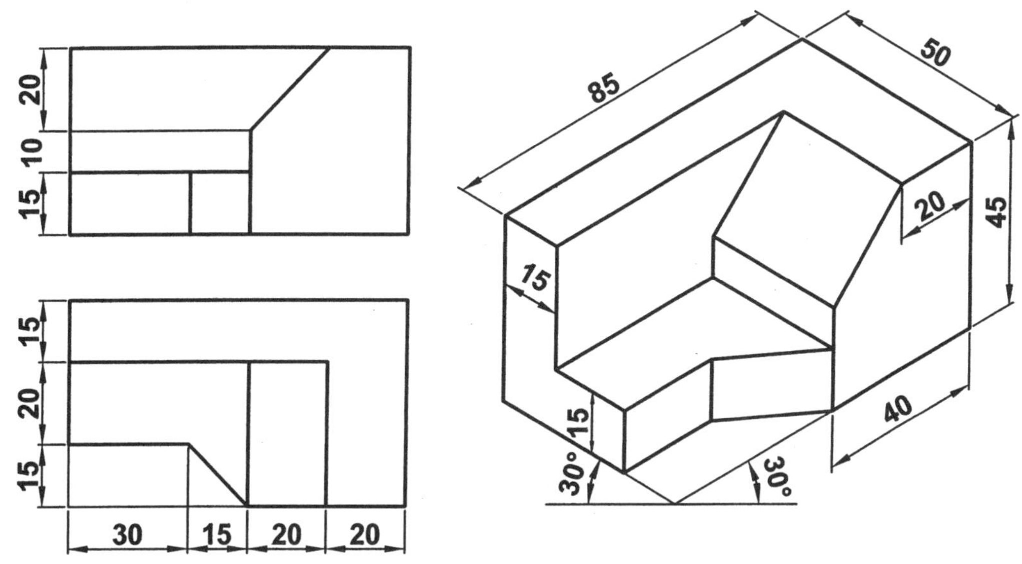

(a) (b)

Fig.12.33

Construction (Fig. 12.33(b)):

1. Enclose the top view in a rectangle.
2. Draw the isometric view of this rectangle.
3. Locate all the corner points on the isometric view of the rectangle.
4. Draw vertical line through all the corners whose lengths may be obtained from the front view.
5. Erase the invisible lines and complete the required isometric view.

Problem 12.30 (Fig. 12.34): *The orthographic projections of a bearing block are shown in Fig. 12.34(a). Draw its isometric view.*

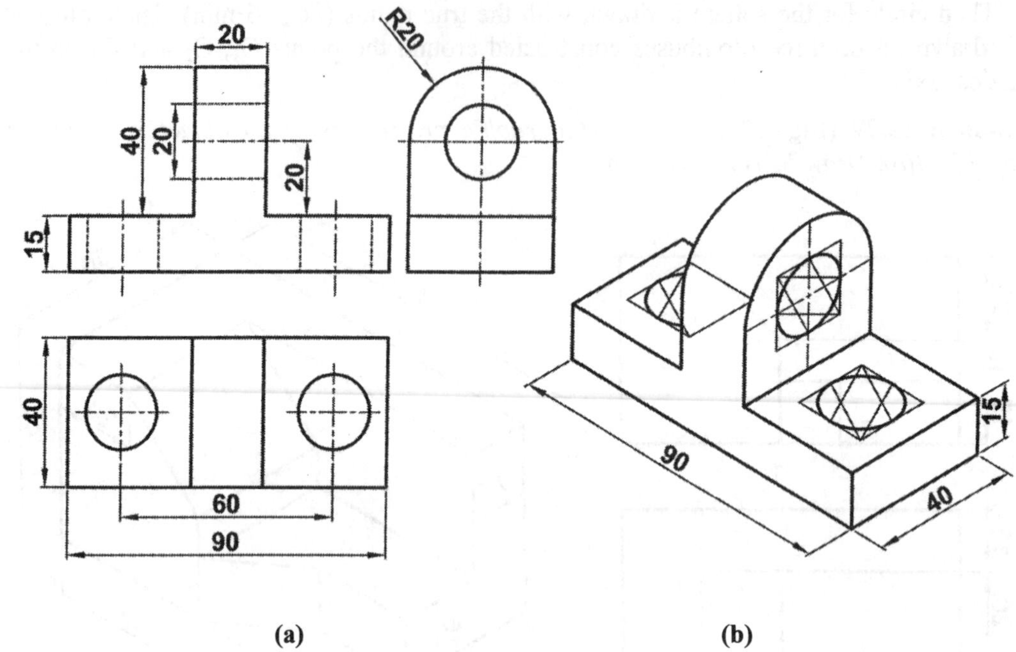

Fig.12.34

Construction (Fig. 12.34(b)):

1. Draw the isometric view of a rectangle of 90 × 40 mm.
2. Draw vertical lines from all the corners whose lengths are obtained from the front view.
3. Complete the isometric view as shown in Fig. 12.34(b).

Problem 12.31 (Fig. 12.35): *The front view and top view of a casting are shown in Fig. 12.35(a). Draw its isometric view.*

See Fig. 12.35(b)

Problem 12.32 (Fig. 12.36): *The orthographic projections of a model of steps are shown in Fig. 12.36(a). Draw its isometric view.*

See Fig. 12.36(b)

Problem 12.33 (Fig. 12.37): *The front view and top view of an object are shown in Fig. 12.37(a). Draw its isometric view.*

See Fig. 12.37(b)

Problem 12.34 (Fig. 12.38): *The front view and top view of an I-beam are shown in Fig. 12.38(a). Draw its isometric view.*

Construction (Fig. 12.38(b)):

Enclose the I-beam in a rectangular box and locate its corner points and complete the isometric view as shown in Fig. 12.38(b)

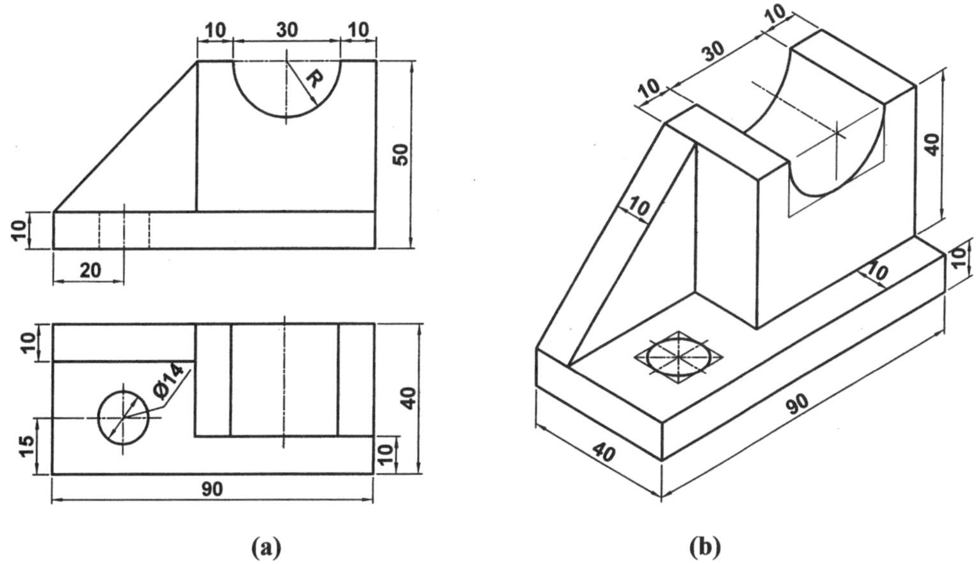

Fig. 12.35

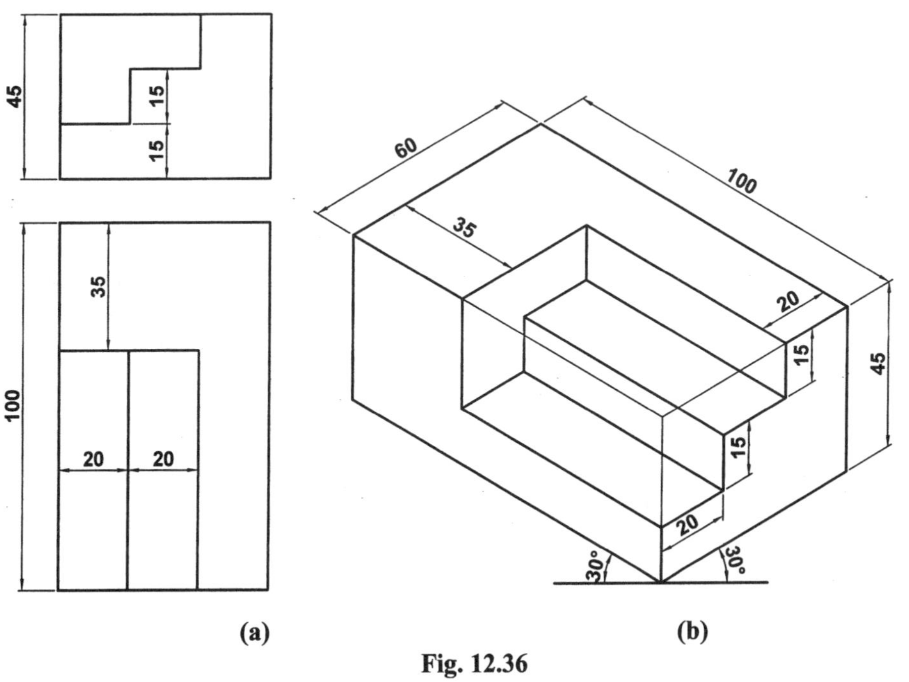

Fig. 12.36

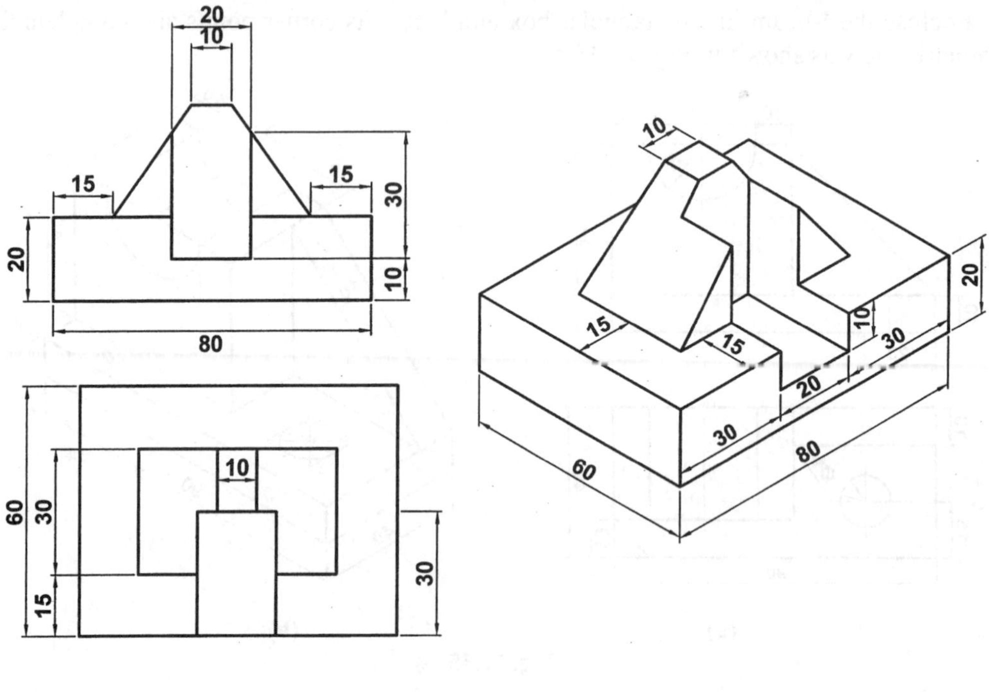

(a) (b)

Fig. 12.37

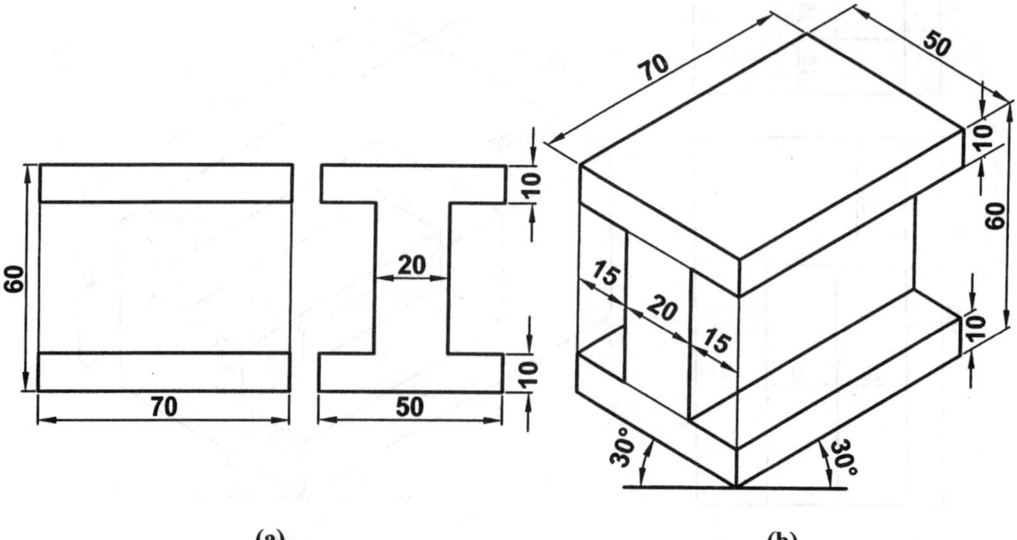

(a) (b)

Fig. 12.38

Exercises

1. Draw an isometric view of an equilateral triangle of 50 mm side when it represents
   (a) a front view (b) a top view
2. Draw an isometric view of a square of 40 mm side when it is considered to be
   (a) parallel to the V.P. (b) parallel to the H.P.
3. Draw an isometric view of a circular lamina of 40 mm diameter when it represents
   (a) a front view (b) a top view
4. A hexagonal plane of 30 mm side, has a circular hole of 30 mm diameter exactly in its centre. Draw its isometric view when it is considered to be
   (a) parallel to the V.P. (b) parallel to the H.P.
5. The top view of a rectangular plane has length 60 mm and breadth 20 mm. Draw its isometric view.
6. A square prism, side of base 30 mm and height 65 mm, is resting on its base on the H.P. Draw its isometric view.
7. Draw the isometric view of a hexagonal prism, side of base 30 mm and axis 70 mm long when it is resting on
   (a) its hexagonal base on the H.P.
   (b) its rectangular face on the H.P. with the axis parallel to both H.P. and V.P.
8. A pentagonal prism, side of base 30 mm and axis 70 mm long is resting on its base on the H.P. with a vertical face parallel to the V.P. Draw its isometric projection.
9. A hexagonal prism, side of base 30 mm and axis 60 mm long has a cylindrical hole of 30 mm diameter axially in the centre. Draw its isometric view.
10. Draw an isometric view of a square pyramid of 30 mm base side and 60 mm height when it resting on its base on the H.P.
11. Draw the isometric view of a hexagonal pyramid having side of base 30 mm and axis 70 mm long when
    (a) its axis is vertical (b) its axis is horizontal
12. Draw the isometric view of a frustum of hexagonal pyramid with a 30 mm base side, 20 mm top side and 35 mm height and resting on its base on the H.P.
13. A frustum of cone with base diameter 50 mm, top diameter 30 mm and height 40 mm, is resting on its base on the H.P. Draw its isometric view.
14. A sphere of 40 mm diameter rests centrally over a cube of 50 mm side. Draw its isometric view.

15. A cylindrical block with 60 mm diameter and 20 mm height is resting on its base on the H.P. A frustum of square pyramid with base side 30 mm and top side 20 mm is kept centrally over the cylindrical block. Draw the isometric view of the combined solid.

16. A frustum of sphere of 50 mm diameter with a frustum circle of 30 mm diameter is kept centrally over a cube of 50 mm side. Draw its isometric view.

17. A cube of 30 mm sides rests on the top of a cylindrical block having 50 mm diameter and 30 mm thickness. The axes of two solids are in a straight line. Draw its isometric view.

18. Figures 12.39 to 12.48 show the orthographic projections of various objects. Draw their isometric views.

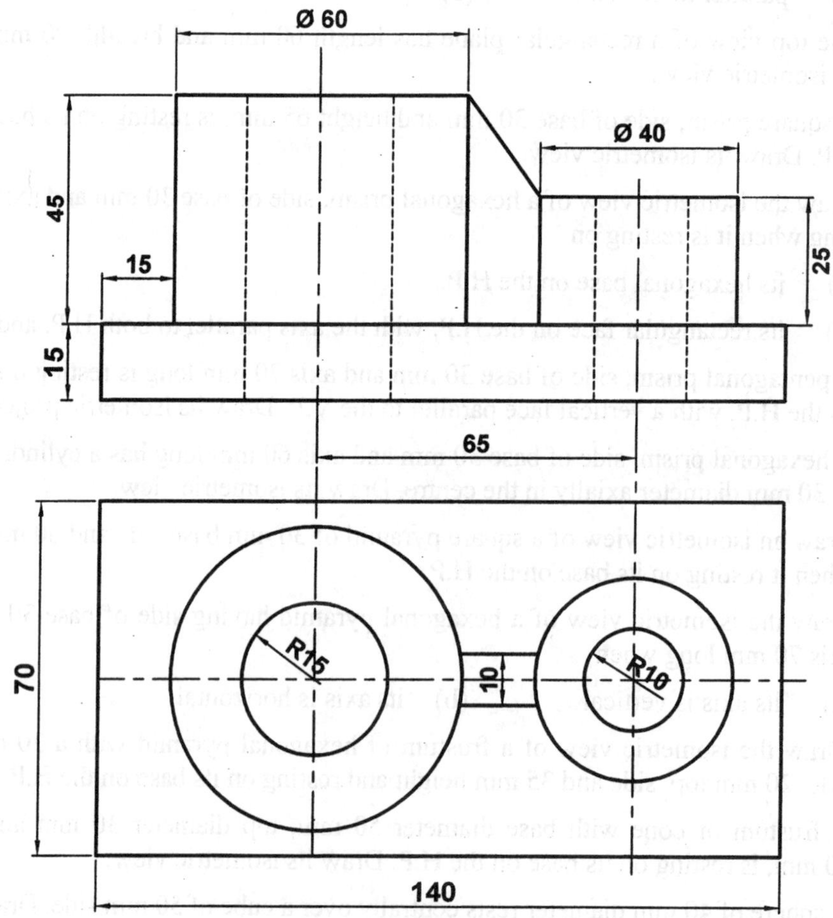

Fig. 12.39

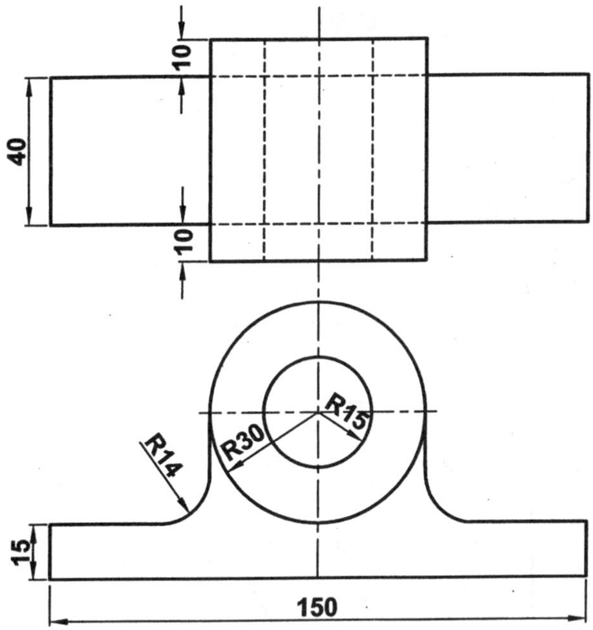

Fig.12.40

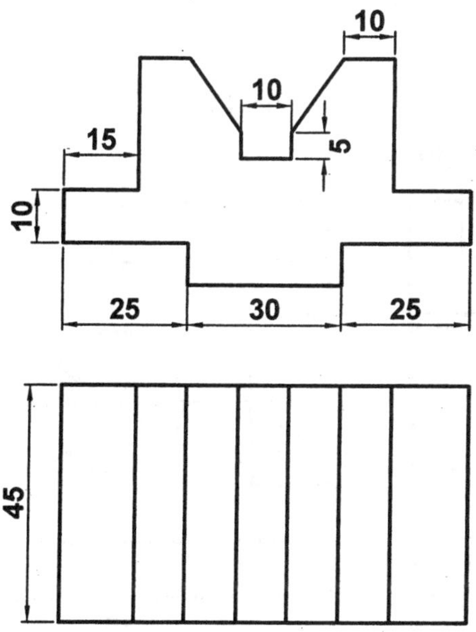

Fig.12.41

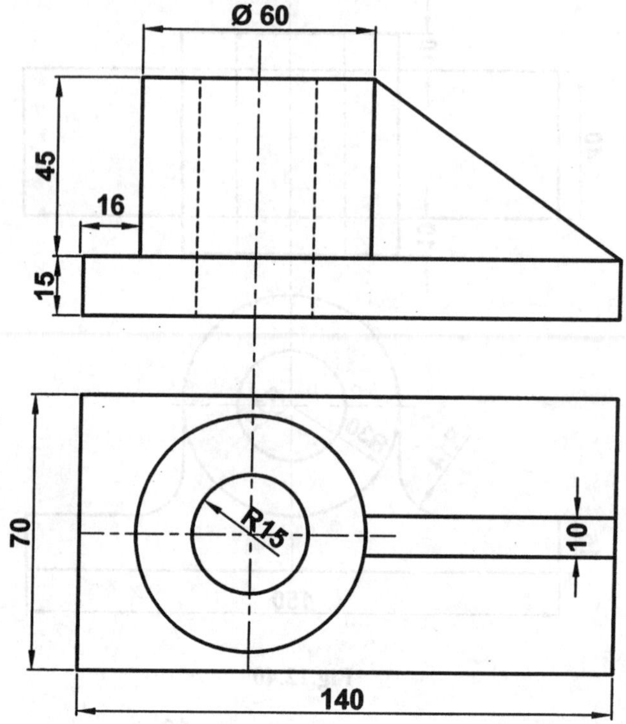

Fig. 12.42

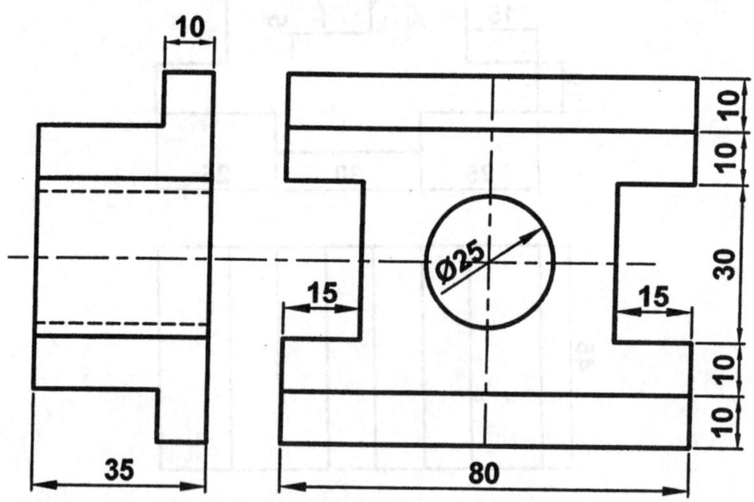

Fig.12.43

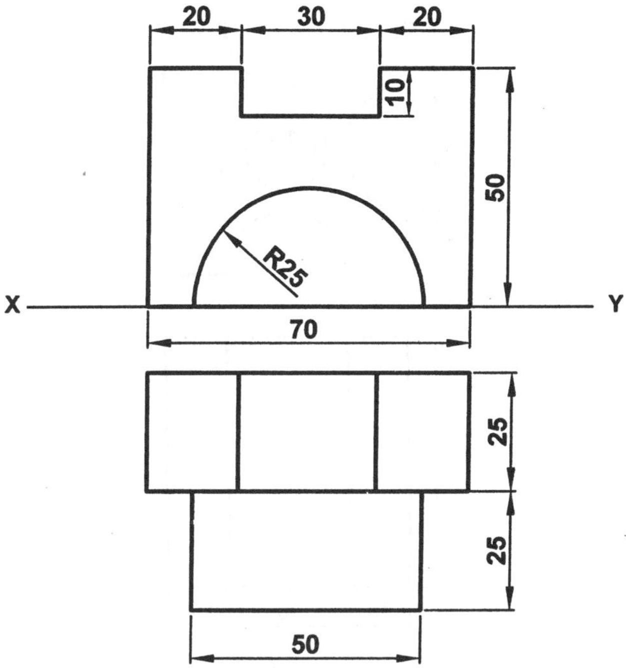

Fig. 12.44

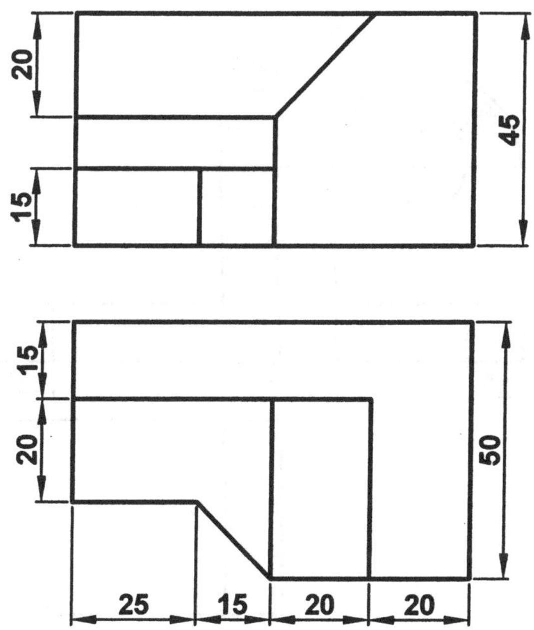

Fig.12.45

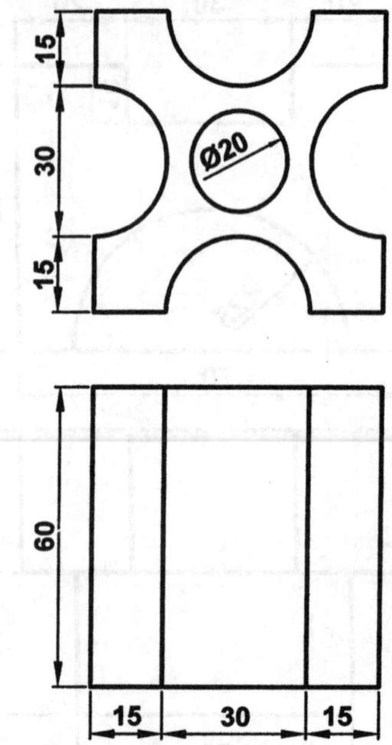

Fig.12.46

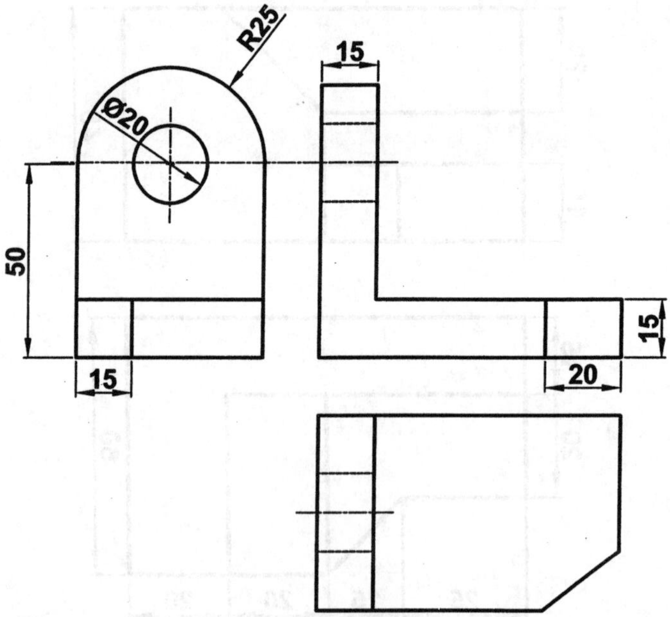

Fig.12.47

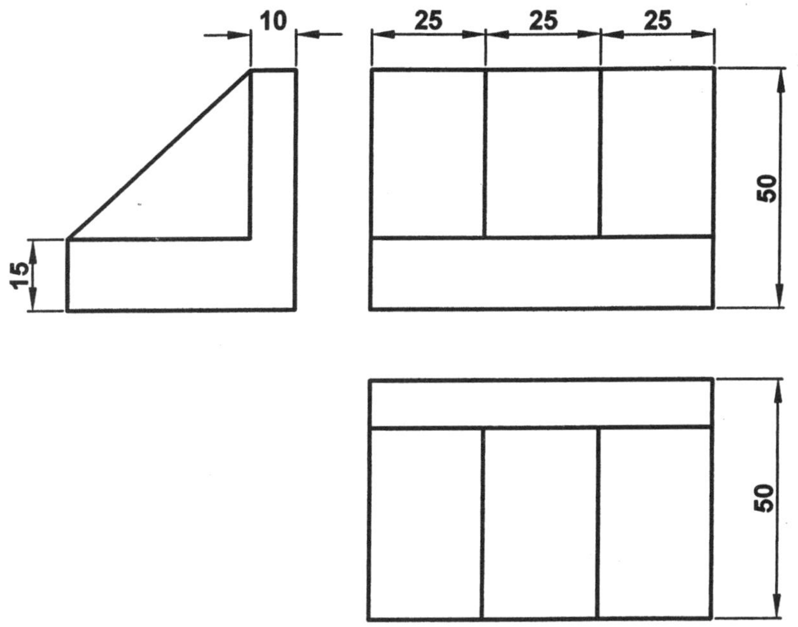

Fig.12.48

CHAPTER 13

Introduction to Computer Aided Drafting

13.1 Computer Aided Drafting (CAD)

Computer aided drafting is the process of making a drawing of an object on the screen (display unit) of a computer with the help of various tools or commands available in the CAD software. Due to its versatile capability and high benefits, modern designers and draft man have adopted CAD in almost all disciplines of engineering. Mechanical engineers use CAD for the design and drawing of machine components, civil engineers use it for preparing plans and layouts of buildings. Electrical engineers prepare layout of power distribution systems using CAD. Similarly it can be used to generate accurate drawings much faster compared to conventional manual drawing, in other engineering fields.

The modern CAD systems make use of interactive computer graphics (ICG) in which computers are employed to create, transform and display data in the form of pictures and symbols. The basic geometrical elements such as points, lines, arcs, circles etc., are used to create an image. It can be modified (enlarged of reduced), moved, rotated and transformed as per the need of the designer.

The basic components of ICG include a hardware, a software and a human designer. The hardware comprises of CPU, input and output devices. A software consists of computer program to perform the specific task. A human designer performs the design process by utilising his knowledge and skill to get the required work output from the system.

13.2 Hardware and Software requirement of CAD System

Windows based computers are usually used for most of the CAD systems. Some CAD systems may require other operating systems such as Unix or Linux. The requirement of system configuration vary with the CAD systems, for example some software may require high speed CPU and RAM of large capacity. Input devices such as keyboard, mouse, track ball, pointing stick, touch pad, input pen, touch screen etc., can be used to enter data, programs, commands etc., to the computer. The processed results(outputs) can be communicated using output devices such as monitors, printers and plotters.

13.3 Advantages of CAD

Computer aided drafting (CAD) offers following advantages over conventional manual drafting.

1. Drawings with higher accuracy can be prepared faster than conventional drafting.
2. It improves drafting productivity.
3. Manpower requirement is reduced.
4. Modification in drawing at any stage is possible.
5. Use of different colours to customize products help to understand the design.
6. No scaling is required to prepare a drawing i.e., drawing can be prepared in full scale.
7. Drawing involving repetition of components can be created faster by simply copying that portion.
8. Drawings can be stored in electronic form and can be preserved for longer duration of time.
9. Visual modelling of any object is possible with CAD.
10. Printing of the drawing can be done to any scale.

13.4 Limitations of CAD

It has following limitations:

1. Since high configuration PCs with CAD software are required, it involves huge investments.
2. Basic knowledge of computer is required to prepare the drawings.

13.5 Software for Creating Drawing/Design

The application software is an interpreter or translator which permits the user to perform the specific task. CAD software provides the facility to create, modify, save and to print or plot the drawing. Following are some commercially available software which can be used to create drawings/design.

1.	AutoCAD	2.	MS Word	3.	MS Power Point
4.	Paint	5.	Page Maker	6.	Corel Draw
7.	Corel CAD	8.	Inventor	9.	Solid Works
10.	Solid Edge	11.	CATIA	12.	Pro-E

Depending upon the user requirement, the suitable software may be used. For example simple drawings can be generated quickly by using MS word, Power Point, Paint, page Maker etc. Corel CAD and AutoCAD may be used for engineering drawing. Solid Works, Solid Edge, CATIA, Pro-E etc., may be used for three dimensional modelling of engineering objects.

13.6 Introduction to AutoCAD

AutoCAD may be used to create two dimensional (2D) and three dimensional (3D) drawings of objects. This section presents some basic features of AutoCAD to draw 2D drawings. For other details, AutoCAD user Manual or other learning resources may be referred.

13.7 Beginning with AutoCAD 2009

When AutoCAD icon is clicked on, a startup menu appears on the screen which offers following four options.

1. **Open a drawing:** By clicking this option any existing (previously saved) file can be opened.
2. **Start from scratch:** It creates a new drawing using either imperial or metric default unit.
3. **Use a template:** It creates a new drawing using settings defined in the available templates.
4. **Use a wizard:** This creates a new drawing using the setting which the user specify either in quick or advanced wizard option.

On selecting one of the above four options, AutoCAD screen appears which provides drawing environment where one can draw, edit, transform or save and print the drawings.

13.8 Drawing Environment (Space)

AutoCAD provides following two drawing environments for creating and laying out the drawing.

1. Model Space
2. Layout (Paper) Space

In model space, we create drawing (model) in full scale (i.e., 1:1) in the area known as model space without considering size and layout of printing paper or sheet. If two dimensional drawing has only one view, it can be drawn entirely in the model space.

The paper space is a sheet layout environment where we can specify the size of drawing sheet in which multiple views can be arranged. Dimensions, title block and other notes can also be added. In a layout space, there can be a number of independent drawings called as viewports. A view port can display a view of drawing at any scale. In the paper space, it is possible to arrange several viewports for printing or plotting in the desired style of layout.

13.9 AutoCAD Workspace

Before beginning of any drawing, it is required to set workspace as follows.

Click on AutoCAD Main Menu ⇒ **Tools** ⇒ **Workspace**

Following options are displayed

1. 2 D Drafting & Annotation
2. 3 D Modelling
3. AutoCAD Classic

For creating 2 D drawings, AutoCAD classic workspace can be used conveniently by selecting the third option. Fig. 13.1 shows AutoCAD classic workspace screen which has following features.

1. **Drawing Area:** The area in the middle of the screen available for creating drawing is known as drawing area.
2. **Tool Bars:** To select various commands, tool bars such as Title Bar, Menu Bar, Standard Tool Bar, Property Tool Bar, and Tool Palettes appear on the screen as shown in Fig. 13.1.
3. **WCS Icon:** The symbol consisting of two arrows in the bottom left corner of the screen is known as WCS (World Coordinate System) icon. This icon shows positive X and Y direction.
4. **Crosshairs Cursor:** A pair of two crossed lines on the screen is known as crosshairs cursor. In its centre, a square box is superimposed. This is called as pick up box which helps to select/pickup the object on the screen.
5. **Command Line:** A pair of windows at the bottom is called command line, where the required command is typed using keyboard or can directly be accessed from the tool bar.

6. **Status Bar:** It appears at the bottom of the screen. X, Y coordinates of the cursor position is displayed at the left. In the middle of this bar, Icons for snap, grid, ortho, polar etc., are displayed and at the right side, the tray icons appear.
7. **Tool Palettes:** Tool palettes can be viewed at the right side of the screen from the menu bar. Various commands can also be selected from this palettes.

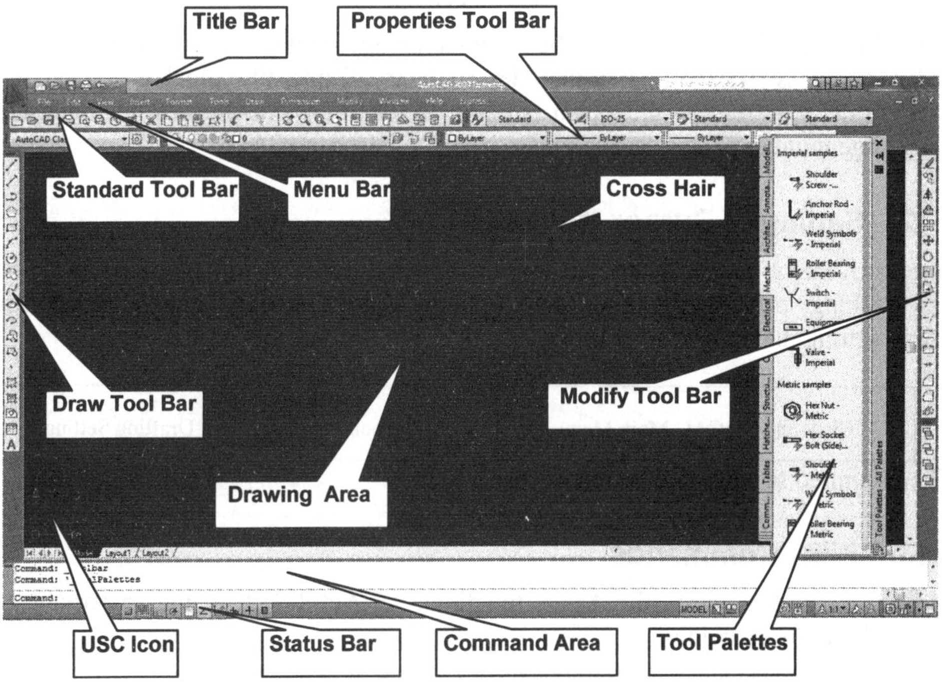

Fig. 13.1 AutoCAD classic Workspace

13.10 Setting of Drawing Space

1. Units Command: Before the beginning of a new drawing, it is important to setup its size and units as per following steps.

Click on AutoCAD Main Menu ⇨ **Format** ⇨ **Units**

This will display a Units dialog box where following parameters are required to be supplied.

(a) Length (b) Angle (c) Insertion scale (d) Direction

2. **Limits Command:** This creates an invisible artificial boundaries for the drawing.
3. **Scale:** Scale is usually chosen at the time of printing or plotting to arrange the drawing within the printing area. The required scale can be set through the plot scale section in the page setup dialog box.

13.11 Sheet Layout for Printing/Plotting

The final output of the drawing can be obtained on any required size of sheet layout. The standard layout is a rectangle that includes boundary, frame, title block etc. It can also be created as per requirement and saved as template for future drawings or it can be selected from a large number of standard templates available in AutoCAD.

13.12 Setting of Cursor Movement

Cursor movement can be restricted for locating the points using snap and grid setting, polar tracking, ortho and object snap options. Above settings can be done through drafting setting dialog box as follows:

Click on AutoCAD Main Menu ⇒ Tools ⇒ Drafting Settings

Then required settings can be made.

13.13 Utility Commands

Utility commands control the basic functions of AutoCAD. Some important utility commands are listed below.

1. NEW : Creates a new drawing file.
2. OPEN : Opens an existing file.
3. CLOSE: Closes the active drawing.
4. SAVE : Saves the current drawing.
5. EXIT : Comes out of the AutoCAD.

13.14 Basic Drawing Commands

In this section some basic commands for creating a drawing, are explained. These commands can be accessed by one of the following methods depending upon the user preference.

(a) By clicking the Menu Browser > Draw > Required TAB (Such as Line, circle etc.,)

(b) Using Command line from the classic workspace, for example Command line: Circle

(c) By clicking the required TAB from the Draw Tool Bar

(d) From Tool palettes

Note: *After typing the required information on the command line, press ENTER.*

13.14.1 POINT Command

The POINT command places a point at the required position in the drawing.

Problem 13.1: *Locate a point whose coordinates are (50, 60).*

Solution: Open the classic workspace of AutoCAD and follow the steps as below:

Command: Point

Current point modes: PDMODE = 0 PDSIZE = 0.0000

Specify a point: 50, 60

Note: *A point can also be located directly by clicking the cursor at the point of crosshair.*

13.14.2 LINE Command

This command is used to draw a line segment of any desired length. A line can also be drawn by one of the following three methods:

(i) Absolute Co-ordinate System

This system is used when the coordinates of end points of a line are known.

Problem 13.2: *Draw a line segment AB whose end coordinates are A(50,50) and B(120,60).*

Solution:

Command : Line

Specify first point or [Undo] : 50, 50

Specify next point or [Undo] : 120, 60

Specify next point or [Undo] : 60, 120

(ii) Relative Rectangular Co-ordinate System (@X distance, Y distance)

Relative co-ordinate values are based on the last point located. This system is used when distance of a point with respect to the previous point is known. Following example will help to understand this system.

Problem 13.3: *Draw a square of 70 mm side using LINE command.*

Command Steps (Fig. 13.2):

Command : Line

Command: line

Specify first point: 50, 50

Specify next point or [Undo]: @70, 0
Specify next point or [Undo]: @0, 70
Specify next point or [Close/Undo]: @-70, 0
Specify next point or [Close/Undo]: C

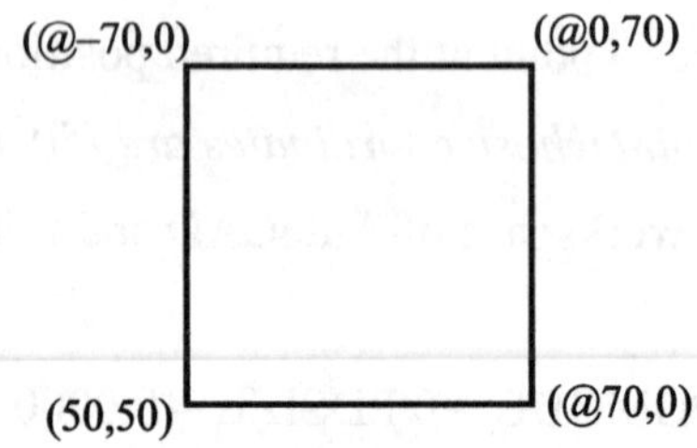

Fig. 13.2

Problem 13.4: *State the series of AutoCAD command steps to draw a rectangle of size 100 mm × 65 mm with the help of LINE command.*

Solution:

Command: Line
Specify first point: (select any suitable point using mouse or supplying coordinates)
Specify next point or [Undo]: @100, 0
Specify next point or [Undo]: @0, 65
Specify next point or [Close/Undo]: @-100, 0
Specify next point or [Close/Undo]: C

Problem 13.5 (Fig. 13.3): *State the series of AutoCAD command steps to reproduce the object shown in Fig. 13.3 with the help of line command using rectangular coordinate system.*

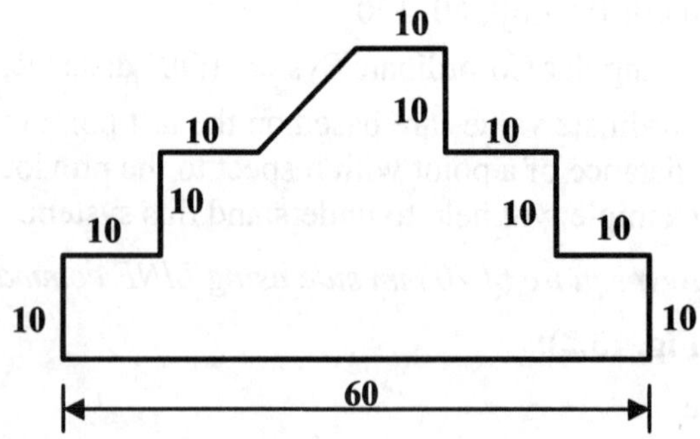

Fig. 13.3

Solution (Fig. 13.3):

Command : Line

Specify first point: (select any suitable point using mouse or supplying coordinates)

Specify next point or [Undo]: @ 60, 0

Specify next point or [Undo]: @ 0, 10

Specify next point or [Close/Undo]: @–10, 0

Specify next point or [Undo]: @ 0,10

Specify next point or [Close/Undo]: @–10, 0

Specify next point or [Undo]: @ 0, 10

Specify next point or [Close/Undo]: @–10, 0

Specify next point or [Undo]: @ –10, –10

Specify next point or [Close/Undo]: @–10, 0

Specify next point or [Undo]: @ 0,–10

Specify next point or [Close/Undo]: @–10, 0

Specify next point or [Undo]: @C

(iii) Relative Polar Co-ordinate System (@distance < angle)

This system uses a distance and an angle with reference to the initial point to locate a point. Consider the following example to understand the process of drawing using this system. Angles are measured in anti-clockwise direction, taking 0° towards right.

Problem 13.6 (Fig. 13.4): *Draw a square of 70 mm side using LINE command with relative polar co-ordinate system.*

Solution (Fig. 13.4):

Command : Line

Command: line Specify first point: 50, 50

Specify next point or [Undo]: @70<0

Specify next point or [Undo]: @70<90

Specify next point or [Close/Undo]: @70<180

Specify next point or [Close/Undo]: C

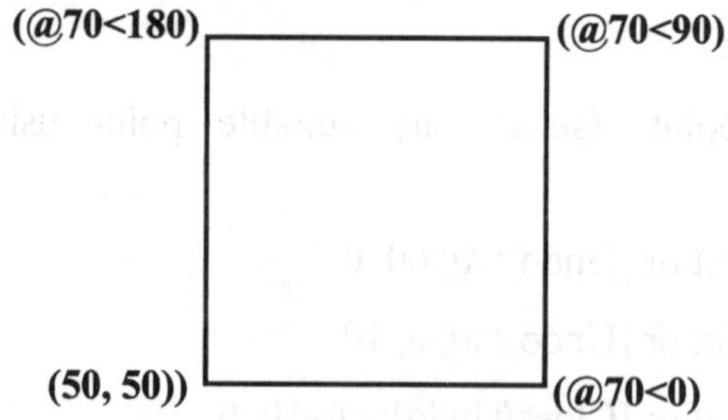

Fig. 13.4

13.14.3 RECTANGLE or RECTANG Command

This command is used to draw rectangles directly without the use of LINE command. With this command one can specify rectangle parameters (length, width, rotation) and control the type of corners (such as square, fillet or chamfer). On accessing this command, following command prompt appears:

Specify first corner point or [Chamfer/Elevation/Fillet/Thickness/Width]:

Here we have six options for drawing a rectangle

1. To draw a simple rectangle: Click on the drawing area and follow the command prompt.
2. To create chamfer at all the four corners: Type C and Enter.
3. To draw a rectangle at a specified distance from the XY plane: Type Elevation (or E) and Enter.
4. To make all the four corners rounded: Type Fillet (or F) and Enter.
5. To provide required thickness to the rectangle: Type Thickness (or T) and Enter.
6. To specify thickness of the line of rectangle: Type Width (or W) and Enter.

Problem 13.7 (Fig. 13.5): *Draw a rectangle having its length and width as 100 mm and 70 mm respectively.*

Solution (Fig. 13.5):

Command: Rectangle

Specify first corner point or [Chamfer/Elevation/Fillet/Thickness/Width]: (By mouse click on the screen)

Specify other corner point or [Area/Dimensions/Rotation]: D

Specify length for rectangles <10.0000>: 100

Specify width for rectangles <10.0000>: 70

It will generate a rectangle as shown in Fig. 13.5.

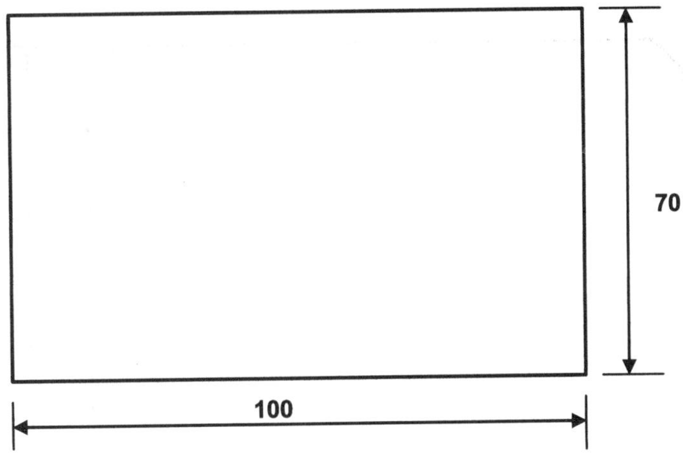

Fig. 13.5

Problem 13.8 (Fig. 13.6): *Draw a rectangle of size 100 mm × 70 mm having chamfer of 5 mm at all corners.*

Solution (Fig. 13.6): Following the command steps as below, a rectangle is drawn as shown in Fig. 13.6

Command: Rectangle

Specify first corner point or [Chamfer/Elevation/Fillet/Thickness/Width]: C

Specify first chamfer distance for rectangles <0.0000>: 5

Specify second chamfer distance for rectangles <5.0000>: 5

Specify first corner point or [Chamfer/Elevation/Fillet/Thickness/Width]: By mouse

Specify other corner point or [Area/Dimensions/Rotation]: D

Specify length for rectangles <10.0000>: 100

Specify width for rectangles <10.0000>: 70

Specify other corner point or [Area/Dimensions/Rotation]:

13.14.4 CIRCLE Command

The CIRCLE command is used to generate circles. This Command offers following options.

(i) Centre-radius/diameter option

(ii) Two-point option

(iii) Three-point option

(iv) Tangent-tangent-radius option

Consider following problems to understand method of drawing a circle using different options.

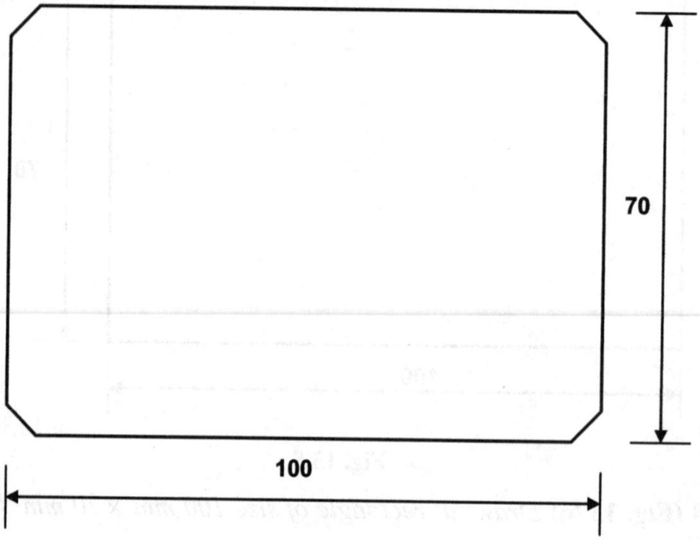

Fig. 13.6

Problem 13.9 (Fig. 13.7): *Draw a circle with centre (100,100) and diameter 50 mm using AutoCAD.*

Solution (Fig. 13.7):

Command: Circle

Specify centre point for circle or [3P/2P/Ttr (tan tan radius)]: 100, 100

Specify radius of circle or [Diameter]: D

Specify diameter of circle: 50

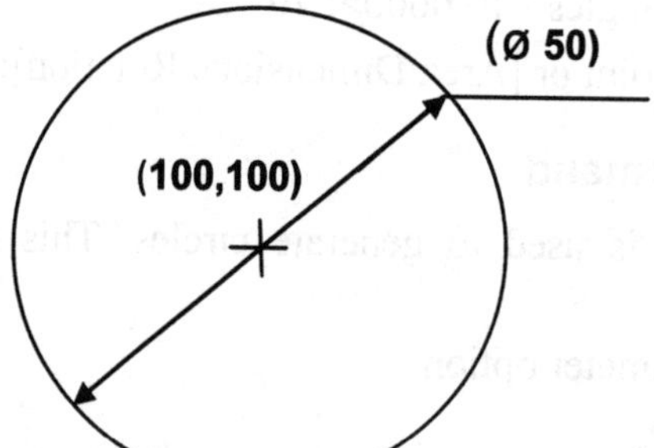

Fig. 13.7

Problem 13.10 (Fig. 13.8): *Draw a circle passing through three points whose coordinates are (100,100), (200, 205), (300,100) using AutoCAD.*

Solution (Fig. 13.8):

Command: Circle

Specify centre point for circle or [3P/2P/Ttr (tan tan radius)]: 3P

Specify first point on circle: 100, 100

Specify second point on circle: 200, 205

Specify third point on circle: 300, 100

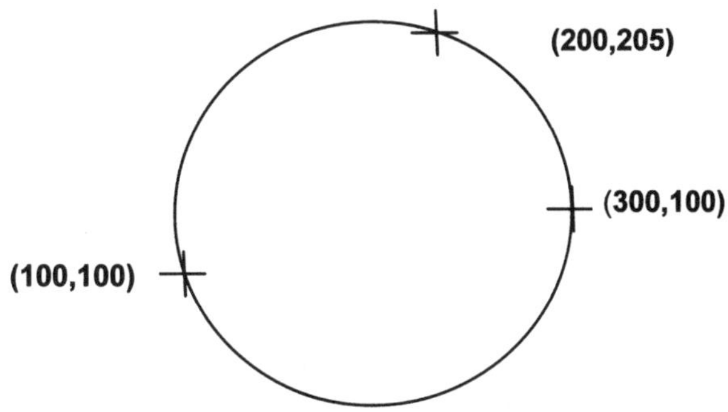

Fig. 13.8

Problem 13.11 (Fig. 13.9): *Draw a circle passing through two points (ends of a diameter) whose coordinates are (195,185), (195,145), using AutoCAD.*

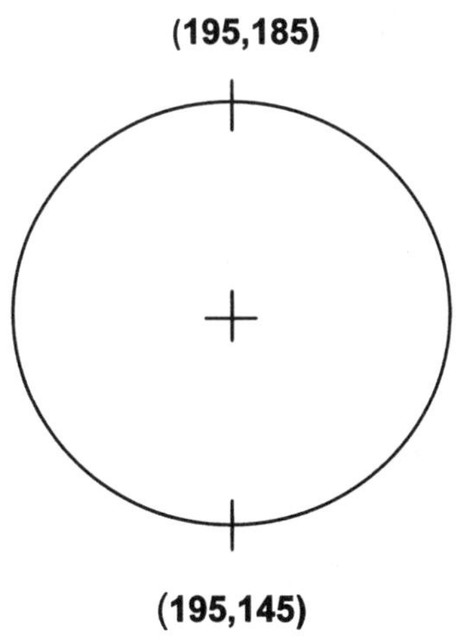

Fig. 13.9

Solution (Fig. 13.9):

Command: Circle

Specify centre point for circle or [3P/2P/Ttr (tan tan radius)]: 2P

Specify first end point of circle's diameter: 195, 185

Specify second end point of circle's diameter: 195, 145

Problem 13.12 (Fig. 13.10): *Draw a circle of 70 mm diameter touching two lines as tangents.*

Solution (Fig. 13.10):

Command: Circle

Specify centre point for circle or [3P/2P/Ttr (tan tan radius)]: Ttr

Specify point on object for first tangent of circle: (*Pick a point on line 1*)

Specify point on object for second tangent of circle: (*Pick a point on line 2*)

Specify radius of circle: 40

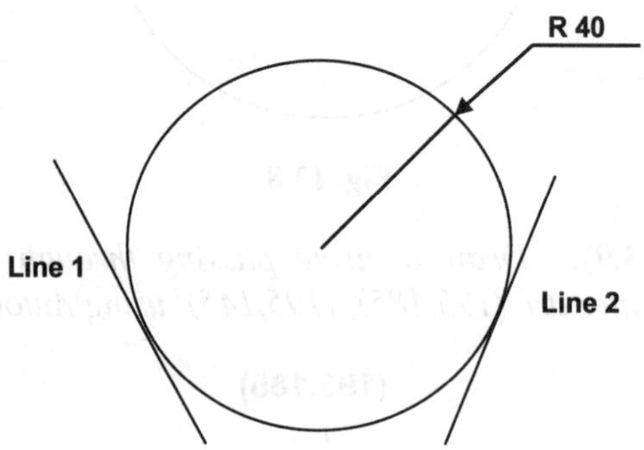

Fig. 13.10

13.14.5 ARC Command

ARC command is used to draw an arc. An arc is also considered as partial circle or a curve specified by centre and radius. The ARC command offers following options to draw the arcs.

1. Three point method
2. Start point-centre point-end point
3. Start point-centre point-length of chord
4. Start point-end point-angle of inclusion
5. Start point-end point-direction
6. Start point-centre point-angle of inclusion
7. Start point-end point-radius

Problem 13.13 (Fig. 13.11): *Draw an arc passing through P_1, P_2 and P_3 using AutoCAD* ***three point*** *option.*

Solution (Fig. 13.11):

Command: Arc

Specify start point of arc or [Centre]: (Pick P_1)

Specify second point of arc or [Centre/End]: (Pick P_2)

Specify end point of arc: (Pick P_3)

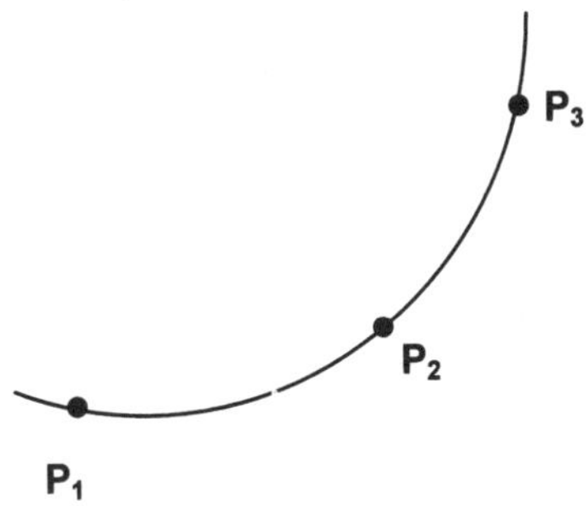

Fig. 13.11

Problem 13.14 (Fig. 13.12): *Draw an arc using* ***start point-centre point-end point*** *option of ARC command.*

Solution (Fig. 13.12):

Command: Arc

Specify start point of arc or [Centre]: (Pick start point)

Specify second point of arc or [Centre/End]: C

Specify centre point of arc: (Pick centre)

Specify end point of arc or [Angle/chord Length]: (Pick end point)

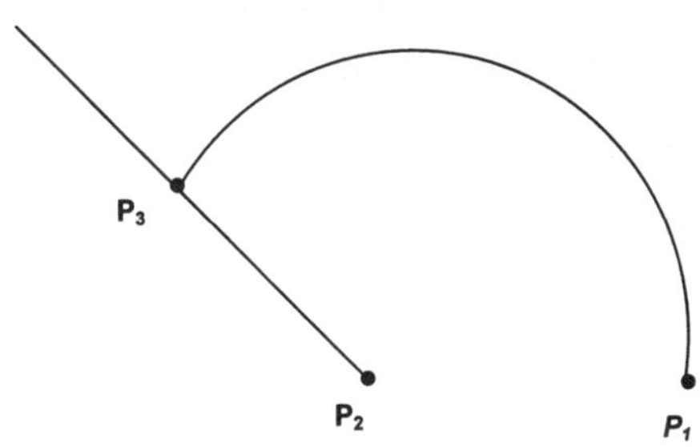

Fig. 13.12

Problem 13.15 (Fig. 13.13): *Draw an arc using **start point-centre point-length of chord** option of ARC command.*

Solution (Fig. 13.13):

Command: Arc

Specify start point of arc or [Centre]: (Pick start point)

Specify second point of arc or [Centre/End]: C

Specify centre point of arc: (Pick end point)

Specify end point of arc or [Angle/chord Length]: L

Specify length of chord: 65

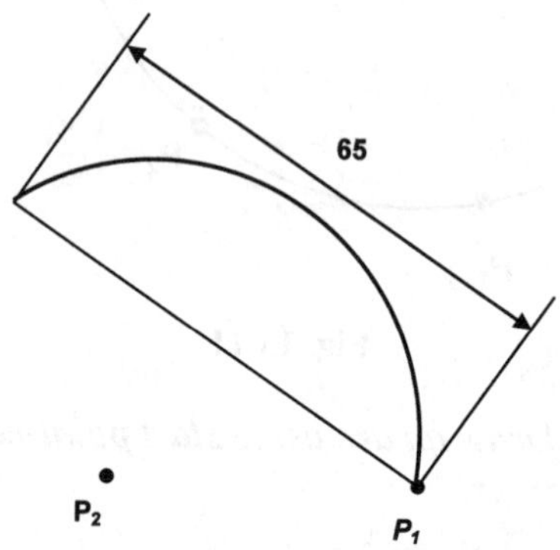

Fig. 13.13

Problem 13.16 (Fig. 13.14): *Draw an arc using **start point-end point-angle** option of ARC command.*

Solution (Fig. 13.14):

Command: Arc

Specify start point of arc or [Centre]: (Pick start point)

Specify second point of arc or [Centre/End]: E

Specify end point of arc: (Pick end point)

Specify centre point of arc or [Angle/Direction/Radius]: A

Specify included angle: 120

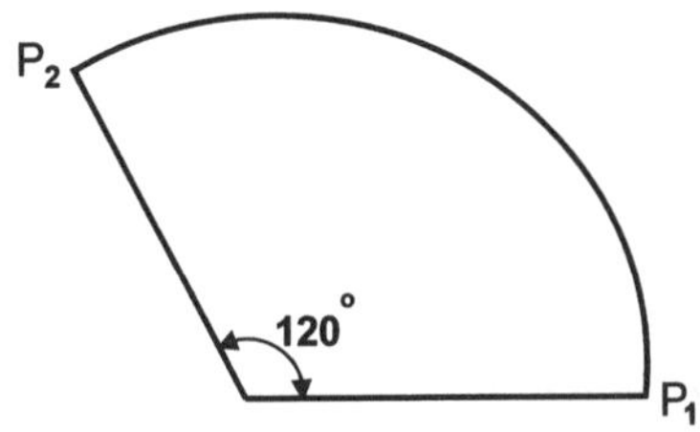

Fig. 13.14

Problem 13.17 (Fig. 13.15): *Draw an arc using **start point-end point-direction** option of ARC command.*

Solution (Fig. 13.15):

Command: Arc

Specify start point of arc or [Centre]: (Pick start point)

Specify second point of arc or [Centre/End]: E

Specify end point of arc: (Pick end point)

Specify centre point of arc or [Angle/Direction/Radius]: D

Specify tangent direction for the start point of arc: (Pick tangent direction)

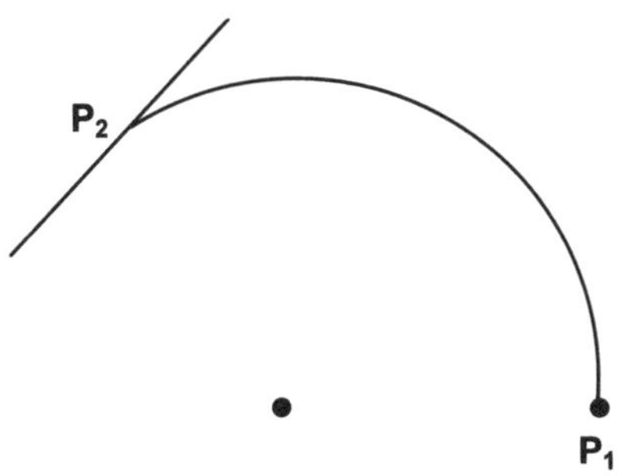

Fig. 13.15

Problem 13.18 (Fig. 13.16): *Draw an arc using **start point-end point-radius** option of ARC command.*

Solution (Fig. 13.16):

Command: Arc

Specify start point of arc or [Centre]:/(Pick start point)

Specify second point of arc or [Centre/End]: E

Specify end point of arc: (Pick end point)

Specify centre point of arc or [Angle/Direction/Radius]: R

Specify radius of arc: 40

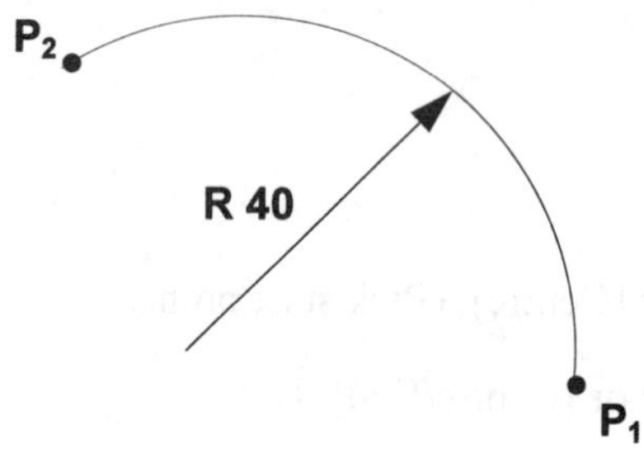

Fig. 13.16

13.14.6 ELLIPSE Command

The ELLIPSE command is used to draw an ellipse using following options.

(i) Axis end points option

(ii) Centre-axis end point option

(iii) Arc option (This option is beyond the scope of present book)

Problem 13.19 (Fig. 13.17): *Draw an ellipse using major axis end points A and B and minor axis end point C using AutoCAD ELLIPSE command.*

Solution (Fig. 13.17):

Command: Ellipse

Specify axis endpoint of ellipse or [Arc/Centre]: (Pick point A)

Specify other endpoint of axis: (Pick point B)

Specify distance to other axis or [Rotation]: (Click point C)

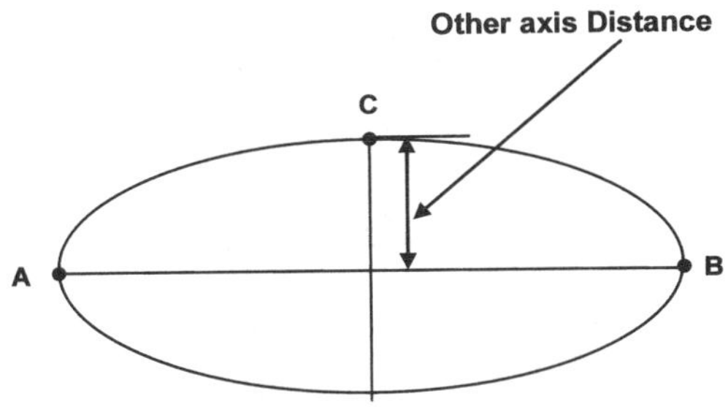

Fig. 13.17

Problem 13.20 (Fig. 13.17): *Draw an ellipse using centre point, major axis end point and minor axis end point.*

Solution (Fig. 13.17):

Command: Ellipse

Specify axis endpoint of ellipse or [Arc/Centre]: C

Specify centre of ellipse: (Pick centre point)

Specify endpoint of axis: (Pick point B)

Specify distance to other axis or [Rotation] (Pick point C)

13.14.7 POLYGON Command

POLYGON command is used to draw a regular polygon of sides ranging from 3-1024. Using this command a polygon can be drawn in the following three ways.

(i) Inscribed in Circle Option

Using this option, a polygon is drawn such that its vertices lie on the circumference of an imaginary circle.

Problem 13.21 (Fig. 13.18): *Draw a regular pentagon inscribed in a circle of 50 mm radius.*

Solution (Fig. 13.18):

Command: Polygon

Enter number of sides <4>: 5

Specify centre of polygon or [Edge]: Pick a point C

Enter an option [Inscribed in circle/Circumscribed about circle] <I>: I

Specify radius of circle: 50

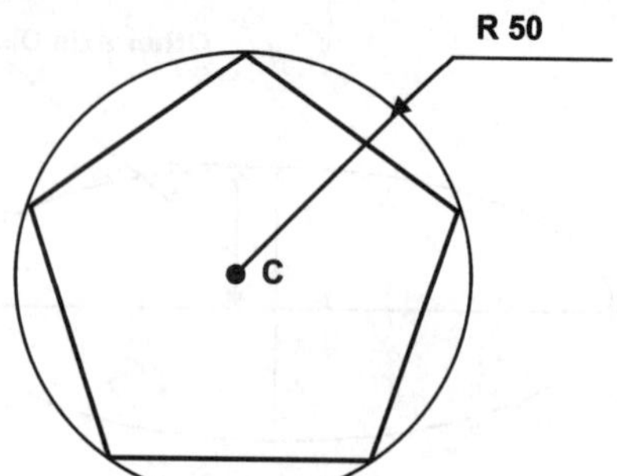

Fig. 13.18

(ii) **Circumscribed about the Circle Option:** Using this option, AutoCAD creates a polygon such that mid points of all the sides touch the circumference of an imaginary circle.

Problem 13.22 (Fig. 13.19): *Draw a regular hexagon circumscribed about a circle of 50 mm radius.*

Solution (Fig. 13.19):

Command: Polygon

Enter number of sides <5>: 6

Specify centre of polygon or [Edge]: Pick a point C

Enter an option [Inscribed in circle/Circumscribed about circle] <I>: C

Specify radius of circle: 50

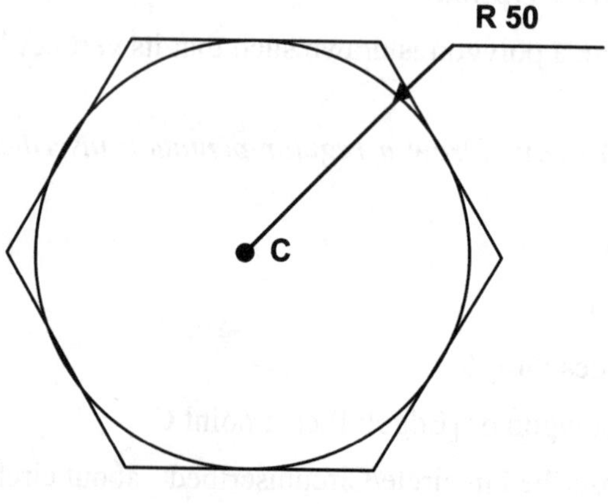

Fig. 13.19

(iii) Edge Option: In this option, a side is defined by fixing two end points. AutoCAD draws rest of its sides.

Problem 13.23 (Fig. 13.20): *Draw a regular hexagon of 40 mm sides having one of its edges horizontal.*

Solution (Fig. 13.20):

Command: Polygon

Enter number of sides <6>: 6

Specify centre of polygon or [Edge]: E

Specify first endpoint of edge: (Pick a point A)

Specify second endpoint of edge: 40

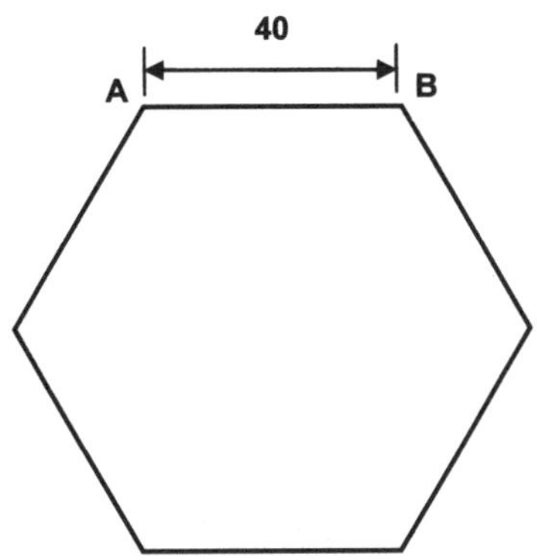

Fig. 13.20

13.14.8 PLINE Command

Poly-line represents a continuous segment of line or arc taken as a single entity. Such line can be drawn using PLINE command. Using this command straight, curved or tapered line segments of desired width can be drawn. Following are some examples which explain the use of PLINE command.

Problem 13.24 (Fig. 13.21): *Draw a straight line AB 100 mm long having width equal to 2.2 mm.*

Solution (Fig. 13.21):

Command: Pline

Specify start point: (Pick a point A)

Current line-width is 0.0000

Specify next point or [Arc/Halfwidth/Length/Undo/Width]: W

Specify starting width <0.0000>: 2.2

Specify ending width <2.2000>: 2.2

Specify next point or [Arc/Half width/Length/Undo/Width]: L

Specify length of line: 100

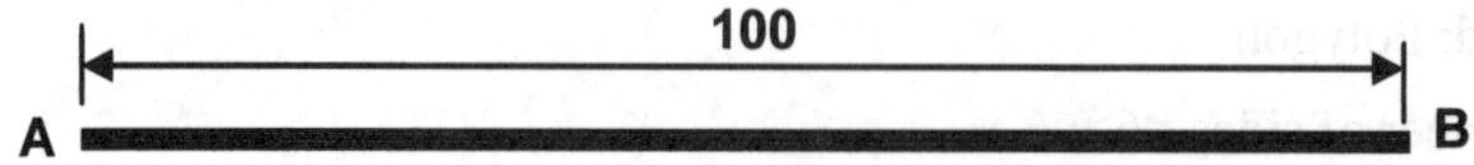

Fig. 13.21

Problem 13.25 (Fig. 13.22): *Draw an arc from point A to B with centre C and having width 1.5 mm using PLINE command.*

Solution (Fig. 13.22):

Command: Pline

Specify start point: (Pick a point A)

Current line-width is 0.0000

Specify next point or [Arc/Halfwidth/Length/Undo/Width]: W

Specify starting width <0.0000>: 1.5

Specify ending width <2.2000>: 1.5

Specify next point or [Arc/Halfwidth/Length/Undo/Width]: A

Specify endpoint of arc or

[Angle/CEnter/Direction/Halfwidth/Line/Radius/Second pt/Undo/Width]: CE

Specify center point of arc: (Pick a point C)

Specify endpoint of arc or [Angle/Length]: (Pick a point B)

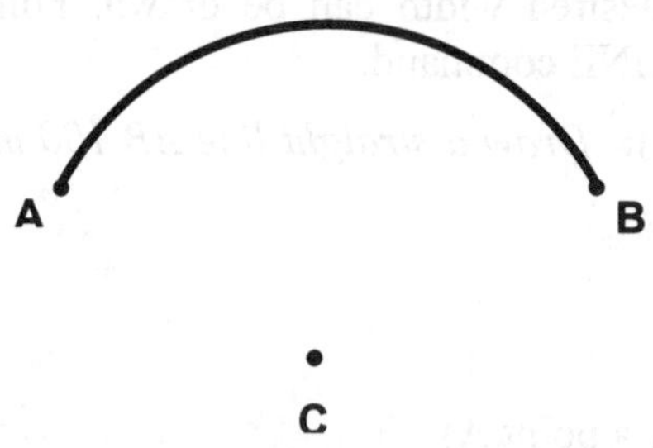

Fig. 13.22

13.14.9 SPLINE Command

This command is used to draw smooth curve that passes through or near to specified points. We can control the maximum distance between the B-spline curve and fit points by changing the value for the fit tolerance using SPLINEDIT. The command steps for drawing a smooth curve through six points are shown as below. We can also fit a smooth curve by using fit tolerance.

Command: Spline

Specify first point or [Object]: (Pick point 1)

Specify next point: (Pick point 2)

Specify next point or [Close/Fit tolerance] <start tangent>: (Pick point 3)

Specify next point or [Close/Fit tolerance] <start tangent>: (Pick point 4)

Specify next point or [Close/Fit tolerance] <start tangent>: (Pick point 5)

Specify next point or [Close/Fit tolerance] <start tangent>: (Pick point 6)

Specify start tangent: (Select first tangent)

Specify end tangent: (Select end tangent)

Using above command steps, following curve (Fig. 13.23) is drawn.

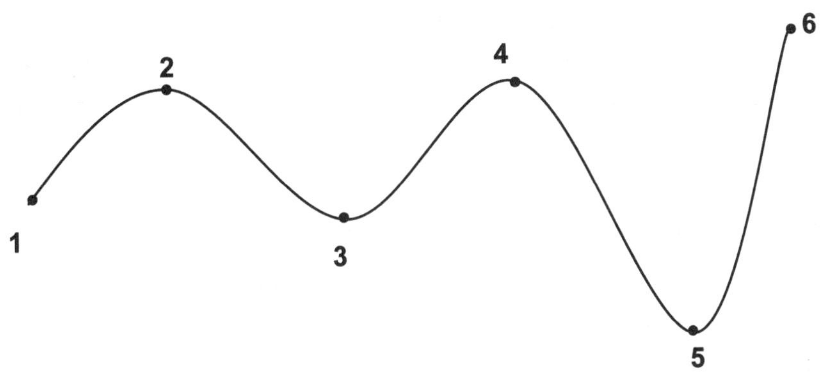

Fig. 13.23

13.14.10 XLINE Command

The XLINE command is used to draw lines that extend to infinity in both the directions. Such lines can be used as construction lines, reference lines etc. This command offers following options:

(i) By Specifying the location of the infinite line using two points through which it passes. Following command steps will create Xline through the specified point.

Command: Xline

Specify a point or [Hor/Ver/Ang/Bisect/Offset]: (Pick a point)

Specify through point: (Pick another point) and ENTER

(ii) **Hor** option creates a horizontal xline passing through a specified point as per following command steps.

Specify a point or [Hor/Ver/Ang/Bisect/Offset]: Hor

Specify through point: (Pick a point and press ENTER)

(iii) **Ver** option creates a vertical xline passing through a specified point as per following command steps.

Specify a point or [Hor/Ver/Ang/Bisect/Offset]: Ver

Specify through point: (Pick a point and press ENTER)

(iv) **Ang** option creates an xline at a specified angle. Following command steps will create a xline at 45° to the horizontal.

Command: Xline

Specify a point or [Hor/Ver/Ang/Bisect/Offset]: Ang

Enter angle of xline (0) or [Reference]: 45

Specify through point: (Pick a point and press ENTER)

(v) **Bisect** option creates a xline that passes through the selected angle vertex and bisects the angle between the first and second line as per following command steps.

Command: Xline

Specify a point or [Hor/Ver/Ang/Bisect/Offset]: Bisect

Specify angle vertex point: (Pick vertex or given angle)

Specify angle start point: (Pick a point on first line)

Specify angle end point: (Pick a point on second line and press ENTER)

(vi) **Offset** creates a xline parallel to another object. For example, following command steps are required to draw a line parallel to and 15 mm away from a given object.

Command: Xline

Specify a point or [Hor/Ver/Ang/Bisect/Offset]: Offset

Specify offset distance or [Through] <15.0000>: 15

Select a line object: (Select an object)

Specify side to offset: (Click left or right side of the object)

13.14.11 RAY Command

This command creates a line that starts at a point and continues to infinity. Following command steps are required to draw a line that starts from a point 1 and passes through point 2.

Command: Ray

Specify start point: (Click at point 1)

Specify through point: (Click at point 2 and press ENTER)

13.15 Commands for Editing and Modification

Once a drawing is created, it is possible to edit or modify or change dimensions of its elements. This can be done using various editing commands as discussed below.

13.15.1 ERASE Command

This command is used to remove a part of drawing or complete drawing.

To erase an object, first select the object using the pick box (cursor) or crossing window method, then press click on erase tab on the modify tool bar or press delete on key board.

Multiple objects can also be selected which can be removed by using the same method as discussed above. The command steps are shown as follows:

Command: Erase

Select objects: 1 found

Select objects: 1 found, 2 total

Select objects: Press Enter or Delete

13.15.2 UNDO Command

This command undoes the most recent operation performed. We can undo more than one command at one time. Following command steps will undo two recently performed operations.

Command: Undo

Enter the number of operations to undo or [Auto/Control/Begin/End/Mark/Back]

<1>: 2 Enter

13.15.3 REDO Command

REDO command is used just to reverse the effect of a single UNDO command.

It must immediately follow the UNDO command.

13.15.4 CHANGE Command

This command is used to modify the properties of object like colour, elevation, thickness etc. Following command steps will modify the property of the selected object.

Command: Change

Select objects: 1 found

Select objects: Enter

Specify change point or [Properties]: P

Enter property to change [Color/Elev/LAyer/LType/ltScale/ LWeight/Thickness /Material/Annotative]: (Enter the required option)

13.15.5 MIRROR Command

This command is used to copy symmetrical image about any required axis. For such object we only require to draw just one half of the object then other half can be drawn by mirroring the first half. Command steps are shown as follows (Fig. 13.24):

Command: Mirror

Select objects: Specify opposite corner: 5 found

Select objects: (By pick box method or specifying opposite corners using mouse)

Specify first point of mirror line: (Pick point1)

Specify second point of mirror line: (Pick point 2)

Erase source objects? [Yes/No] <N>: N (Press Enter)

13.15.6 OFFSET Command

The OFFSET command is used to draw parallel lines, concentric circles, Rectangle in rectangle etc. For example, following command steps will draw a circle inside or outside the given circle at offset distance of 15 mm (Fig.13.25). Similarly parallel lines at required offset, can be drawn using this command.

Command: Offset

Specify offset distance or [Through/Erase/Layer] <15.0000>: 15

Select object to offset or [Exit/Undo] <Exit>: (Select the object)

Specify point on side to offset or [Exit/Multiple/Undo] <Exit>: (Pick the point inside or outside the given circle)

Select object to offset or [Exit/Undo] <Exit>: Enter

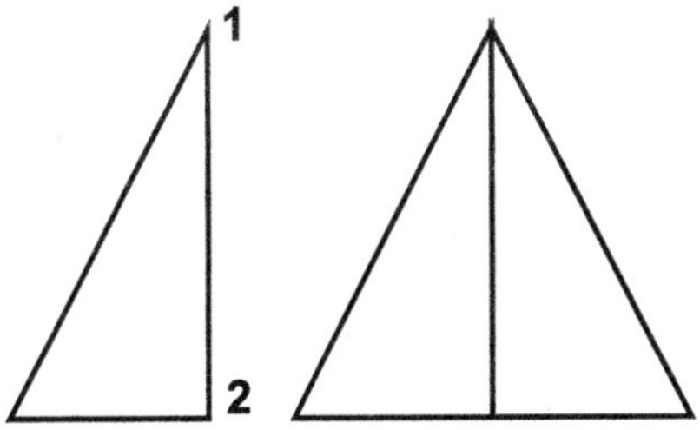

Fig. 13.24

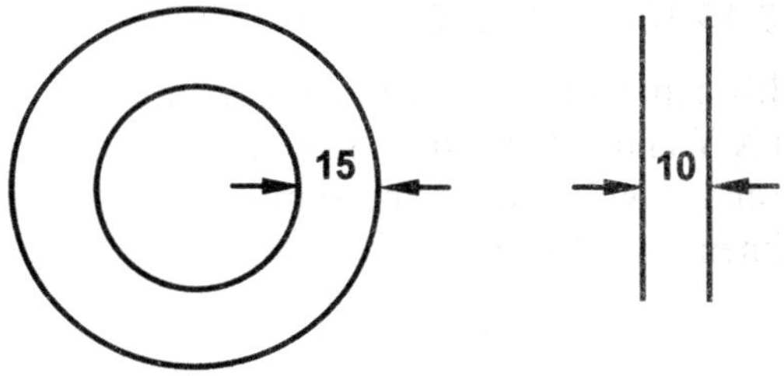

Fig. 13.25

13.15.7 ARRAY Command

This command is used to make multiple copies of any object in a regular pattern i.e., in rectangular or polar array. When Array command is invoked, it displays an array dialog box where one can set the pattern of array, number of rows, number of columns, offset distances etc., and can get the preview before obtaining the final drawings.

13.15.8 MOVE Command

This command is used to move the selected components of a drawing to the specified location. For Example:

Command: Move

Select objects: 1 found (Click on the object to be moved)

Select objects: (Press ENTER)

Specify base point or [Displacement] <Displacement>: Specify second point or

<use first point as displacement>: (Select second point on the screen)

Here we can use displacement option also, where command prompt asks for the coordinates or the distance of the new location.

13.15.9 ROTATE Command

This command is used to rotate an object around a base point by a specified angle. For example, following command steps will rotate an object by 330° anticlockwise.

Command: Rotate

Current positive angle in UCS: ANGDIR = counterclockwise ANGBASE = 0

Select objects: 1 found

Select objects: (Press Enter)

Specify base point: (Select the base point)

Specify rotation angle or [Copy/Reference] <330>: 330 (Press Enter)

13.15.10 SCALE Command

This command is used to enlarge or reduce an object to the required scale factor equally in X,Y and Z directions. Scale factor greater than 1 will enlarge and scale factor between 0 and 1 will reduce the dimensions. For example, to enlarge an object to double its dimensions, following command steps are required.

Command: Scale

Select objects: 1 found

Select objects: (Press Enter)

Specify base point: (Select base point)

Specify scale factor or [Copy/Reference] <1.0000>: 2 (For enlarging) (Press Enter)

13.15.11 STRETCH Command

The Stretch command is used to stretch or compress the selected object. Here it is important to select the object using crossing window method only. On selecting this command, command prompt appears as follows:

Command: Stretch

Select objects to stretch by crossing-window or crossing-polygon...

Select objects: Specify opposite corner: 3 found

Select objects: (Press ENTER)

Specify base point or [Displacement] <Displacement>: D

Specify second point or <use first point as displacement>:

13.15.12 TRIM Command

Trim command is used to trim objects at a cutting edge defined by other objects. The objects that can be trimmed include arcs, circles, lines, elliptical arcs, open 2D or 3D polylines, rays and splines. For trimming the object, following command prompt appears:

Command: Trim

Current settings: Projection = UCS, Edge = None

Select cutting edges ...

Select objects or <select all>: 1 found

Select objects: (Press ENTER)

Select object to trim or shift-select to extend or

[Fence/Crossing/Project/Edge/eRase/Undo]:

It can also be used with Fence, Crossing, Project, Edge and Undo sub options.

13.15.13 EXTEND Command

This command is used to extend an object to meet the another object. The objects that can be extended include lines, open 2D and 3D polylines, the rays, arcs and elliptical arcs.

Command: Extend

Current settings: Projection = UCS, Edge = Extend

Select boundary edges ...

Select objects or <select all>: Specify opposite corner: 2 found

Select objects: (Press ENTER)

Select object to extend or shift-select to trim or

[Fence/Crossing/Project/Edge/Undo]: (Select the object)

It can also be used with Fence, Crossing, Project, Edge and Undo sub options.

13.15.14 BREAK Command

The BREAK command is used to remove the selected part of an object between any two points. When it is invoked, the command prompt appears as follows:

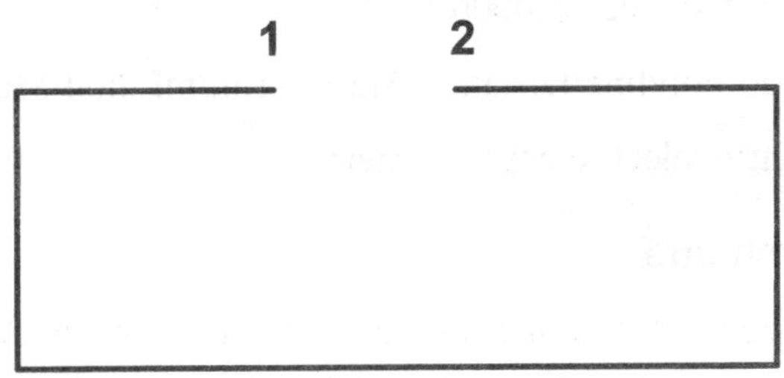

Fig. 13.26

Command: Break

Select object: (Select first point on the object)

Specify second break point or [First point]:

Point or option keyword required. (Select the second point)

13.15.15 JOIN Command

The action of this command is just reverse of Break command. It is used to join objects to form a single unbroken object. This can be used to join lines, poly-lines, arcs, elliptical arcs, splines and helixes. To join two parts of a line, command steps appear as follows.

Command: Join

Select source object: (Select the first part of a line)

Select lines to join to source: 1 found (Select the second part of the line)

Select lines to join to source: 1 line joined to source

13.15.16 CHAMFER Command

This command is used to level the edges of an object. For this two distances from the corner are to be specified and then select line 1 and line 2. For example following command steps will create chamfer at the corner of a right angle as shown in Fig. 13.27.

Command: Chamfer

(TRIM mode) Current chamfer Dist1 = 5.0000, Dist2 = 5.0000

Select first line or [Undo/Polyline/Distance/Angle/Trim/mEthod/Multiple]: D

Specify first chamfer distance <5.0000>: 5

Specify second chamfer distance <5.0000>: 5

Select first line or [Undo/Polyline/Distance/Angle/Trim/mEthod/Multiple]:

Select second line or shift-select to apply corner:

13.15.17 FILLET Command

This command is used to connect two lines or edges by an arc which is tangent to the two connecting lines or edges. For example, to create the fillet of radius10 mm at two right angled lines as shown in Fig.13.28, following command steps appear.

Command: Fillet

Current settings: Mode = TRIM, Radius = 0.0000

Select first object or [Undo/Polyline/Radius/Trim/Multiple]: R

Specify fillet radius <0.0000>: 10

Select first object or [Undo/Polyline/Radius/Trim/Multiple]:

Select second object or shift-select to apply corner:

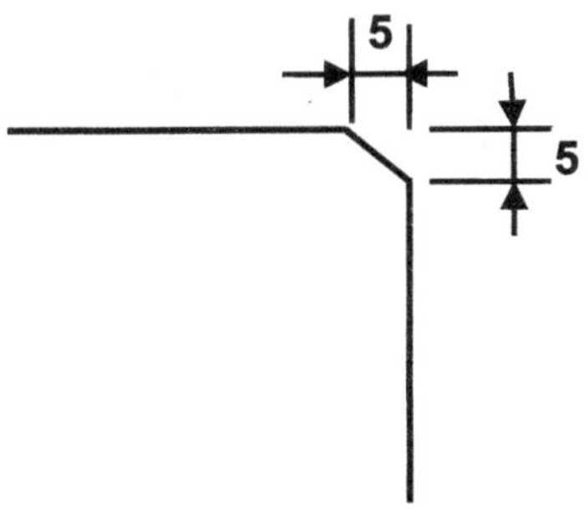

Fig. 13.27

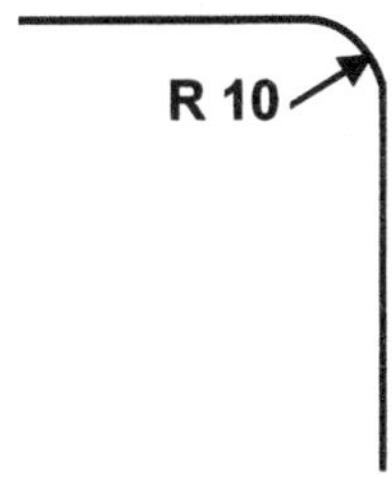

Fig. 13.28

13.15.18 EXPLODE Command

This command is used to break a compound object into its component objects when it is required to modify its components separately. Objects that can be exploded include blocks, polylines, and regions, among others.

Command: Explode

Select objects: 1 found

Select objects: Press Enter

13.15.19 LENGTHEN Command

This command increases or decreases the length of a selected object. It does not affect the close objects. The command prompt appear as follows:

Command: Len

Select an object or [Delta/Percent/Total/Dynamic]:

Current length: 66.6001

Select an object or [Delta/Percent/Total/Dynamic]: P

Enter percentage length <100.0000>: 120

Select an object to change or [Undo]:

13.15.20 EXTRUDE Command

It creates a 3D solid or surface by extruding a 2D object.

Command: Extrude

Current wire frame density: ISOLINES = 4

Select objects to extrude: 1 found

Select objects to extrude:

Specify height of extrusion or [Direction/Path/Taper angle] <54.2288>: Enter

13.15.21 HATCH Command

This command is used to fill the object with different colours and patterns. This command allows selection of required pattern, scale of pattern and angle of patterns with the help of a dialog box.

Command: Hatch

Pick internal point or [Select objects/remove Boundaries]: Selecting everything...

Selecting everything visible...

Analyzing the selected data...

Analyzing internal islands...

Pick internal point or [Select objects/remove Boundaries]:

Fig. 13.29

13.16 Dimensioning Commands

Three types of dimensioning such as linear, radial and angular, can be provided using following commands.

(i) DIMLINEAR

This is used to measure horizontal or vertical distances between two points.

(ii) DIMALIGNED

This is used to dimension inclined edges of an object.

(iii) DIMRADIUS

It draws radial dimensions depending on the size of the circle or arc and position of the cursor.

(iv) DIMDIAMETER

It draws a diameter dimension depending on the size of the circle.

(v) DIMCENTER

It draws a centre mark for arcs or circles.

(vi) DIMANGULAR

It draws an angular dimension.

(vii) DIMBASELINE

It draws a series of dimensions measured from the same base line.

(viii) DIMCONTINUE

This command is used to draw dimensions that add up to the total measurements. This is also known as chain dimensioning.

13.17 TEXT Command

The TEXT command is used to draw a single line text on the drawing with the required height and orientation.

Command: Text

Current text style: "Standard" Text height: 2.5000 Annotative: No

Specify start point of text or [Justify/Style]: (Pick a start point)

Specify height <2.5000>: (Press ENTER)

Specify rotation angle of text <0>: (Press ENTER)

13.18 Projection Problems on CAD

In engineering drawing, following types of projections are used.

1. Orthographic Projection
2. Isometric Projection
3. Oblique Projection
4. Perspective Projection

These projections have been described in Chapter 5.

1. **Orthographic Projection:** When the projectors are drawn parallel to each other and perpendicular to the plane of projection, such projection is called orthographic projections. In this method, various views such as front, top and side views of the object are drawn.

   AutoCAD provides facility to draw the orthographic projections of any object conveniently using cross hair. ORTHO command can be used to restrict the movement of cursor to horizontal and vertical directions which helps to draw horizontal and vertical lines. DIST command allows measurement of distance during preparing drawing.

2. **Isometric Projection:** The projection obtained on a plane when the projectors are parallel but are inclined at 30° to the plane of projection, such projection is known as isometric projection. An isometric view represents all three dimensions i.e.,

length, breadth and height. This method is based on measurement of dimensions along X, Y and Z axis. X and Y axes are assumed to be inclined at 120° and Z axis is parallel to Y axis of the screen. In this method, projectors are parallel but inclined at 30° to the plane of projection. The true length is reduced by a factor of 0.815 which is called isometric length.

Following AutoCAD commands can be used conveniently to draw isometric drawing.

(i) SNAP Command: This restricts the movement of cursor to the specified intervals.

(ii) ISOPLANE Command: This command specifies the current isometric plane.

3. **Oblique Projections:** This also represents all three dimensions of the object in a single drawing but its one face is drawn parallel and other adjacent face is inclined at an angle of 45° to the plane of projection. In AutoCAD, oblique projections are drawn in the same way as in isometric projection.

4. **Perspective Projections:** This is also a three dimensional representation of an object on a plane as it appears at human eye from a given distance. In orthographic projection, the projectors are parallel to each other but in perspective projections, they converge to a point of sight.

AutoCAD provides facility to draw perspective projections of any object using a camera and view angle of object.

13.19 Typical Problems

Problem 13.26 (Fig. 13.30): *The front view of a V-Block is shown in Fig. 13.30. Reproduce the same using AutoCAD and write a series of command steps.*

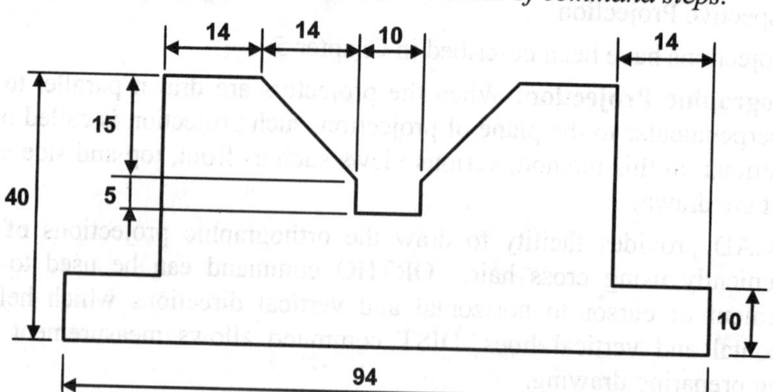

Fig. 13.30

Solution (Fig. 13.30):

Command: Line

Specify first point: 0,0 ↵ (Assuming lower left corner to be at (0,0)

Specify next point or [Undo]: @0, 10 ↵

Specify next point or [Undo]: @14, 0 ↵

Specify next point or [Close/Undo]: @0, 30 ↵

Specify next point or [Close/Undo]: @14, 0 ↵

Specify next point or [Close/Undo]: @14, –15 ↵

Specify next point or [Close/Undo]: @0, –5 ↵

Specify next point or [Close/Undo]: @10, 0 ↵

Specify next point or [Close/Undo]: @0, 5 ↵

Specify next point or [Close/Undo]: @14, 15 ↵

Specify next point or [Close/Undo]: @14, 0 ↵

Specify next point or [Close/Undo]: @0, –30 ↵

Specify next point or [Close/Undo]: @14, 0 ↵

Specify next point or [Close/Undo]: @0, –10 ↵

Specify next point or [Close/Undo]: @–94, 0 ↵

Specify next point or [Close/Undo]: C ↵

Problem 13.27 (Fig. 13.31): *State a series of command steps required to reproduce the Fig. 13.31 using AutoCAD rectang, chamfer, break, and line commands.*

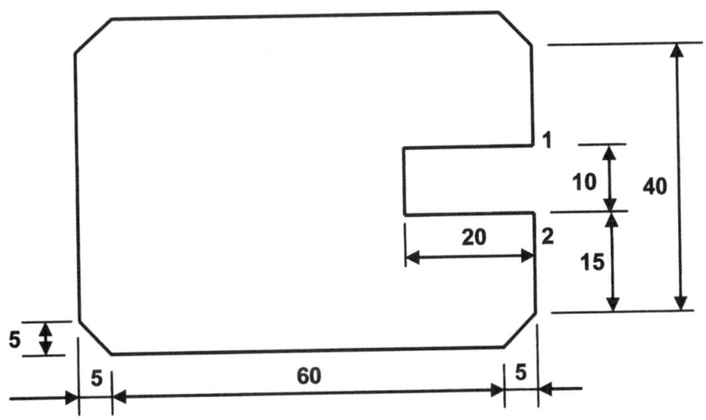

Fig. 13.31

Solution (Fig. 13.31):

Step 1: Draw a rectangle of 70 × 50 mm with the chamfer of 5 mm at the corners.

Command: Rectang ↵

Current rectangle modes: Chamfer = 5.0000 × 5.0000

Specify first corner point or [Chamfer/Elevation/Fillet/Thickness/Width]: C ↵

Specify first chamfer distance for rectangles <5.0000>: 5 ↵

Specify second chamfer distance for rectangles <5.0000>: 5 ↵

Specify first corner point or [Chamfer/Elevation/Fillet/Thickness/Width]:

Specify other corner point or [Area/Dimensions/Rotation]: D ↵

Specify length for rectangles <70.0000>: ↵

Specify width for rectangles <50.0000>: ↵

Specify other corner point or [Area/Dimensions/Rotation]:

Step 2: Break the side of the rectangle from point 1 to 2.

Command: Break ↵

Select object: (Select the rectangle)

Specify second break point or [First point]: F ↵

Specify first break point: (Pick point 1)

Specify second break point: (Pick point 2)

Step 3: Draw internal lines.

Command: Line

Specify first point: (Pick point 1)

Specify next point or [Undo]: @ –20, 0 ↵

Specify next point or [Undo]: @0, –10 ↵

Specify next point or [Close/Undo]: @20, 0 ↵

Problem 13.28: *An isometric view of a combined solid is shown in Fig. 13.32. State a series of command steps required to draw its front and top views using AutoCAD.*

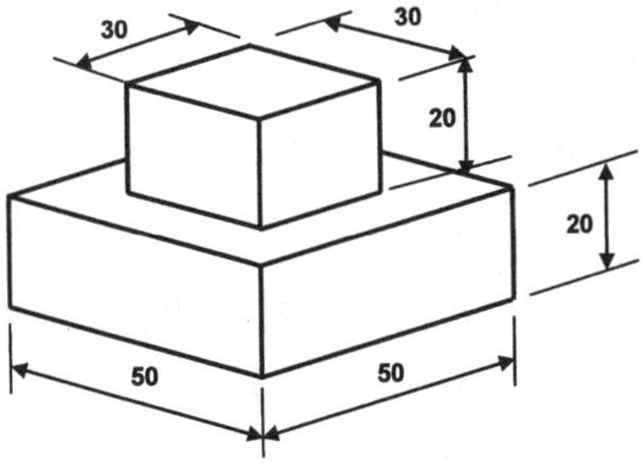

Fig. 13.32

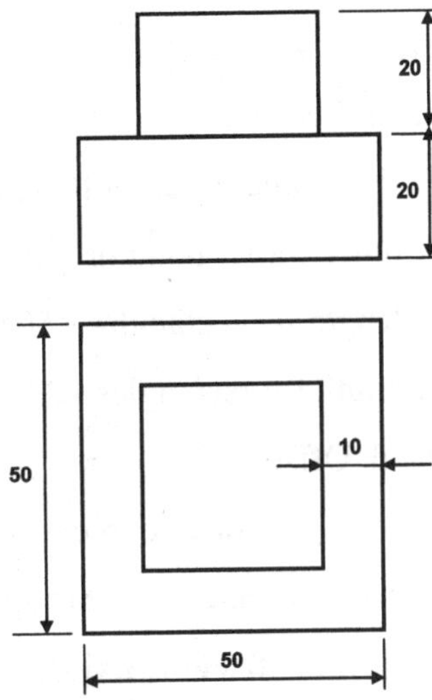

Fig. 13.33

Solution (Fig. 13.33)

To Draw Front View:

Command: Line

Specify first point: 0, 0 ↵ (Assuming lower left corner at (0, 0)

Specify next point or [Undo]: @50, 0 ↵

Specify next point or [Undo]: @0, 20 ↵

Specify next point or [Close/Undo]: @–50, 0 ↵

Specify next point or [Close/Undo]: @0, –20 ↵

Specify next point or [Close/Undo]: *Cancel*

Command: Line

Specify first point: 10, 20 ↵

Specify next point or [Undo]: @0, 20 ↵

Specify next point or [Undo]: @30, 0 ↵

Specify next point or [Close/Undo]: @0, –20 ↵

Specify next point or [Close/Undo]: *Cancel*

To Draw top view:

Command: Rectang

Specify first corner point or [Chamfer/Elevation/Fillet/Thickness/Width]:10,10 ↵

Specify other corner point or [Area/Dimensions/Rotation]: D ↵

Specify length for rectangles <50.0000>: 50 ↵

Specify width for rectangles <20.0000>: 50 ↵

Command: Offset

Current settings: Erase source – No Layer = Source OFFSETGAPTYPE=0

Specify offset distance or [Through/Erase/Layer] <1.0000>: 10 ↵

Select object to offset or [Exit/Undo] <Exit>: (Select the square)

Specify point on side to offset or [Exit/Multiple/Undo] <Exit>: (Click inside the square)

Select object to offset or [Exit/Undo] <Exit>: E ↵

Exercises

1. What is CAD ? List out five advantages of CAD as compared to conventional drawing.
2. Give limitations of manual drawing and discuss salient features of CAD.
3. What is a CAD software? Name any three software commonly used for CAD.
4. What do you mean by drawing entity? Name any five drawing entities.
5. Name the commands used for the following:
   (a) Dimensioning a drawn object.
   (b) Constraining the lines drawn in horizontal and vertical direction only.
   (c) Saving a drawn object.
   (d) Creating a bevel surface at the corner
   (e) Displaying the object at a specified scale factor.
   (f) Filling the drawn object with different pattern or colour
6. Name various methods of locating a point in CAD and explain any one of them.
7. Explain any four methods of drawing an arc in AutoCAD.
8. Mention any two methods by which a circle can be drawn using AutoCAD.
9. Write the steps for drawing a cylinder of 20 mm base and 50 mm height.

10. Explain various commands for transformation of an object. (Hint: Move, Mirror, Scale, Rotate etc., can be used as transformation command)
11. State any five significant EDIT commands used in CAD software.
12. Mention a few Utility commands and state their respective basic functions.
13. Prepare ellipse using four different methods in AutoCAD.
14. What is a poly-line? How a poly-line and a poly-arc are constructed using AutoCAD ?
15. Explain the use of Hatch command in AutoCAD.
16. Write the steps used in erasing the drawing object using AutoCAD.
17. What is the use of Polar and Rectangle Array ? Explain.
18. Explain methods of drawing chamfer and fillet at the corner of an object.
19. Write the steps for drawing a pentagon of 55 mm side.
20. State the series of AutoCAD command steps to draw a rectangle of size 100mm × 65mm.
21. State the series of AutoCAD command steps to draw two concentric circles of 50 mm and 40 mm diameters.
22. Draw the front view and top view of an object shown in Fig. 13.34

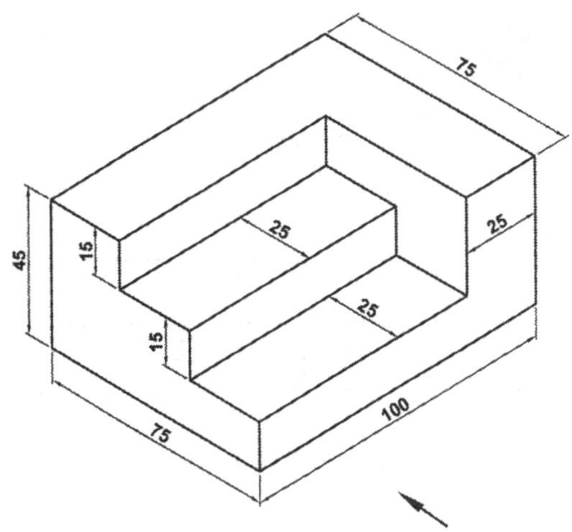

Fig. 13.34

23. Reproduce the isometric view of an object shown in Fig. 13.35 using AutoCAD.

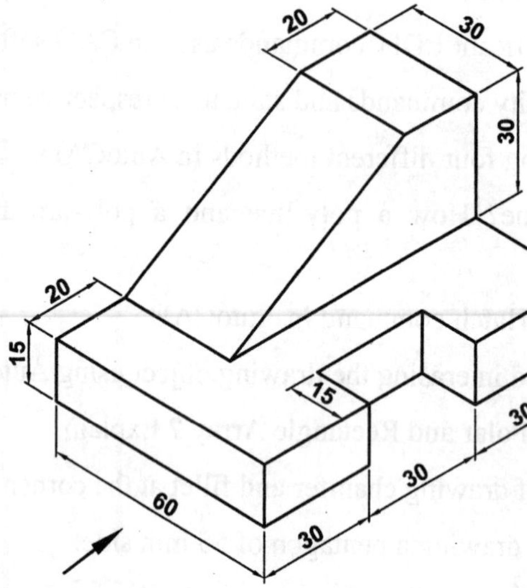

Fig. 13.35

24. Write the series of command steps required to draw an object shown in Fig. 13.36 using AutoCAD.

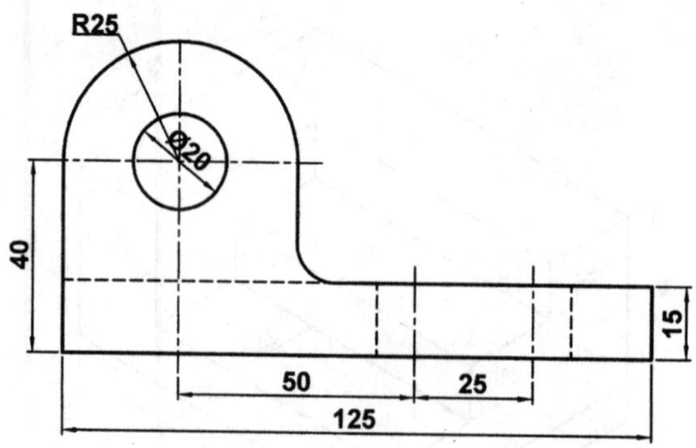

Fig. 13.36

Index

J

L

M

N

O

P

T

U

X